AF717985

SCOPE 56

Global Change: Effects on Coniferous Forests and Grasslands

SCOPE 36: Acidification in Tropical Countries, 1988, 424 pp
SCOPE 37: Biological Invasions: A Global Perspective, 1989, 528 pp
SCOPE 38: Ecotoxicology and Climate with Special Reference to Hot and Cold Climates, 1989, 432 pp
SCOPE 39: Evolution of the Global Biogeochemical Sulphur Cycle, 1989, 224 pp
SCOPE 40: Methods for Assessing and Reducing Injury from Chemical Accidents, SGOMSEC 6, 1989, 320 pp
SCOPE 41: Short-Term Toxicity Tests for Non-Genotoxic Effects, SGOMSEC 4, 1990, 353 pp
SCOPE 42: Biogeochemistry of Major Worlds Rivers, 1991, 356 pp
SCOPE 43: Stable Isotopes: Natural and Anthropogenic Sulphur in the Environment, 1991, 472 pp
SCOPE 44: Introduction of Genetically Modified Organisms into the Environment, 1990, 224 pp
SCOPE 45: Ecosystem Experiments, 1991, 296 pp
SCOPE 46: Methods for Assessing Exposure of Human and Non-Human Biota, SGOMSEC 5, 1991, 448 pp
SCOPE 47: Long-term Ecological Research. An International Perspective, 1991, 312 pp
SCOPE 48: Sulphur Cycling on the Continents: Wetlands, Terrestrial Ecosystems and Associated Water Bodies, 1992, 345 pp
SCOPE 49: Methods to Assess Adverse Effects of Pesticides on Non-Target Organisms, SGOMSEC 7, 1992, 264 pp
SCOPE 50: Radioecology After Chernobyl, 1993, 367 pp
SCOPE 51: Biogeochemistry of Small Catchments: A Tool for Environmental Research, 1993, 432 pp
SCOPE 52: Methods to Assess DNA Damage and Repair: Interspecies Comparisons, SGOMSEC 8, 1994, 257 pp
SCOPE 53: Methods to Assess the Effects of Chemicals on Ecosystems, SGOMSEC 10, 1995, 440pp
SCOPE 54: Phosphorus in the Global Environment: Transfers, Cycles and Management, 1995, 480pp
SCOPE 55: Functional Roles of Biodiversity: A Global Perspective, 1996, 516pp
SCOPE 56: Global Change: Effects on Coniferous Forests and Grasslands, 1996, 480pp
SCOPE 57: Particle Flux in the Ocean, 1996, 396pp

Funds to meet SCOPE expenses are provided by contributions from SCOPE Committees, an annual subvention from ICSU (and through ICSU, from UNESCO), an annual subvention from the French Ministère de l'Environnement, contracts with UN Bodies, particularly UNEP, and grants from Foundations and industrial enterprises.

SCOPE 56

Global Change: Effects on Coniferous Forests and Grasslands

Edited by

A. I. BREYMEYER

Polish Academy of Sciences, Poland

D. O. HALL

King's College London, UK

J. M. MELILLO

The Ecosystem Center, USA

G. I. ÅGREN

Swedish University of Agricultural Sciences, Sweden

Published on behalf of the
Scientific Committee on Problems of the Environment (SCOPE)
of the International Council of Scientific Unions (ICSU)
and the United Nations Environment Programme (UNEP)

ICSU

UNEP

by

JOHN WILEY & SONS

Chichester · New York · Brisbane · Toronto · Singapore

Published in 1996 by John Wiley & Sons Ltd
Baffins Lane, Chichester,
West Sussex PO19 1UD, England

National 01243 779777
International (+44) 1243 779777
e-mail (for orders and customer service enquiries): cs-books@wiley.co.uk
Visit our Home Page on http://www.wiley.co.uk
or http://www.wiley.com

Other Wiley Editorial Offices

John Wiley & Sons, Inc., 605 Third Avenue,
New York, NY 10158-0012, USA

Jacaranda Wiley Ltd, Park Road, Milton,
Queensland 4064, Australia

John Wiley & Sons (Canada) Ltd, 22 Worcester Road,
Rexdale, Ontario M9W 1L1, Canada

John Wiley & Sons (Asia) Pte Ltd, 2 Clementi Loop £02-01,
Jin Xing Distripark, Singapore 129809

Library of Congress Cataloging-in-Publication Data
Global change: effects on coniferous forests and grasslands / edited
by A. I. Breymeyer ... [et al.].
p. cm. —(SCOPE; 56)
Includes bibliographical references and index.
ISBN 0-471-96837-4
1. Climatic changes. 2. Forest microclimatology. 3. Forest ecology. 4. Conifers—Ecology.
5. Grasslands-ecology. 6. Carbon cycle (Biogeochemistry) I. Breymeyer, A. I. (Alicja I.),
1932–11. Series: SCOPE report: 56.
SD390.7.C55G58 1996
574.5′ 2642—dc20 96-25987
CIP

British Library Cataloguing in Publication Data
A catalogue record for this book is available from the British Library

ISBN 0-471-96837-4

Typeset in 10/12pt Times from disks supplied by the editors by Vision Typesetting, Manchester.
Printed and bound in Great Britain by Biddles Ltd, Guildford, Surrey.
This book is printed on acid-free paper responsibly manufactured from sustainable forestation, for which at least two trees are planted for each one used for paper production.

International Council of Scientific Unions (ICSU)
Scientific Committee on Problems of the Environment (SCOPE)

SCOPE is one of a number of committees established by a non-governmental scientific organization, the International Council of Scientific Unions (ICSU). The membership of ICSU includes representatives of 94 academies of science, 23 scientific unions and 30 scientific associates. ICSU has established 26 interdisciplinary bodies to bring together activities which include the interests of several unions.

SCOPE was established by ICSU in 1969 as one such interdisciplinary body. Its attention and scientific programme are directed to existing and potential environmental issues – either global or shared by several nations – which are in urgent need of interdisciplinary syntheses.

The mandate of SCOPE is to assemble, review, and assess the information available on human-made environmental changes and the effects of these changes on people; to assess and evaluate the methodologies of measurement of environmental parameters; to provide an intelligence service on current research; and by the recruitment of the best available scientific information and constructive thinking to establish itself as a corpus of informed advice for the benefit of centres of fundamental research and of organizations and agencies operationally engaged in studies of the environment. It acts at the interface between the science and decision-making spheres, providing advisors, policy-planners and decision-makers with analytical tools to promote sound management and policy practices.

At present, representatives of 38 member countries and 22 international unions and scientific committees participate in the work of SCOPE, with scientists from around the world contributing their time and expertise to the projects in the scientific programme. SCOPE is governed by a General Assembly, which meets every three years. Between such meetings its activities are directed by the Executive Committee.

R. E. MUNN
Editor-in-Chief
SCOPE Publications

Executive Director: V. Plocq-Fichelet
Secretariat: 51 boulevard de Montmorency
75016 Paris, France

Contents

List of Contributors

John D. Aber
Complex Systems Research Center, Institute for the Study of Earth, Oceans, Space, University of New Hampshire, Durham, NH 03824, USA

G. I. Ågren
Department of Ecology and Environmental Research, Swedish University of Agricultural Sciences, PO Box 7072, S-750 07 Uppsala, Sweden

Joaquim Azcon-Bieto
Department Fisologia Vegetal, Universitat de Barcelona, 08028 Barcelona, Spain

Bjorn Berg
Dept. of Forest Soils, PO Box 7001, Swedish University of Agricultural Science, S-750 07 Uppsala, Sweden

E. Bosatta
Department of Ecology and Environmental Research, Swedish University of Agricultural Sciences, PO Box 7072, S-750 07, Uppsala, Sweden

Alicja I. Breymeyer
Institute of Geography and Spatial Organization, Polish Academy of Sciences, Krakowskie Przedmiescie 30, PL 00-927 Warsaw, Poland

I. C. Burke
Department of Forest Sciences and Natural Resource Ecology Laboratory, Colorado State University, Fort Collins, CO 80523, USA

M. B. Coughenour
Natural Resource Ecology Laboratory, Colorado State University, Fort Collins, CO 80523, USA

W. Cramer
Potsdam Institute of Climate Impact Research (PIK), Postfach 60 12 03, D-14473, Potsdam, Germany

Bert G. Drake
Smithsonian Environmental Research Center, PO Box 28, Edgewater, MD 21037, USA

A. D. Friend
Institute of Terrestrial Ecology, Bush Estate, Penicuik, Midlothian H26 0QB, UK

E. Garcia Moya
Centro de Botanica, Colegio de Postgraduados, Chapingo, Mex. 56230, Mexico

T. G. Gilmanov
Department of Vertebrate Zoology and Ecology, Moscow State University, 119899 Moscow, Russia

M. Gonzalez-Meler
Smithsonian Environmental Research Center, PO Box 28, Edgewater, MD 21037, USA

Stith T. Gower
University of Wisconsin-Madison, Department of Forestry, 12 Russell Labs, 1630 Linden Drive, Madison, WI 53706, USA

D. O. Hall
Division of Life Sciences, King's College London, Campden Hill Road, London W8 7AH, UK

S. W. Humphries
Department of Biology, University of Essex, Colchester CO4 3SQ, UK

E. R. Hunt
School of Forestry, University of Montana, Missoula, MT 59812, USA

Dale Johnson
Desert Research Institute, Biological Sciences Center, 7010 Dandini Boulevard, Reno, NV 89512, USA

Apinan Kamnalrut
Faculty of Natural Resources, Prince of Songkla University, Hatyai 90110, Thailand

D. W. Kicklighter
The Ecosystems Center, Marine Biological Laboratory, Woods Hole, MA 02543, USA

J. I. Kinyamario
Department of Botany, University of Nairobi, P.O. Box 30197, Nairobi, Kenya

T. B. Kirchner
Natural Resource Ecology Laboratory, Colorado State University, Fort Collins, CO 80523, USA

M. U. F. Kirschbaum
Division of Forestry, CSIRO, PO Box 4008, Canberra, ACT 2600, Australia

T. G. F. Kittel
Natural Resource Ecology Laboratory, Colorado State University, Fort Collins, CO 80523, USA

W. K. Lauenroth
Departments of Range Science and Natural Resource Ecology Laboratory, Colorado State University, Fort Collins, CO 80523, USA

R. Leemans
Global Change Department, Dutch National Institute of Public Health and the Environment (RIVM), PO Box 13720 BA Bilthoven, The Netherlands

S. Linder
Department of Ecology and Environmental Research, Swedish University of Agricultural Sciences, PO Box 7072, S-750 07 Uppsala, Sweden

S. P. Long
Department of Biology, University of Essex, Colchester, CO4 3SQ, UK

A. D. McGuire
The Ecosystems Center, Marine Biological Laboratory, Woods Hole, MA 02543, USA

R. E. McMurtie
Department of Biological Science, The University of New South Wales, Kensington, NSW 2033, Australia

J. M. Melillo
The Ecosystems Center, Marine Biological Laboratory, Woods Hole, MA 02543, USA

J-C Menaut
Laboratoire d' Ecologie, Ecole Normale Supérieure, CNRS-URA 258, 46 rue d'Ulm, 75230 Paris, Cedex 05, France

T. R. Moore
Dept. of Geography and Centre for Climate and Global Change Research, McGill University, 805 Sherbrooke St.W., Montreal, H3A 2K6, Canada

D. S. Ojima
Natural Resource Ecology Laboratory, Colorado State University, Fort Collins, CO 80523, USA

C. P. Osborne
Department of Biology, University of Essex, Colchester CO4 3SQ, UK

W. J. Parton
Natural Resource Ecology Laboratory, Colorado State University, Fort Collins, CO 80523, USA

W. M. Pulliam
Natural Resource Ecology Laboratory, Colorado State University, Fort Collins, CO 80523, USA

R. J. Raison
CSIRO, Division of Forestry, PO Box 4008, Queen Victoria Terrace, Canberra ACT 2600, Australia

E. B. Rastetter
The Ecosystems Center, Marine Biological Laboratory, Woods Hole, MA 02543, USA

M. G. Ryan
USDA Forest Service, Rocky Mountain Experiment Station, 240 West Prospect Street, Fort Collins, CO 80526-2098, USA

O. E. Sala
Department of Ecology, Faculty of Agronomy, University of Buenos Aires, Av. San Martin 4453, Buenos Aires 1417, Argentina

D. S. Schimel
Climate System Modelling Program, NCAR, PO Box 3000, Boulder, CO 80303, USA

R. J. Scholes
Division of Water, Environment and Forest, CSIR, PO Box 395, Pretoria 0001, South Africa

J. M. O. Scurlock
Division of Life Sciences, King's College London, Campden Hill Road, London W8 7AH, UK

T. R. Seastedt
Institute of Arctic and Alpine Research, University of Colorado, Campus Box 450, Boulder, CO 80309, USA

G. R. Shaver
The Ecosystems Center, Marine Biological Laboratory, Woods Hole, MA 02543, USA

J. G. van Minnen
Global Change Department, Dutch National Institute of Public Health and the Environment (RIVM), PO Box 13720 BA Bilthoven, The Netherlands

J. A. van Veen
DLO Research Institute for Plant Protection, PO Box 9060, 6700 GW Wageningen, The Netherlands

Preface

In 1987 SCOPE launched a project to help improve our understanding of ecosystem and climate coupling. This study focuses on the effects of climate change on production and decomposition in coniferous forests and in both temperate and tropical grasslands.

It was necessary to investigate whole plant responses to simultaneous changes in atmospheric CO_2 content, temperature regime and water availability to document direct and indirect effects of these environmental changes on ecosystem processes. Ecosystem models linking plant and soil response to climate change were reviewed. Diagnostic and predictive models were designed to work at multiple space and time scales for effective extrapolation and analysis of climate change scenarios.

Two major international interdisciplinary workshops were convened in the USA (1988) and in the UK (1990) while several smaller thematic workshops and modelling workshops were organised in the UK, USA, Sweden and Poland. A final synthesis meeting in Sweden (1992) brought together the information collected over the course of the project.

This volume critically assesses the present state of our knowledge, reviews extant forest and grasslands models, and develops foundations for the design of diagnostic and predictive models, as well as identifying plans for future research on ecosystem response to global change. Carbon flow and storage are emphasised, although the flow and storage of key nutrients such as nitrogen, sulphur and phosphorus are also considered when deemed necessary to understand changes in the carbon budget of coniferous forests and grasslands.

The support and financial assistance provided by the following is gratefully acknowledged. Without their assistance this project would not have been possible: A. W. Mellon Foundation, National Science Foundation, and Electric Power Research Institute in the USA; the Royal Society and King's College London in the UK; in Sweden the Council for Forestry and Agricultural Research, Council for Planning and Co-ordination of Research, Environmental Protection Agency, Natural Science Research Council, Swedish University of Agricultural Sciences, Wenner-Gren Center Foundation for Scientific Research; in Poland the Academy of Sciences (Scientific Committee Man and Environment); and the United Nations Environmental Programme (UNEP).

D. O. Hall
September 24, 1996

1 Executive Summary

J. M. MELILLO,[1] D. O. HALL[2], G. I. ÅGREN[3]
[1]*The Ecosystems Center, Marine Biological Laboratory, Woods Hole, USA*
[2]*Division of Life Sciences, King's College London, UK*
[3]*Department of Ecology and Environmental Research, Swedish University of Agricultural Sciences, Uppsala, Sweden*

1.1 INTRODUCTION

Human activities are rapidly changing the global environment. They are causing increases in atmospheric concentrations of carbon dioxide and other greenhouse gases; in turn, these increases are thought to be a primary force for global climate change. By the middle of the next century, the mean global temperature is expected to be higher than it has been for millions of years. In addition, changes in rainfall patterns and increased storm intensities are predicted.

A rapidly changing global climate will certainly affect the fundamental ecological processes of the planet. Exactly how ecosystems will respond to climate change is not well known. Predicting the effects of climate change on ecosystems is complicated by interactions among the effects of concurrent changes in atmospheric carbon dioxide concentration, available water, and temperature on a number of key ecosystem processes such as primary production and decomposition.

A major challenge is to understand the fate of man-induced sources of emissions of greenhouse gases. For carbon dioxide there are presently two principal sources: tropical land use changes and combustion of fossil fuels plus cement production. Some of this carbon dioxide remains in the atmosphere, some is absorbed in the oceans, and some disappears due to an increased plant production. Expanding the 1994 IPCC assessment Schimel (1995) gives the following average balance sheet for the period 1980–1989:

Sources (10^{15} $g\,yr^{-1}$)	
Tropical land use change	1.6
Fossil fuel combustion and cement production	5.5
Sinks (10^{15} $g\,yr^{-1}$)	
Regrowth of Northern Hemisphere forests	0.5
Increased terrestrial plant production	1.6
(From carbon dioxide fertilization	1.0)

Global Change: Effects on Coniferous Forests and Grasslands.
Edited by A. I. Breymeyer, D. O. Hall, J. M. Melillo and G. I. Ågren

UNEP

(From nitrogen (N) fertilization	0.6)
Net uptake in oceans	2.0
Increase in the atmosphere	3.2

This budget is not balanced and there are uncertainties of 50% or more for all terms except the increase in the atmosphere and fossil fuel combustion which are reasonably accurately known. The requirement on accuracy in the terms in the terrestrial budget has to be seen against the background of the terrestrial net primary production of 60×10^{15} g yr^{-1}.

To improve our understanding of ecosystem responses to climate change, SCOPE, the Scientific Committee on Problems of the Environment, sponsored a project on the effects of climate change on production and decomposition in coniferous forests and grasslands. The project had three objectives: (1) a critical assessment of our understanding of the effects of climate on production and decomposition processes in coniferous forests and grasslands, temperate and tropical; (2) a review of extant forest and grassland ecosystem models, and an evaluation of their usefulness for studies of the effects of climate change on production and decay processes; and (3) the development of a foundation for the design of diagnostic and predictive models to describe the effects of climate change on forests and grasslands.

1.2 ORGANIZATION OF THE STUDY

The report contains five major sections. In the executive summary chapter we briefly review the highlights of the volume and identify future experimental and modelling needs.

Section two comprises four chapters, each of which reviews what we know about key aspects of the carbon (C) budgets in major coniferous forest and grassland biomes: boreal forests; temperate coniferous forests; tropical savannas, woodlands and grasslands; temperate grasslands. Several of these chapters discuss the usefulness of studying ecosystems along climatic gradients as part of a global change research program. All point to key research areas for a global change research agenda.

The third section of the book deals with 'ecosystem physiological responses' to global change. Each of the five chapters in this section reviews how an important ecological process might respond to changes in either climate, atmospheric carbon dioxide, or both. The five processes considered are photosynthesis, plant respiration, C allocation in plants, litter decomposition and soil organic matter dynamics.

Section four of the book summarizes the results of work done within the SCOPE project with both grassland and forest models. In a total of four chapters, grassland and forest model structures are reviewed and compared. Model runs for both biome types for various climate change scenarios are reported and analyzed. Grassland models used in this work are CENTURY and GRASS. The forest models used are BIOMASS, BIOME-BGC, CENTURY (forest version), HYBRID, MBL-GEM, PnET-CN, and Q. The focus of both the grassland and forest modelling activities is primarily site-specific. Regional and global modelling activities are reviewed in section five.

Comprised of two chapters, section five deals with two aspects of ecological response to global change; biogeochemical response and biogeographical response. The biogeochemical response includes C and nutrient cycling, while the biogeographical response includes the shifting of species from place to place with climate change. One important conclusion to be drawn from the final section is the importance of coupling biogeochemistry and biogeography models for climate change research. The coupling would not only provide insights into ecological responses to climate change but also into the feedbacks between the terrestrial biosphere and the climate system.

This book reports the state of knowledge as of 1993 but no major developments have occurred during the time between the writing of this report and its final publication date that change the report's conclusions.

1.3 FOREST AND GRASSLAND ECOSYSTEMS

1.3.1 Boreal forests

The boreal forest covers approximately 12×10^6 km^2 or 17% of the earth's land surface. There is now a reasonable data base at the regional and subcontinental scales of the biomass of the boreal forest, though several areas (e.g. Eurasia) still appear to be poorly represented. There is less certainty about the annual productivity of the boreal forest, owing to the more limited number of sites at which measurements have been made, potential differences in measurement techniques (e.g. increment vs whole-ecosystem C budget) and year-to-year variability, as well as the C stored in the soil. Storage of C in boreal forest soils is highly variable and may be a major area of uncertainty in the C storage of the biome.

Fire is recognized as an important control on the structure and functioning of the boreal forest. Few quantitative data are available for changes in C storage and fluxes in the whole ecosystem at the time of burning and in post-fire succession. Water stress on shallow soils is another factor that will increase in importance under a warmer climate. Differences in response between the northern and the southern margins of the boreal forests are likely.

Attempts have been made to calculate the overall C budgets for boreal forest sites. The calculations consider both fire and human impacts. Studies from Canada and Europe suggest that boreal forests may be a substantial sink of C (50–250 Tg C yr^{-1}). The identification of evidence for enhanced boreal tree growth in response to increasing carbon dioxide concentrations since the mid-nineteenth century deserves further attention. This is likely to appear as a positive residual when growth patterns associated with climate change, stand dynamics, etc., are taken out. However, the correlation between rising atmospheric temperatures and carbon dioxide concentrations means that the statistical separation of these two effects will be a difficult task.

Modelling of C budgets in boreal forests ranges from basic ecophysiological models, through whole-tree and ecosystem models that include nutrient availability and uptake components. These models need further verification against data from a wider range of boreal forest ecosystems. There is also a great need to couple physiological and stand dynamic models to better simulate forest dynamics at scales larger than the individual stand.

The direct effect of elevated atmospheric carbon dioxide concentrations on sapling and seedling growth has been established for several boreal forest tree species, but because it is so difficult and costly to perform environmentally relevant experiments on mature trees sapling and seedling experiments have a limited applicability to field conditions. Given the costs of manipulating whole forests, evaluation of the carbon dioxide effect of the boreal forest may have to be based on modelling and dendrological records searching for enhanced tree growth beyond that derived from changes in climate.

1.3.2 Temperate coniferous forests

The area of temperate coniferous and mixed forests is about 2.5×10^6 km^2. Although the data on aboveground biomass and aboveground net primary productivity (NPP) for temperate coniferous forests are numerous, there are few complete C budgets for these ecosystems. This is largely due to the lack of information on fine root production and respiration. Estimates of fine root production range from 30 to 710 g C m^{-2} yr^{-1}.

Aboveground litter inputs and their decay are among the most well studied aspects of the C budgets for temperate coniferous forests. Both climate and initial litter quality are important controls of litter decomposition. Useful predictive models of aboveground litter decay have been developed based on these controls.

Temperate coniferous forests are currently thought to be net C sinks. Many of these forests are accumulating C as they recover from disturbance. In addition, C accumulation may be stimulated by the high N deposition from the atmosphere that many of these forests are subject to, and the increase in atmospheric carbon dioxide concentration.

Nitrogen is considered to be the most common limiting nutrient in temperate coniferous forests. Three major mechanisms by which enhanced N availability increases aboveground net production are: (1) increased net photosynthesis; (2) increased leaf area index (LAI); and (3) reallocation of C from fine roots and mycorrhizae to aboveground tissues.

1.3.3 Tropical savannas, woodlands and grasslands

The contribution of nonforested tropical ecosystems to the global C cycle has received scant attention in the past. Tropical grasslands, savannas and drought-deciduous woodlands occupy at least 11% of the global land surface. If the definition of tropical grass-containing communities is broadened to include hot semiarid shrublands and desert grassland, the cover increases to over 20% of the global land surface. These types form a vegetation continuum with varying proportions of grasses and woody plants, which can only be separated into discrete types by applying arbitrary limits. Savannas, which in most classifications form the largest single class on this continuum, are a tropical vegetation type in which the primary production is approximately equally shared by trees and grasses. They occur in regions with a hot, wet season of 3–9 months' duration, with a warm dry period of little plant growth for the rest of the year. Fires occur every 1 to 10 years during this dry season.

Net annual C fixation in tropical tree–grass systems is about 7.6×10^{15} g C yr^{-1}, with a possible range between 3.2 and 10.8×10^{15} g C yr^{-1}. This is about half of the net annual C fixation attributed to tropical forests. The main controls on C fixation are water availability, nutrient availability and vegetation composition and structure. Between 0.9 and 1.7×10^{15} g C of this fixed C is returned to the atmosphere annually as the products of combustion (over 95% as carbon dioxide). Assuming steady-state conditions the rest returns via the respiration of decomposer organisms and herbivores. However, with our present state of knowledge globally it is unclear whether the C stock in savanna vegetation and soil is increasing or decreasing, since opposing processes predominate in different areas of the world.

The total C stock in tropical grasslands, savannas and woodlands is about 135×10^{15} g C, 80% in the soil and 20% in woody plants. The relatively low average soil C density (6.7 kg C m^{-2}) is due to the sandy nature of many savanna soils, the high soil temperatures, alternating wet-and-dry cycles and frequent fires. The tropical savannas, grasslands and woodlands have a large potential for either sequestration or release of C, depending on future land management practices, climate and atmospheric composition.

There is no direct information on the impact of rising carbon dioxide concentrations on the primary production, decomposition and herbivory in tropical grasslands and savannas. There is a need for several secure, well-documented long-term ecological research sites in tropical grasslands and savanna regions. Information about the extent, frequency and behavior of vegetation fires in this particular fire-prone system is presently inadequate, but will improve substantially within the next few years as research currently in progress is published. The dynamics and mechanisms of the interaction between woody plants and grasses in mixed tree–grass systems are being investigated in an ongoing SCOPE project. Perhaps the largest uncertainty in understanding and predicting the C budgets for the tropical woodlands, savannas and grasslands is the extent, degree and nature of land use changes, along with the human processes which drive it.

1.3.4 Temperate grasslands

Temperate grasslands are located mainly at mid-latitudes of Asia and North and South America where large changes in climate and land use have occurred or are expected to occur in the near future. They occupy approximately 8.3×10^6 km^2. One of the most important characteristics of these ecosystems is that the major C store is located below ground. The size of the belowground C pool is related to temperature, whereas the size of the aboveground C pool is related to annual precipitation.

Primary production is the major C input to grassland ecosystems and its rate is controlled by annual precipitation with belowground production equalling or exceeding the aboveground production, but estimates of its magnitude are uncertain. However, the slope of the production/precipitation relationship is steeper across space than across time. This suggests that the short-term response of these ecosystems to an increase in precipitation is constrained by its structure. Changes in structure are mediated by processes which range from leaf expansion, and recruitment to migration, and which have characteristic response times. Decomposition is the major C output in these grasslands and it is primarily controlled by temperature although soil texture and litter quality play some role. Regional analysis suggests that soil organic C accumulation is more controlled by output (decomposition) than by input (primary production).

The three forces driving global change are related to changes in land use, atmospheric composition, and climate. A large fraction of the temperate grasslands has already been converted into croplands and the loss of C resulting from cultivation appeared to be much larger than the loss resulting from 50 years of the expected climate change. The direct response of grasslands to increases in carbon dioxide concentration is controlled by water and nutrient availability with responses being largest in nutrient-rich, and water-limited systems. Finally, climate change will result in an increase in temperature and an increase or decrease in precipitation depending upon the region selected and the global circulation model consulted. The temperature increase will result in an increase of the decomposition rate. An increase in precipitation will result in an increase of production which may or may not offset the increase in the decomposition rate. An increase in temperature which coincides with a decrease in precipitation will certainly result in a severe C loss.

Concern over losses of soil organic matter and net C release to the atmosphere has resulted in new research on cultivation practices over the past several decades. One of the objectives has been the conservation of soil organic matter which has resulted in crop production methods that either minimize tillage or eliminate it altogether. Preliminary results suggest that these new methods have the potential to conserve or, in some cases, to increase soil C storage. A recent trend in many developed countries of converting cropland back to the original ecosystem will create a new and potentially large category of recovering or successional ecosystems. In the temperate grassland regions in the Northern Hemisphere, such conversions will have important effects on regional C budgets.

1.4 ECOSYSTEM PHYSIOLOGICAL RESPONSES

1.4.1 Photosynthesis in atmospheric and climate change

Photosynthesis provides the driving step to the biogeochemical global C cycle. While much is known about the direct effects of temperature, water vapor pressure deficit and nutrient supply on this process, until recently little has been known about the effects of elevated carbon dioxide concentration over the long term. The instantaneous effects of atmospheric carbon dioxide concentration on photosynthesis are well understood with the mechanisms being highly conserved across terrestrial vegetation; this has provided a sound platform for the development of mechanistic models that are valid at scales from the cell to the landscape. Exploration of the mechanisms shows that it is unreasonable to describe the effects of rising carbon dioxide on photosynthetic C-uptake by use of a simple multiplier, such as the β-factor. Interactions with temperature, N and water can all markedly modify the degree and even the direction of change. This is further complicated by the effects of growth at elevated atmospheric carbon dioxide concentration on the photosynthetic apparatus. Although most of the 'down-regulation' of photosynthesis may be explained as an artifact of the pot size used or the measurements made, there are species differences and clear interactions with nutrient supply. Understanding these species differences and nutrient effects will be critical to the development of sound models for the prediction of the long-term changes of photosynthetic C-uptake by coniferous forests and grasslands.

Theoretical considerations suggest that the relative stimulation of both light-limited and light-saturated C_3 photosynthesis resulting from elevated carbon dioxide concentration will increase with temperature. These are consistent with field observations. The conserved nature of the mechanism of C_3 photosynthesis and its temperature relations provides a strong basis for the development of canopy and vegetation scale models across biomes. The validity of this approach will depend largely on the significance and understanding of second-order effects determined by nutrient limitations, water availability and acclimation, which may alter light-saturated photosynthesis.

1.4.2 Rising atmospheric carbon dioxide and plant respiration

There is now abundant evidence that dark respiration in foliage and possibly in other plant tissues is reduced by approximately 30% for a doubling of atmospheric carbon dioxide. The primary effect is immediate and reversible and this effect appears to be altered by changes in tissue composition in some species.

In slow growing woody perennial species in which soluble carbohydrates do not accumulate, respiration appears to be inhibited and there does not seem to be the level of engagement of the alternative oxidase that occurs in fast growing species including crops.

Although there is evidence that the inhibitory effect of elevated carbon dioxide on respiration extends to nonphotosynthetic microbes involved in decomposition, this would only have practical meaning for a high carbon dioxide world if these decomposers were normally exposed to atmospheric levels of carbon dioxide. It is hard to imagine that the high concentrations of carbon dioxide found in soil could be substantially altered by doubling of atmospheric carbon dioxide. Thus, the rates of soil respiration are not likely to be influenced by increasing atmospheric carbon dioxide except as related to changes in the properties of decomposing plant tissues.

If increasing carbon dioxide concentrations reduce the rates of respiration in all plant tissues and in microbes and fungi as well, then it should also be expected that there would be a significant increase in the C accumulating in ecosystems. This would result from the combined effects of stimulation of photosynthesis and reduction in respiration.

While the results of the effect of elevated carbon dioxide on respiration have been found to vary widely, even small effects would have a significant impact on global C budgets. For this reason, additional study of the role of increasing atmospheric carbon dioxide in respiration should be given a high priority for research.

1.4.3. Carbon and nutrient allocation

The term ‘allocation’ refers to the ways in which plants distribute element and energy resources among the various plant functions including growth, reproduction, maintenance, storage, defense and uptake of additional resources. While the basic patterns of allocation have been described in relation to the environment for many ecosystems, the physiological mechanisms controlling allocation are poorly understood. Nonetheless, a unified theory of allocation is now being developed. It can be represented as a circular set of interactions resulting in positive feedbacks between site quality, plant response, and future site quality. The power of the theory is that it unifies allocation with other ecosystem processes including photosynthesis, foliar longevity, herbivory and element cycling in a single interactive whole.

1.4.4 Elevated carbon dioxide, litter quality and decomposition

Litter from plants grown in elevated carbon dioxide often has a reduced N concentration. Sometimes the litter also has slightly increased concentrations of structural compounds such as lignin, but this is not always the case. Litter from plants grown in elevated carbon dioxide often decomposes more slowly than litter from plants grown at ambient carbon dioxide. This slower decay rate could reduce the rate of nutrient cycling and so feed back negatively to plant productivity. If warming accompanies increases in atmospheric carbon dioxide, the temperature increase will likely increase decay rate. This increase in the rate of decay and nutrient cycling may offset the negative feedback to plant productivity associated with the carbon dioxide-induced decrease in litter quality.

1.4.5 Soils, a source or sink of carbon

Detrital inputs to the soil are estimated at about 60×10^{15} g yr^{-1} and decomposition at 50–60 $\times 10^{15}$ g yr^{-1}. There is no a priori reason to suspect that litter and soil organic matter are in a steady-state condition at present, given the large changes in land use patterns that continue to occur at a global scale and the continuous input of C to the global cycle from fossil fuel burning. One study of soil C accumulation rate suggested that, on a worldwide basis, approximately 0.4×10^{15} g C yr^{-1}, or less than 1% of primary production, escapes decomposition and is added to a soil C store of 1500×10^{15} g. This conclusion is, however, still open for question and more research is needed.

Given the uncertainties associated with C balances on a global scale, let alone on a plot scale, an input–output imbalance of 5–10% could easily escape detection, and imbalances of this magnitude would make soils either a sizeable source or sink for C. With respect to the future, it does not seem possible at this time to predict the balance between the forces that increase soil organic matter accumulation under climate change and those that decrease accumulation.

1.5 MODELLING

1.5.1 Grassland models

The SCOPE Grasslands Modelling Group developed computer models of grassland productivity and nutrient cycling. Rather than attempting to review all existing models, the group decided to concentrate on further validation and development of the CENTURY ecosystem model, a model of terrestrial biogeochemistry based on the effect of climate, human management (fire, grazing), and soil properties on plant productivity and decomposition. The grassland version of CENTURY was tested using observed data from 11 temperate and tropical grasslands around the world. In order to better simulate climatic responses of photosynthetic productivity in a range of grasslands, the CENTURY model was also compared with GRASS, a more ecophysiologically based model of grass productivity which has been applied to both tropical and temperate grasslands. The objective of this effort was twofold: (a) to establish a method to compare ecosystem dynamics across a broad range of environmental factors which influence grassland dynamics and, (b) to evaluate the sensitivity of different grassland communities to changes in climatic patterns as projected by various climate change scenarios.

With respect to the first objective, the results show that CENTURY can simulate soil C and N levels to within 25% of the observed values for a diverse set of soils. Peak live biomass and annual plant production can be simulated within ±25% of the observed values for natural, as well as, burned, fertilized and irrigated grassland sites where annual precipitation ranged from 220 mm to over 1500 mm. Seasonal live biomass can be generally predicted to within ±50% of the observed values. The model underestimated the seasonal live biomass in extremely high plant production years at two of the Russian sites. A comparison of CENTURY model results with statistical models showed that the CENTURY model had higher r^2 values than the statistical models. A comparison of CENTURY with GRASS indicated that the two models behaved in a similar fashion and that no significant difference was determined between the two models in this prediction of green live biomass, annual net productivity and peak live biomass. Data and calibrated model results from this study are useful for analysis and description of grassland C dynamics across different soil, N inputs, and climatic gradients, and will be useful for validating other grassland models, including physiologically-based models.

Climate and carbon dioxide perturbations were modelled with CENTURY for 31 grassland sites, both temperate and tropical, representing 7 ecoregions of the world. The model was validated against actual field data for 16 of these sites. The GRASS model was tested under current climate conditions for a site in Colorado, USA and in Kenya. The projected climate change scenarios of the two general circulation models (GCMs) showed similar temperature changes, but contrasted in their predicted changes in precipitation for certain of these ecoregions, notably the temperate steppes and the humid savannas.

Temperature and precipitation change alone resulted in an increase in total above- and belowground production for the mesic regions (humid temperate and Mediterranean), mainly attributable to increased N mineralization, and decreased plant production in the cold desert steppe regions. Soil organic matter decreased in all the mesic and colder regions, due to increased decomposition. In line with most GCM predictions, the tropical savanna regions were affected the least. Climate-driven redistribution of grassland regions was not modelled, but the world area of grasslands is expected to increase in the future or at least remain constant.

Carbon dioxide increase alone resulted in a significant increase in production in all regions, with the greatest proportional increase in soil organic matter in tropical savanna regions. When combined with predicted climate change, carbon dioxide had an additive effect, tending to ameliorate climate change effects. The net effect of climate change and carbon dioxide was a significant increase in NPP in mesic regions (attributable to N mineralization) as well as in dry savannas, with little or no net change in cold desert steppe or humid tropical regions. Overall, soil organic matter showed a decrease, especially in temperate steppes and cold desert steppes due to stimulation of decomposition by both climate change and carbon dioxide, but tropical savanna and humid savanna regions were actually soil C sinks, regardless of GCM scenario.

Climate change alone predicts a C loss of 3–4 $\times 10^{15}$ g after 50 years of climate change. However, the carbon dioxide enhancement effect, amounting to 2×10^{15} g over the same time, results in a smaller net loss of 1–2 $\times 10^{15}$ g over 50 years. These numbers are substantially lower than previous estimates of Jenkinson and co-workers (1991).

1.5.2 Forest models

The SCOPE Forest Modelling Group compared seven different site-specific forest models under current climate and under future climate scenarios. The models used were BIOMASS, BIOME-BGC, CENTURY (forest version), HYBRID, MBL-GEM, PnET-CN, and Q. These models represent the state of the art for understanding the influence of climate on element cycling, C storage, and the interactions between plant and soil processes in temperate conifer forests. The models differ in time scale, linkages between C, N, and water, representation of heterogeneity, detail of photosynthesis, allocation, and decomposition. The models have been applied to radiata pine stands in Australia and Scots pine stands in Sweden both for validation against existing stand records and for predictions under future climate scenarios. The ability of the models to reproduce to current conditions depends to a large extent on how constrained they are by the initial conditions.

For current and future climates there were major differences among model predictions of important storages and fluxes. Because model predictions for altered climates differed, large uncertainties still remain about the response of temperate conifer ecosystems to changes in air temperature and atmospheric carbon dioxide. Model predictions differed for: (1) the amount of C stored over a stand's lifetime; (2) the absolute response of productivity and C to increased temperature and carbon dioxide; and (3) the magnitude of the response relative to the same model's predictions under current climate. Modelling growth and C storage for the expected life of a stand using a mechanistic approach remains problematic. Therefore, our current knowledge, as incorporated into these mechanistic models, is not sufficient to provide an unambiguous answer to the question of how C storage in specific forest stands will respond to altered climate.

Models predicted that growing forests are strong sinks for C. Additionally, most models did predict that productivity would increase for growing forests under elevated carbon dioxide and all models estimated that C storage would increase. Model predictions also showed that changes in fluxes and C storage that normally occur over a forest's lifetime are far greater than those expected under elevated carbon dioxide or increased temperature.

Four critical areas for future research were identified. First, the model predictions differed for the increased temperature scenario, because the models made different assumptions about how temperature affects productivity. In some models the major controls were on gas exchange and in some models on nutrient mineralization. Second, because the models made different assumptions about the long-term controls over productivity, model predictions for stand development disagreed. These factors are in general poorly understood. Third, the biogeochemical feedbacks constraining plant response to elevated carbon dioxide are difficult to handle. Finally, there is a critical need for flux measurements made at the ecosystem level, and an understanding of the acclimation processes.

1.5.3 Global-scale biogeochemistry modelling

The terrestrial ecosystem model (TEM) was used to estimate NPP, vegetation C, and soil C for the potential global distribution of grasslands and conifer forests under conditions of contemporary carbon dioxide and climate. For grasslands, the TEM estimates an area-average NPP of 267 g C m^{-2} yr^{-1}, vegetation C of 413 g C m^{-2}, and soil C to 1 m of 9160 g C m^{-2}. The estimates for conifer forests are NPP of 378 g C m^{-2} yr^{-1}, vegetation C of 14 600 g C m^{-2}, and soil C of 11 800 g C m^{-2}. TEM was also used to investigate potential responses of NPP and C pools to doubled carbon dioxide and associated climate changes predicted by GCMs. For high- and low-temperature scenarios applied to grasslands, TEM predicts NPP increases of 27 and 18%, vegetation C increases of 36 and 27%, and soil C increases of 4.0 and 6.8%, respectively. For high- and low-temperature scenarios applied to conifer forests, TEM predicts NPP increases of 33 and 27% and vegetation C increases of 34 and 23%, respectively. Soil C in conifer forests decreased by 0.1% for the high-temperature scenario, but increased 2.9% for the low-temperature scenario.

The absolute responses of soil C in grasslands and conifer forests are less when TEM is calibrated for soil C to 20 cm depth rather than 1 m; the response of NPP, vegetation C, and the proportional response of soil C are insensitive to the calibration depth. These results demonstrate that it is important to identify at the calibration site the actively decomposing soil C that is appropriate to the time frame of interest. The responses of TEM for climate change are generally consistent with the results of other models that have examined potential responses of C cycling to climate change, but there is little information available for comparison to changes in both carbon dioxide and climate.

1.5.4 Global-scale biogeography modelling

Assessments of C cycling in coniferous forests and grasslands should be based on a description of land cover with realistic geography and adequate resolution. Such description can be derived from a global geographic data base of existing land cover types. Some global data bases for land cover are described in terms of their underlying assumptions and classification principles. These data bases differ from each other for many regions. We conclude that there is an urgent need for a new assessment of current global land cover for future assessments of global change on ecosystems.

Global ecosystem and land cover distribution is largely driven by climate and human land use. The climatic component is represented by a variety of equilibrium models or climate classifications, differing from each other in terms of inclusion of biophysical processes. Classifications based on annual means of climatic parameters generally have a poor fit. The performance improves considerably if seasonality and an adequate soil moisture balance are taken into account. Recent plant functional type models represent the most realistic and robust patterns of climate-driven ecosystem distribution, and they are therefore valuable tools to determine current and future ecosystem distributions for global change studies. Different applications of global vegetation models are reviewed and discussed.

Global land cover bases and bioclimatic classifications or models have different shortcomings. Nevertheless, a pairwise comparison between data bases from either of these groups may lead to insights into the limitations of either the data base or the model. If so, we can start to derive some general principles which should be included in a model used for global change studies. The discrepancy between different data sets can perhaps be explained by the large differences in the actual classifications. The conclusion is that we have only very approximate comprehension of actual global land cover. This conclusion is further strengthened by the fact that even in the statistical, tabular-based country censuses such as FAO-agricultural and forestry yearbooks, large differences occur between the data.

1.6 FUTURE RESEARCH

This SCOPE project has pointed to several areas of future research.

1.6.1 Reduction of uncertainties in data bases

For all biomes investigated it is recognized that there is a lack of reliable data bases of the extent and rate of production. In some cases well-documented studies of a few sites exist but their representativeness can be questioned. Long-term studies are required of representative sites where trends in behavior can be detected. Short-term flux studies of carbon dioxide and water for tests of our mechanistic interpretations of system properties are also required.

Those mechanisms which are not well understood at present are respiration and allocation. Belowground storage and processes which result in C sequestration and losses require special attention because of the increasing awareness of their importance and their sensitivity to external manipulations.

1.6.2 Reduction of uncertainties regarding key processes

The SCOPE study shows that key areas of divergence between the forest models arise from the formulation of the effects of carbon dioxide on water and nutrient use, on allocation of NPP to different plant components (explicitly or implicitly), and on long-term coupling of C and N storage. The models also differ in the degree to which canopy processes are coupled to the atmosphere and the feedbacks which result from this coupling. It is crucial that modellers and experimentalists communicate new results suggesting mechanisms not included in extant models.

1.6.3 Modularization of models

From the model comparison activity, we concluded that converting all of the models to a modular structure will facilitate model comparisons. A modular structure would make it possible to exchange modules describing key ecological processes such as photosynthesis among models. By doing this we could pinpoint the conceptual differences among models and we could explore the system-level consequences of these differences.

1.6.4 Validation of models

One way to begin to narrow the range of possible responses and to determine which models are most accurate is to collect appropriate measurements from natural ecosystems that can confirm or validate the projections of the models. Certain measurements can and should be made in the near future. In particular, these include flux measurements of C and water exchange over large areas of major ecosystem types. Measurements over the seasonal cycle and, periodically, over several years to determine fluxes during years with different temperatures and precipitation amounts are essential. These data can be used to determine whether measured evapotranspiration and net C exchange are consistent with values predicted by particular models. It is also important to recognize that the ability of a model to simulate current ecological conditions does not validate its ability to accurately simulate responses to future climate change. Model sensitivity to different climates may be examined by simulating ecological responses to Holocene climates or more distant past warm periods for which paleoecological data exist.

1.6.5 Development of models of transient ecological responses

From both a scientific and policy perspective it is critical that we develop models that incorporate transient dynamics and make real-time predictions about the patterns of ecological change. This is not a trivial task, in part because of the numerous aspects of ecological response (e.g. vegetation dieback, migration, succession, soil development) that must be incorporated and because the time constants for different responses can vary widely. Several modelling groups are working on transient models, and we simply want to convey to others the critical importance of this work.

1.7 CONCLUSION

At the start of this SCOPE project there was some doubt as to whether we could usefully catalyze cooperation between empiricists and modellers in a way which would progress research in the field of climate change. Fortunately the research was ahead of its time and initiated research which has contributed significantly to the role of coniferous forests and grasslands in global change. During the project continuous inputs were made by the participants into numerous IGBP and IPCC studies which provided a more solid scientific basis to these studies than would otherwise have been the case.

The review of the forest and grassland ecosystem models provides a unique comparison of models and sites around the world. While the outcome was not as perfect as we initially wished, the exercise has shown the limitations of model comparisons and the problems of using a single model to compare imperfect data from many different sites. Nevertheless, the conclusions with regard to terrestrial C sources and sinks and the role which various physiological processes in plants and soils play in such C fluxes are of considerable importance in predicting global change effects.

As a result of the research undertaken in the project we feel confident in highlighting those areas of research which should receive priority in the future. There is little doubt that such future research should enhance active collaboration between experimentalists and modellers so as to reduce the distinction between such disciplines. This is the way forward to help solve global change research problems.

1.8 REFERENCES

Jenkinson, D. S., Adams, D. E. and Wild, A. (1991) Model estimates of CO_2 emissions from soil in response to global warming. *Nature* **351**, 304–306.

Schimel, D. S. (1995) Terrestrial ecosystems and the carbon cycle. *Global Change Biology* **1**, 77–81.

2 The Carbon Budget of Boreal Forests: Reducing the Uncertainty

T. R. MOORE
Department of Geography and Centre for Climate and Global Change Research, McGill University, Montreal, Canada

2.1 INTRODUCTION

The boreal forest covers approximately 12×10^6 km^2, or 17%, of the land's surface, but it plays important roles in resource utilization in nordic countries, in the storage and cycling of global C and in the terrestrial response to climatic change. Although described in only two words, and deemed to be relatively sparse in plant species diversity, this biome contains a great diversity in structure and function related to variations in climate (e.g. dry–wet, cold–cool), hydrology (xeric–hydric), soil type (nutrient rich–poor) and disturbance (e.g. damage by insects and fires). Thus, a critical issue is to adequately describe this diversity at a global scale and to develop models which will simulate the function, structure and dynamics of the contemporary boreal forest. In addition, the models should provide a vehicle for the examination of potential changes brought about by perturbations, such as increased frequency of disturbance, increased atmospheric inputs and climate change.

In the past five years, attention on the C budget of boreal forests has been heightened by suggestions of the role of this biome in atmospheric chemistry. These include the linkage between seasonal CO_2 concentrations in Alaska and boreal forest photosynthesis (d'Arrigo *et al.* 1987) and the controversial suggestion by Tans *et al.* (1990) that mid-latitude land surfaces may act as a significant sink of CO_2. The latter modelling suggested a major uptake of CO_2 in cold deciduous forests and with boreal forests acting as a small source of CO_2. Very recently, enhanced tree growth and C storage has been reported from European forests (Kauppi *et al.* 1992), and Apps and Kurz (1992) calculated that Canadian forests, dominated by the boreal, are a sink for C. The BOREAS project attempts to address some of these issues in a North American context (BOREAS Science Steering Committee 1990) and Shugart *et al.* (1992) have recently published a volume describing advances in the modelling of boreal forests. As a comparison with an adjacent ecosystem, an assessment of the response of arctic tundra

Global Change: Effects on Coniferous Forests and Grasslands.
Edited by A. I. Breymeyer, D. O. Hall, J. M. Melillo and G. I. Ågren

UNEP

ecosystems to climatic change concluded that carbon/nutrient interactions are of critical importance to an understanding of the functioning of this ecosystem and its response to global change (Shaver *et al.* 1992).

In this chapter, I address three issues: (1) an evaluation of our knowledge of the processes and storage of C in boreal forests and the environmental controls on these; (2) a review of the application of models to simulate boreal forest C storage and dynamics; and (3) an examination of the potential effects of climatic changes on the boreal forest.

2.2 STORAGE AND DYNAMICS OF C IN BOREAL FORESTS

2.2.1 Storage

The storage of C in phytomass and soils within the boreal forest should be the easiest parameter to measure. As could be anticipated from its diversity, estimates of above-ground biomass within the boreal forest cover over an order of magnitude, ranging from < 1 to > 10 kg C m^{-2} (Figure 2.1). Development of volume or mass tables, primarily for commercial forestry, allow the ready estimation of biomass. Not surprisingly, emphasis has been placed on the more productive or commercially and ecologically interesting sites.

Uncertainties are introduced when attempts are made to scale up measurements from individual plots to regional, continental or global scales. Botkin and Simpson (1990), having defined the boreal forest in terms of mean July temperatures, conducted a statistically based assessment of above-ground biomass in the North American boreal forest. Their results, expressed in terms of 12 major regions, suggested an average of 1.9 ± 0.4 kg C m^{-2} and a total of 9.7 ± 2.3 Pg C (Figures 2.1 and 2.2). These figures are substantially less than many previous estimates, but similar to a Forestry Canada estimate (Bonnor 1985; Apps and Kurz 1992; Kurz *et al.* 1992). As appears to be common with many environmental measures, extrapolating plot measurements to spatial scales up to global frequently involves major uncertainties, particularly the representativeness of the sites selected for the point measurements, with a distinct temporal change in estimates (Figure 2.3).

Less agreement is found in estimates of total organic C in the boreal forest system, which combines above- and below-ground phytomass as well as soil organic C. Schlesinger (1977) quotes an average soil organic C content of 15 kg C m^{-2} for boreal forests, equivalent to 24% of the global soil C pool, whereas Kurz *et al.* (1992) estimate a value of 11.8 kg C m^{-2} for the eastern and western sections of the Canadian boreal forest.

Consequently, there is a wide range of above:below ground ratios, as might be expected given the diversity of the environmental conditions of the forest and its history. The process of paludification is common in boreal environments, contributing to the storage of large amounts of soil organic matter. As above-ground

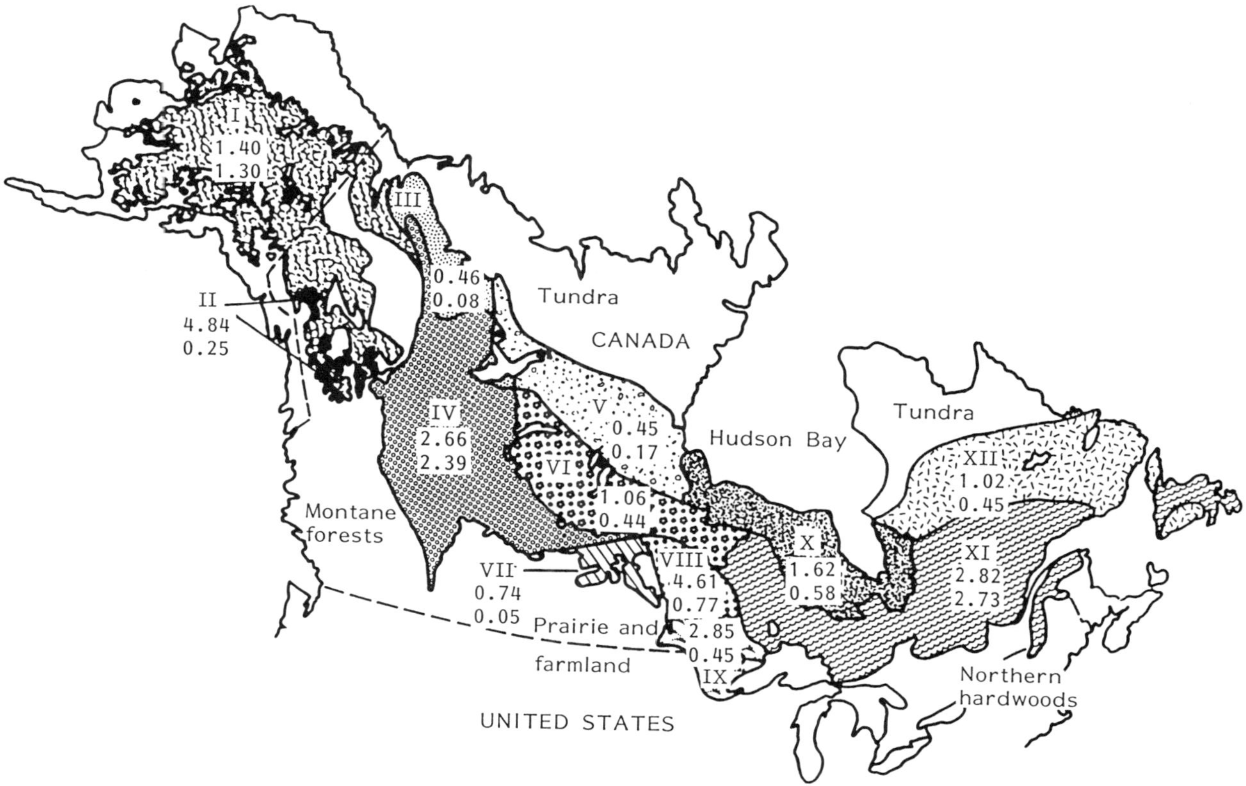

Figure 2.1 Map of biomass in the North American boreal forest, in terms of mean value (upper figures in kg C m^{-2}) and total of each of 12 regions (lower figures in Pg C). Adapted from Botkin and Simpson (1990)

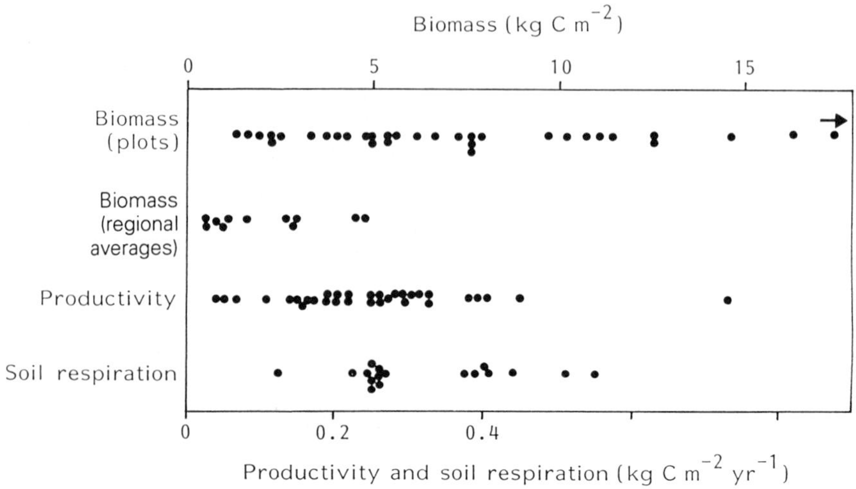

Figure 2.2 Collation of values of boreal forest biomass by plot or regional average, productivity and forest floor respiration. Data derived from DeAnglis *et al.* (1981), Raich and Schlesinger (1992) and Viereck *et al.* (1983)

biomass and vegetation may be a weak indicator of soil type and depth in some regions, there is still major uncertainty in the estimates of C stored in boreal forest soils and this component may be the least reliable in boreal C pools (e.g. Chapin and Matthews 1993). Biomass:soil C ratios vary enormously, but also show significant differences in regional or global estimates. Above:below ground ratios generally range from 0.2 at the northern limit of the boreal forest to 1.0 or more at the southern limit (e.g. Adams *et al.* 1990; Prentice and Fung 1990; Apps and Kurz 1992). The danger in all these studies is that original estimates, which may be biased, tend to be used in later studies, lending credibility to the original, flawed estimates.

These sets of figures have been used in estimations of the role of the boreal forest, and other ecosystems, in storing C in the recent past. However, not surprisingly, there can be major discrepancies, partially associated with differences in ecological definitions used. For example, Adams *et al.* (1990) and Prentice and Fung (1990) have both estimated the changes in C storage associated with vegetation change since the maximum glaciation, about 18 000 yr BP. They used boreal forest (or its approximate equivalent) biomass and soil C figures of 8 and 16 kg C m^{-2} (Prentice and Fung 1990) and 7 and 9 kg C m^{-2} (Adams *et al.* 1990). Whereas Adams *et al.* (1990) estimated that the boreal forest or its equivalents accumulated about 100 Pg C since 18 000 yr BP, Prentice and Fung (1990) estimated about 245 Pg C. In an era when critical attention is being paid to the issue of uncertainty in global change predictions, the spatially averaged data base of C storage in the boreal forest could be improved, as well as an attempt made to use comparable ecological units.

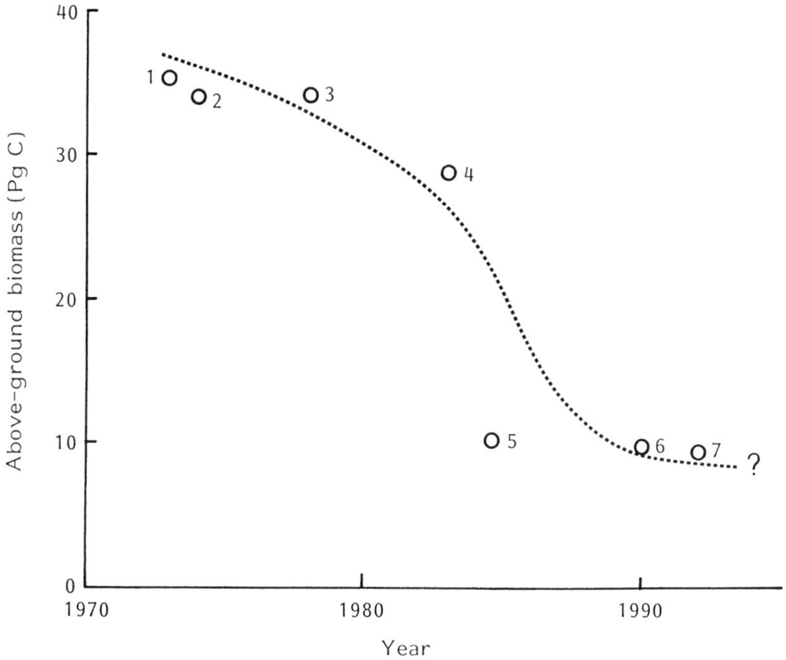

Figure 2.3 Temporal variation in the estimation of the North American boreal forest biomass. Adapted from Botkin and Simpson (1990). Key: (1) Whittacker and Likens (1973); (2) Olson *et al.* (1978); (3) Atjay *et al.* (1979); (4) Olson *et al.* (1983); (5) Bonnor (1985); (6) Botkin and Simpson (1990); (7) Apps and Kurz (1992)

2.2.2 Productivity

The physiological processes contributing to the dynamics of C in boreal forests, such as photosynthesis, respiration, C allocation, etc., have been dealt with elsewhere and will not be examined here. Bonan and Shugart (1989) have reviewed our knowledge and understanding of the environmental factors which influence processes controlling forest structure, dynamics and vegetation patterns (Figure 2.4). Most attention has focused on the above-ground portion of the forest, compared to the activities going on beneath the soil surface. Root and mycorrhizal production and turnover are difficult to measure, yet form an important component of the C cycle of boreal forests.

Net primary production estimates for boreal forests, derived from estimates of increment growth detritus production, range from 0.06 to 0.8 kg C m^{-2} yr^{-1} (Figure 2.3). The issue arises as to how typical these figures are for the boreal forest biome, given the uncertainty concerning the representativeness of biomass estimates.

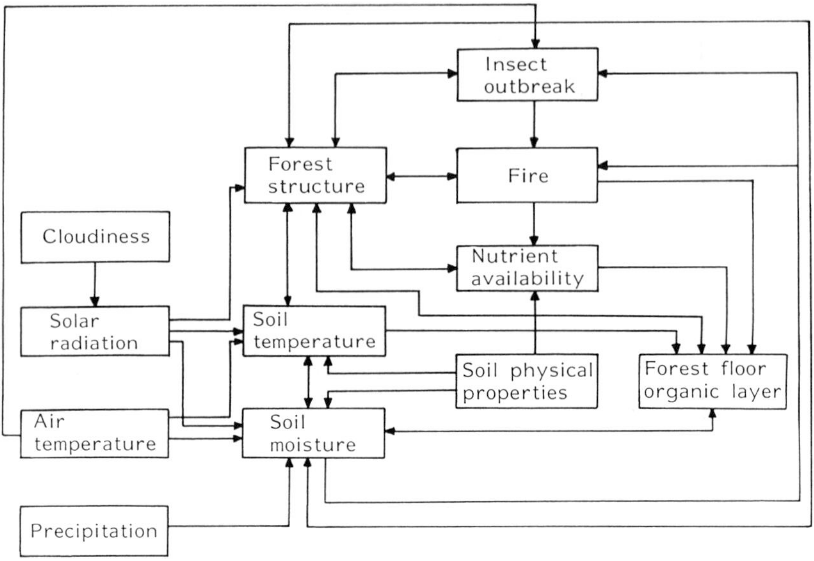

Figure 2.4 Processes controlling boreal forest structure and pattern (adapted from Bonan and Shugart 1989)

Measurements of CO_2 flux from the soil give some indication of the strength of the below-ground respiration component. Annual respiration fluxes for boreal forest soils range from 120 to 550 g C m^{-2} yr^{-1}, with a mean of 322 g C m^{-2} yr^{-1} (Figure 2.3; Raich and Schlesinger 1992). Work in Alaska elucidated the control that environmental variables, such as temperature and moisture content, play in this flux of CO_2 (e.g. Schlentner and van Cleve 1985). Perhaps surprisingly, fire appears to have a minimal effect on this flux, with soil respiration 84–120% of pre-fire values, probably through the confounding effects of root respiration decreases, soil temperature increases, and enhanced nutrient availability (e.g. Weber 1985; Gordon *et al.* 1987).

An alternative approach to measurements of C budget in forests is to measure the annual net exchange of CO_2 with the atmosphere, which combines plant metabolic and soil activities. There is, as far as I am aware, only one study of whole-ecosystem CO_2 flux by a boreal forest. Wofsy *et al.* (1991) reported a growing season (early June–late August) uptake of 65 g C m^{-2} from a mature subarctic spruce–lichen woodland near Schefferville, northern Qubec. This compares with an uptake of CO_2 of 250–350 g C m^{-2} yr^{-1} at the deciduous Harvard forest in the northeastern USA, 15 g C m^{-2} during the growing season in Alaskan tundra and 90 g C m^{-2} during the wet season by an Amazonian forest (Table 2.1). The pattern of uptake in the Harvard forest differed significantly from idealized functions in global models, in terms of both absolute uptake (greater

Table 2.1 Whole-ecosystem exchange of CO_2 with the atmosphere (Wofsy *et al.* 1991, 1993; Wofsy, pers. comm.)

Ecosystem	Period	CO_2 exchange (g C m^{-2})
Subarctic woodland	June–August	−65
Alaskan tundra	June–August	−15
Temperate deciduous forest	Year-round	−300
Tropical forest (Amazon)	Wet season	−90

than predicted) and seasonal pattern (greater uptake in midsummer and greater release in autumn).

Attention has also focused on the ability of remote sensing to link with vegetation parameters to allow an estimation of the seasonal and spatial patterns of plant production. Spectral reflectance, or its remotely sensed correlate, spectral radiance, has commonly been employed in studies of vegetation and atmospheric chemistry (e.g. Tucker *et al.* 1986) and in net primary productivity (e.g. Goward *et al.* 1987). The normalized difference vegetation index (NDVI), combining red light absorption by chlorophyll with green leaf infra-red reflectance, shows a strong correlation with biomass or leaf area index for grasslands and agricultural crops. However, as noted by Ranson and Williams (1992), the complex canopy structure and variability of underlying surface of many boreal forests reduces the direct utility of remote sensing techniques in providing spatially averaged and temporal patterns of biomass, leaf area index or productivity, though remotely sensed information is an important input into models of forest processes over landscape scales. The BOREAS project combining remote sensing and process measurements at boreal forest sites in northern Manitoba and central Saskatchewan in 1993–94 should shed further light on the utility of remote sensing attributes (BOREAS Science Steering Committee 1990).

2.2.3 The effect of disturbance

Disturbance is an important control on the distribution, structure and function of the boreal forest, particularly in terms of fire and insect damage. Stocks (1991) has collated data on the area of burned forest in Canada, Sweden and Finland, and estimates an average annual burn area of 5.6×10^{10} m^2, for 1980–89. There is a large uncertainty in the area of burn in the former Soviet Union, but it has tentatively been estimated at 3×10^{10} m^2 yr^{-1}.

Most of the fires are of a small size (in Canada, 90% are $< 4 \times 10^2$ m^2), but the annual area of burn is primarily formed from a few very large ($> 2 \times 10^6$ m^2) fires. Fire regime can be characterized in terms of the intensity of the fire and its frequency. Fire intensity and temperature are related to the combination of fuel and meteorological conditions, with ranges in intensity of 10–100 000 kW m^{-2} (van Wagner 1983) and temperature of 200–600 °C (e.g. Smith and Sparling 1966).

Table 2.2 Estimates of consumption (%) of biomass and forest floor during fire in the boreal forest, based on data in MacLean and Wein (1980) and Dyrness and Norum (1983)

	Fire intensity		
	Light	Moderate	Severe
Foliage	60	80	100
Twigs	20	40	60
Wood	10	30	50
Roots	50	70	90
Forest floor	21	63	76

Fire frequencies in the boreal forest decrease from the south to the north, with rotation periods ranging, for example, from 130 yr for forests in southern Quebec to several thousand years for northern forest tundra (Cogbill 1985; Payette *et al.* 1989).

During fire, C contained in the phytomass is detritus are converted to gases, primarily CO_2, CO and CH_4 and to particulate matter such as smoke and ash, which may be removed from the site. Quantitative data on the proportion of C lost during fire are largely absent, but, in general, losses depend on fuel level, substrate type and moisture content as well as fire characteristics. During a fire 20–70% of the biomass and forest floor C are combusted (Table 2.2). Of the gases formed during combustion, CO_2 dominates (an average of 89%) with smaller amounts of CO (9%) and CH_4 and non-methane hydrocarbons (2%) (Cofer *et al.* 1991). An additional 20 g of smoke ash per kilogram biomass burned can be formed (Radke *et al.* 1991).

Based on pre- and post-burn vegetation and forest floor C content, fire can release, almost instantaneously, between 0.2 and 7 kg C m^{-2} in boreal forests. In addition, fire can be indirectly responsible for further releasing C from decomposing plant remains and forest floor after the fire, as well as the probable increased productivity and thus C fixation by plants in the post-fire succession. Few data are available for post-fire accumulation of C in biomass and soil (Johnson 1992). Stocks (1991), using an average fuel consumption of 2.5 kg m^{-2} in boreal forests, and the global average of burned area in the boreal forest of 5.6×10^{10} m^2, estimated that about 70 Tg of C would be released annually to the atmosphere from boreal forest fires. This figure is a relatively small proportion of the estimated global total of 3.5–4.0 Pg C yr^{-1} from all biomass burning sources (Andreae 1991).

Insects, although they play a profound role in the structure and landscape pattern in boreal forests (e.g. Holling 1992), probably play a less important direct role than fire in the C budget. Although insect attacks, for example by defoliators such as spruce budworm, may produce large areas of killed trees, rates of boreal forest plant tissue decomposition, except for the most nutrient-rich and lignin-

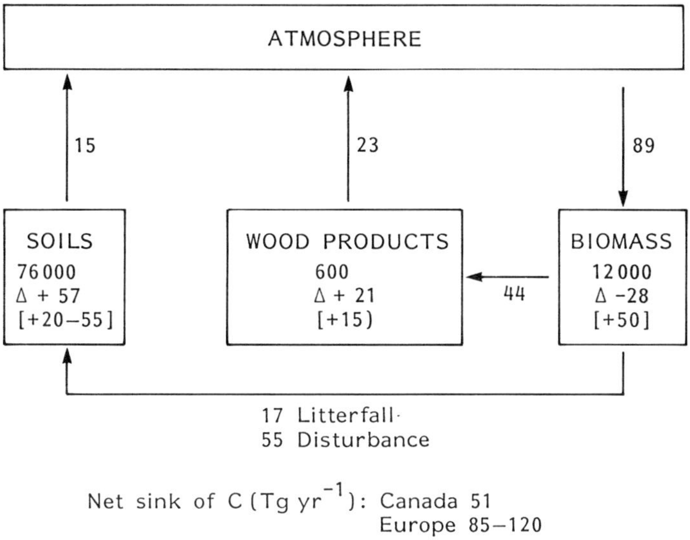

Figure 2.5 An overall C budget for the Canadian forest, 1986, adapted from Apps and Kurz (1992). Figures in brackets are C storage values for similar components of the European forests, based on Kauppi *et al.* (1992)

poor, are slow so that release of C as CO_2 through post-disturbance decomposition is probably minor.

An attempt has recently been made to integrate our knowledge of the different components of the C budget of the contemporary boreal forest, the subcontinental scale. Apps and Kurz (1992) described 41 ecoclimatic regions (dominated by the boreal forest) across Canada in terms of biomass, maturity class and productivity and forest soil C dynamics for three major terrestrial C pools–biomass, forest products and soils (Figure 2.5). Disturbances, including wildfire, insect damage, clear-cut logging with and without slash burning and partial logging, are also included in terms of C storage and dynamics. They integrated these measurements for 1986. Phytomass in the forest was estimated to be 12 Pg C (12×10^{15} g), with a further 76 Pg C in forest soils (including standing dead and downed trees) and 135 Pg C in peatlands. During 1986, live and dead trees, coarse wood and debris were estimated to decrease by 28 Tg C, forest products increase by 21 Tg C and storage in soils (including dead biomass) increased by 57 Tg C. Thus, during 1986, Apps and Kurz (1992) estimated that the forest sector acted as a net sink of 51 Tg C.

In Europe, although recognizing that forest decline and death may be important in some areas, Kauppi *et al.* (1992) used forest resource surveys to estimate an annual build-up in forests over the period 1971–90 of 50 Tg C yr^{-1}. In addition, they estimate that accumulation of other forest biomass and wood products may add a further 20–55 and 15 Tg C yr^{-1}, respectively. These are of the same order as those derived for Canadian forests by Apps and Kurz (1992). In the Canadian context, management (harvesting and fire regimes) is a controlling factor, whereas in Europe Kauppi *et al.* (1992) suggest that enhanced tree growth is derived from fertilization brought about by increased atmospheric deposition of nutrients, especially N. Combined (*ca.* 250 Tg C yr^{-1}), these two sinks for atmospheric CO_2 may form about 20% of the estimated 'missing' C in the global budget and lower than the estimate of 400–2600 Tg C yr^{-1} released from land use changes in the tropics (e.g. Tans *et al.* 1990; Dale *et al.* 1991).

2.3 MODELLING THE C BUDGET OF THE BOREAL FOREST

In the past decade, major advances have been made in the application of models to productivity, C storage, population dynamics and nutrient cycling and decomposition in the boreal forest. Physiological-based models, their structure, application and development, have been reviewed by Ågren *et al.* (1991), and some applications are discussed in this volume (see Chapters 13 and 14). Models range in their scale, from the cell and individual organism to the whole plant/soil system, and range in their time steps from $<$ 1 day to months. Variations also occur in their driving variables from relatively simple, physiological models, driven by CO_2, radiation and water, through to those that incorporate nutrient cycling with a strong interaction between the plant and soil and the processes of decomposition, nutrient release and plant uptake. They also vary in their composition, allowing changes in species composition as well as productivity, and in their spatial scale from individual trees to the landscape.

Since the Ågren *et al.* (1991) review, two other models have been applied to the boreal forest. Bonan (1990a, b, 1991, 1992) combined species change and environmental/biogeochemical models which were initially applied in the simulation of stand characteristics and distribution in Alaska. Since then, the model has been applied to other North American boreal forest stands. The ecophysiological model dealing with CO_2 fluxes, based on photosynthesis, plant productivity, soil respiration and decomposition, was applied to 23 stands in the Fairbanks area (Bonan 1991). Only one of the stands had a positive simulated net CO_2 exchange (i.e. a source). When expressed as the mean for stand type, C uptake ranged from 105 to 731 g C m^{-2} yr^{-1}; however, this is larger than the observed dry matter production in these forests (Table 2.3).

Rastetter *et al.* (1991, 1992) have developed a model applicable to a variety of ecosystems, from forest to tundra, based on a detailed knowledge of the physiological processes and C and N cycles, though without the ability for major successional changes. In some cases, however, there is a dearth of information for some ecosystems for the model to run effectively (e.g. Rastetter *et al.* 1991). There

Table 2.3 Simulated and observed C flux in forest stands near Fairbanks, Alaska (from Bonan 1991)

Stand type	Simulated C flux ($g\ C\ m^{-2}\ yr^{-1}$)	Observed plant production
Black spruce	105	58
White spruce	302	188
Birch	7	241
Aspen	498	290
Balsam poplar	731	283

is still the need for the inclusion of a fire component, important in driving species, C storage and nutrient changes in the boreal forest.

A characteristic of many of these models is the ability to simulate a change in environmental conditions, such as temperature, precipitation and atmospheric CO_2 concentration, thereby allowing a simulation of conditions under 2 × CO_2 climate change scenarios, or through the increased addition of N by atmospheric deposition.

The extrapolation of models from the scale of the stand to landscapes, regions and continents also requires further attention. King *et al.* (1989) used the ABISKO II tundra and CONIFER forest models to simulate CO2 exchange within the region from 64 - 90°N. The simulation produced annual CO_2 exchange patterns which were similar to that of another terrestrial biosphere model and an atmospheric tracer transport model, though there were some discrepancies. The model predicted that the tundra acted as a sink and the boreal forest as a source of CO_2 (-6 and $+79\ g\ C\ m^{-2}\ yr^{-1}$, respectively), suggesting that this latitudinal belt was a source of CO_2.

2.4 THE EFFECT OF CLIMATIC CHANGE ON THE C BUDGET OF THE BOREAL FOREST

2.4.1 Predicted changes in climate associated with greenhouse gases

The development of general circulation models (GCMs) has spawned a series of predictions of changes in climatic parameters, such as temperature, precipitation and soil moisture. A first generation of models was used in the first IPCC assessment of climate change (Houghton *et al.* 1990). More recent models have involved a transient increase in CO_2 concentrations, rather than an instantaneous one, as well as improved atmosphere–ocean coupling. In addition, emission scenarios have been changed (Wigley and Raper 1992) resulting in a plethora of models of climate changes under doubled CO_2 conditions. Nevertheless, the models show similar overall patterns of change in climate. The range of changes predicted for the boreal forest regions of North America and Eurasia in

Table 2.4 Range of climate parameters anticipated to change in the boreal forest regions, associated with a doubling of CO_2 in GCMs (from Houghton *et al.* 1990)

Climate change	Model parameter	CCC	GFDL	UKMO
Summer (JJA)	N. America	+3 to +5	+3 to +7	+3 to +7
temperature (°C)	Eurasia	+3 to +7	+3 to +7	+3 to +7
Winter (DJF)	N. America	+5 to +9	+3 to +9	+3 to +9
temperature (°C)	Eurasia	+3 to +7	+5 to +7	+5 to +10
Summer (JJA)	N. America	−0.5 to +1.5	−0.5 to +0.5	−0.5 to +0.5
precipitation (mm d^{-1})	Eurasia	−0.5 to +1.5	−0.5 to +0.5	+0.5
Winter (JDF)	N. America	−0.5 to +1.5	+0.5	+0.5
precipitation (mm d^{-1})	Eurasia	−0.5 to +0.5	+0.5	+0.5
Summer soil	N. America	−2 to +2	−2 to +1	−2 to +1
moisture (cm)	Eurasia	−2 to +2	−2 to +1	−1 to +1

CCC = Canadian Climate Centre.
GFDL = General Fluid Dynamics Laboratory.
UKMO = UK Meteorological Office.

winter and summer temperature and precipitation and soil moisture for the Canadian Climate Centre, General Fluid Dynamics Laboratory and UK Meteorological Office models are summarized in Table 2.4.

The models suggest that we can anticipate significant changes in temperature and precipitation in the boreal forest by the end of the next century. Winter temperature shows the greatest change, increasing by about 6 °C (except for Alaska), with smaller increases in summer temperatures, perhaps of 4 °C. In both cases, the warming is greater at night than in the day. Precipitation shows an extremely variable pattern, but is unlikely to change in so pronounced a fashion as temperature. Although some areas of the boreal forest show predicted increases in soil moisture in some of the models, there is a general decrease of about 1 cm. It is important to realize that these generalizations hide considerable variability between the models and between the various regions occupied by the boreal forest at present. Temperature increases in the winter are generally greatest at the northern margin of the forest, whereas summer increases often show a maximum at the southern margin, where the reduction in soil moisture is also the greatest. Thus, any attempt to use these climate change predictions must take into account this regional variability and its effect on forest structure and functioning.

2.4.2 Direct effect of CO_2 fertilization on boreal forest growth

There is indisputable evidence that there has been a significant increase in the atmospheric concentration of CO_2 from about 270 ppmv 150 yr ago to about 350 ppmv at present. Through the use of fossil fuels and changes in land use, it is predicted that this concentration will double to about 700 ppmv by the middle to late twenty-first century, though recent projections suggest a reduced rate of growth of CO_2 concentrations (Wigley and Raper 1992). Thus, it is important to evaluate the effect of this increased concentration on the activities of boreal forest from scales of the individual cell, leaf and plant to whole forests and from short- to long-term time scales, though there will be important interactions with factors, such as nutrient and water availability, affecting tree growth.

Eamus and Jarvis (1989) reviewed the evidence for CO_2 fertilization effects on temperate trees and forests and Bazzaz (1990) has addressed similar issues for a broader range of natural vegetation. All boreal forest species have a C_3 photosynthesis pathway. Although several experiments have shown an increase in photosynthesis and growth in response to elevated (usually doubled or larger) CO_2 concentrations, these results may be of limited applicability to field conditions. These limitations include generally young trees, experiments of short duration (*ca.* four months), small rooting volume and unrealistically high availability of moisture and nutrients. Many studies have shown an acclimation to elevated CO_2 concentrations, usually within a year. Growth response for conifers was generally positive, with a median increase of *ca.* 40% (Eamus and Jarvis 1989). Increased growth may lead to increased root exudation of carbohydrates, thereby leading to possibly increased rhizosphere and mycorrhizal effects and increased uptake of nutrients. Using simulation models under $2 \times CO_2$ conditions, Shugart (1984) suggested that important effects were faster gap closure and succession, rather than effects on the species distribution of biomass.

Elevated CO_2 concentrations in field experiments with natural systems led to increased plant production over three years in C_3 species in a salt marsh (Curtis *et al.* 1989), but no significant response was observed in arctic tundra species, after a short-lived increase in C accumulation (Shaver *et al.* 1992). Eamus and Jarvis (1989) suggest that there is a strong probability that CO_2 may be partly responsible for increased growth in temperate forests in Europe, though the elevated CO_2 concentrations have been accompanied by enhanced N fertilization from the atmosphere (Kauppi *et al.* 1992).

One technique to check for this response, particularly in natural forests where historical stand data are unavailable, is through dendroecology, in particular, ring width and character, which would be related to biomass (Fritts and Swetnam 1989). Although sophisticated techniques have been developed in dendrochronology, most of the attention has been directed at those forests which respond to specific changes in environmental conditions, such as climate change, acid precipitation and other atmospheric pollutants. LaMarche *et al.* (1984) attributed increased ring widths in subalpine conifers to rising concentrations of atmospheric CO_2, beyond that expected from changes in climate. However, further research by Graumlich (1991) and Kienst and Luxmoore (1988) has shown that, for the most part, tree-ring width variations can be adequately explained by variations in climatic parameters. Increased water use efficiency arising from elevated CO_2 concentrations would also lead to improved tree growth in drought-susceptible sites. Research at the Lamont-Doherty Observatory has found no trend in tree-ring data that cannot be explained by climate change at several North American sites (Jacoby and d'Arrigo 1989; d'Arrigo and Jacoby 1993), nor at sites in subarctic Qubec (Payette, pers. comm.). Increased ratios between C and nutrients in plant tissues, especially C:N ratios, arising from the CO_2 fertilization effect, may result in slower rates of litter decomposition.

The identification of evidence for enhanced boreal tree growth in response to increasing CO_2 concentrations since the mid nineteenth century deserves further attention, at sites which are typical of the boreal forest, not at those that have special ecological or palaeoecological significance. This is likely to appear as a positive residual when growth patterns associated with climate change, stand dynamics, etc., are taken out, as has been observed at a few sites by Graumlich (1991). However, the correlation between rising atmospheric temperatures and CO_2 concentrations means that statistical separation of these two effects on tree-rings will be a difficult task.

2.4.3 Modelling climate change effects on C budgets of the boreal forest

There are two ways of approaching the effect of climate change on the C budget of boreal forests. One approach utilizes the relationship between vegetation associations and climate parameters, established at the present time and then translated to climate change scenarios. One example of this approach has been shown by Smith *et al.* (1992a, b), who used the Holdridge life zone classification scheme of natural vegetation types and their relationship to precipitation, temperature and potential evapotranspiration. The distribution of the present-day distribution of the life zones was mapped, from contemporary climatic patterns, and then compared to the pattern produced when four $2 \times CO_2$ GCM climate change scenarios were employed.

All four models predict a northward migration of the vegetation types of the boreal forests, displacing tundra and, in turn, being displaced by hardwood forests and grasslands to the south. There was a reduction in the area occupied by the boreal forests, but the four GCM scenarios varied in the scale of this reduction: from a present area of 15×10^{12} m^2, reductions to between 9 and 14×10^{12} m^2 were predicted (Figure 2.6). Using Botkin and Simpson's (1990) average values, this would result in a reduction in boreal forest (as we know it) biomass from 29 to 18–27 Pg C. In addition to the assumed stability of vegetation types under changed climate, and the ability of plants to migrate with changes in the spatial pattern of climate, the results of this study show that a major limitation is the discrepancy in climate changes predicted by the GCMs, even though they are in agreement with the overall pattern of change. Using ecoclimatic provinces and discriminant analysis within Canada, Rizzo and Wiken (1992) provide a

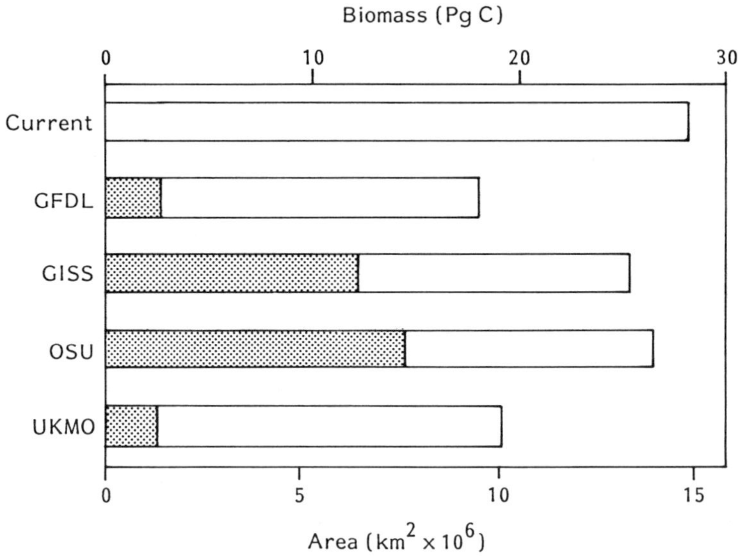

Figure 2.6 Changes in extent of boreal forest predicted from Holdridge life zone relocation, based on four GCMs (from Smith *et al.* 1992a, b). The shaded section represents the area currently occupied by the boreal forest and remains so under the changed climatic conditions of the senarios. The open section represents the area currently occupied by other biomes, which are predicted to change to the boreal forest under the changed climatic conditions. Estimates of change in biomass are based on the average value for the North American boreal forest estimated by Botkin and Simpson (1990)

further example, using the GISS GCM model, which suggests that there will be a reduction in the area of boreal ecoprovince from 28.9 to 14.9% of the land area, and a decrease from 20.1 to 7.8% in the subarctic ecoprovince. Advances in this area will be made through improvements in the reliability of GCM predictions, especially through atmosphere–ocean coupling and the recognition of greater regional differences in climatic change. The utilization of global vegetation models and distributions based on a stronger consideration of plant physiology and physiognomy and soils, as well as climate, will also result in better predictions of the ecological effects of climatic change (Prentice *et al.* 1992).

A second approach is to use process-based forest ecosystem simulation models to determine changes in species composition and biomass and soil organic matter under changed climatic conditions. One example of this is presented by Smith *et al.* (1992b) for seven sites in the North American boreal forest (Table 2.5). The models were run under present-day climatic conditions for 500 yr, then a linear change of climate under 2 × CO_2 scenarios was applied over a 70 yr period, and the changes in species composition and C storage predicted. Again, there are major differences in the response from four common GCM models, suggesting

Table 2.5 Changes in above-ground biomass at seven North American boreal forest sites under four GCM scenarios, 150 yr after creation of new climatic conditions (from Smith *et al.* 1992b)

Site	Species	Above-ground biomass (kg m^{-2})				
		Present-day	GISS	GFDL	OSU	UKMO
Alaska	Various	6	2	0	1	0
Alaska	Black spruce	4	7	4	6	4
N. Quebec	Black spruce	4	11	12	11	12
Central Quebec	Black spruce	10	9	0	10	0
Newfoundland	Black spruce	12	3	0	8	0
Newfoundland	Fir	15	15	13	16	0
New Brunswick	Jack pine	9	8	0	9	0

that the GCMs are a major limitation in the application of these techniques. There is an incompability in the resolution scales of most GCMs (200–500 km) with that of landscape-induced spatial variations in forest type (1–10 km).

A further example of the range of responses utilizing GCM-based climate change scenarios with forest dynamics models is provided by Pastor and Post (1988). They used an early version of the UKMO GCM and a forest succession model that uses climate, soil moisture and N availability to drive the model (Pastor and Post 1986). Present-day species composition, biomass and soil organic matter were included with a run of 200 yr, then 2 × CO_2 climate change effects introduced linearly over a 100 yr period, with the model being run for a further 200 yr under the changes. Two soil types were part of the simulation, a silty clay loam with an available water capacity of 18.3 cm m^{-1} and a sand with an available water capacity of 10 cm m^{-1}. Their results are summarized in Figure 2.7, for biomass and soil organic matter for 11 sites spanning the North American boreal forest.

The simulations of biomass show an enormous range of changes and reveal the importance of soil available water capacity in determining changes in the southern sections of the boreal forest. In general, only the three northernmost sites (N. Manitoba, N. Quebec and N. Maine) show a consistent increase in biomass, whereas the remainder showed varying degrees of decrease or slight increases. This is a response to the importance of low soil moisture availability in the southern forests, exacerbated under the climate change scenarios, and the response of the northern forests to increased growing degree-days. Less drastic changes were encountered when the loam soil was used in the simulations. The predicted changes in soil organic matter were less than those for biomass, but again a wide range (−43 to +80%). As noted by Pastor and Post (1988), these models illustrate the diversity of response to climate change within the boreal forest, where the dominant controls on tree growth are temperature, moisture or N availability. However, they do not allow for a CO_2 fertilization effect, changed

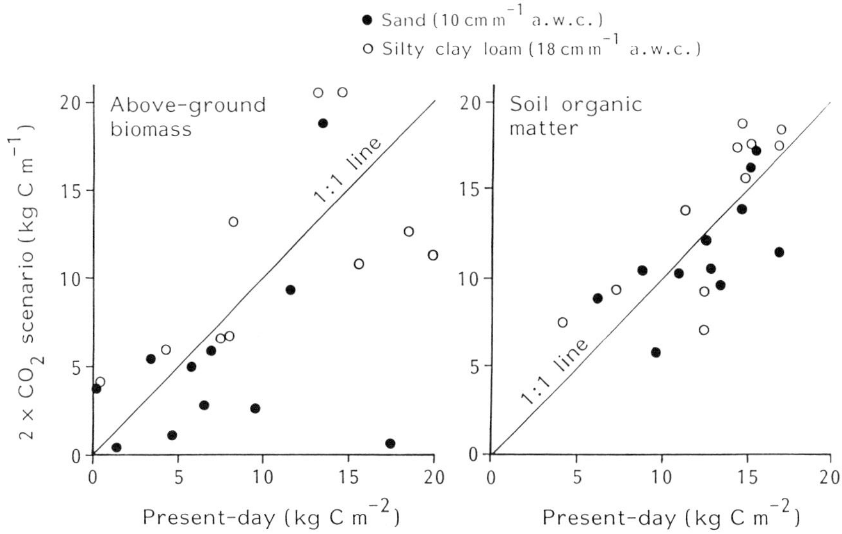

Figure 2.7 Predicted changes in above-ground biomass and soil organic matter based on a $2 \times CO_2$ climate change scenario, with a sand soil (from Pastor and Post 1988). a.w.c. = available water capacity

leaf area, improved water use efficiency, or changed fire frequency and its effect on succession. Using a similar biogeochemical forest model, Rastetter *et al.* (1992) suggested for a hardwood forest that doubling CO_2 concentrations would lead to a biomass increase of 5%, an increase in temperature of 5 °C to 23% and a combination of CO_2 and temperature to a 28% increase in biomass. Soil organic matter would change little. Using a terrestrial ecosystem model and several GCM predictions of $2 \times CO_2$ scenarios, Melillo *et al.* (1993) estimated that global boreal forest net primary productivity will increase by an average of only 4–5%, a major constraint being the availability of nutrients, especially N.

The time is ripe for a comparison of forest models, using the same climatic change scenarios, such as appear in this volume (Chapters 13 and 14). The work so far has shown the uncertainties induced in C changes associated with different GCMs, so now is the time to evaluate the forest models. One indirect comparison can be made: both Pastor and Post (1988) and Smith *et al.* (1992b) used climatic data from the Schefferville region of northern Quebec. The Pastor and Post (1988) study suggests an increase in above-ground biomass from 8–9 kg m^{-2} currently (unrealistically large for well-drained sites) to 12–13 kg m^{-2} under the GISS $2 \times CO_2$ scenario; the Smith *et al.* (1992b) study predicts an increase from 4 kg m^{-2} at present (a realistic number) to 11 - 12 kg m^{-2} under the four GCM scenarios. The agreement may be fortuitous, given the differences in the GCM predictions used, the initial conditions, and the length of the run. One may also question the reality of the models. For example, the Pastor and Post (1988) model used an available water capacity of 10–18 cm m^{-1}. In the spruce–lichen woodlands of subarctic Quebec, the soils are shallow, infertile and cool, with the roots concentrated into the thin organic surface horizons, limiting the available water capacity to perhaps 2 cm. Thus, it is probably a myth to assume that, under $2 \times CO_2$ scenarios, temperature will be all-important in a 'greening' of the northern boreal forest, whereas increased stress on soil moisture may be more important and reduce biomass.

Much less agreement occurs when comparing the northern Maine and New Brunswick sites in the two studies. Pastor and Post (1998) predict increases from 30–32 to 42–46 kg m^{-2}, whereas Smith *et al.* (1992b), using a jack pine forest, predict changes from 9 kg m^{-2} at the present day to 0–9 kg m^{-2} under the GCM scenarios.

2.4.4 Fire frequency

Fire plays an important role in the boreal forest, and it should be expected that changes in climate will produce changes in the frequency of fires. Within Canada, the probability of fire within a forest is represented by the fire weather index, which is a daily or seasonal parameter incorporating fuel moisture content, surface wind, humidity, precipitation and temperature and fuel build-up. Flannigan and van Wagner (1991) have used predicted changes in climate parameters from three GCMs based on doubled CO_2 concentrations to estimate the change in the seasonal severity rating for forest fires for six sites across Canada, compared to the range and mean for a 35 year record.

Combining changes in temperature, precipitation and humidity in the GCM models produced ratios in the seasonal fire severity rating, compared to present-day conditions, ranging from 1.9 to 3.5, with an average of 2.6, and with the Oregon State University and Goddard Institute of Space Science GCMs portraying a more conservative pattern. When modification included only temperature and precipitation changes (leaving relative humidity constant) the ratio of modelled versus present day was 1.5. Flannigan and van Wagner (1991) caution, however, that actual fire frequency will depend not only on the climate parameters leading to a drying, but also ignition frequency and fire control activity. Thus, there is often a poor correlation between the seasonal severity rating and the area burned over a number of years (r^2 0–0.34). This study, then, suggests that climate change as envisaged by the GCMs will increase the potential for fires in the boreal forest of Canada by about 40%, with obvious effects not only on the structure of the forest and the C budget, but also in terms of species and forest type redistribution in response to climate change.

Overpeck *et al.* (1991) suggested that increases in moist air convective penetration predicted by GCMs will lead to increased intensity and frequency of thunderstorms, and thus ignition frequency. Flannigan and van Wagner (1991) note, however, that the burned area in Canada has increased more than 50% since 1970, and it is uncertain whether this is a response to global warming, or natural variability and changes in management strategies. The fire-danger index used in Soviet forestry is presumably amenable to the same response to climate change parameters, as it is based on temperature, humidity and precipitation (Antonovski *et al.* 1992).

Overpeck *et al.* (1991) used the FORENA stand-simulation model with changes in precipitation and temperature predicted in the GCM model and increases in disturbance frequency (0.00–0.01) for the southern boreal forest in Quebec. The results of this simulation show not only a change in forest composition, but also a speeding of the replacement of species, in response to the climate change. Changes in species composition, and thus C storage and dynamics, are also likely to be significantly altered in areas where vegetation patterns are strongly influenced by fire. Thus, increased fire frequency may be expected to have a dramatic effect on the northern tree-line, where features are preserved from warmer conditions in the absence of fires. Increased fire frequency would lead not only to the more rapid destruction of these tree-line relicts, but also, conversely, to the possibility of a more rapid northward extension of the tree-line (Payette 1992; Sirois 1992). Similarly, climate changes will have pronounced effects at the southern, grassland border of the boreal forest (e.g. Clark 1988). Combination of stand dynamics with C storage during fire sequences will provide information on the effect of fire frequency changes on C in the boreal forest.

2.5 CONCLUSIONS

There is now a reasonable data base of the regional and subcontinental scales of the biomass of the boreal forest, though several areas (e.g. Eurasia) still appear to be poorly represented. There is less certainty of the annual productivity of the boreal forest, owing to the more limited number of sites at which measurements have been made, potential differences in measurement techniques (e.g. increment vs whole-ecosystem C budget) and year-to-year variability, as well as the C stored in the soil. Storage of C in boreal forest soils is highly variable and may be a major area of uncertainty in the C storage of the biome.

Remote sensing techniques, utilizing spectral characteristics of the canopy, offer an alternative mechanism for mapping productivity and biomass and their spatial and temporal variability, though reliable methods have yet to be implemented in a wide range of diverse boreal forests.

Although fire is recognized as an important control on the structure and functioning of the boreal forest, few quantitative data are available for changes in C storage and fluxes in the whole ecosystem, during fire and in post-fire succession.

Using spatial extrapolations, attempts have been made to calculate the overall C budget of forests, including fire and human impacts, and studies from Canada and Europe suggest that boreal forests may be a substantial sink of C (50–250 Tg C yr^{-1}).

Modelling of C budgets in boreal forests ranges from basic ecophysiology, through those dealing with whole-tree and ecosystem models with a nutrient availability and uptake component. These models need further verification against a wider range of boreal forest ecosystems and the better meshing of empirical data. There is also a great need to couple physiological and stand dynamic models to better simulate forest dynamics at scales larger than the individual stand.

The direct effect of elevated atmospheric CO_2 concentrations on sapling and seedling growth has been established for several boreal forest tree species, but these experiments have a limited applicability to field conditions, in terms of their duration and environmental conditions. Given the costs of manipulating whole forests, evaluation of the CO_2 effect of the boreal forest will have to be based on modelling and dendrological records searching for enhanced tree growth beyond that derived from changes in climate.

General circulation models have predicted significant changes in temperature, precipitation and available water for boreal forest regions under $2 \times CO_2$ scenarios. However, there are discrepancies in the magnitude and seasonality of the change, therefore when these changes are used in forest models, there are major differences in response of forests associated with the different GCMs. In general, it is anticipated that biomass will increase significantly at the northern margin of the boreal forest, driven by temperature increases, but the response at the southern margin will be much more variable, as influenced by temperature, water and CO_2 availability and the differential forest response. Comparisons of forest model predictions with standard climatic change parameters should be attempted. There is also a need to incorporate changes in fire frequency (predicted to increase significantly under $2 \times CO_2$ scenarios) into forest/climate change models and to couple physiological and stand dynamic models.

2.6 REFERENCES

Adams, J. M., Faure, H., Faure-Denard, L., McGlade, J. M. and Woodward, F. I. (1990) Increases in terrestrial carbon storage from the Last Glacial Maximum to the present. *Nature* **348**, 711–714.

Ågren, G. I., McMurtie, R. E., Parton, W. J., Pastor, J. and Shugart, H. H. (1991) State-of-the-art of models of production–decomposition linkages in conifer and grassland ecosystems. *Ecol. Appl.* **1**, 118–138.

Andreae, M. O. (1991) Biomass burning: its history, use, and distribution and its impact on environmental quality and global change. In: Levine, J. S. (Ed.) *Global Biomass Burning: Atmospheric, Climatic and Biospheric Implications*, pp. 3–21. MIT Press,

Cambridge, Mass.

Antonovski, M .Y., Ter-Mikaelian, M. T. and Furyaev, V. V. (1992) A spatial model of long-term forest fire dynamics and its applications to forests in western Siberia. In: Shugart, H. H., Leemans, R. and Bonan, G. B.(Eds) *A Systems Analysis of the Global Boreal Forest*, pp. 373–403. Cambridge University Press, Cambridge.

Apps, M. J. and Kurz, W. A. (1992) Assessing the role of Canadian forests and forest sector activities in the global carbon balance. *World Res. Rev.* **3**, 333–344.

Atjay, G. L., Ketner, P. and Davigneaud, P. (1979) Terrestrial primary production and phytomass. In: Bolin, B., Degens, E. T., Kempe, S. and Ketner, P.(Eds) *The Global Carbon Cycle*, pp. 129–182. J. Wiley & Sons, Chichester.

Bazzaz, F. A. (1990) The response of natural ecosystems to the rising global CO_2 levels. *Ann. Rev. Ecol. Syst.* **21**, 167–196.

Bonan, G. B. (1990a) Carbon and nitrogen cycling in North American boreal forests. I. Litter quality and thermal regime effects in interior Alaska. *Biogeochemistry* **10**, 1–28.

Bonan, G. B. (1990b) Carbon and nitrogen cycling in North American boreal forests. II. Biogeographic patterns. *Can. J. For. Res.* **20**, 1077–1088.

Bonan, G. B. (1991) Atmosphere–biosphere exchange of carbon dioxide in boreal forests. *J. Geophys. Res.* **96**, 7301–7312.

Bonan, G. B. (1992) A simulation analysis of environmental factors and ecological processes in North American boreal forests. In: Shugart, H.H., Leemans, R. and Bonan, G. B.(Eds) *A Systems Analysis of the Global Boreal Forest*, pp. 404–427. Cambridge University Press, Cambridge.

Bonan, G. B. and Shugart, H. H. (1989) Environmental factors and ecological processes in boreal forests. *Ann. Rev. Ecol. Syst.* **20**, 1–28.

Bonnor, G. M.(1985) *Inventory of Forest Biomass in Canada.* Canadian Forestry Service. Petawawa National Forestry Research Institute, Ont.

BOREAS Science Steering Committee (1990) Charting the boreal forest's role in global change. *EOS* **72**, 33, 35, 40.

Botkin, D. B. and Simpson, L. G. (1990) Biomass of the North American boreal forest. *Biogeochemistry* **9**, 161–74.

Chapin, F. S. and Matthews, E. (1993) Boreal carbon pools: approaches and constraints in global extrapolations. In: Vinson, T. S. and Kolchugina, T. P.(Eds) *Carbon Cycling in Boreal Forest and Sub-Arctic Ecosystems*, pp. 9–20. Proceedings International Workshop, EPA 600/R-93/084.

Clark, J. S.(1988) Effect of climate change on fire regimes in northwestern Minnesota. *Nature* **334**, 233–235.

Cofer, W. R., Levine, J. S., Winstead, E. L. and Stocks, B. J. (1991) Trace gas and particulate emissions from biomass burning in temperate ecosystems. In: Levine, J.S.(Ed.) *Global Biomass Burning: Atmospheric, Climatic and Biospheric Implications*, pp. 203–208. MIT Press, Cambridge, Mass.

Cogbill, C. V. (1985) Dynamics of the boreal forests of the Laurentian Highlands, Canada. *Can. J. For. Res.* **15**, 252–261.

Curtis, P. S., Drake, B. G. and Whigham, D. F. (1989) Nitrogen and carbon dynamics in C_3 and C_4 estuarine marsh plants grown under elevated CO_2 in situ. *Oecologia* **78**, 297–301.

Dale, V. H., Houghton, R. A. and Hall, C. A. S. (1991) Estimating the effects of land-use change in global atmospheric CO_2 concentration. *Can. J. For. Res.* **21**, 87–90.

D'Arrigo, R. D. and Jacoby, G. C. (1993) Tree growth–climate relationships at the northern boreal forest tree line of North America: evaluation of potential response to increasing carbon dioxide. *Glob. Biogeochem. Cyc.* **7**, 525–536.

D'Arrigo, R. D., Jacoby, G.C. and Fung, I.Y. (1987) Boreal forests and atmosphere–

biosphere exchange of carbon dioxide. *Nature* **329**, 321–323.

DeAngelis, D. L., Gardner, R. H. and Shugart, H. H. (1981) Productivity of forest ecosystems studied during the IBP: the woodlands data set. In: Reichle, D. E.(Ed.) *Dynamic Properties of Forest Ecosystems*, pp. 567–672. Cambridge University Press.

Dyrness, C. T. and Norum, R. A. (1983) The effects of experimental fires on black spruce forest floors in interior Alaska. *Can. J. For. Res.* **13**, 879–893.

Eamus, D. and Jarvis, P. G. (1989) The direct effects of increase in the global atmospheric CO_2 concentration on natural and commercial temperate trees and forests. *Adv. Ecol. Res.* **19**, 1–55.

Flannigan, M. D. and van Wagner, C. E. (1991) Climate change and wildfire in Canada. *Can. J. For. Res.* **21**, 66–72.

Fritts, H. C. and Swetnam, T. W. (1989) Dendroecology: a tool for evaluating variations in past and present forest environments. *Adv. Ecol. Res.* **19**, 111–188.

Gordon, A. M., Schlentner, R. E. and van Cleve, K. (1987) Seasonal patterns of soil respiration and CO_2 evolution following harvesting in the white spruce forests of interior Alaska. *Can. J. For. Res.* **17**, 304–311.

Goward, S. N., Kerber, A., Dye, D. and Kalb,V. (1987) Comparison of North and South American biomes from AVHRR observations. *Geocarto International* **2**, 27–39.

Graumlich, L. (1991) Subalpine tree growth, climate and increasing CO_2: an assessment of growth trends. *Ecology* **72**, 1–11.

Holling, C. S. (1992) The role of forest insects in structuring the boreal landscape. In: Shugart, H. H., Leemans, R. and Bonan, G. B. (Eds) *A Systems Analysis of the Global Boreal Forest*, pp. 170–191. Cambridge University Press, Cambridge.

Houghton, J. T., Jenkins, G. J. and Ephraums, J. J. (Eds) (1990) *Climate Change. The IPCC Scientific Assessment*. Cambridge University Press.

Jacoby, G. L. and d'Arrigo, R. D. (1989) Reconstructed northern hemisphere annual temperature since 1871. *Clim. Change* **14**, 39–59.

Johnson, D. W. (1992) Effects of forest management on soil carbon storage. *Water Air Soil Pollut.* **64**, 83–120.

Kauppi, P. E., Mielikäinen, K. and Kuusela, K. (1992) Biomass and carbon budget of European forests, 1971 to 1990. *Science* **256**, 70–74.

Kienst, F. and Luxmoore, R. J. (1988) Tree-ring analysis and conifer growth responses to increased atmospheric CO_2 levels. *Oecologia* **76**, 487–495.

King, A. W., O'Neill, R. V. and DeAngelis, D. L. (1989) Using ecosystem models to predict regional CO_2 exchange between the atmosphere and the terrestrial biosphere. *Glob. Biogeochem. Cyc.* **3**, 337–362.

Kurz, W. A., Apps, M. J., Webb, T. M. and McNamee, P. J. (1992) *The Carbon Budget of the Canadian Forest Sector: Phase I.* Information Report NOR-X-326, Canadian Forestry Service, Edmonton, Alberta.

LaMarche, V. C., Graybill, D. A., Fritts, H. C. and Rose, M. R. (1984) Increasing atmospheric carbon dioxide: tree-ring evidence for growth enhancement in natural vegetation. *Science* **225**, 1019–1021 and **231**, 859–860.

MacLean, D. A. and Wein, R. W. (1980) Simulation of wild fire effects on the nitrogen cycle of a *Pinus banksiana* ecosystem in New Brunswick. *Ecol. Model.* **10**, 167–92.

Melillo, J. M., Kicklighter, D. W., McGuire, A. D., Moore, B., III, Vörösmarty, C. J. and Grace, A. L. (1993) Global climate change and terrestrial net primary production. *Nature* **363**, 234–240.

Olson, J. S., Pfuderer, H. A. and Chan, Y. H. (1978) *Changes in the Global Carbon Cycle and the Biosphere.* ORNL/EIS-109, Oak Ridge National Laboratory, Oak Ridge, Tenn.

Olson, J. S., Watts, J. A. and Allison, L. J. (1983) *Carbon in Live Vegetation of Major World Ecosystems.* ORNL-5862, Oak Ridge National Laboratory, Oak Ridge, Tenn.

Overpeck, J. T., Rind, D. and Goldberg, R. (1991) Climate-induced changes in forest disturbance and vegetation. *Nature* **343**, 51–53.

Pastor, J. and Post, W. M. (1986) Influence of climate, soil moisture and succession on forest carbon and nitrogen cycles. *Biogeochemistry* **2**, 3–28.

Pastor, J. and Post, W. M. (1988) Response of northern forests to CO_2-induced climate change. *Nature* **334**, 55–58.

Payette, S. (1992) Fire as a controlling process in the boreal forest. In: Shugart, H. H., Leemans, R. and Bonan, G. B. (Eds) *A Systems Analysis of the Global Boreal Forest*, pp. 144–169. Cambridge University Press, Cambridge.

Payette, S., Morneau, C., Sirois, L. and Desponts, M. (1989) Recent fire history of the northern Québec biomes. *Ecology* **70**, 656–673.

Prentice, I. C., Cramer, W., Harrison, S. P., Leemans, R., Monserud, R. A. and Solomon, A. M. (1992) A global biome model based on plant physiology and dominance, soil properties and climate. *J. Biogeogr.* **19**, 117–134.

Prentice, K. C. and Fung, I. Y. (1990) The sensitivity of terrestrial carbon storage to climate change. *Nature* **346**, 48–51.

Radke, L. F., Hegg, D. A., Hobbs, P. V., Nance, J. D., Lyons, J. H., Laursen, K. K., Weiss, R. E., Riggan, P. J. and Ward, D. E. (1991) Particulate and trace gas emissions from large biomass fires in North America. In: Levine, J. S. (Ed.) *Global Biomass Burning: Atmospheric, Climatic and Biospheric Implications*, pp. 209–224. MIT Press, Cambridge, Mass.

Raich, J. W. and Schlesinger, W. H. (1992) The global carbon dioxide flux in soil respiration and its relationship to vegetation and climate. *Tellus* **42**B, 81–99.

Ranson, K. J. and Williams, D. L.(1992) Remote sensing technology for forest ecosystem analysis. In: Shugart, H. H., Leemans, R. and Bonan, G. B.(Eds) *A Systems Analysis of the Global Boreal Forest*, pp. 267–290. Cambridge University Press, Cambridge.

Rastetter, E. B., McKane, R. B., Shaver, G. R. and Melillo, J. M. (1992) Changes in C storage by terrestrial ecosystems: how C–N interactions restrict responses to CO_2 and temperature. *Water Air Soil Pollut.* **64**, 327–344.

Rastetter, E. B., Ryan, M. G., Shaver, G. R., Melillo, J. M., Nadelhoffer, K. J., Hobbie, J. E. and Aber, J. D. (1991) A general biogeochemical model describing the responses of the C and N cycles in terrestrial ecosystems to changes in CO_2, climate, and N deposition. *Tree Physiol.* **9**, 101–126.

Rizzo, B. and Wiken, E. (1992) Assessing the sensitivity of Canada's ecosystems to climatic change. *Clim. Change* **21**, 37–55.

Schlentner, R. L. and van Cleve, K. (1985) Relationships between CO_2 evolution from soil, substrate temperature, and substrate moisture in four mature forest types in interior Alaska. *Can. J. For. Res.* **15**, 97–106.

Schlesinger, W. H.(1977) Carbon balance in terrestrial detritus. *Ann. Rev. Ecol. Syst.* **8**, 51–81.

Shaver, G. R., Billings, W. D., Chapin, F. S., Giblin, A. E., Nadelhoffer, K. J., Oechel, W. C. and Rastetter, E. B. (1992) Global change and the carbon balance of arctic ecosystems. *BioScience* **42**, 433–441.

Shugart, H. H. (1984) *A Theory of Forest Dynamics: The Ecological Implications of Forest Succession Models*. Springer-Verlag, New York.

Shugart, H. H., Leemans, L. and Bonan, G. B. (Eds) (1992) *A Systems Analysis of the Global Boreal Forest*. Cambridge University Press, Cambridge.

Sirois, L. (1992) The transition between boreal forest and tundra. In: Shugart, H. H., Leemans, R. and Bonan, G. B. (Eds) *A Systems Analysis of the Global Boreal Forest*, pp. 196–215. Cambridge University Press, Cambridge.

Smith, D. W. and Sparling, J. H. (1966) The temperature of surface fires in jack pine barrens. I. The variation in temperature with time. *Can. J. Bot.* **44**, 1285–1292.

Smith, T. M., Leemans, R. and Shugart, H. H. (1992a) Sensitivity of terrestrial carbon storage to CO_2-induced climate change: comparison of four scenarios based on general circulation models. *Clim. Change* **21**, 367–384.

Smith, T. M., Shugart, H. H., Bonan, G. B. and Smith, J. B. (1992b) Modelling the potential response of vegetation to global climate change. *Adv. Ecol. Res.* **21**, 93–116.

Stocks, B. J. (1991) The extent and impact of forest fires in northern circumpolar countries. In: Levine, J. S. (Ed.) *Global Biomass Burning: Atmospheric, Climatic and Biospheric Implications*, pp. 197–202. MIT Press, Cambridge, Mass.

Tans, P. P., Fung, I. Y. and Takahashi, T. (1990) Observational constraints on the global atmospheric CO_2 budget. *Science* **247**, 1431–1438.

Tucker, C. J., Fung, I. Y., Keeling, C. D. and Gammon, R. H. (1986) Relationships between atmospheric CO_2 variations and a satellite derived vegetation index. *Nature* **319**, 195–9.

Van Wagner, C. E. (1983) Fire behaviour in northern conifer forests and shrublands. In: Wein, R.W. and MacLean, D. A. (Eds) *The Role of Fire in Northern Circumpolar Ecosystems*, pp. 65–80. Wiley, New York.

Viereck, L. A., Dyrness, C. T., van Cleve, K. and Foote, M. J. (1983) Vegetation, soils, and forest productivity in selected forest types in interior Alaska. *Can. J. For. Res.* **13**, 703–720.

Weber, M. G. (1985) Forest soil respiration in eastern Ontario jack pine ecosystems. *Can. J. For. Res.* **15**, 1069–1073.

Whittacker, R. H. and Likens, G. E. (1973) Carbon in the biota. In: Woodwell, G. M. and Pecan, E. V. (Eds) *Carbon and the Biosphere*, pp. 281–302. U.S. National Technical Information Service.

Wigley, T. M. L. and Raper, S. C. (1992) Implications for climate and sea level of revised IPCC emissions scenarios. *Nature* **357**, 293–300.

Wofsy, S. C., Goulden, M. L., Munger, J. W., Fan, S.-M., Bakwin, P. S., Daube, B. C., Bassow, S. L. and Bazzaz, F. A. (1993) Net exchange of CO_2 in a mid-latitude forest. *Science* **260**, 1314–1317.

Wofsy, S. C., Munger, J. W., Bakwin, P. S., Daube, B. C. and Moore, T. R. (1991) Net CO_2 uptake by northern woodlands. *Trans. Amer. Geophys. Union* **72**, 76.

3 Carbon Budget: Temperate Coniferous Forests

A. I. BREYMEYER[1], B. BERG[2], S. T. GOWER[3] and D. JOHNSON[4]

[1]*Institute of Geography and Spatial Organization, Polish Academy of Sciences, Warsaw, Poland*
[2]*Department of Forest Soils, Swedish University of Agricultural Science, Uppsala, Sweden*
[3]*University of Wisconsin-Madison, Department of Forestry, Madison, USA*
[4]*Desert Research Institute, Biological Sciences Center, Reno, USA*

3.1 INTRODUCTION

This chapter is divided into five sections. The first section, 3.2, authored by Gower, Johnson and Berg, describes a general approach to site-specific C budgets for temperate coniferous forests and it considers one site in Sweden in some detail. The second section, 3.3, authored by Breymeyer, reviews recent literature on coniferous forests' C budgets in Europe and considers how a climate gradient affects litter production and decay in these forests. The next section, 3.4, authored by Berg, reviews soil organic matter budgets for a set of forest sites in Sweden and also for a transect from northern Sweden to central Europe. The final section, authored by Gower and Johnson, reviews C balance in coniferous forests of temperate North America.

3.2 CARBON BUDGETS–EXAMPLES OF A MODEL AND EMPIRICAL MEASURES

Although aboveground biomass and net primary production (ANPP) data for temperate coniferous forests of North America and the world are numerous (Cannell 1982), there are very few complete C budgets for forest ecosystems (Table 3.1). This is largely due to a lack of information on soil C flux and allocation of C to fine root production and respiration.

The major components of a forest C budget can be estimated as follows:

$$C_{veg} = A - R_{wood} - R_{leaf} - R_{root} - D_{wood} - D_{leaf} - D_{root} \quad (1)$$

Global Change: Effects on Coniferous Forests and Grasslands.
Edited by A. I. Breymeyer, D. O. Hall, J. M. Melillo and G. I. Ågren

UNEP

Table 3.1 Carbon content and fluxes for temperate conifer forests

Carbon content or flux	*Pinus taeda* 16 yrs (1)	*Pinus elliotti* 7 yrs (2)	*Pinus elliotti* 27 yrs (2)	*Abies amabilis* 23 yrs (3)	*Abies amabilis* 180 yrs (3)	*Pseudotsuga menziessii*[a] >450 yrs (4)
Carbon content ($g\ C\ m^{-2}$)						
Aboveground						
wood	4 542	327	1 749	1 692	20 262	34 698
foliage	265	77	297	614	974	558
understory	—	162	64	203	94	325
Belowground						
coarse roots	1 120	180 [b]	1 132 [b]	741 [b]	5 971 [b]	7 064 [b]
fine roots	[c]	121 [d]	311 [d]	305 [d]	763 [d]	564 [d]
understory	—	399	570	136	17	—
Detritus						
forest floor	760	406	1 652	2 390	7 475	2 558
woody	—	0	0	3 939	11 467	10 730
Mineral Soil	—	10 624	6.448	12 754	15 834	6 552
Carbon fluxes ($g\ C\ m^{-2}\ yr^{-1}$)						
Aboveground N	878	170	487	323	157	382
Belowground NPP	101	71	208	396	618	145
Aboveground detritus	873	119	230	82	98	193
Belowground detritus	466	107	226	477	603	184
Autotropic respiration	2 068	—	—	120	550[e]	6 667
Heterotropic respiration	694	113	135	156	197	756
Gross primary productivity	—	—	—	941	1 204	7 722
Net ecosystem productivity	—	—	—	—	—	149

[a]Watershed average.
[b]>5 mm diameter.
[c]Included in coarse roots.
[d]<5 mm diameter.
[e]Calculated from Figure 1 from Ryan (1990).
Sources: (1) Kinerson *et al.* (1977); (2) Gholz *et al.* (1986); Ewel *et al.* (1987a, b); (3) Grier *et al.* (1981); Vogt *et al.* (1980, 1982, 1983, 1990); Harmon *et al.* (1986); Edmonds (1987); Ryan (1990); (4) Grier and Logan (1977).

where C_{veg} is the annual accumulation of C in the vegetation, A is net daytime canopy photosynthesis by the over- and under story vegetation, R_{wood} is respiration for woody tissue, R_{leaf} is night respiration only, since most chamber measurements of photosynthesis are net photosynthesis, R_{root} is root respiration and D_{wood}, D_{leaf}, and Droot are detritus inputs of dead wood, leaf and root tissue, respectively. The annual change in soil C (C_{soil}) can be described as

$$C_{soil} = D_{wood} + D_{leaf} + D_{root} - R_{soil} \qquad (2)$$

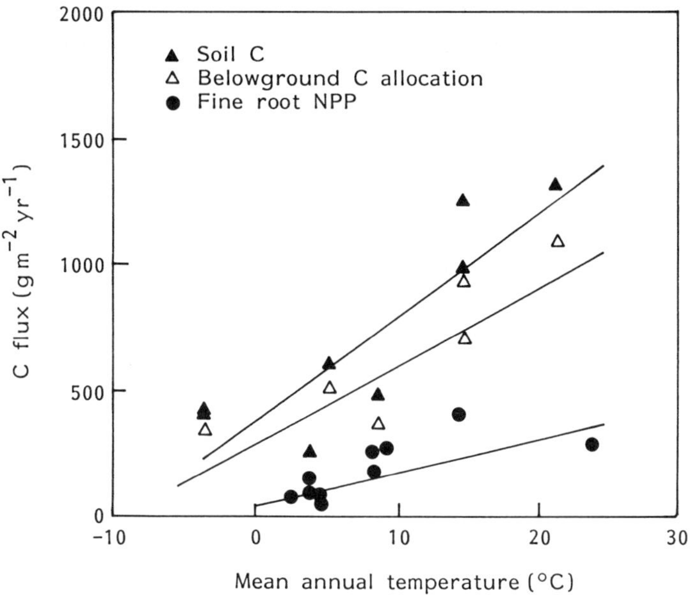

Figure 3.1 Relationship between C belowground allocation, fine root production and mean annual temperature. After Gower *et al.* (1994b)

where R_{soil} is microbial or heterotrophic respiration. Total CO_2 flux from the soil surface can be described as

$$R_s = R_{soil} + R_{root} \tag{3}$$

however, it is extremely difficult to measure R_{soil} and R_{root} separately. The net change in C in a forest ecosystem (vegetation + soil) or net ecosystem productivity is equal to the sum of C_{veg} and C_{soil} and can be described by combining equations (1) and (2) to yield equation (4):

$$C_{veg} + C_{soil} = A - R_{wood} - R_{leaf} - R_s \tag{4}$$

Soil surface CO_2 flux (R_s) for conifer forests ranges from 259 to 1314 g $Cm^{-2}yr^{-1}$ (Raich and Nadelhoffer 1989) and is positively correlated to mean annual temperature (Figure 3.1). It is difficult to partition total soil CO_2 flux into R_{soil} and R_{root}; however, soil surface CO_2 flux data for control (R_{soil} + R_{root}) and trenched (R_{soil}) plots in warm temperate (*Pinus elliottii*) and cold temperate (*P. resinosa*) pine plantations suggests that R_{root} comprises 62 and 35% of R_s, respectively (Ewel *et al.* 1987b; Gower, unpublished data). Nakane *et al.* (1983) estimated that R_{root} comprised 47–51% of R_s in a 80-year-old *P. densitlora* forest in Japan.

Fine root production for temperate conifer forests ranges from 30 g $Cm^{-2}yr^{-1}$

for a *Picea rubens* forest in Ontario, Canada to 710 g $Cm^{-2}yr^{-1}$ for an *Abies amabilis* forest in Washington, USA. Although fine root production is an important component of forest C budgets, it is still one of the least understood processes. Assuming C_{soil} in equation (2) is near zero in the short term (e.g. year-to-year) in forests, then the total allocation of C to roots (root turnover and root respiration) can be estimated as $R_{soil} - D_{leaf} + D_{wood}$ (cf. Raich and Nadelhoffer 1989). Rates of soil C accumulation (C_{soil}) are, in fact, typically smaller than litterfall or soil respiration rates. For example, Schlesinger (1990) calculated an average long-term soil C accumulation rate of 2.4 ± 0.7 g $Cm^{-2}yr^{-1}$, compared to average litterfall rates of 62–319 g $Cm^{-2}yr^{-1}$ and soil respiration rates of 259–1314 g $Cm^{-2}yr^{-1}$ for conifer forest ecosystems (Raich and Nadelhoffer 1989). However, forest floor accumulation rates can be of the order of 95 g $Cm^{-2}yr^{-1}$ (e.g. Turner 1981), making the assumption of negligible soil and litter C accumulation invalid in aggrading forest ecosystems. Using the C balance approach, Gower *et al.* (1994b) reported that total C allocated to roots was positively correlated to mean annual temperature; this relationship supports the previously reported relationship between fine root NPP (determined by sequential coring method) and mean annual temperature for pine forests (Figure 3.1) (Gower *et al.* 1994b). The difference between the two regression lines should approximate the amount of C allocated to coarse root NPP and root respiration.

The availability of nitrogen, the most common limiting nutrient in temperate conifer forests (Gessel *et al.* 1973; Johnson *et al.* 1981), influences the assimilation and allocation of C. Three major mechanisms by which N availability increases aboveground net production are: (1) increased net photosynthesis (Smolander and Oker-Blom 1990; but see Teskey and Whitehead 1994); (2) increased leaf area index (LAI) (Vose and Allen 1988; Gholz *et al.* 1991; Gower *et al.* 1992); and (3) reallocation of C from fine roots and mycorrhizae to aboveground tissues (Keyes and Grier 1981; Vogt *et al.* 1990; Gower *et al.* 1992).

3.2.1 One site – the SWECON site, general background and site data

We have selected a Scots pine forest of the former Swedish Coniferous Forest Project (SWECON) to create a dynamic budget for the surface organic soil layers, also known as the F and H layers. The background data and site history are very favourable which makes it possible to use it as a starting point for budgets in a regional perspective.

Amounts of organic matter measured in the stand and some flows (Figure 3.2) give a static organic-matter budget for the forest at an age of about 120 years. The numbers given are based on a good set of measurements and the original compilation of data was published by Andersson *et al.* (1980). Litterfall data were measured for the period 1973–80 (cf. Flower-Ellis 1985; Berg *et al.* 1993). Root production was measured and estimates were made for root-litter formation as well as for that from the shrubs and mosses. Decomposition studies were long

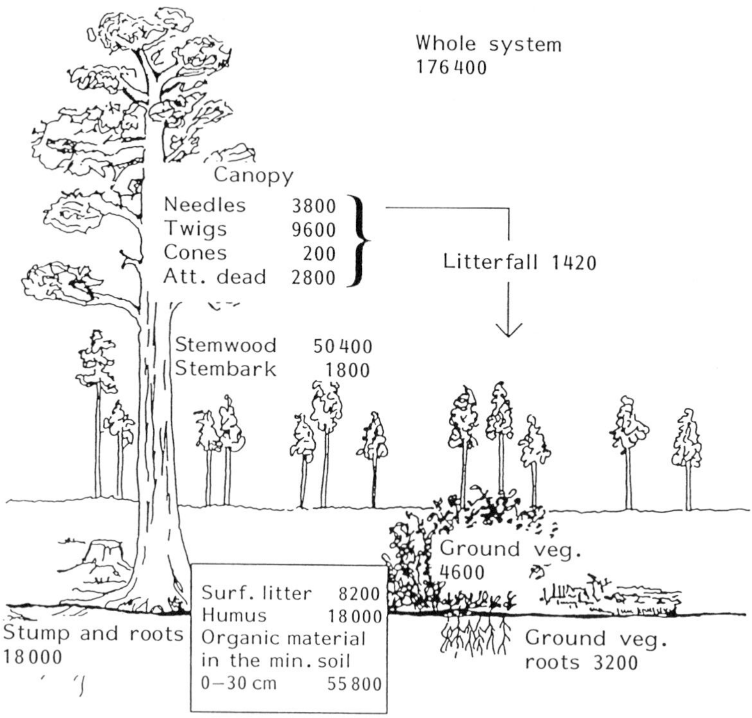

Figure 3.2 A budget for organic matter in a mature (120-year-old) Scots pine monoculture (SWECON site). Based on data from Andersson *et al.* (1980). Units are in kg of organic matter per ha. Att. – attached; Surf. – surface; min. – mineral; and veg. – vegetation

term, using different litter components, such as needles (e.g. Berg *et al.* 1982) and root litter (Berg 1984) as well as other ones. Also the chemical composition of different litter components was investigated (Berg 1981). With this background information it appears reasonable to try to create a budget for organic matter (C).

The stand was a 140-year-old (in 1993) Scots pine monoculture located at Jädraås in central Sweden (60°49′N; 16°30′E) at an altitude of 185 m a.s.l. on a flat area of deep sand sediments. The mean annual precipitation is 609 mm and the mean annual temperature is +3.8 °C. The forest has been described in earlier papers (e.g. Axelsson and Bråkenhielm 1980; Berg *et al.* 1982).

A very loose A_{00} horizon, interwoven with living mosses and lichens, covers a combined A_{01}–A_{02} horizon of 5–10 cm. This latter organic layer was very well separated from the mineral soil. The parent mineral material as well as the whole soil is considered to be very poor in essential nutrients. In the middle part of the last century the forest at this site was burned completely, including the existing organic layers, leaving coal at the bottom of today's organic layer.

3.3 EUROPEAN TRANSECT STUDIES ON ORGANIC MATTER PRODUCTION/DECOMPOSITION

3.3.1 Trends of coniferous organic matter accumulation in the light of new European literature

Cannell *et al.* (1992) reviewed the new information on C flux and storage in forests of Europe. Coniferous forests comprise about 50% of European forest. In comparison with broad-leaved forests they cover a smaller area (except in the extreme northern latitudes), but their C density is high. In central Europe they store large amounts of C in foliage (12 t ha^{-1}) and litter (45 t ha^{-1}). The total amount of C currently stored in the biomass and litter in these coniferous forests is 2.8 Gt. Since the last glacial maximum reforestation followed the ice sheet retreat at the rate of 0.01 million ha per year, equivalent to 7×10^8 kg annually. From the Middle Ages until the beginning of this century, Europe was deforested due to conversion of forests to agricultural land. The average rate of deforestation is evaluated at 0.2–0.4 million ha yr^{-1}; thus, the forest was eliminated 20–40 times faster than the rate of forest increase during deglaciation. In recent decades, intensive afforestation is noted in Europe (land-use changes, CO_2 and N fertilization) and EU forests are now a net sink for C.

Adams *et al.* (1990) indicate that the amount of organic C in vegetation and soils of the globe has more than doubled (from 0.96 to 2.3 Gt) during the present interglacial (last 10 000 years). The class of forest called by the authors 'dry open pine woodland' exactly doubled its C storage (from 7 to 14.1 Gt) during this time.

Every year there is a massive exchange of C between the atmosphere and temperate vegetation. The amount of C in vegetation and soils is around two to three times as much as that in the atmosphere and is particularly important as a relatively rapidly responding pool.

Kauppi *et al.* (1992) surveyed the growth patterns and stock of a third of European forest through the 1970s and calculated that forest resources increased as much as 25% (Figure 3.3). According to forest surveys in European countries, growing stock increased in Europe by 12% between 1971 and 1980. The increase was projected to continue at the same rate in the 1980s yielding a 25% larger growing stock in 1990 than in 1971. The fertilization effect of pollutants override the adverse effects, and forest biomass increased from 70 to 105 million tonnes of C accumulated annually during the 1970s and 1980s in European forests. These results suggest that the European forest is a C sink in the global C budget.

Brubaker and Graumlich (1989) found a similar tendency of increased net primary production (NPP) for coniferous forests in the Pacific Northwestern USA during the past 100 years (Figure 3.4). The authors conclude that NPP increased by 60% over the period 1880–1979; the natural variability of productivity is great and controlled primarily by variations in summer temperature.

Klimo (1992) provided the data which can be used for the following calculation of organic matter budget in a Norway spruce forest (stand of *Picea abies* L.Karst

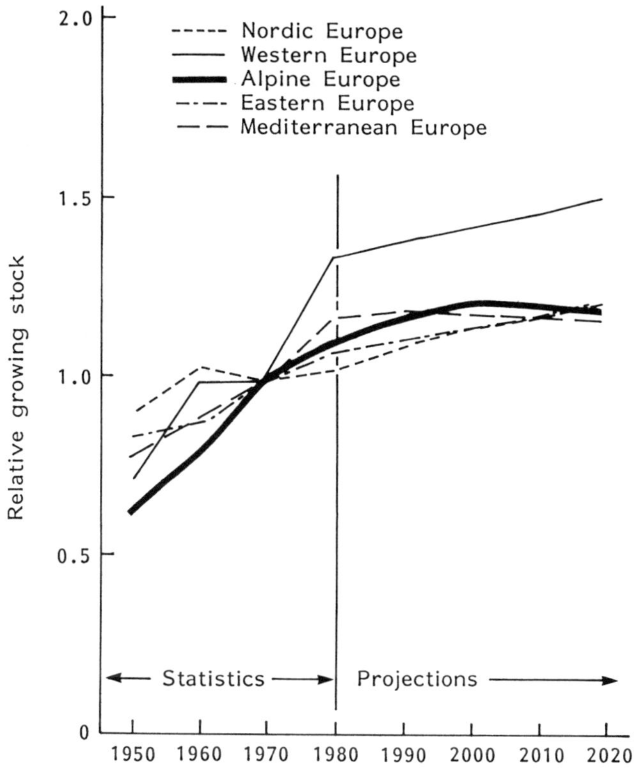

Figure 3.3 Relative change of growing stock based on the best available information from different regions in Europe. The value for 1970 has been adjusted to 1.0. After Kauppi *et al.* (1992)

established around 1900 in the Czech Drahonska Uplands): dry matter of aboveground parts of trees equals 273.43 t ha^{-1}; dry matter of underground parts of trees – 40.50 t ha^{-1}; total 319.93 t ha^{-1}. Underground parts of spruce trees do not exceed 13% of the whole tree biomass.

Above-ground biomass (273.43 t ha^{-l}) produces litterfall, which, measured over 10 years (1977–88), fluctuates from 3.39 to 6.76 t ha^{-l} yr^{-1}. The highest value of litterfall was registered in the year 1986 when *Cephalia abieties* L., the fir pest, occurred. In the other 'normal' years the sequence of organic matter portions in this trophic chain are as follows in units of t ha^{-l} yr^{-1}: (1) ABVG biomass of trees, 273.43; (2) litterfall, 4.55; (3) herbivore feces fall (average year), 0.156; (4) herbivore feces fall (*Cephalia* year), 0.676.

On the basis of physiological studies we can assume that food digestibility of herbivorous invertebrates oscillates around 30%; consequently, the biomass of their food is 43% higher from registered feces biomass and equals 0.222 (average

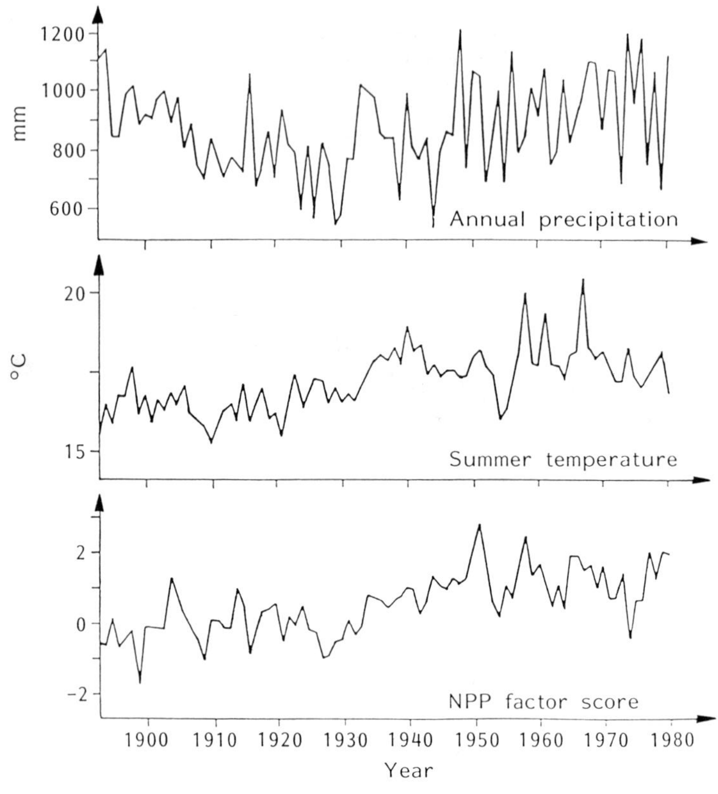

Figure 3.4 Brubaker and Graumlich (1989) studies on 80 years time series of NPP, annual precipitation and summer temperature. NPP factor score obtained from the principal component analysis of the NPP measurements at four forest sites (Seattle, Washington)

year)–0.966 (*Cephalia* year) t $ha^{-1} yr^{-1}$. Thus, in spruce forest more than 0.2–1.0 t ha^{-1} of plant matter is annually consumed by canopy herbivores; we say 'more than' because in this calculation the amount of organic matter utilized for costs of herbivorous life is not considered. The metabolism of small invertebrates is intensive and its energetic costs are high. Moreover, Funke and Roth-Holzapfel (1991) have found that the invertebrates of a spruce forest accumulate the elements K, Mg, Ca, Cu, Zn, Cr, Ni, Se, Sb, Hg and Cd from their food. Most elements are accumulated by primary consumers, some by saprophages and predators. The accumulation factor (the ratio of element concentrations in animals to those in food) can be as high as 4, 10, 15 or even 33. So important elements accumulated in invertebrates' bodies must influence the allocation and flows of organic matter in the forest system and its parts.

In summary, the quoted literature suggests that coniferous forests accumulated organic C during recent decades and were a C sink in its global budget. The sink is a consequence of a combination of regrowing forests, CO_2 fertilization and N fertilization from poluted rain.

3.3.2 Response of coniferous litter production/decomposition to a climate gradient along a north–south transect

Breymeyer conducted studies (1991a, b, 1993) on production and decomposition of organic matter along a north–south transect of pine forests distributed from northern Sweden to central Poland. Temperature was considered to be the main climatic factor influencing both of these ecological processes. In this study, litterfall was treated as a production index.

The transect extends between 50 and 68° N, or approximately 18° latitude. Efforts were made to choose the stands in such a way that longitudinal deviations were avoided: in the central part of Europe, the shifting of the transect to the east or west would mean changing the effect of continentalism. The difference in temperature between the most northern and southern stand is approximately 8.4 °C. All forests studied on this transect occur on sandy soils; near the ground, they shelter shrub berries and mosses. Admixtures of the other tree species are insignificant. The studies were carried out in the years 1986–88; some characteristics of all sampled stands are presented in Table 3.2.

The most northern stand, Tornetrask, is outside the acknowledged range of pine forests; it is in the basin of the large Tornetrask Lake, *ca.* 200 km north of the Arctic Circle, at a height of about 300 m a.s.l. Mean long-term temperatures for this area range from −1.0 to −0.5 °C (at the Tornetrask climatic station −0.7 °C), precipitation ranges from 300 to 1000 mm annually (at Tornetrask 470 mm). In 1986 the mean annual temperature and precipitation in Tornetrask averaged −1.10 °C and 476 mm, respectively. The Tornetrask study site is situated several hundred metres from the edge of a lake; fog and groundfrosts often occur in this area. Soils were derived from poor, acid, glacial–fluvial sands originating from the weathered material of siliceous rocks (Sonesson 1979). This is the edge of the forest; the undulating terrain around the lake is covered by bogs and birch tundra (*Betula pubescens* Ehrh. f. Tortuosa Led.). Among the rather homogeneous low birch tundra occur stands of *Pinus sylvestris* L.forest, most frequently in the vicinity of the Tornetrask railway station and in the Abisko Valley. These are thought to be relict stands of pine, as years with sufficiently warm temperatures for this species occur very seldom. However, our studies indicate that in stands of pine forest around Tornetrask there are numerous young trees. Sonesson (1979) came to the same conclusion. Selected for the study was one of the largest stands of pine forest (about 10 000 m^2). The age of the oldest pines is estimated at 80 years. The undergrowth observed were: *Betula nana*, *Empetrum hermaphroditum*, *Vaccinium victicoides* and *V. myrtylli*, *Losselueria procumbens* and *Artous alpina*. It seems that this forest is in good condition and

Table 3.2 North–south transect studies of litter budget in coniferous forests. A. Litterfall; B. Decomposition; C. Organic matter lost annually from forest floor (Breymeyer 1991a, b)

A

Stand geographic position, temperature (Yr)		Litterfall		
		G m^{-2} y^{-1}	% needles	% woody parts
68°14′N	−0.7 °C	58.4	55.2	42.5
60°49′N	3.8 °C	155.8	46.9	52.7
52°20′N	7.2 °C	232.7	56.5	26.4
52°22′N	7.8 °C	238.1[a]	59.0	28.0
Index I:	LIT 52°22′N / LIT 68°14′N	4.08	1.07	0.66
Index II[b]: Rate of change per 1°C		21.1		

[a]After Jozefaciukowa (1975).
[b]Calculated as ratio of differences between extreme litter masses and temperatures.

B

Stand geographic position, temperature (yr)		Yearly decomposition rates (%)			
		Mixed litter	Needles	Branches	Cones
68°14′N	−0.7 °C	12.7	14.6	4.8	2.9
60°49′N	3.8 °C	18.4	28.0	5.7	9.2
54°10′N	6.7 °C	23.0	36.9	13.6	13.9
52°41′N	6.6 °C	23.1	34.9	19.2	17.1
50°30′N	7.6 °C	27.8	40.2	12.7	11.8
Index I:	DEC 50°30′ / DEC 68°14′	2.15	2.75	2.64	4.0
Index II: Rate of change per 1°C		1.80	3.08	0.95	1.06

C

Stand geographic position, temperature (yr)		Loss of OM from forest floor (gm^{-2} y^{-1}
68°14′N	−0.7 °C	7.41
60°49′N	3.8 °C	28.66
52°41′N – 51°20′N	6.6 °C–7.2 °C	53.75
50°30′N–52°22′N	7.6 °C–7.8 °C	66.19
Index I: LOS 50°30′N / LOS 68°14′N		8.93
Index II: Rate of change per 1°C		6.99

ought to be functioning similarly to other pine forest ecosystems. The next stand is the pine forest in Jädraås, in the central part of Sweden. The five stands from Poland studied in the transect are situated in Borecka Forest, in Mierzwice and Litewniki forests, in the Bialowieza Forest (Czerlonka stand) and in two pine forests in central Poland (Kampinos and Pinczow stands).

Litterfall was caught on 10–20 surfaces; they were smooth circular areas, cleared of vegetation and covered by sand, each 0.1 m^2. In such a way litterfall was gathered from the ground surface and additionally the production of the undergrowth was observed. Once a year, the plant litter was collected and transported to the laboratory, sorted, dried at a temperature of 90 °C, and weighed.

The decomposition of organic matter was evaluated by the litter bag method. Litter produced at a site was decomposed there. Three types of litter were used in the decay studies; pine needles, mixed litter (needles, leaves and so on in the proportions in which they fell to the ground), and pine cones. The litter bags were made of nylon netting with a mesh size of 1 mm; 10 g of material was put into each litter bag.

The comparison of litterfall and decomposition rates for the *P. sylvestris* forest along the whole transect, from Tornetrask to Pinczow, is presented in Table 3.2. Litterfall was lowest in Tornetrask (55–61 g m^{-2}). This litter also decayed the slowest, decomposing at an average rate of 11–14% per year. In Jädraås, in central Sweden, the litterfall was twice as large, and the decay rate was also higher, although it did not change as greatly as litterfall. Mixed litter (composition identical to that which falls on the bottom of the forest) has a decomposition rate ranging from 11% per year in the north to 28% per year in the south. Pine needles decay most rapidly–from 14% per year in the north to 40% per year in the south. The biggest differences were observed in the decay rates of pine cones. Pine cones increase their decomposition rates as much as five-fold after shifting 18° south from Tornetrask. On average, the decomposition of mixed litter doubles, of needles and small twigs almost triples, and of cones, quadruples.

Five summary points can be drawn from this transect study. First, the decay rate of cones is very low, requiring about 30 years for complete decomposition. Second, the decomposition of cones reacts most to temperature changes along the transect stretching from 50 °N to 68 °N. Third, the decomposition of litter includes at least two phases: fast initial decay of carbohydrates and slow decay of lignin and cellulose, while warming will make a difference in the ratio of decay in both phases, it will more strongly stimulate the slow lignin–cellulose phase. Fourth, litterfall for the northern stands was comprised of a greater percentage of lignified, slowly decaying woody tissue and fifth, as one moves south, the composition of litter changes: it contains more remains of soft, green parts of plants (Breymeyer 1991a, b).

Table 3.3 Relationship between annual total litterfall and needle litterfall as dependent on stand age in Scots pine monoculture stands at the SWECON site, Jädraås. Basic data from Axelsson and Bråkenhielm (1980), Berg *et al.* (1993) and Flower-Ellis (1985). Significance levels to the coefficient (b); n.s. = not significant

Stand age (yr)		b	a	r^2	n	$p<$
18	Total litter	89.11	−1041.6	0.716	7	0.05
	Needle litter	54.83	−456.7	0.506	7	n.s.
55	Total litter	128.78	−5450.0	0.622	10	0.01
	Needle litter	79.81	−3273.2	0.517	10	0.05
120	Total litter	56.22	−5377.6	0.573	10	0.05
	Needle litter	75.12	−832.2	0.358	10	n.s.
18, 55	Total litter	37.39	11.1	0.837	17	0.001
	Needle litter	21.99	211.9	0.773	17	0.001
18, 55, 120	Total litter	4.76	1284.6	0.099	27	n.s.
	Needle litter	1.44	1011.6	0.018	27	n.s.

3.4 A BUDGET APPROACH FOR SOIL ORGANIC MATTER IN SCOTS PINE STANDS–FROM ONE SITE TO A REGION

In this attempt to set up a budget for organic matter we first use information from a site supplying very good data: this is the Jädraås SWECON site described above as an example of the empirical C budget. Using this approach we attempt to generalize our observations on soil organic matter accumulation to a region.

Two younger stands (18 and 55 years old) on the same nutrient-poor ground were investigated for litterfall. Further data on these are given by Axelsson and Bråkenhielm (1980).

3.4.1 Litter production

3.4.1.1 Trends in total litterfall

Litterfall was observed for 7–10 years in three Scots pine stands, initially 18, 55 and 120 years of age (Table 3.3). Some trends were evident within stands, and there were clear differences between the different-aged forests. A detailed analysis of annual litterfall as well as of single litter components such as needles, cones, branches, and fine litter has been given by Berg *et al.* (1993) and by Flower-Ellis (1985) (cf. Figure 3.2). Total litterfall in all three stands increased with stand age (cf. Berg *et al.* 1993). The stands thus represented the age periods of 18–25 years, 55–65 years and 120–130 years, giving a relatively good distribution among ages.

The methods used by Flower-Ellis (1985) to determine total litterfall was very much adapted to litter type and the reader is referred to Flower-Ellis (1985) and Flower-Ellis and Olsson (1978) for a full description.

3.4.1.2 Linear relation for litterfall vs stand age

Total litterfall was positively correlated to stand age, using data for the Jädraås site for the two younger stands only (18 and 55 years old). For example, for the 18-year-old stand, total litterfall was positively correlated to age. The 55-year-old stand also showed a significant, positive relationship for the 10 years during which it was followed (Table 3.3). These data suggest that the amount of litter was still increasing. However, this trend seems not to be an effect of age alone, but includes a temporal trend which is common to all three stands (Berg *et al.* 1993). The old stand did not follow a similar pattern. At an age of 120 years, it may no longer have been expanding its needle mass but there was no indication that this was declining during the period of observation.

3.4.1.3 Litter decomposition–maximum decomposition–limit values

For most litter types it has been observed that the decomposition rate decreases as decomposition proceeds and the more easily metabolized components are decomposed and the recalcitrant compounds remain (Berg *et al.* 1982, 1987). For some litter types it has even been observed that the decomposition proceeds at such a low rate that is not detectable.

Howard and Howard (1974) presented estimated final values for mass loss of some leaf litter types that were decomposing in the absence of soil fauna. Based on their approach, they found that asymptotic functions described their data well. The interpretation of their finding was that in the absence of soil animals, the decomposition could stop or at least come close to a halt and reach a limit value well below 100% mass loss. Using data from a nutrient-poor forest, almost free from soil animals, Berg and Ekbohm (1991) tested long-term mass-loss values for such limits. They found that for seven, chemically very different, litter types, limit values could be estimated ranging from about 50% remaining, down to none for one specific litter type. A later study (Berg and Ekbohm 1993) gave results that confirmed the tendency for some litter types to reach certain, specific limit values. They found when comparing 12 different sets of decomposing needle litters (Scots pine and lodgepole pine) that there was a considerable homogeneity in limit values between species. For Scots pine needles an average amount of 11% remained.

3.4.2 Method of estimation

In their study, Howard and Howard (1974) compared several statistical models for mass-loss data and preferred an asymptotic nonlinear model with three parameters. Later, Berg and Ekbohm (1991) modified their function somewhat and obtained

$$\text{m.l.} = m(1 - \mathrm{e}^{-kt.m}) \tag{5}$$

where m.l. is the accumulated mass loss (in %) and t is time in days. The parameter m represents the asymptotic level which the accumulated mass loss will ultimately reach, and the parameter k is the decomposition rate at the beginning of the decay.

3.4.2.1 A possible interpretation of the phenomenon of limit values

Howard and Howard (1974) used an experimental environment, with no soil animals present. This may very well be a good basic condition for the phenomenon as the mechanical treatment of the litter is minimized or nonexistent. The main published results are from systems which were either free from soil animals or had a soil fauna with a very low activity toward litter, at least in the early decomposition stages. It is thus possible that the asymptotes reflect a stage at which microbial decomposition comes to a halt and some other agent is needed for the process to continue. However, even if the phenomenon as such may be connected to the absence of soil animals, the observation that there is a considerable range of limit values suggests that a low level of soil animals is a necessary condition though it is not necessarily the only one. Berg *et al.* (1995) used existing data and examined the relationships between chemical composition and limit values for decomposition of litters. They observed a highly significant, negative relationship to initial N concentration ($r = -0.706$; $n = 28$; $p < 0.001$). In this analysis data from both field studies were used and they seemed to fit the same line. The higher the initial N level, the more organic material should be recalcitrant and thus be stored.

There may be a causal relationship. As decomposition of the litter proceeds, the recalcitrant component lignin is enriched. Before its decomposition has started and afterwards, lignin is subject to chemical transformations such as the chemical incorporation of N in a first step of humification (cf. Nömmik and Vahtras 1982). Such chemical incorporation may result in structures that are not necessarily broken by soil microorganisms.

The microflora able to decompose lignin completely is limited and those fungi and bacteria able to do this are called 'white rot'. The composition of the microbial community is very dependent on the chemical quality of the litter input into a system. This includes not only the availability of the energy source but also of the concentrations of different nutrients. The interaction of nutrients may be complex and although, for example, a good supply of N may stimulate the initial decomposition rate of litter, the same supply of N may have a reversed effect in somewhat later stages. Properties like these will affect the selection of decomposing organisms including 'white rot'. One example was given above and another one is the observation by Keyser *et al.*(1978) that a white-rot fungus did not synthesize its lignin-degrading enzyme system in the presence of low-molecular, N-rich compounds such as ammonium and amino acids. Fungi with such properties would naturally be hampered in their development when growing on a highly decomposed, lignin-rich litter which also is rich in N. Because a good number of white-rot fungi have this property (P. Ander, pers. comm.) it follows that the effect of high N in the litter could hamper lignin degradation and consequently, the decomposition of litter in several soil systems.

When testing the relationship between initial concentrations of manganese and the limit values for all available decomposition data, we observed a highly significant, positive relation ($r = 0.757; n = 17; p < 0.001$). This nutrient has been suggested by some scientists to promote lignin decomposition by stabilizing the enzymes of the lignin degrading system and by increasing the production of the enzyme (Perez and Jeffries 1992; cf. Berg *et al.* 1995). Although these causal relationships are not clear, we may still see that there was a clear connection, and we cannot exclude this as a possible factor that may determine limit values for decomposition.

3.4.2.2 Soil organic matter data for the stand

In a measurement on the amounts of soil organic matter in the 120-year-old stand, Staaf and Berg (1977) estimated–with the use of large diameter samplers–the amount of 'humus' in the combined A_{01}–A_{02} layer to be 1.54 kg ha^{-1}. This estimate did not include visible litter remains.

3.4.2.3 Budget

From this site we have good access to data for total litter input to the mature forest. Using this information and knowledge of an approximate linear relationship (Table 3.3) between litter input and stand age we estimated the annual litter input for the period 0–120 years. We assumed that the canopy development stopped after about 100 years although there was no real canopy closure and that the litter input thus became constant and similar to the estimated value of 0.280 kg $m^{-2}yr^{-l}$. This figure was based on long-term measurements and includes litter formation from shrubs, mosses and roots (Andersson *et al.* 1980).

With regard to the importance of the chemical composition of the litter, it is essential to emphasize that the litter produced from the pines was chemically similar including all parts of the trees (with the exception of the rough bark) and the main part of the shrub litter had a chemical composition that did not deviate much (Berg 1981). In addition, litter formation from the pines dominated the litter input in the stand, and we may thus assume that the decomposition process in the late stages was similar among the litter types and on the average leaving recalcitrant or stabilized remains of about 11% of the input. This figure was estimated for this stand as based on a separate determination. We further assumed that in this stand, 10 years are required before the litter has decomposed far enough to lose its structure enough to become called soil organic matter. Based on this assumption we estimated a total accumulation of 1.82 kg m^{-2}

which compares favorably with the measured value of 1.54 kg m^{-2} (Staaf and Berg 1977). When the former stand at this site burned in the middle of the nineteenth century, almost all of the surface organic matter (L + F + H layers) was burned, leaving just coal remains, a fact that allows us to suggest this comparison. The difference between the estimated and the determined amount was 15%.

In this regard, we may consider whether we thus have a method to estimate the accumulation of soil organic matter in forests in other soil types. This kind of approach is, of course, sensitive to the accuracy of litter input data (both root litter and litterfall data) as well as to the limit value for litter decomposition.

3.4.3 Trends in a region from the Arctic Circle to the arch of the Alps and the Carpathians

3.4.3.1 Litterfall – a transect of Scots pine sites across Scandinavia

In an attempt to construct a regional model we compared the annual fall of needle litter as well as that of total litter, to stand age, basal area, site index and latitude (Tables 3.4 and 3.5).

The litterfall data used, originating from 33 Scots pine sites, was based on three main studies. Litterfall was measured from 3 to 10 years at each site. The geographical range for these sites varied from north of the Arctic Circle (66° 32′ N) to south Sweden (58° 07′ N) which covers about half of the north/south distribution of Scots pine in Europe. For all of these sites information was presented about needle litterfall (Albrektsson 1988; Johansson 1986; Flower-Ellis 1985), and about half of them also had total litterfall measured (Johansson 1986; Flower-Ellis 1985). A compilation of these data was made by Berg *et al.* (1993).

Mean annual needle litterfall ranged from 530 to 2312 kg ha^{-1}. The lowest amounts were found for nutrient-poor sites (sediment soil) in the north while litterfall was commonly greater for the more nutrient-rich sites with till deposits or clay soils. Among sites of similar fertility, needle litterfall was lower for sites located in the north than for sites at a more southern latitude. This is evident when needle litterfall at two extreme sites, at latitudes 66° 32′ N and 58° 04′ N, are compared. These sites had nearly identical site qualities and basal areas (17.5 and 18.3 m^2 ha^{-1}, respectively) but the needle litterfall at the northern site (608 kg ha^{-1} yr^{-1}) was only about one-third of the amount obtained at the site located farther to the south (1571 kg ha^{-1} yr^{-1}).

The mean annual total aboveground litterfall (larger twigs and branches excluded as limited by the method–70 cm litter trap) varied from 735 to 4198 kg ha^{-1}. The proportion of needle material in the total litterfall, which comprises needles and fine litter, varied between 33 and 75% in the stands studied. However, in most of the stands the needle litter fraction accounted for about 50–60% of the total annual litterfall. According to Bonnevie-Svendsen and Gjems (1957) and Mälkönen (1974), needle litterfall is approximately 75% of the total litterfall in Scots pine stands. Their values are higher than ours, despite the fact that larger

Table 3.4 Simple and multiple linear relations for average annual amounts of needle litterfall (kg ha^{-1}) as dependent on latitude, stand age, site index and basal area in Scots pine stands in Scandinavia. All available average values for Scots pine were used (data from Albrektsson 1988; Flower-Ellis 1985; Johansson 1986; cf. Berg *et al.* 1993). Levels of significance refer to the coefficient (*b*); n.s. stands for not significant

	b	*a*	r^2	*n*	$p<$
Simple linear relations					
Latitude	−147.16	10 526.81	0.48	33	0.001
Site index	93.49	−676.41	0.42	33	0.001
Age	−9.01	2 105.20	0.27	33	0.01
Basal area	46.32	281.91	0.26	33	0.01
Multiple linear relations					
Site index plus	75.55	−924.01	0.49	33	0.001
Basal area	25.77				0.001
Latitude plus	−140.60	9 030.30	0.71	33	0.001
Basal area	42.42				0.001
Latitude plus	−122.17	7 487.37	0.73	33	0.001
Site index plus	25.31				n.s.
Basal area	36.05				0.001
Age plus	−4.54	8 710.80	0.62	33	0.001
Altitude plus	−2.07				
Latitude	−107.51				

Table 3.5 Relationships between total litterfall (kg ha^{-1}) as dependent on latitude, stand age, site index and basal area for Scots pine stands in Sweden. All data from Flower-Ellis (1985); Johansson (1986); cf. Berg *et al.* (1993). Levels of significance refer to the coefficient (*b*); n.s. stands for not significant

	b	*a*	r^2	*n*	$p<$
Simple linear regressions					
Latitude	−212.51	15 387.38	0.39	17	0.01
Age	−19.28	3 587.65	0.19	17	n.s.
Site index	153.40	−1154.73	0.50	17	0.01
Basal area	102.68	132.71	0.52	17	0.001
Multiple linear regressions					
Latitude plus	−152.70		0.71	17	0.01
Basal area	83.39	9 924.33			0.01
Site index plus	93.60		0.64	17	0.05
Basal area	66.63	−1224.16			0.05
Latitude plus	−130.86		0.77	17	0.05
Site index plus	70.47				n.s.
Basal area	59.01	7 501.89			0.05

twigs and branches are not included in our figures. Although the methods used in the two transect studies were not ideal for e.g. cones and branches we have used data for estimates in a budget (below) to indicate trends.

3.4.3.2 Linear relations of litterfall in Scandinavia to stand age, latitude, basal area, and site index

Stand age: When we compared all available data for needle litter to stand age we obtained an r^2 value of 0.27 ($n=33$; ns) for needle litter (Table 3.4). For total litter the r^2 value was 0.19 ($n=17$: ns) (Table 3.5).

Latitude: We obtained a negative relationship between latitude and litterfall (both total litterfall with $r^2=0.39$; $n=17$; $p<0.01$) (Table 3.4) and needle litterfall ($r^2=0.48$; $n=33$; $p<0.001$) (Table 3.5). This may be compared with the results of Albrektson (1988), who also found a negative relationship for needles ($r^2=0.46$; $n=16$).

Basal area: A comparison of litterfall to basal area gave a coefficient of determination (r^2) of 0.52 for total litterfall ($n=17$; $p<0.001$) (Table 3.5). When all available data for needle litterfall were combined (Table 3.4), basal area gave a highly significant, but poorer relationship ($r^2=0.26$; $n=33$; $p<0.01$).

Site index: Site index values were available for all the Scandinavian sites. All needle litter data combined gave an r^2 value of 0.42 ($n=33$; $p<0.001$;Table 3.4). Total litter gave an r^2 value of 0.50; $n=17$; $p<0.01$ (Table 3.5).

Multiple linear relations for litterfall of both needles and total litter, using all Scandinavian data: Relationships on latitude and site index were highly significant ($n=33$) and gave r^2 values of 0.48 and 0.42 respectively ($p<0.001$). In multiple linear relationships, the combined factors gave significant or highly significant relationships with r^2 values from 0.49 for site index plus basal area, to 0.71 for latitude plus basal area. A three-variable model that included latitude, site index and basal area, gave an r^2 value of 0.73, thus explaining about three-quarters of the variation in needle litterfall (Table 3.4). For total litterfall an r^2 value of 0.77 was reached when latitude, site index and basal area were included (Table 3.5).

Decomposition of litter and limit values: Also for the Scots pine needle litter decomposing in the transect of Johansson (1986) limit values were estimated for the whole climatic range of sites (Johansson, pers. comm.). Although some trends in limit values apparently were dependent on climate and the nutrient status of the sites (Berg, Johansson, unpubl.) they have so far not been observed to be of a magnitude to be considered in the present budget estimate. We have therefore used the value of 11% litter remaining, estimated for the Jädraą006s site for the whole transect range.

Accumulation of soil organic matter on a regional level–network studies: With the use of the presented data and by using the equations of Tables 3.4 and 3.5 we estimated the accumulation of organic matter layers for a larger region. We have preferred to illustrate this with the use of a change in latitude from, say, northern Scandinavia to central Germany or south Poland.

Let us take first needle litter and then total litter. Needle litterfall would be expected to range from a low value of about 810 kg ha^{-1} for forests at 66° N to

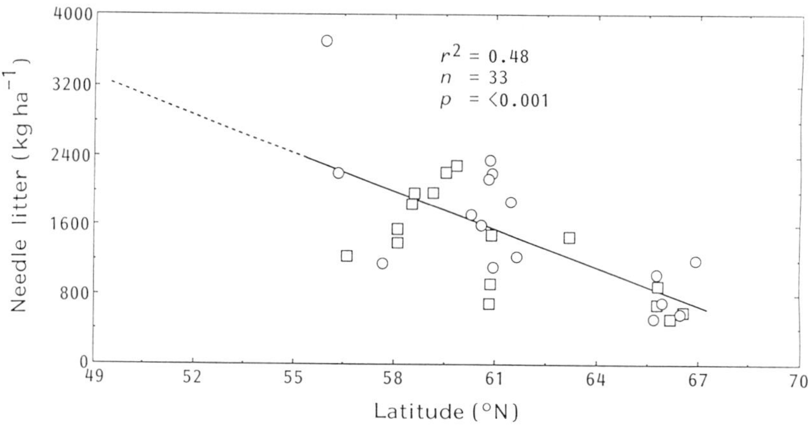

Figure 3.5 A linear regression giving needle litterfall as compared to latitude. The dashed line gives the extrapolation of data to central Europe. Circles = total litter; squares = needles

3316 kg ha^{-1} at 49° N (cf. Figure 3.5). These data suggest that soil organic matter may accumulate about four times faster in the south just due to the needle component. For total litter we would have a figure of 3.5 times as fast, assuming that the limit values would be similar in both cases and that the stands would have similar nutrient status.

If we extend the calculations and use the fact that the stand in the south would grow faster and reach a maximum canopy in 40 years as compared to 110 years in the north we can extend the estimates. We use the simple model, suggesting a linear increase in litterfall and a remaining difficult-to-decompose component of 11%. We may thus estimate that in the northern stand the needle component in litterfall would have added about 4900 kg ha^{-1} to the soil organic matter layer in 110 years. In the south the stand would have added 7500 kg ha^{-1} in 40 years, in both cases until the canopies reached a maximum.

3.4.3.3 *Concluding remarks and recommendations*

The extension of Scandinavian data to the northern half of continental Europe lacks validation data. Our results indicate a considerable difference in the growth rate of the soil organic matter layer as a consequence of a difference in latitude, and the magnitude may seem alarmingly large with respect to the possibility of a corresponding storage of nutrients.

The example given above is valid for Scots pine stands. We may expect more pronounced differences between northern and southern latitude forests for other coniferous species, e.g. spruce and also for deciduous trees. We would therefore make the following recommendations:

- It appears self-evident that a good basic knowledge about lignin-degrading organisms and their ecology is vital for further progress in the research field of soil organic matter dynamics.
- As budgets of dead organic matter make up the basic starting point for budgets of nutrients in forest stands some basic approaches are necessary: (1) well-coordinated litter input measurements in selected stands in transects; and (2) a generalization of the concept of limit values as well as an understanding of the causal relationships for their magnitude.

3.5 NORTH AMERICAN CONIFEROUS FORESTS AND STUDIES ON CARBON BALANCE

Temperate conifer forests are common in the western half of North America, across extreme moisture and temperature gradients from west to east resulting from the Cascade, Sierra Nevada and Rocky Mountains. The greater abundance of evergreen conifers over broad-leaved deciduous forests is particularly striking in the Pacific Northwest where, except for the early successional species such as red alder (*Alnus rubra*) and occasional patches of aspen (*Populus tremuloides*), conifers dominate the landscape. The unique environmental conditions of the Pacific Northwest, namely dry, warm days and cool nights in the summer and mild, moist winter days are largely responsible for the dominance of evergreens in the Pacific Northwest (Waring and Franklin 1979). Old-growth Douglas fir in the western hemlock zone are unique in that biomass accumulation is as great or greater than anywhere else in the world and second-growth Douglas fir forests are among the most productive in the world (Waring and Franklin 1979; Long 1981). In the eastern USA, conifer forests are common after disturbance or on nutrient-poor soils. The mild climate, long growing season and infertile soils common to the southeastern USA favor evergreen conifer species (Hicks and Chabot 1985). Both western and eastern conifer forests are commonly nitrogen deficient, and productivity can be greatly enhanced by N fertilization (Gessel *et al.* 1973; Johnson *et al.* 1981; Vose and Allen 1988; Gower *et al.* 1994a).

3.5.1 Carbon balance and stand development

Deforestation and natural disturbances are important influences on the global C cycle because ecosystem energetics are altered by disturbance. Despite our incomplete understanding of how forest C budgets change during stand development, it is useful to infer patterns based on the limited data. The annual net C change or net ecosystem productivity (NEP) of a forest is defined by equation (4). Aboveground net primary production (ANPP) for conifer forests follows a sigmoid curve during early stand development and often exhibits a decrease in many overmature conifer forests (see Sprugel 1985; and Gower *et al.* 1994a, for reviews). The exact reason for the decline is not precisely known, but decreased nutrient availability (van Cleve and Viereck 1981; Miller 1981; Gholz *et al.* 1985) or reduced stomatal conductance due to inefficient transport of water to the canopy in older trees (Ryan and Waring 1992) may in part explain the decrease in ANPP. Few data for belowground net primary production (BNPP) are available, but it appears maximum BNPP coincides with canopy closure (Vogt *et al.* 1986). A third factor that may help explain decreased NPP in overmature conifer forests is increased autotrophic respiration costs and constant or decreased net canopy photosynthesis. This hypothesis was first proposed by Kira and Shidei (1967), but few empirical data are available to rigorously examine this hypothesis, and the data that are available are often the result of extrapolating a few chamber measurements to the stand. Autotrophic respiration, as a percentage of gross primary productivity, either does not differ between overmature and aggrading forests (Sprugel 1984) or is slightly greater for overmature forests (Ryan and Waring 1992; Table 3.1, this study).

Immediately after a forest is harvested or destroyed, GPP (Gross Primary Production) is very low because LAI is small or near zero; the duration of the period of low LAI and GPP is related to the severity of the disturbance. The GPP increases rapidly during the early stages of stand development and reaches a long-term steady state near canopy closure (Sprugel 1985). During the period when GPP is very low, heterotrophic respiration reaches a maximum because of improved substrate and increased soil temperature and moisture conditions. Consequently, NEP is negative for some period immediately after disturbance and reaches a maximum near or before canopy closure when the foliage:woody biomass ratio is highest (Gower *et al.* 1994b) and resources are not strongly limiting. Although NEP is hypothesized to approach zero in overmature forests (Sprugel and Bormann 1981), few data are available to test this hypothesis. Grier and Logan (1977) reported that NEP for five old-growth conifer forest communities averaged 331 g Cm^{-2} yr^{-1} (Table 3.1), but given the uncertainty in many of their measurements it is difficult to ascertain if these estimates of NEP differed from zero.

Despite our tenuous understanding of how the C budget changes during stand development, it is clear that any attempt to balance the global C budget must consider the effects of natural and anthropogenic disturbances on forest C budgets. Johnson (1992) reviewed the literature and found that losses of soil C after harvesting and reforestation were small in the vast majority of studies. However, Harmon *et al.* (1990) calculated that converting Douglas fir in the Pacific Northwest from old-growth to second growth forests released $1.5–1.8 \times 10^{12}$ g C as CO_2 to the atmosphere. Forests in the southeastern USA may have been a regional C source last century due to large-scale deforestation (Delcourt and Harris 1980). There is increasing evidence that northern latitude, temperate and boreal forests are currently an important C sink (Tans *et al.* 1990).

3.5.2 Oregon transect study

In the Pacific Northwest USA, very steep climatic gradients exist because of the orographic effect from the maritime influence of the Pacific Ocean and the high mountain ranges. Annual precipitation can exceed 3000 mm yr^{-1} for coastal *Picea sitchensis* forests to less than 200 mm yr^{-1} for Juniperus woodlands on the east slopes of the Cascades. Along this large climatic gradient, Grier and Running (1977) found that leaf area index (LAI) was positively correlated to site water balance, which they defined as precipitation minus open pan evaporation. Gholz (1982) also reported a similar relationship along the same Oregon transect and that ANPP was positively correlated to LAI. More recently, Runyon *et al.* (1993) reported that ANPP was positively correlated ($r^2 = 0.82$) to annual intercepted photosynthetic active radiation (IPAR) and this relationship was even stronger ($r^2 = 0.98$) when they accounted for the percentage of annual IPAR that was not used to assimilate CO_2 because of environmental constraints on photosynthesis. The positive relationship between ANPP and IPAR reported by Runyon *et al.* (1993) and others (e.g. Linder *et al.* 1985; Cannell *et al.* 1987; Daugherty *et al.* 1992) may be a useful tool to scale stand-level measurements of net primary production to larger scales, especially since IPAR has been shown to be positively related to the remotely sensed vegetation index (normalized difference vegetation index – NDVI) (Bartlett *et al.*1990; Goward and Huemmrich 1992; Waring *et al.* 1993).

3.6 REFERENCES

Adams, J. M., Faure, H., Faure-Denard, L., McGlade, J. M. and Woodward, F. I. (1990) Increases in terrestrial carbon storage from the Last Glacial Maximum to the present. *Nature* **348**, 711–714.

Albrektsson, A. (1988) Needle litterfall in stands of *Pinus sylvestris* L. in Sweden, in relation to site quality, stand age, and latitude. *Scand. J. For. Res.* **3**, 333–342.

Andersson, F., Axelsson, G., Lohm, U., Perttu, K. and Ågren, G. (1980) *Skogen som miljö. Naturvetenskapliga Forskningsrådets årsbok 1979 - 1980. Forska för livet*, pp. 142–179. Naturvetenskapliga forskningsrādet, Stockholm. (In Swedish.)

Axelsson, B. and Brākenhielm, S. (1980) Investigation sites of the Swedish Coniferous Forest Project. Biological and physiographical features. *Ecol. Bull.* (Stockholm) **32**, 25–64.

Bartlett, D. S., Whiting, G. J. and Hartman, J. M. (1990) Use of vegetation indices to estimate intercepted solar radiation and net carbon dioxide exchange of a grass canopy. *Remote Sens. Environ.* **30**, 115–128.

Berg, B. (1981) Litter decomposition studies within SWECON. Data on weight loss and organic chemical composition. Swed. Conif. For. Proj., Internal Report No. 104.

Berg, B. (1984) Decomposition of root litter and some factors regulating the process. Long-term root litter decomposition in a Scots pine forest. *Soil Biol. Biochem.* **16**, 609–617.

Berg, B., Berg, M., Flower-Ellis, J. G. K., Gallardo, A., Johansson, M.-B., Lundmark, J.-E.

and Madeira, M. (1993) Amounts of litterfall in some european coniferous forests. In: Breymeyer, A. (Ed.) *Proceedings from Scope Seminar, Conference Papers 18. Geography of Carbon Budget Processes in Terrestrial Ecosystems*, pp. 123–146. Szymbark, 17–23 Aug. 1991.

Berg, B. and Ekbohm, G. (1991) Litter mass-loss rates and decomposition patterns in some needle and leaf litter types. Long-term decomposition in a Scots pine forest. VII. *Can. J. Bot.* **69**, 1449–1456.

Berg, B. and Ekbohm, G. (1993) Decomposing needle litter in lodgepole pine (*Pinus contorta*) and Scots pine (*Pinus silvestris*) monocultural systems. Is there a maximum mass loss? *Scand. J. For. Res.* **8**, 457–465.

Berg, B., Hannus, K., Popoff, T. and Theander, O. (1982) Changes in organic-chemical components of needle litter during decomposition. Long-term decomposition in a Scots pine forest. 1. *Can. J. Bot.* **60**, 1310–1319.

Berg, B., McClaugherty, C., Virzo De Santo, A., Johansson, M.-B. and Ekbohm, G. (1995) Decomposition of forest litter and soil organic matter – a mechanism for soil organic matter buildup? *Scand. J. For. Res.* (in press).

Berg, B., Staaf, H. and Wessen, B. (1987) Decomposition of Scots pine needle litter of differing nutrients concentrations. *Scand. J. For. Res.* **2**, 399–415.

Bonnevie-Svendsen, C. and Gjems, O. (1957) Amount and chemical composition of the litter from larch, beech, Norway spruce and Scots pine stands and its effect on the soil. *Medd. Nor. skogsforsöksves.* **14**(48), 115–168.

Breymeyer, A. (1991a) Search for geographic scale regularities in ecosystem processes. In: Esser, G. and Overdieck, D. (Eds) *Modern Ecology: Basic and Applied Aspects*, pp. 751–771. Elsevier, Amsterdam.

Breymeyer, A. (1991b) Comparative analysis of organic matter transformations in coniferous forests in Europe. In: Nakagoshi, N. and Golley, F.B. (Eds) *Coniferous Forest Ecology from an International Perspective*, pp. 161–177. SPB Academic Publishing bv, The Hague.

Breymeyer, A. (1993) Production/decomposition of organic matter in forest; studies in one climatic region and in climatic gradient. In: Breymeyer, A. (Ed.) *Geography of Organic Matter Production/Decomposition in Terrestrial Ecosystems.* Conference Papers 17, 147–174, Warsaw.

Brubaker, L. B. and Graumlich, L. J. (1989) 100-year records of forest productivity at high elevations in Western Washington, USA. In: Noble, R. D., Martin, J. L. and Jensen, K. F. (Eds) *Air Pollution Effects on Vegetation*, pp. 49–56. USDA–Forest Service, USEPA, Washington.

Cannell, M. G. R. (1982) *World Forest Biomass and Primary Production Data.* Academic Press, New York.

Cannell, M. G. R., Dewar, R. C. and Thornley, J. H. M. (1992) Carbon flux storage in European forests. In: Teller, A., Mathy, A. and Jeffers, J. N. R. (Eds) *Responses of Forest Ecosystems to Environmental Changes*, pp. 256–271. Elsevier, Amsterdam.

Cannell, M. G. R., Milne, R., Sheppard, L. J. and Unsworth, M. H. (1987) Radiation interception and productivity of willow. *J. Appl. Ecol.* **24**, 261–278.

Daugherty, C. S. T., Gallo, K. P., Goward, S. N., Prince, S. D. and Kustas, W. P. (1992) Spectral estimates of absorbed radiation and phytomass production in corn and soybean canopies. *Remote Sens. Environ.* **39**, 141–152.

Delcourt, H. R. and Harris, W. F. (1980) Carbon budget of the southeastern U.S. biota: analysis of historical change in trend from source to sink. *Science* **210**, 321–322.

Edmonds, R. L. (1987) Decomposition rates and nutrient dynamics in small-diameter woody litter in four ecosystems in Washington, USA. *Can. J. For. Res.* **17**, 499–509.

Ewel, K. C., Cropper, W. P. Jr and Gholz, H. L. (1987a) Soil CO_2 evolution in slash pine plantations. I. Changes through time. *Can. J. For. Res.* **17**, 325–329.

Ewel, K. C., Cropper, W. P. Jr. and Gholz, H. L. (1987b) Soil CO_2 evolution in slash pine plantations. II. Importance of root respiration. *Can. J. For. Res.* **17**, 330–333.

Flower-Ellis, J. G. K. (1985) Litterfall in an age series of Scots pine stands; summary of results for the period 1973–1983. *Swed. Univ. Agric. Sci., Dep. Ecol. Environ. Res. Rep.* **19**, 75–94.

Flower-Ellis, J. G. K. and Olsson, L. (1978) Litterfall in an age series of Scots pine stands and its variation by components during the years 1973–1976. Swed. Univ. Agric. Sci., Swed. Conif. For. Proj., *Technical Report 15*, Uppsala.

Funke W. and Roth-Holzapfel, M. (1991) Animal-coenoses in the 'spruce forest' ecosystem (Protozoa, Metazoa-invertebrates): indicators of alterations in forest-ecosystems. In: Esser, G. and Overdieck, D. (Eds) *Modern Ecology: Basis and Applied Aspects*, pp. 579–600. Elsevier, Amsterdam.

Gessel, S. P., Cole, D. W. and Steinbrenner, E. C. (1973) Nitrogen balances in forest ecosystems of the Pacific Northwest. *Soil Biol. Biochem.* **5**, 19–34.

Gholz, H. L. (1982) Environmental limits on above-ground net primary production, leaf area and biomass in vegetation zones of the Pacific Northwest. *Ecology* **63**, 469–481.

Gholz, H. L., Fisher, R. F. and Pritchett, W. L. (1985) Nutrient dynamics in slash pine plantation ecosystems. *Ecology* **66**, 647–659.

Gholz, H. L., Hendry, L. C. and Cropper, W. P. Jr (1986) Organic matter dynamics of fine roots in plantations of slash pine (*Pinus elliottii*) in north Florida. *Can. J. For. Res.* **16**, 529–538.

Gholz, H. L., Vogel, S. A., Cropper, W. P. Jr, McKelvey, K., Ewel, K. C., Teskey, R. O. and Curran, P. J. (1991) Dynamics of canopy structure and light interception in *Pinus elliottii* stands of north Florida. *Ecol. Monogr.* **61**, 33–51.

Goward, S. N. and Huemmrich, K. E. (1992) Vegetation canopy PAR absorptance and the normalized difference vegetation index: an assessment using the SAIL model. *Remote Sens. Environ.* **39**, 119–140.

Gower, S. T., Gholz, H. L., Nakane, K. and Baldwin, V. C. (1994a) Production and carbon allocation patterns of pine forests. In: Gholz, H. L., Linder, S. and McMurtrie, R. E.(Eds) *The Structure and Productivity of Pine Forests: A Synthesis. Ecol. Bull.* (in review).

Gower, S. T., Isebrands, J. G. and Sheriff, D. W. (1994b) Carbon allocation and accumulation patterns for conifer forests. In: Smith, W. K. and Hinckley, T. M. (Eds) *Ecophysiology of Conifers.* Academic Press (submitted).

Gower, S. T., Vogt, K. A. and Grier, C. C. (1992) Carbon dynamics of Rocky Mountain Douglas-fir: influence of water and nutrient availability. *Ecol. Monogr.* **62**, 43–65.

Grier, C. C. and Logan, R. S. (1977) Old-growth *Pseudotsuga menziesii* communities of western Oregon watershed: biomass distribution and production budgets. *Ecol. Monogr.* **47**, 373–400.

Grier, C. C. and Running, S. W. (1977) Leaf area of mature northwestern coniferous forests: relation to site water balance. *Ecology* **58**, 893–899.

Grier, C. C., Vogt, K. A., Keyes, M. R. and Edmonds, R. L. (1981) Biomass distribution and above and belowground production in young and mature *Abies amabilis* zone ecosystems of the Washington Cascades. *Can. J. For. Res.* **11**, 155–167.

Harmon, M. E., Ferrel, W. K. and Franklin, J. F. (1990) Effects on carbon storage of conversion of oldgrowth forests to young forests. *Science* **247**, 699–702.

Harmon, M. E., Franklin, J. F., Swanson, F. J., Sollins, P., Gregory, S. V., Lattin, J. D., Anderson, N. H., Cline, S. P., Aumen, N. G., Sedell, J. R., Lienkaemper, G. W., Cromack, K. Jr and Cummins, K. W. (1986) Ecology of coarse woody debris in temperate ecosystems. *Adv. Ecol. Res. 15*, 133–302.

Hicks, D. J. and Chabot, B. F. (1985) Deciduous forests. In: Chabot, B. F. and Mooney, H. A. (Eds) *Physiological Ecology of North American Plant Communities*, pp. 257–277. Chapman & Hall, New York, NY.

Howard, P. J. A. and Howard, D. M. (1974) Microbial decomposition of tree and shrub leaf litter. *Oikos* **25**, 311–352.

Johansson, M.-B. (1986) Decomposition rates of Scots pine needle litters in relation to site properties and litter quality. In: Chemical composition and decomposition pattern of leaf litters from forest trees in Sweden with special reference to methodological aspects and site properties. Ph.D. thesis, Swed. Univ. Agric. Sci., Rep. No. 56, 154 pp, Uppsala.

Johnson, D. W. (1992) Effects of forest management on soil carbon storage. *Water Air Soil Pollut.* **64**, 83–120.

Johnson, D. W., Cole, D. W., Bledsoe, C. S., Cromack, K. Jr, Edmonds, R. L., Gessel, S. P., Grier, C. C., Richards, N. and Vogt, K. A. (1981) Nutrient cycling in forests of the Pacific Northwest. In: Edmonds, R. L. (Ed.) *Analysis of Coniferous Forest Ecosystems in the Western United States*, pp. 186–232. Hutchinson Ross, Stroudsburg, Pa.

Józefaciukowa, W. (1975) Variation of the fall rate of plant debris from trees in the association *Vaccinio-myrtylli-Pinetum typicum* (Kobenza 1930). Br.-Bl. et Vlieger 1939 in the Kampinos National Park. *Ekol. Pol.* **23**, 93–101.

Kauppi, P. E., Mielikäinen, K. and Kuusela, K. (1992) Biomass and carbon budget of European forests, 1971 to 1990. *Science* **256**, 70–74.

Keyes, M. R. and Grier, C. C. (1981) Above- and below-ground net production in 40-year-old Douglas-fir stands on low and high productivity sites. *Can. J. For. Res.* **11**, 599–605.

Keyser, P., Kirk, T. K. and Zeikus, I. G. (1978) Ligninolytic enzyme system of *Phanerochaete chrysosporium*: synthesized in the absence of lignin in response to nitrogen starvation. *J. Bacteriol.* **135**, 790–797.

Kinerson, R. S., Ralston, C. W. and Wells, C. G. (1977) Carbon cycling in a loblolly pine plantation. *Oecologia* **29**, 1–10.

Kira, T. and Shidei, T. (1967) Primary production and turnover of organic matter in different forest ecosystems of the western Pacific. *Jap. J. Ecol.* **17**, 70–87.

Klimo, E. (1992) The spruce forest ecosystems in Czechoslovakia. In: Teller, A., Mathy, A. and Jeffers, J. N. R. (Eds) *Responses of Forest Ecosystems to Environmental Changes*, pp. 256–271. Elsevier, Amsterdam.

Linder, S., McMurtrie, R. E. and Landsberg, J. J. (1985) Growth of eucalypts: a mathematical model applied to *Eucalyptus globulus*. In: Tigerstedt, P. M. A., Puttonen, P. and Koski, V. (Eds) *Crop Physiology of Forest Trees*, pp. 117–126. Helsinki University Press.

Long, J. N. (1981) Productivity of western coniferous forests. In. Edmonds, R. L. (Ed.) *Analysis of Coniferous Forest Ecosystems in the Western United States*, pp. 89–125. Hutchinson Ross, Stroudsburg, Pa.

Mälkönen, E. (1974) Annual primary production and nutrient cycle in some Scots pine stands. *Commun. Inst. For. Fenn.* **84**, 1–87.

Miller, H. G. (1981) Forest fertilization: some guiding concepts. *Forestry* **54**, 157–167.

Nakane, K., Yamamoto, M. and Tsubota, H. (1983) Estimation of root respiration rate in a mature forest ecosystem. *Jpn. J. Ecol.* **33**, 397–408.

Nömmik, H. and Vahtras, K. (1982) Retention and fixation of ammonium and ammonia in soils. In: Stevenson, F. J. (Ed.) *Nitrogen in Agricultural Soils*, pp. 123–171. Agronomy Monographs, No. 22. Agron. Soc. Am., Madison, Wis.

Perez, J. and Jeffries, T. W. (1992) Roles of manganese and organic acid chelators in regulating lignin degradation and iosynthesis of peroxidases by *Phanerochaete chrysosporium*. *Appl. Environ. Microbiol.* **58**, 2402–2409.

Raich, J. W. and Nadelhoffer, K. J. (1989) Belowground carbon allocation in forest ecosystems: global trends. *Ecology* **70**, 1346–1354.

Runyon, J., Waring, R. H., Goward, S. N. and Welles, J. M. (1993) Environmental limits on above-ground production: observations from the Oregon transect. *Ecol. Appl.* **4**(2), 226–237.

Ryan, M. G. (1990) Growth and maintenance respiration in stems of *Pinus contorta* and *Picea engelmannii. Can. J. For. Res.* **20**, 48–57.

Ryan, M. G. (1991) A simple method for estimating gross carbon budgets for vegetation in forest ecosystems. *Tree Physiol.* **9**, 255–266.

Ryan, M. G. and Waring, R. H. (1992) Maintenance respiration and stand development in a subalpine lodgepole pine forest. *Ecology* **73**, 2100–2108.

Schlesinger, W. H. (1990) Evidence from chronosequence studies for a low carbon-storage potential of soils. *Nature* **348**, 232–234.

Smolander, H. and Oker-Blom, P. (1990) The effect of nitrogen content on the photosynthesis of Scots pine needles and shoots. *Ann. Sci. For.* **46**, 473–475.

Sonesson, M. (1979) Abisko Scientific Research Station: environment and research. *Holarct. Ecol.* **2**, 279–283.

Sprugel, D. G. (1985) Natural disturbance and ecosystem energetics. In: Pickett and White, P. (Eds) *The Ecology and Natural Disturbance and Patch Dynamics*, pp. 335–455. Academic Press, New York.

Sprugel, D. G. and Bormann, F. H. (1981) Natural disturbance and the steady state in high altitude balsam fir forests. *Science* **211**, 390–393.

Staaf, H. and Berg, B. (1977) A structural and chemical description of litter and humus in the mature Scots pine stand at Ivantjarnsheden. Swed. Conif. For. Proj. Int. Rep. No. 65, Uppsala.

Tans, P. P., Fung, I. Y. and Takahashi, T. (1990) Observational constraints on the global atmospheric CO_2 budget. *Science* **247**, 1431–1438.

Teskey, R. O. and Whitehead, D. (1994) Photosynthesis and carbon gain by pine. In: Gholz, H. L., Linder, S. and McMurtrie, R. E. (Eds) *The Structure and Productivity of Pine Forests: A Synthesis. Ecol. Bull.* (in press).

Tinker, P. B. (1992) Session report V INPUT/OUTPUT. In: Teller, A., Mathy, A. and Jeffers, J. N. R. (Eds) *Responses of Forest Ecosystems to Environmental Changes*, pp. 256–271. Elsevier, Amsterdam.

Turner, J. (1981) Nutrient cycling in an age sequence of western Washington Douglas-fir stands. *Ann. Bot.* **48**, 159–169.

Van Cleve, K. and Viereck, L. A. (1981) Forest succession in relation to nutrient cycling in boreal forests of Alaska. 1. In: West, D. C., Shugart, H. H. and Botkin, D. B. (Eds) *Forest Succession: Concepts and Applications*, pp. 185–211. Springer-Verlag, New York.

Vogt, K. A., Edmonds, R. L., Antos, G. C. and Vogt, D. J. (1980) Comparisons between carbon dioxide evolution, ATP concentrations and decomposition in red alder, Douglas-fir, western hemlock, and Pacific silver fir ecosystems in western Washington. *Oikos* **35**, 72–79.

Vogt, K. A., Grier, C. C., Meier, C. E. and Edmonds, R. L. (1982) Mycorrhizal role in net primary production and nutrient cycling in *Abies amabilis* ecosystems in western Washington. *Ecology* **63**, 370–380.

Vogt, D. A., Grier, C. C., Meier, C. E. and Keyes, M. R. (1983) Organic matter and nutrient dynamics in forest floors of young and mature *Abies amabilis* stands, as affected by fine root input. *Ecol. Monogr.* **53**, 139–157.

Vogt, K. A., Grier, C. C. and Vogt, D. J. (1986) Production, turnover, and nutrient dynamics of above and below-ground detritus of world forests. *Adv. Ecol. Res.* **15**, 303–377.

Vogt, K. A., Vogt, D. J., Gower, S. T. and Grier, C. C. (1990) Carbon and nitrogen interactions for forests ecosystems. In: Proc. of above- and below-ground interactions in forest trees in acidified soils. *Air Pollution Research Report* **32**, 203–235.

Vose and Allen (1988) Leaf area, stemwood growth and nutritional relationships in loblolly pine. *For. Sci.* **34**: 547–563.

Walter, H. (1979) *Vegetation of the Earth and Ecological Systems of the Geo-biosphere.* Springer-Verlag, New York.

Waring, R. H. and Franklin, J. F. (1979) Evergreen conifer forests of the Pacific Northwest. *Science* **29**, 1380–1386.

Waring, R. H., Runyon, J., Goward, S. N., McCreight, R., Yoder, B. and Ryan, M. G. (1993) Developing remote sensing techniques to estimate photosynthesis and annual forest growth across a steep climatic gradient in western Oregon, U.S.A. *Stud. For. Suec.* **191**, 33–42.

4 The Carbon Budget of Tropical Savannas, Woodlands and Grasslands

R. J. SCHOLES[1] **and D. O. HALL**[2]
[1]*Division of Water, Environment and Forest Technology, CSIR, Pretoria, South Africa*
[2]*Division of Life Sciences, King's College London, UK*

4.1 INTRODUCTION

The extensive zone between the evergreen tropical forests and the mid-latitude deserts is occupied by vegetation types containing both trees and grasses, in varying proportions (Figure 4.1). The significance of this zone to the global environment is based on five factors:

(a) The vast extent of the area (a conservative estimate is 16.1 million km^2, or 11.5% of the global land surface) makes it a significant component of global element budgets and gaseous fluxes;
(b) The seasonally dry climate permits extensive annual biomass burning;
(c) The proportion of trees and grass is inherently unstable, and small changes in climate or land-use practice can lead to rapid changes in biomass and soil properties;
(d) It lies mostly within the area of the developing world where population pressure and land-use changes will be greatest in the next few decades. The agro-pastoral communities in savannas are highly dependent on the natural vegetation, which is vulnerable to degradation;
(e) The biodiversity of plants and animals in these areas is greater than is usually recognized.

4.1.1 Definitions of grasslands, savannas and related formations

The importance of the tropical tree–grass systems has been obscured in the past by the vagueness of vegetation type definitions. The woodier savanna variations are often grouped with tropical forests (for instance, in the FAO deforestation statistics), while the grassier forms are lumped with temperate grasslands. The

Global Change: Effects on Coniferous Forests and Grasslands.
Edited by A. I. Breymeyer, D. O. Hall, J. M. Melillo and G. I. Ågren

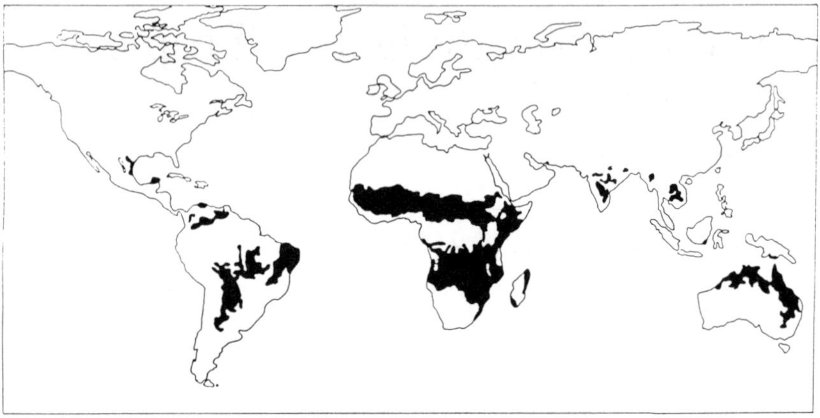

Figure 4.1 The approximate global distribution of tropical grasslands, savannas and woodlands, derived from Olson *et al.* (1983). The black areas represent savannas and are a combination of Olson categories: 32 (drought-deciduous forest), 43 (savannas), 48 (primarily *Eucalyptus*) and 59 (thorn and succulent bush). The details of this map are far from accurate, but there is no consistently better map at a global scale. The Olson *et al.* (1983) data have been used throughout this chapter to maintain internal consistency

problem arises because tropical drought-deciduous woodlands, savannas, shrublands, thickets and grasslands form a seamless continuum which can only be divided into distinct structural types by applying arbitrary limits (Figure 4.2). In this chapter we have followed the following naming conventions:

(a) Forests have complete tree canopy cover and three or more overlapping vegetation strata;
(b) Woodlands have 50–100% tree canopy cover by trees, and a sometimes sparse, but always significant gramineous layer;
(c) Savannas have 10–50% cover by woody plants, and in the unexploited state, a well-developed grass layer;
(d) Grasslands have less than 10% tree cover.

There are also a variety of other formations which are related to, or subsets of, the above broad categories. Grasslands with 2–10% tree cover are often referred to as wooded grasslands. Low-statured (<3 m tall) savannas are often called bushlands or shrublands, while low-statured multi-stemmed woodlands or forests are called thicket or scrub forest. If the woody plants are very short (<1 m) the formation is usually called a dwarf shrubland.

Of the non-forest types, tropical grasslands have the smallest extent (probably less than 10% of the total area). Completely treeless tropical grasslands are confined to the following situations: soils which are intermittently waterlogged; areas with frequent (annual or biannual) intense fires, typically on fertile soils;

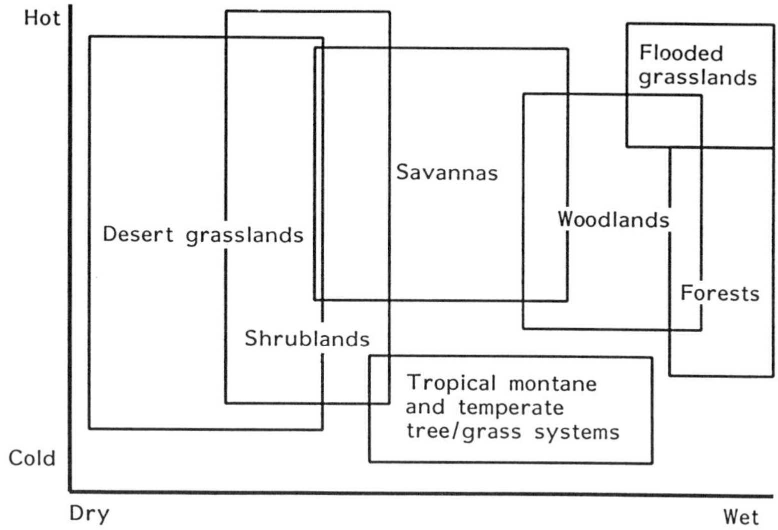

Figure 4.2 Grasslands, savannas and woodlands are part of a continuum of vegetation types on moisture and temperature gradients. The divisions on this continuum are arbitrary, which leads to great variations in the estimates by different authorities of the area occupied by each type

soils which contain elements toxic to trees (typically metals); areas subject to frequent frosts (i.e. high altitudes); areas receiving very low rainfall (<100 mm yr^{-1}); and areas cleared of trees by human activity.

Woodlands and savannas are characterized by having an annual dry season of sufficient duration and intensity that most of the woody plants shed their leaves, and the grasses dry out. This accumulation of dry fuel permits fires every few years; and the fires help prevent complete domination by the woody plants. Tropical forests are generally evergreen and not prone to fires.

We used 30° N and 35° S as the limits of the tropical formations, based on the latitudinal distribution of woodlands, savannas and grasslands. The latitudinal distribution of these classes, inferred from the global vegetation data set of Olson *et al.* (1983) is illustrated in Figure 4.3. The bulk of the tropical grasslands, savannas and woodlands fall between 30° S and 20° N. The strict geographical definition of 'the tropics', being the area between 23.5° N and 23.5° S, has little biological meaning. The key difference between tropical ecosystems and their temperate analogues is the degree to which processes such as primary production and litter decay are constrained by low temperatures: in tropical systems, low temperatures are at best a secondary constraint.

The high-altitude tropical grasslands share many characteristics with the temperate grasslands, including dominance by species with a C_3 photosynthetic

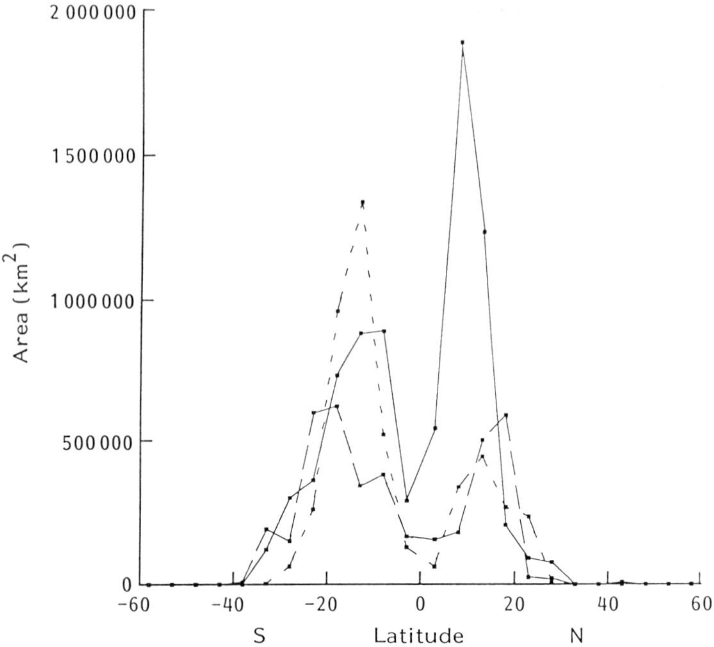

Figure 4.3 The latitudinal distribution of tropical woodlands, savannas and grasslands derived from the Olson *et al.* (1983) data. The solid line represents Olson's class 43 ('savannas'), the long-dashed line is class 59 (thorn and succulent woodlands) and the short-dashed line is classes 32 (drought-deciduous tropical forests) and 48 (mainly Australian *Eucalyptus* and *Acacia* woodlands). Note the bimodal distribution on either side of the equator, extending to 35° S and 30° N. The equatorial region is mainly occupied by evergreen forest. At the higher latitudes, savannas are generally replaced by semi-desert shrublands and grasslands

pathway and a relatively high soil organic carbon content. They should be grouped ecologically with temperate rather than tropical grasslands, since low temperatures are an important constraint. Subtropical and high-altitude savannas also experience a degree of temperature constraint at the beginning and end of the growing season, but it is small relative to the water constraint. This is because the low temperatures typically occur during the dry season, where no growth is possible in any case. Frost can be an ecologically significant factor in tropical savannas, on the infrequent occasions when unseasonal rainfall coincides with low night temperatures.

The surface area, carbon store and net primary productivity of tropical woodlands, savannas and grasslands as estimated by previous studies and calculated according to the above criteria, are shown in Table 4.1.

Table 4.1 The surface area, carbon store and NPP of tropical woodlands, savannas and grasslands, as estimated by various sources. DM = oven dry matter.

Source and vegetation type	Area (10^6 km^2)	Biomass Pg DM[a]	Biomass kg DM m^{-2}	NPP Pg DM yr^{-1} [a]	NPP g DM m^{-2}yr^{-1}
Ajtay *et al.* (1979)	22.5	[b]	[b]	[b]	[b]
Whittaker and Likens (1973)	15.0 Savannas	60	4 (0.2–15)	13.5	900 (200–2000)
	8.5 Woodland and shrubland	50	6 (2–20)	6	700 (250–1200)
This study (based on Olson *et al* 1983 classes 32, 43, 48 and 59)[c]	16.1	58.7[d]	3.6 (0.9–21)	16.9[e]	1050 (380–3570)
Comparative values: Whittaker and Likens (1973), quoted in Ajtay *et al.* (1979)					
Tropical forest	17	765		37.4	
Global terrestrial vegetated area	149.3	1837		117.5	

[a]Pg = 10^{15} g.
[b]Not calculated in this study.
[c]32 = drought-deciduous tropical forest and woodland; 43 = Savanna; 48 = *Eucalyptus* or *Acacia*, quabracho, saxual; 59 = succulent and thorn woods.
[d]The total ecosystem carbon store is equivalent to 134.5 Pg DM, including carbon in the soil.
[e]See Table 4.5. Assumes 45% carbon content.

4.2 FACTORS CONTROLLING CARBON FLUXES IN TROPICAL GRASSLANDS AND SAVANNAS

4.2.1 Primary production

Net primary production (NPP) is defined here as the sum of net increases in total plant standing crop over a given period, plus losses due to litterfall, root death and herbivory during the same period. In practice, the available data seldom satisfy this definition perfectly. It is often not possible to determine from the sources whether turnover losses and herbivory have been included in the estimate. The estimates given below are therefore generally conservative. We have excluded all estimates based simply on end-of-season standing crop. The sources quoted have measured NPP over a period of at least a year, and in most cases, several years.

Three main factors control primary production in tropical woodlands, savannas and grasslands:

(a) water availability to the plants;
(b) nutrient availability to the plants;
(c) vegetation structure (particularly the tree–grass ratio, the plant density and cover, and the amount of standing dead material).

4.2.1 *Water availability*

Water availability is the relationship between water supply and water demand, seen from the plant point of view. Supply and demand have independent seasonal patterns, so although availability can be expressed as an annual mean, it must be calculated using a daily, weekly or monthly water balance model which takes into account precipitation, air temperature, radiation, humidity, wind, soil texture, rooting depth, vegetation attributes and landscape position.

Precipitation in the lowland tropics is almost entirely in the form of rainfall, most of which occurs as short-duration, high-intensity convectional storms. The near-tropical zone occupied by savannas and woodlands is climatically characterized by:

(a) stable, descending airmasses for protracted periods of the year, leading to highly seasonal rainfall, sometimes bimodally distributed;
(b) high continentality (much of the area is far from the ocean);
(c) very high solar radiation, since the cloud cover is relatively low.

These conditions result in generally low humidity and high daytime temperatures, and thus very high rates of evaporation. Annual potential evaporation substantially exceeds annual rainfall. Therefore comparing production between tropical and temperate regions on the basis of equal annual rainfall alone is misleading, since rainfall is less effective in the tropics. A better approach (still imperfect, since it disregards soil and plant details) is to relate production to a water availability index calculated as the annual sum of the monthly ratio of rainfall to open-water evaporation. The water availability index (WAI) is calculated as the sum over 12 months

$$\mathrm{WAI} = \sum_{i=12} d_i R_i/E_i \quad R_i/E_i \leqslant 1.0$$

where d_i is the number of days in month i, R_i is the monthly rainfall, and E_i is the monthly potential evaporation. Constraining R/E to less than 1.0 is equivalent to saying that evaporation cannot exceed rainfall and that stored soil water is not carried over between months. This is a conservative assumption, taken in the absence of detailed soil information. The climate data were from Leemans and Cramer (1990). Open-water potential evaporation was calculated using the Priestley–Taylor equation (Rosenberg *et al.* 1983, using a coefficient of 1.36, appropriate for semi-arid environments). Net radiation was calculated from the

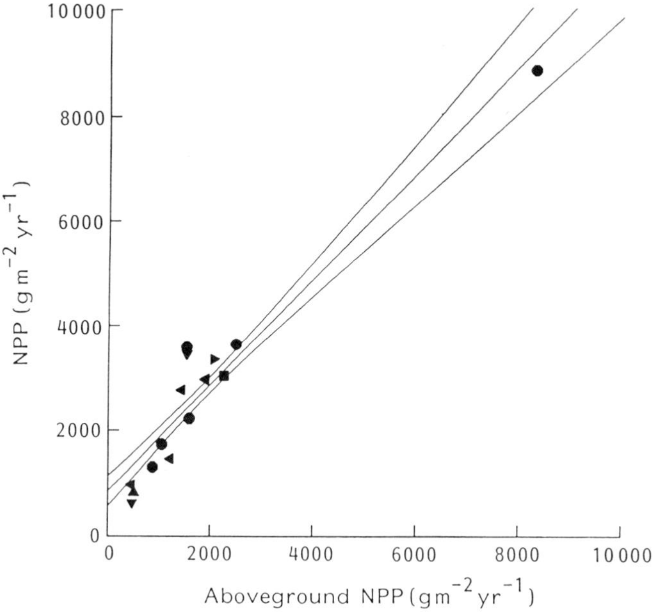

Figure 4.4 The relationship between total NPP and aboveground NPP for sites in Table 4.2 which report both aboveground and belowground NPP. The equation for this relationship is total NPP = 853 + 1.01 × aboveground NPP ($n = 15$, $F = 141.3$, $p < 0.001$, $r^2 = 0.91$). This relationship was used to calculate the total NPP for sites reporting only aboveground NPP, for inclusion in Figure 4.5. The meaning of the symbols is explained in the caption of Figure 4.5

solar geometry and cloud cover (McMurtrie 1993). The sources of the NPP data are in Table 4.2. The sites for which both belowground NPP and aboveground NPP data were available were used to develop an equation for total NPP as a function of aboveground NPP (Figure 4.4), and this equation was used to predict total NPP at the sites which reported aboveground NPP only. This equation predicts that belowground production accounts for between 5 and 70% of the total NPP, the proportion decreasing with increasing water availability. The average proportion of belowground NPP is high, but in line with recent estimates by other techniques (Raich and Nadelhoffer 1989).

The NPP predicted by the relationship shown in Figure 4.5 was calculated for a large number of randomly selected locations in each of the Olson *et al.* (1983) vegetation classes representing tropical woodlands, savannas and grasslands (Figure 4.6 and Table 4.3). The relations describing the upper and lower data envelopes in Figure 4.5 were used to estimate the likely error range.

Slope, soil texture and depth control the fraction of water which leaves the ecosystem 'productively' (i.e. through transpiration) and 'unproductively' (runoff, drainage and evaporation from the soil). Where the soil is deep and rainfall is

Table 4.2 The primary production reported for tropical grasslands, savannas and woodlands

Structura class site name	Latitude	Longitude	Aboveground ($g\ DM\ m^{-2}\ yr^{-1}$)		Belowground ($g\ DM\ m^{-2}\ yr^{-1}$)		Source
			Trees	Grass	Trees	Grass	
Drought-deciduous woodlands							
Lamto, Côte d'Ivoire	6°13′ N	5°02′ W	633	1450	37	633	Menaut and Cesar (1979)
Mokwa, Nigeria	9°18′ N	5°04′ E	378				Collins (1977)
Savannas							
Nylsvley, South Africa	24°42′ S	28°42′ E	282	157	187	325	Scholes and Walker (1993)
Lamto, Côte d'Ivoire	6°13′ N	5°02′ W	55	1610	5	1900	Menaut and César (1979)
			137	1280	13	1330	
			277	1610	23	1040	
Niono, Senegal	14°18′ N	6°0′ W		225			De Ridder *et al.* (1982)
Tropical grasslands, known to be derived from forests, woodlands or savannas							
Towoomba, South Africa	24°50′ S	28°15′ E		141			Donaldson *et al.* (1984)
Makaholi, Zimbabwe	19°48′ S	30°48′ E		146			Ward and Cleghorn (1964)
Nuanetsi, Zimbabwe	21°24′ S	30°48′ E		261			Kelly *et al.* (undated)
Klong Hoi Kong, Thailand	6°0′ N	100°56′ E		1595		625	Long *et al.*(1992)[a]
Uttar Pradesh, India	24°18′ N	83°0′ N					Pandey and Singh (1992)
Ranitali			5200	6450	600	6750	
Hathinala			2300	7150	300	7200	
Telburva			4300	4900	400	4600	
Charleville, Australia	26°24′ S	146°12′ E		520		290	Christie (1978)
Tropical grasslands, thought to be controlled by fire, browsing or flooding							
Nairobi, Kenya	1°0′ N	36°49′ E		881		431	Long *et al.* (1989)
Udaipur, India	25°30′ N	72°24′ E		180			Vyas *et al* .(1972)
Jodphur, India	26°18′ N	73°06′ E		108			Gupta *et al.* (1972)
Delhi, India	28°54′ N	77°12′ E		798			Vashney (1972)
Ejura, Ghana	7°30′ N	3°30′ E		870			Greenland & Nye (1959)

Kurukshetra, India	9°58′ N	76°51′ E		2500		1131	Rajvanshi and Gupta (1985)
Calabozo, Venezuela	8°48′ N	67°27′ W		478		146	Sarmiento (1984)
Manaus, Brazil	3°20′ S	60°0′ W		9418		507	Long *et al.* (1989)
Montecillos, Mexico	19°28′ N	98°28′ W		1063		678	Long *et al* .1989)
Lamto, Côte d'Ivoire	6°13′ N	5°02′ W		1540		2040	Menaut & César (1979)
Plantations, mostly exotic trees							
Kurukshetra, India	29°58′ N	76°51′ E	1547	248			Rajvanshi and Gupta (1985)
Gandi Nagar, India	23°12′ N	77°6′ E	2270		757		Gurumurti *et al.* (1986)
Mudigere, India	13°12′ N	75°36′ E	1872				Swaminath (1988)
Nigeria	10°30′ N	7°18′ E	1337				Kadeba and Aduayi (1985,6)

[a]Long *et al.* (1989,1992) used a variety of methods for calculation of NPP. The sum-of-positive increments method is quoted here, which gives the third highest estimate on their data. The highest estimate, which includes consideration of decay between sampling periods, may be a better estimate, but was not used because that approach was not applied in any other study.

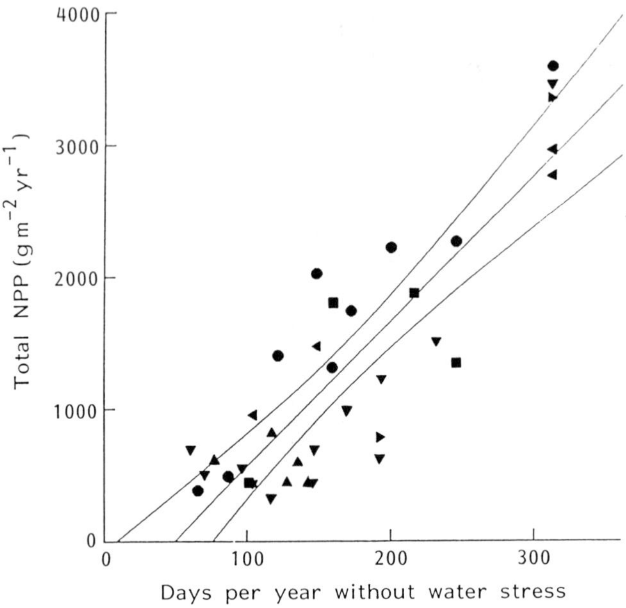

Figure 4.5 The relationship between total NPP and an index of water availability for tropical grasslands and tree–grass mixtures, with 95% confidence limits. The water availability index (WAI) can be thought of as the number of days per year on which growth is not water limited. The derivation of WAI is discussed in the text. The least-squares best fit line has the equation NPP = 11 × WAI − 539 ($n = 37, F = 108.6, p < 0.001, r^2 = 0.75$). The data are from Table 4.2, and a number of other studies which did not satisfy the criteria for inclusion in Table 4.2, for instance because they refer to regions rather than sites. The symbols denote different structural types: [○] grasslands; [△] derived grasslands; [▽] grassy savannas; [◁] savannas; [▷] woodlands; [□] plantations of exotic trees in savanna areas

much less than evaporation, less water is lost 'unproductively' from sandy than from clayey soils. Where the rainfall is higher and the soils shallower, sandy soils lose water through drainage and runoff, while clay soils benefit from their higher water-holding capacity and nutrient availability (see below). In moist and mesic systems, therefore, clay soils support a higher NPP than sandy soils receiving the same rainfall, but the reverse is true in arid areas. The crossover occurs around 500 mm yr^{-1} (Noy-Meir 1973).

The implicit assumption in much of the literature on water limitation of growth is that the mechanism of limitation is the reduction in stomatal conductance resulting from stomatal closure during periods of water stress. This reduces CO_2 assimilation and therefore dry matter production. The weakness in this argument is that there is little evidence to suggest that the rate of carbon assimilation constrains plant growth in the tropics. Photosynthetic carbon assimilation substantially exceeds the amount of carbon which appears as dry matter (Scholes

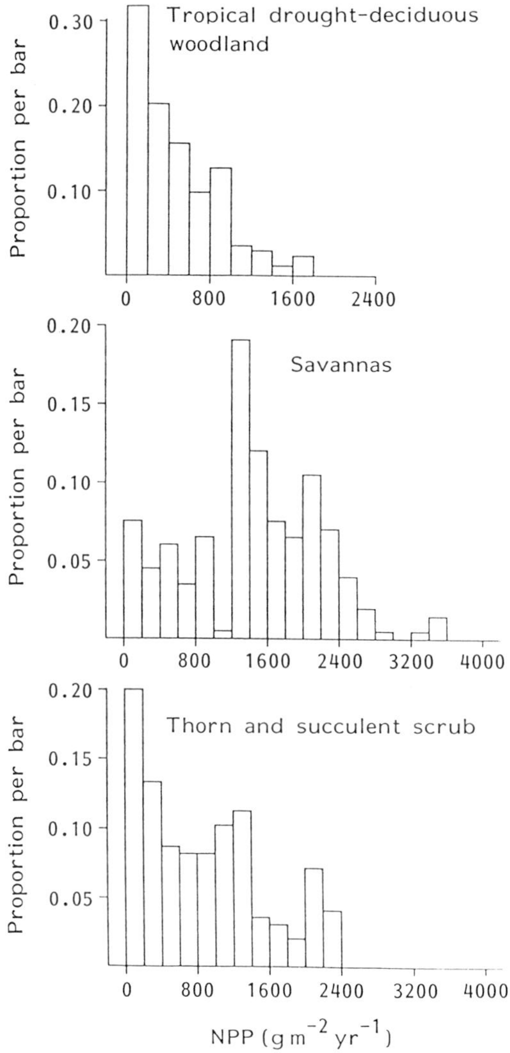

Figure 4.6 The probability distribution of total NPP for tropical woodlands, savannas and thorn scrub, calculated using the regression equation developed in Figure 4.5, the Olson *et al.* (1983) vegetation map and a global climate data set (Leemans and Cramer 1990). Total NPP includes trees and grass, above ground and below ground. The bars represent the fraction of all locations in the given type which have NPP in the range indicated on the *x*-axis

Table 4.3 Calculated net primary productivity in tropical woodlands, and savannas. The estimates are based on the relationship shown in Figure 4.5, along with its upper and lower data envelopes, applied to sites randomly drawn from the areas classified as having those vegetation characteristics, using a global climate data set. The 'best estimate' is based on the regression line defined in Figure 4.5. The 'lower limit' is based on the line which forms the lower envelope to the data in the figure, and the 'upper limit' applies the line defining the upper envelope

Vegetation class (after Olson *et al.* (1983)	*N*[a]	Total NPP (g DM m^{-2} yr^{-1})		
		Best estimate	Lower limit	Upper limit
Drought-deciduous woodlands	173	1263	462	1789
Savannas	199	1426	681	1941
Eucalypt	111	733	186	1242
Thorn and succulent 'woods'	195	856	289	1370

[a]Number of locations at which the NPP was simulated.

and Walker 1993). The high carbon:nutrient ratios of tropical grasses, in particular, suggest that the availability of nutrients is the major production constraint, rather than the rate of carbon assimilation.

A more convincing case can be made that water availability controls plant growth via its influence on nutrient availability. The mineralization of nitrogen and phosphorus in litter and soil, the transport of solutes to the root, and the extension of roots into the soil are all under the direct control of soil water content. This line of argument is supported by fertilization experiments and by comparisons of NPP on adjacent soils with different fertility status. In both these cases productivity is increased without an increase in water availability.

In reality, primary production in savannas (and other vegetation types) is probably controlled and constrained by a number of partially interdependent factors, either acting simultaneously or sequentially. The relative importance of individual factors shifts with time; the point is that it must not simply be assumed that the process of carbon assimilation is all-important, and that the stomate is the locus of its control by water availability.

4.2.1.2 Soil fertility and plant nutrition

Over half of the savannas and tropical grasslands worldwide occur on extremely old land surfaces, mostly formed from acid, crystalline igneous parent material. As a result, the soils are sandy and deficient in many nutrients. The small amount of clay is predominantly in minerals such as kaolinite and sesquioxides, which have a low surface area and low ion exchange capacity. Nitrogen and phosphorus play a central role in limiting primary production in tropical savannas (Bremen and de Wit 1983). Deficiencies of other elements in the natural vegetation are usually only obvious once nitrogen and phosphorus constraints have been alleviated. In the case of nitrogen, deficiency may be due to losses of nitrogen through repeated burning ('pyrodenitrification'); in the case of phosphorus it is usually associated with low supplies of weatherable phosphorus coupled with high phosphorus fixation. Secondary production is also constrained by nitrogen, phosphorus and trace elements. There is a striking difference in large mammal biomass between savannas on fertile and infertile soils (East 1984; Fritz and Duncan 1993).

Widespread nitrogen deficiency may seem odd, given that the ultimate nitrogen source is the atmosphere, which is the same over fertile and infertile soils. Although leguminous plants are a conspicuous (frequently dominant) component of savannas, there is very little nitrogen fixation by those on infertile soils, possibly because of deficiencies of cofactors such as molybdenum and phosphorus (Zietsman *et al.* 1988). The nitrogen cycle is dominated by mineralization of soil organic matter, which is in turn controlled by soil water content, soil temperature and the amount of readily mineralizable soil organic nitrogen present. The soil organic matter content of infertile savannas is low, for reasons discussed below. The principal pathway of nitrogen loss from savanna ecosystems is pyrodenitrification (Scholes and Walker 1993), which is high due to the combination of frequent fires and the accumulation of decomposition-resistant fuel.

The phosphorus supply in these old, weathered soils is controlled by two factors: the rate of microbial mineralization of organic phosphorus; and the sorption of inorganic phosphorus by the abundant aluminium and iron oxides in the soil. Large tracts of virgin tropical savanna and grassland remain on soils of this type, because of their unsuitability for low-input agriculture. Some ingenious traditional methods (such as *chitemene* ash-fertilization agriculture in Zambia; Lawton 1982; Stromgaard 1985) have evolved to use these soils. Advances in agricultural technology, notably the ability to supply high rates of lime, nitrogen and phosphorus fertilizer, make the infertile savannas highly attractive for agricultural expansion, such as occurred in the *cerrado* of Brazil during the 1980s.

In addition to the extensive areas of old, weathered, acidic soils, there are significant areas of basic intrusive and extrusive geology in the tropics. These not only have a more favourable nutrient content to begin with, but retain nutrients better as well. The basic igneous rocks weather to produce predominantly 'high activity' clays, with a large surface area and high surface charge. The soils formed from these materials store more carbon and nitrogen than soils formed from acid crystalline materials (Jones 1973). They also tend to occur on younger land surfaces, and therefore have a greater reserve of weatherable nutrients and have had less time to lose nutrients through leaching.

Due to the more favourable nutrient status, more phytomass is produced per unit of water transpired by ecosystems based on soils resulting from these recent, basic igneous parent materials, but they are more drought-prone in arid areas due to the high clay content of the soil. The areas of fertile soil are the focus of agriculture in the developing world. The only uncultivated patches are usually in extremely arid areas or in national parks. Where they have not been planted to crops they are heavily grazed (and therefore seldom burned) and have a sparse tree cover due to wood collection for domestic purposes.

The most dramatic difference between the infertile and fertile tropical ecosystems lies in the degree of herbivory. Whereas less than 10% of the NPP of tropical grasslands, savannas and woodlands on infertile soils is typically consumed by herbivores, up to 80% is consumed on fertile soils (Drent and Prins 1987; Scholes and Walker 1993). As a consequence, fire is less frequent and intense on grazed fertile soils than on infertile soils, since a fuel load does not accumulate. The high acceptability of the forage to herbivores is the main reason why most of the famous savanna game parks occur on these soils. The reasons for this difference are discussed below under herbivory.

Soils derived from sedimentary and calcareous rocks, and from alluvium and windblown sands, are also extensive in savanna regions. They have a fertility status and NPP intermediate to the above extremes.

4.2.1.3 Vegetation structure

The primary productivity of grasses in tropical mixed tree–grass systems is strongly reduced by small increases in the tree biomass (e.g. Walker *et al.* 1972). While a dense grass cover is able to suppress young trees (those still within the grass canopy and flame zone), it has little effect on mature trees (Knoop and Walker 1984). This competitive asymmetry is a recipe for structural instability. Where the herbivory is moderate (or intense but intermittent, such as in the migratory systems of East Africa), the combination of occasional fires and browsing keeps fertile savannas open or sparsely treed. Consistently heavy grazing by domestic stock combined with the exclusion of indigenous browsing mammals and fire leads to a rapid invasion by woody plants, culminating in a dense thicket with little grass cover (e.g. van Vegten 1984; Harrington and Johns 1990). This is known as 'bush encroachment' and is most common in cattle and sheep ranching areas on fertile soils, with a low human population. In densely populated areas, wood harvesting leads to decreased woody biomass, accompanied by high grazing pressure on the grass layer.

The inverse relationship between tree leaf biomass and grass production is typically non-linear: the steepest decline occurs while the woody plant biomass is low (Figure 4.7). The degree of curvature decreases as the productive potential of the site (i.e. water and nutrient availability) increases, suggesting that it is related to the degree of belowground competition (Scanlan and Burrows 1990). A few cases have been reported where the productivity of the grass layer increases with initial increases in tree cover, before declining at higher levels of cover. These cases may be explained in terms of the higher nutrient supply and lower evaporative demand immediately below the tree canopy (Belsky *et al.* 1989).

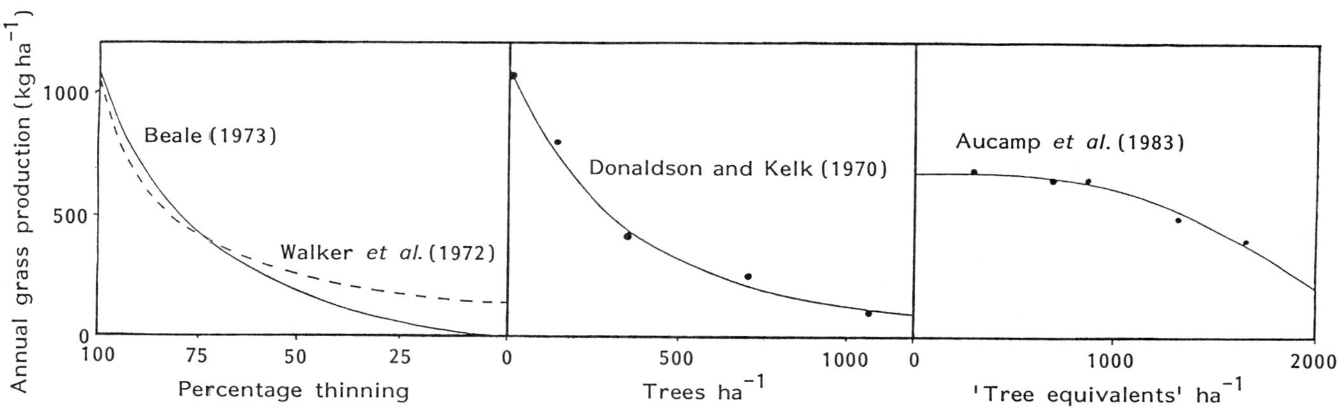

Figure 4.7 Some published relationships between the quantity of trees in a savanna and the annual grass aboveground NPP. Almost all the studies show concave inverse relations; the inverse convex relationship is unusual

Reviews of savanna NPP have generally given the impression that they are unproductive systems (Bourliere and Hadley1970; Murphy 1977), especially when compared to plantations of exotic trees in the same environment (Jackson and Ojo 1971). This is partly due to the small number of studies and the application of inadequate methods of estimating NPP: in particular the use of end-of-season standing crops, and the failure to account for belowground NPP (Long *et al.* 1989). More than half of all savannas and woodlands are on infertile soils, and all are by definition water limited to some degree. These factors do constrain savanna productivity (and cause a large fraction of NPP to shift belowground), but the inherent capacity for production by savanna plants is not markedly lower than other systems. If the upper range of data in Figure 4.5 (i.e. non-nutrient-limited systems) is projected to a water availability of 365 days (i.e. no water limitation) the predicted annual NPP is 4000 g m^{-2}, which is well within the average-to-high range for other natural ecosystems not subject to strong water or nutrient constraints (Hall *et al.* 1993). Moist, fertile tropical grasslands achieve productivities which rival the highest published crop or natural ecosystem values (Long *et al.* 1989).

Although plants with the C_4 photosynthetic system generally have higher instantaneous peak rates of photosynthesis, there is no statistically significant difference in the total NPP achievable by plants with C_3 (tree) or C_4 (grass) photosynthetic systems, when grown under identical environments (Snaydon 1991). In dry areas the total aboveground NPP in mixed tree–grass ecosystems is usually higher than in pure grasslands derived from them through the removal of woody plants, despite the large increase in grass production which accompanies this action. The explanation for higher total production by partly wooded ecosystems in dry areas appears to lie in the greater leaf area duration of trees, permitted by their larger internal storage of water, nutrients and carbohydrates, as well as their deeper roots (Scholes and Walker 1993).

Primary production by tropical grasses typically increases by about 20% in the season after a fire (San José and Medina 1975; Grossman *et al.* 1981), although the reverse has also been recorded (Smith 1960; Brockington 1961). Where there is little grazing, production increases are partly due to the removal of the self-shading effect of old, dead standing material. The return of nutrients in the ash and changes in the soil microclimate increase the availability of nutrients, and therefore increase both quality and quantity of grass produced. The long-term consequence of frequent fire may be to reduce NPP, due to nutrient depletion. Grazing on nutrient rich sites has the same effects as fire: moderately grazed grasslands have greater NPP than ungrazed or excessively grazed grasslands (McNaughton 1983).

Soil fertility, water availability, fire, herbivory and vegetation structure interact to determine the fraction of NPP which is allocated belowground. It is typically higher in grasses (40–80%) than in woody plants (20–60%), and increases with nutrient and water stress, fire and herbivory. In the extreme (but widespread) case of savannas on deep, sandy soils, up to 80% of the plant biomass, and about 60% of the NPP, may be belowground (Rushworth 1978).

4.2.2 Carbon loss to the atmosphere

4.2.2.1 Controls on decomposition

The overriding control on the decomposition rate of plant residues in tropical grasslands and savannas is water availability. The chemical composition of the litter, and to a lesser extent its physical structure, control its decay rate during the periods when it is moist. The chemical composition of woody plant leaf litter is significantly different from that of grass litter: tree litter typically has a higher content of nitrogen, lignin and secondary compounds (Scholes and Walker 1993). The chemistry of savanna tree litters also varies substantially between trees growing on fertile soils and those growing on infertile soils (Scholes and Walker 1993). Leaves from the latter group are heavily defended with secondary chemicals, and can take several years to decompose in the field. Grass litter decomposes faster than tree litter provided it is in contact with the soil surface; however grass litter frequently remains attached to the plant, where it decays gradually until burned, grazed or trampled.

The lignin:nitrogen indices which have been useful in predicting litter decay rates in temperate regions frequently fail when applied in the tropics (e.g. Palm and Sanchez 1990). The main reason appears to be the high content of secondary plant compounds, such as tannins and other phenolics, in the litter. It is suggested that warm temperatures, which favour microbial attack, and the high incidence of mainly insect herbivory promote high concentrations of secondary chemicals in tropical plants. A low litter decay rate is a consequence of these defensive compounds, rather than the principal reason for their synthesis.

Soil fauna, notably termites and earthworms, process a large fraction of the dead organic matter in some locations and at some times, but in general only a small fraction (<5%) of the ecosystem respiration results directly from detritivores (Lavelle *et al.* 1992; Scholes and Walker 1993). Their major impact is to move surface litter into the soil and fragment it, thereby speeding up decomposition and preventing them from being burned.

4.2.2.2 Fire

Even in annually burned savannas, a surprisingly small fraction (<20%) of the NPP ends up being consumed by fire (Scholes and Walker1993). In triennially burned savannas, this fraction decreases to about 5%. This is because a large part of the NPP is protected from fire, being underground or in the tree biomass, or having decayed in the interval between fires. Furthermore, most savanna fires are very patchy and incomplete, due to discontinuous fuel, dissected topography and variation of meteorological conditions between day and night. Some estimates of gaseous fluxes to the atmosphere from biomass burning in savannas are substantially too large, since they assume too large a fraction of the aboveground biomass is combusted, or they overestimate the fuel load or fire frequency (Menaut *et al.* 1991). About 40% of the total savanna area is too dry to accumulate substantial fuel loads except in high-rainfall years, or are too heavily grazed to burn fiercely.

Hao *et al.* (1990) estimate that 3.69 Pg DM is consumed annually by fires in savannas, of which 2.43 Pg is in Africa. This is close to the 2.52 Pg estimated by Delmas *et al.* (1991) for African savannas, but both are probably overestimates. Hao *et al.* (1990) base their calculation on a mean fuel load of 6.6 Mg ha^{-1}, a burn efficiency of 83%, and an annual area burned in Africa of about 450 million ha (out of 1000 million ha of African savannas). Delmas *et al.* (1991) estimate an area burned of 470 million ha, a burning efficiency of 80% and a mean fuel load of 6.1 Mg ha^{-1}. These estimates translate to an average of 112 g C m^{-2} yr^{-1}, or 1.68 Pg C for all the savannas, using the Hao *et al.* (1990) estimate of global savanna area (15 million km^2). Menaut *et al.* (1991) provide a more resolved calculation for West Africa, in which the different types of savannas, fuel loads and fire frequencies are taken into account. For an area of 2.27 million km^2, they estimate an annual emission of 123 Tg C. This works out to only 54 g C m^{-2} yr^{-1}.

Seiler and Crutzen (1980) calculated that vegetation burning in savannas amounts to 1.19 Pg DM yr^{-1}, or 17% of the global total for all forms of biomass burning. Hao *et al.* (1990) estimated 3.69 Pg DM yr^{-1}, or 59% of all biomass burning calculated by their method. We estimate that the true emission from savanna burning is somewhere between 1.94 Pg DM yr^{-1} (0.87 Pg C yr^{-1}), derived from an extrapolation of the Menaut *et al.* (1991) rate using our savanna extent estimate, and 3.69 Pg DM yr^{-1} (1.66 Pg C yr^{-1}) taken from Hao *et al.* (1990). Even allowing for some overestimation, fire in tropical grasslands and savannas (particularly in Africa) remains an important global source of CO, CH_4, N_2O, NO_x and tropospheric O_3.

4.2.2.3 *Herbivory*

The fraction of NPP which is consumed by herbivores in tropical grasslands and savannas, and the type of herbivores involved, are strongly dependent on the soil fertility status (East 1984). The nitrogen content of the grasses growing on the old, infertile land surfaces is below the threshold (about 1% nitrogen) for efficient ruminant digestion for most of the year. Consequently, these areas have a low mammalian herbivore biomass, dominated by large-bodied herbivores (>200 kg). Grasslands on fertile soils remain above this threshold all year, and therefore support a large number and diversity of grazers, many quite small-bodied (<100 kg). In both fertile and infertile savannas, a large (perhaps even dominant) part of the herbivory is by insects. In the grass layer, these are mainly locusts and grasshoppers. The trees of infertile savannas are periodically defoliated by lepidopteran larvae (Scholes and Walker 1993).

Although the fraction of production consumed by mammals in infertile savannas is small, herbivory has an important consequence on global warming through the production of methane by enteric fermentation. This source of methane is of the same order of magnitude as that from burning of tropical grasslands and savannas (Schütz *et al.* 1990).

4.3 CARBON STOCKS IN TROPICAL WOODLANDS, SAVANNAS AND GRASSLANDS

4.3.1 Biomass

The underlying factors which control the biomass in unharvested tropical woodlands, savannas and grasslands are water and nutrient availability. The pattern of biomass distribution in the plane defined by these two factors is complex, since a large number of modifying factors alter the basic pattern (Figure 4.8). Especially at the moist and fertile end of the spectrum, the control is mainly exerted through fire. There are two broad trends: in general, the low-fertility soils support greater tree biomass than the high-fertility soils, especially in arid areas; and generally the woody biomass increases with increasing water availability, up to the point at which the soils are saturated for several months a year, above

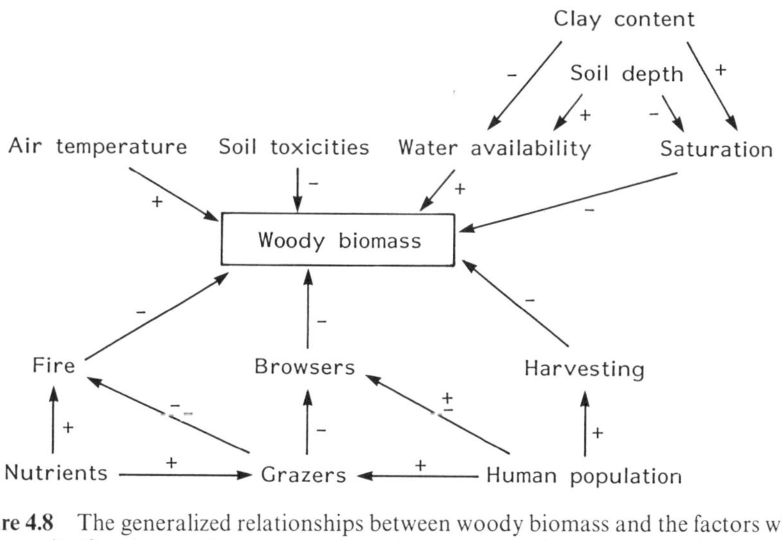

Figure 4.8 The generalized relationships between woody biomass and the factors which influence it. If an interaction has a positive sign, it means that an increase in the causal factor results in an increase in the affected factor. Note that there are several loops containing both positive and negative interactions. This contributes to the fundamental instability of woody biomass in savannas and the existence of multiple quasi-stable states

Table 4.4 Biomass standing stocks (DM) reported for tropical savannas, woodlands and grasslands. The site descriptions are those used by the authors of the studies. They have been regrouped according to our definitions and judgement of woodland, savanna and grassland, which frequently differs from the definition applied by Olson *et al.* (1983).

	Aboveground biomass (t ha^{-1})				
Site	Wood	Tree leaf	Grass/ herbs	Roots (t ha^{-1})	Source
Drought-deciduous forests: mean total aboveground biomass = 133 t ha^{-1} s.d. = 76					
Kurukshutra, India	53.12	3.29	1.33		Rajvanshi and Gupta (1985)
Dalbergia sissoo dry forest					
Brigalow, Australia *Acacia harpephylla* dry forest.	158.1	11	0	40.7	Moor *et al.* (1967)
Oro Forest Reserve, Nigeria	132				Fatubarin (1984)
Woodlands: mean total aboveground biomass = 62 t ha^{-1} s.d. = 28					
Kasapa, Zaire *Brachystegia* spp.	101.7	4.82			Malaisse *et al.* (1975)
Makaholi, Zimbabwe *Brachystegia* spp.	44.6	2.29			Ward and Cleghorn (1964)
Zambia *Brachystegia* spp.	40				Endean (1967)
Marondera, Zimbabwe *Brachystegia* spp.	39.73	2.74	0.1		Frost (pers. comm.)[a]
Dukwe, Botswana	81.9				Tietema (1993)
Savannas: mean total aboveground biomass = 18 t ha^{-1} s.d. = 8					
Sengwa, Zimbabwe	22.5	0.53			Guy (1981)
Cobar, NSW, Australia	36	3.2	1		Harrington and Johns (1990)
Nylsvley, South Africa	15.54	0.748	1.5	7.1[b]	Rutherford (1982)
Burkea africana					Scholes and Walker (1993)
Waterberg, Namibia *Burkea africana*	22.3				Rutherford (1982)
Klaserie, South Africa *Colophospermum mopane*	20.84	0.8	0.5		Scholes (1988)
Klaserie, South Africa *Acacia nigrescens*	5.05	0.664	1		
Klaserie, South Africa	10.5	0.759	0.5		

Table 4.4 (*continued*).

Site	Aboveground biomass ($t\ ha^{-1}$) Wood	Tree leaf	Grass/ herbs	Roots ($t\ ha^{-1}$)	Source
Combretum apiculatum					
Kruger Park, South Africa	15.4	1.5	0.5		Dayton (1978)
Combretum apiculatum					
South east Zimbabwe	19.7				Kelly and Walker (1976)
Colophospermum mopane					
Hwange, Zimbabwe	3.65	3.21	1.23	34.4	Rushworth (1978)
Terminalia sericea					
Khakhea, Botswana	8.05				Tietema (1993) Reported as fresh mass. Multiplied by 0.7 to get dry mass
Acacia and Combretum					
Kang, Botswana	18.06				
Morwa Forest Reserve	17.5				
Acacia tortilis	30.87				
Morwa Hill, Botswana					
Croton and Combretum					
Dikeletsane, Botswana	*32.97*				
Acacia and Combretum					

[a]PGH Frost, Dept of Biological Sciences, University of Zimbabwe, Harare, Zimbabwe.
[b]Sum of coarse roots (4.6) plus the mean annual fine root biomass (2.5).

which the tree biomass drops abruptly. Firewood collection, tree harvesting and charcoal burning are important biomass-reducing factors in populous savannas.

At this stage there are insufficient data to extrapolate these trends on a global basis. The best that can be done is to associate mean biomass estimates with the structural categories defined in one of the global vegetation databases, and multiply by the extent of that category. Table 4.3 contains data of this type, organized according to the categories in Olson *et al.* (1983). From the sparse belowground biomass data that are available, it is estimated that belowground biomass is 20% of the total biomass (both on a DM basis) in drought deciduous forests, 25% in moist, broad-leaved woodlands and savannas, and 30% in the arid savannas which comprise most of the 'thorn woods' category of Olson *et al.* (1983). Note that the fraction of the total biomass which is belowground is much smaller than the fraction of the NPP which occurs belowground (see section on primary production). Applying these estimates to the mean aboveground biomasses in Table 4.4, and assuming that DM is 45% carbon, the vegetation carbon density in dry forests is about 7.47, in woodlands is 3.74, and dry savannas is 1.13 kg C m^{-2}.

4.3.2 Soil carbon

It is accepted wisdom that savannas have a low soil organic carbon (SOC) content by comparison with tropical forests (Ahn 1970; Askew *et al.* 1970; Kalpage 1974; Montgomery and Askew 1983) or temperate grasslands. This is generally true, but not necessarily so (Sanchez *et al.* 1982). Savannas, like all vegetation types, exhibit a range of soil carbon contents. At the woody end, where they grade into seasonally dry forests, the organic matter content is as high as in moist tropical forests (see later). Tropical montane grasslands, hydromorphic grasslands, grasslands on vertisols or andosols and tropical forests have relatively high SOC, while desert grasslands, sandy savannas and old fields have a very low SOC. With careful management of grazing and soil fertility, the SOC under pastures derived from tropical forests can exceed that of the original forest soil, although the total ecosystem carbon store will remain lower due to the substantial reduction in aboveground biomass which occurs when trees are removed (Lugo and Brown 1993).

The SOC in tropical grasslands, savannas and woodlands increases with increasing soil clay content, rainfall, tree cover and decreasing temperature (Birch and Friend 1956; Jenny and Raychaudhuri 1960), and is higher on soils dominated by high-activity clays (such as smectites and allophanes) than low-activity clays (kaolinite, aluminium and iron oxides: Jones 1973).

The generally low SOC in savannas may be due due to one or more of the following factors:

(a) The high maximum temperatures lead to large soil respiration losses, since respiration is exponentially related to temperature;
(b) Many savanna soils are low in clay content, and the clays are of the low-activity type which do not stabilize large amounts of carbon;
(c) Some carbon is lost to the ecosystem due to burning;
(d) The alternation of wetting and drying favours decomposition over production, since soil respiration can continue at water potentials below which primary production has ceased (this is despite the higher belowground allocation in arid areas);
(e) Primary production in many savannas is low, due to low rainfall and soil fertility.

High densities of earthworms occur in some moist savannas. Epigeic earthworms increase the rate of turnover of SOC, and could therefore reduce the SOC content (Lavelle *et al.* 1992). Termites dominate the soil fauna of savannas receiving less than 800 mm yr^{-1} rainfall. In exceptional years, they can consume up to half of the aboveground primary production, although 2–5% is more

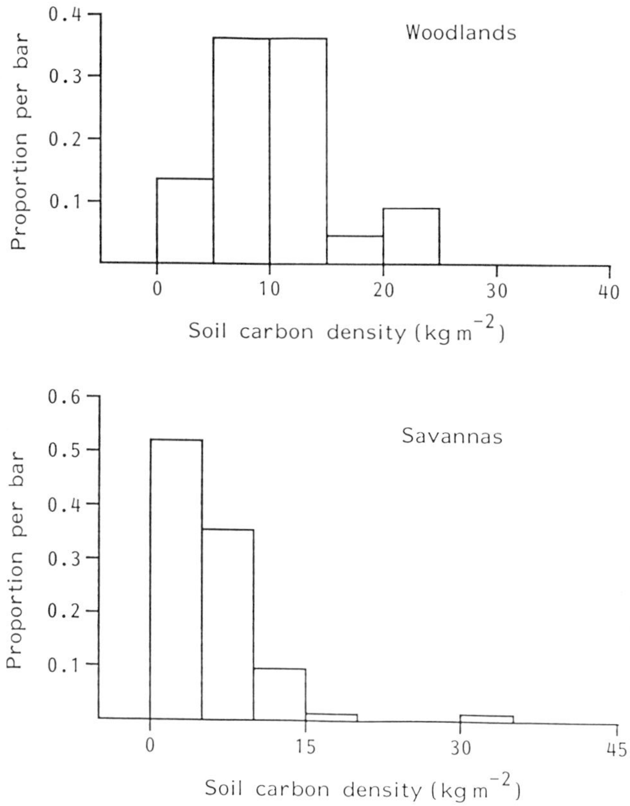

Figure 4.9 The carbon density of 78 soil profiles under tropical savanna and 22 under tropical dry woodland. Data from Zinke *et al.* (1986). The bars represent the fraction of the soil profiles recorded as being under that vegetation which have a carbon density in the range given on the x-axis

typical (Lavelle *et al.* 1992; Scholes and Walker 1993). There is a suggestion from one study that exclusion of termites results in slight increases in SOC and aboveground litter accummulation (Parker *et al.* 1982).

This review made use of a global soil carbon database (Zinke *et al.* 1986) to arrive at estimates of soil carbon in tropical grasslands, savannas and woodlands. The data were corrected for bulk density, and are expressed as kg C m^{-2} to a depth of 1 m. Only in the Olson ecosystem categories of tropical savannas (i.e. tall-grass, moist, infertile savannas) and seasonally dry tropical woodlands were there sufficient recorded profiles in the database for a useful analysis (78 and 22 respectively). Woodlands have a mean and standard deviation soil carbon density of 11.8 $\pm$ 5.34 kg C m^{-2} (very close to the 10.57 kg C m^{-2} average for tropical moist forests) and the savanna mean is 5.65 $\pm$ 4.60 kg C m^{-2} (Figure 4.9).

Five records for tropical thorn woodland gave a mean carbon density of 5.4 $\pm$ 1.9 kg C m^{-2}, and three warm semi-arid woodlands 10.2 $\pm$ 1.3.

Cultivation of tropical grassland and savanna soils rapidly leads to large reductions in the SOC (Nye and Greenland 1960). It appears the decomposition-resistant fractions which make up a large part of carbon in temperate grassland soils, contribute a much smaller part in tropical grasslands (Scholes *et al.* 1992). Typically, about half of the SOC in tropical woodlands, grassland or savanna soils converted to croplands is lost in 20 years, whereas the average decay rate in temperate grassland soils is slower.

4.3.2.1 Carbon stored in tropical savannas and woodlands

The information detailed in Tables 4.3 and 4.4 is summarized in Table 4.5, after weighting by the aerial extent of the soil and vegetation types, and the degree of disturbance. The degree of cultivation and harvesting is a rough estimate, based on the experience of the authors. Note that the vegetation descriptions in Table 4.4 are those used by the authors of the studies cited, and do not necessarily match the definitions used in Olson *et al.* (1983). For instance, the south-central part of Africa is decribed as drought-deciduous forest in Olson *et al.* (1983), but is described as a woodland by African ecologists. Some judgement has therefore been exercised in assigning vegetation carbon densities to the Olson *et al.* (1983) categories in Table 4.5.

4.3.2.2 The potential for carbon sequestration in savannas

More than three-quarters of the ecosystem organic carbon in woodlands and savannas is in the soil (Table 4.5). The spatial distribution of SOC in savannas is notably patchy: it is substantiallly higher beneath the tree canopies than between them (Campbell *et al.* 1988; Scholes and Walker 1993). Therefore, removal of trees leads to an overall decline in SOC over a period of years. Permitting the trees to become larger and denser (for example, through excluding fires) increases both the biomass carbon store and the SOC (Trapnell 1959; Moore 1960; Jones *et al.* 1990). The upper limit of biomass plus soil stored carbon should be equivalent to the current carbon density of the tropical woodlands (14.9 kg C m^{-2}), or on average about double the current savanna soil carbon store (6.7 kg C m^{-2}). An estimate of the upper limit of the carbon sequestration potential of tropical savannas is therefore the product of the area which is currently not woodland (11.5×10^{12} m^2) and the difference in carbon density (8.2 kg C m^{-2}), about 94.3 Pg C. This sequestration could be achieved over a period of about 50 years, by excluding fire from savannas, giving an annual rate of around 2 Gt, or about one-quarter of the annual anthropogenic carbon emissions to the atmosphere as CO_2. Although this is a tempting option it is a risky one. Given the seasonally dry climate, it is nearly impossible to exclude fire reliably from savannas. When fire

Table 4.5 Total ecosystem carbon stock and turnover in tropical grasslands, savannas and woodlands

			Carbon in vegetation[c]		Carbon in soils[d]			
Category	Area[a]	Disturbance[b](%)	Density (g C m^{-2})	Total[g] (Pg C)	Density (g C m^{-2})	Total[h] (Pg C)	Total Carbon[e] (Pg C)	NPP[f] (Pg C yr^{-1})
Drought-deciduous woodland	4616005	10	3735	15.5	11 180	49.03	64.55	2.36
Savannas	6696927	10	1125	6.78	5 650	35.95	42.73	3.87
Eucalyptus and saxual	902483	40	1 125	0.61	5 600	4.04	4.65	0.18
Thorn and succulent 'woods'	3915149	20	1125	3.52	5400	19.03	22.55	1.21
Total	16130564			26.4		108.1	134.48	7.61

[a]From Olson *et al.* (1983), as represented in the digital data set distributed by EPA/NOAA.
[b]Defined as complete removal of original vegetation, and a reduction in soil carbon of 50%. Estimated by the authors.
[c]From Table 4.4.
[d]See text and Figure 4.6. Derived from Zinke *et al.* (1986) database.
[e]Soil plus vegetation carbon.
[f]From Table 4.3, multiplied by the undisturbed area, assuming dry matter is 45% carbon.
[g]Undisturbed area * (vegetation carbon density).
[h]Undisturbed area * (soil carbon density) + disturbed area * (soil carbon density/2).

inevitably occurs, much of the stored carbon will be returned to the atmosphere. In the meantime, the ecology of the savannas will have been significantly altered. One of the long-term consequences of regular savanna burning is a small but steady sequestration of carbon in the form of elemental carbon, which is virtually inert (Seiler and Crutzen 1980).

4.4 CARBON FLUXES IN TROPICAL WOODLANDS, SAVANNAS AND GRASSLANDS

The NPP model (Figure 4.5) based on the data collated in Table 4.2 and reported in Tables 4.3 and 4.5, predicts that the gross annual flux of carbon from the atmosphere into tropical grasslands, woodlands and savannas via photosynthesis is between 3.16 and 10.84 Pg C, with a best estimate of about 7.61 Pg C. Between 0.54 and 1.7 Pg C is returned to the atmosphere through burning, and a small amount (less than 0.5 Pg C) through enteric digestion. Assuming steady-state conditions, which is clearly an oversimplification, the remainder returns through decomposition in the litter layer and soil.

The net annual carbon flux from tropical grasslands and savannas is assumed to have been zero before 1800, when significant European colonization of the tropical regions began. Since that time there has probably been a small flux of carbon to the atmosphere due to woodland clearing and cultivation, and possibly a reverse flux due to bush encroachment and higher productivity as a result of rising atmospheric CO_2. There are two opposing land-use changes occurring in tropical tree–grass systems. The first is towards the removal of trees, mostly due to cultivation of the soil and for use as domestic fuelwood. This trend occurs in areas of dense human settlement. The extent of conversion has been limited by the inherent low fertility of the moist savannas, and the unsuitability of the dry savannas for rainfed crops. Where infrastructural development can support high-input agriculture, and the rainfall is above 600 mm, conversion of savannas to cropland can be expected at a large scale in the next decade. An example of this trend has occurred in the *cerrado* of Brazil, where the use of lime and phosphate fertilizers has allowed millions of hectares of savanna to be converted to croplands during the 1980s. There are large areas of similar climate and soils in Central and Southern Africa, where the same technology could be applied.

The second trend occurs in areas of low human population, usually low rainfall, and pastoral use. Particularly on the more fertile soils, sustained heavy grazing and the exclusion of fire and browsers have resulted in an increase in woody plant density. This trend, known as 'bush encroachment', has been particularly evident in South and East Africa, Australia, Argentina and the southern United States (Walker *et al.* 1981).

There are no reliable data to indicate the relative importance of these two processes, the first of which releases carbon, while the second sequesters carbon.

The future trend of increasing human population and agricultural technology is likely to favour carbon release.

4.5 RESEARCH NEEDS

There is no direct information on the impact of rising CO_2 concentrations on primary production, decomposition and herbivory in tropical grasslands and savannas. The amount of data on biomass, soil carbon and productivity is still small compared to some less extensive vegetation formations, but is gradually improving. There is a need for several secure, well-documented long-term ecological research sites in tropical grassland and savanna regions. Information about the extent, frequency and behaviour of vegetation fires in this particularly fire-prone system is inadequate, but will improve substantially within the next few years as research currently in progress is published. The dynamics and mechanisms of the interaction between woody plants and grasses in mixed tree–grass systems are a research theme which has been adopted for a future SCOPE project. Perhaps the largest single uncertainty in understanding and predicting the carbon budgets for the tropical woodlands, savannas and grasslands is the extent, degree and nature of land-use change, along with the human processes which drive it.

4.6 ACKNOWLEDGEMENTS

The calculations in this chapter would not have been possible without access to the global soil carbon database, made available by the Oak Ridge National Laboratory, and the Global Ecosystems Database, a joint US Environmental Protection Agency and National Oceanic and Atmospheric Administration/National Geophysics Data Centre programme.

4.7 REFERENCES

Ahn, P. M. (1970) *West African Soils*. Oxford University Press, Oxford.

Ajtay, G. L., Ketner, P. and Duvigneaud, P. (1979) Terrestrial primary production and phytomass. In: Bolin, B., Degens, E. T., Kempe, S. and Ketner, P. (Eds) *The Global Carbon Cycle*, pp. 129–182. SCOPE Vol. 13. Wiley & Sons, New York.

Askew, G. P., Moffat, D. J., Montgomery, R. F. and Searle, P. L. (1970) Interrelationships of soil and vegetation in the savanna–forest boundary zone of north-eastern Mato Grosso. *Geogr. J.* **136**, 370–376.

Aucamp, A. J. Dankwerts, J. E., Teague, W. R. and Venter, J. J. (1983) The role of *Acacia karroo* in the false thornveld of the Eastern Cape. *J. Grassl. Soc. Sthn. Afr.* **8**, 151–154.

Beale, I. F. (1973) Tree density effects on yields of herbage and tree components in south-west Queensland mulga (*Acacia aneura* F. Meull.) scrub. *Tropical Grasslands* **7**, 135–142.

Belsky, A. J., Amundson, R. G., Duxbury, J. M., Riha, S. J., Ali, A. R. and Mwonga, S. A. (1989) The effects of trees on their physical, chemical and biological environments in a semi-arid savanna in Kenya. *J. Appl. Ecol.* **26**, 1005–1024.

Birch, H. F. and Friend, M. T. (1956) The organic matter and nitrogen status of East African soils. *J. Soil Sci.* **8**, 156–167.

Bourliere, F. and Hadley, M. (1970) The ecology of tropical savannas. *Annu. Rev. Ecol. Syst.* **1**, 125–151.

Bremen, H. and de Wit, C. T. (1983) Rangeland productivity and exploitation in the Sahel. *Science* **221**, 1341–1347.

Brockington, N. R. (1961) Studies of the growth of *Hyparrhenia*-dominant grassland in Northern Rhodesia. II. The effect of fire. *J. Br. Grassl. Soc.* **16**, 54–64.

Campbell, B. M., Swift, M. J., Hatton, J. and Frost, P. G. H. (1988) Small-scale vegetation pattern and nutrient cycling in *miombo* woodland. In: Verhoeven, J. T. A., Heil, G. W. and Werger, M. J. A. (Eds) *Vegetation Structure in Relation to Carbon and Nutrient Economy*, pp. 69–85. SPD Academic Publishing, Amsterdam.

Christie, E. K. (1978) Ecosystem processes in semi-arid grasslands. I. Primary production and water use of two communities possessing different photosynthetic pathways. *Aust. J. Agric. Res.* **29**, 773–787.

Collins, N. M. (1977) Vegetation and litter production in a southern Guinea savanna, Nigeria. *Oecologia* **28**, 163–175.

Dayton, B. R. (1978) Standing crops of dominant *Combretum* species at three browsing levels in the Kruger National Park. *Koedoe* **21**, 67–76.

Delmas, R. A., Loudjani, P., Podaire, A. and Menaut, J.-C. (1991) Biomass burning in Africa: an assessment of annually burned biomass. In: Levine, J.S. (Ed.) *Global Biomass Burning: Atmospheric, Climatic and Biospheric Implications*, pp. 126–132. MIT Press, Cambridge, Mass.

De Ridder, N., Stroosnijder, L., Cisse, A. M. and van Keulen, H. (1982) *Productivity of Sahelian Rangelands*. Vol. 1. Dept of Soil Science and Plant Nutrition, University of Wageningen, The Netherlands.

Donaldson, C. H. and Kelk, D. M. (1970) An investigation of the veld problems of the Molopo area. 1. Early findings. *Proc. Grassl. Soc.* **5**, 50–57.

Donaldson, C. H., Rethmann, G. and Grossman, D. (1984) Long-term nitrogen and phosphorus application to veld. *J. Grassl. Soc. South. Afr.* **1**, 27–32.

Drent, R. H. and Prins, H. H. T. (1987) The herbivore as prisoner of its food supply. In: van Andel, J. *et al.* (Eds) *Disturbance in Grassland*, pp. 131–147. Dr. W. Junk Publishers, Dordrecht.

East, R. (1984) Rainfall, soil nutrient status and biomass of large African savanna mammals. *Afr. J. Ecol.* **22**, 245–270.

Endean, F. (1967) The productivity of miombo woodland in Zambia. *For. Res. Bull.* **14**, Government Printer, Lusaka, Zambia.

Fatubarin, A. (1984) Biomass estimates for the woody plants in a savanna ecosystem in Nigeria. *Trop. Ecol.* **25**, 208–213.

Fritz, H. and Duncan, P. (1993) Large herbivores in rangelands. *Nature* **364**, 292–293.

Greenland, D. J. and Nye, P. H. (1959) Increases in the carbon and nitrogen contents of tropical soils under natural fallows. *J. Soil Sci.* **10**, 284–299.

Grossman, D., Grunow, J. O. and Theron, G. K. (1981) The effect of fire with and without subsequent defoliation on the herbaceous layer of *Burkea africana* savanna. *Proc. Grassl. Soc. South. Afr.* **16**, 117–120.

Gupta, R. K., Sharm, S. and Sharm, S. K. (1972) Aboveground productivity of grasslands at Jodphur, India. In: Golley, P. M. and Golley, F. B. (Eds) *Tropical Ecology with Emphasis on Organic Production*, pp. 75–93. Univ. of Georgia, Athens, Ga.

Gurumurti, K., Bhandari, H. C. S. and Dhawan, M. (1986) Studies of yield, nutrients and energy conversion efficiency in energy plantations of *Acacia nilotica*. *J. Tree Sci.* **5**, 36–42.

Guy, P. R. (1981) Changes in biomass and productivity of woodlands in the Sengwa wildlife research area, Zimbabwe. *J. Appl. Ecol.* **18**, 507–519.

Hall, D. O., Scurlock, J. M. O., Bolhar-Nordenkampf, H. R., Leegood, R. C. and Long, S. P. (1993) *Photosynthesis and Production in a Changing Environment*. Chapman & Hall. London.

Hao, W. M., Lui, M. H. and Crutzen, P. J. (1990) Estimates of annual and regional release of CO_2 and other trace gases to the atmosphere from fires in the tropics, based on FAO statistics for the period 1975–1980. In: Goldammer, J. G. (Ed.) *Fire in the Tropical Biota: Ecosystem Processes and Global Challenges*, Ecol. Stud. **84**, 440–462. Springer-Verlag, Berlin.

Harrington, G. N. and Johns, G. G. (1990) Herbaceous biomass in a Eucalyptus savanna woodland after removing trees and/or shrubs. *J. Appl. Ecol.* **27**, 775–787.

Jackson, J. K. and Ojo, G. O. A. (1971) Productivity of natural woodland and plantations in savanna zones of Nigeria. *Nigerian J. For.* **1**, 53–90.

Jenny, H. and Raychaudhuri, S. P. (1960) *Effect of Climate and Cultivation on Nitrogen and OM Reserves in Indian Soils*. Indian Council of Agricultural Research, New Delhi.

Jones, C. L., Smithers, N. L., Scholes, M. C. and Scholes, R. J. (1990) The effect of fire frequency on the organic components of a basaltic soil in the Kruger National Park. *South Afr. J. Plant Soil* **7**(4), 236–238.

Jones, M. J. (1973) The organic matter content of the savanna soils of West Africa. *J. Soil Sci.* **24**, 42–53.

Kadeba, O. and Aduayi, E. A. (1985) Litter production, nutrient recycling and litter accumulation in *Pinus caribaea* Morelet var *hondurensis* stands in the northern Guinea savanna of Nigeria. *Plant Soil* **26**, 197–206.

Kadeba, O. and Aduayi, E. A. (1986) Dry matter production and nutrient distribution in a *Pinus caribaea* stand planted in a subhumid tropical savanna site. *Oikos* **46**, 237–242.

Kalpage, F .S. C. F. (1974) *Tropical Soils*. St Martin's Press, Macmillan, New York.

Kelly R. D., Schwimm, W. F. and Barnes, D. L. (no date) Bush clearing trial at Nuanetsi. *Annual Report of the Rhodesian Dept of Specialist Services*, Harare, Zimbabwe. About 1975.

Kelly, R. D. and Walker, B. H. (1976) The effects of different forms of land use on the ecology of a semi-arid region of South Eastern Rhodesia. *J. Ecol.* **64**, 555–576.

Knoop, W. T. and Walker, B. H. (1984) Interactions of woody and herbaceous vegetation in two savanna communities at Nylsvley. *J. Ecol.* **73**, 235–253.

Lavelle, P., Blanchart, E., Martin, A., Spain, A. V. and Martin, S. (1992) Impact of soil fauna on the properties of soils in the humid tropics. In: *Myths and Science of Soils of the Tropics. SSSA Special Publication* **29**, pp. 157–185. Soil Sci. Soc. Am., Madison.

Lawton, R. M. (1982) Natural resources of *miombo* woodland and recent changes in agricultural and land-use practices. *For. Ecol. Manage.* **4**, 287–297.

Leemans, R. and Cramer, W. P. (1990) The IIASA database for monthly values of temperature, precipitation and cloudiness on a global terrestrial grid. Working paper, August 1990. WP-90-41, Biosphere Dynamics Project, International Institute for Applied Systems Analysis, A-2361, Laxenburg, Austria.

Long, S. P., Garcia Moya, E., Imbamba, S. K., Kamnalrut, A., Piedade, M. T. F., Scurlock, J. M. O., Shen, Y. K. and Hall, D. O. (1989) Primary productivity of natural grass ecosystems of the tropics: a reappraisal. *Plant Soil* **115**, 155–166.

Long, S. P., Jones, M. B. and Roberts, M. J. (Eds) (1992) *Primary Productivity of Grass Ecosystems of the Tropics and Sub-tropics.* Chapman & Hall, London.

Lugo, A. E. and Brown, S. (1993) Management of tropical soils as sinks or sources of atmospheric carbon. *Plant Soil* **149**, 27–41.

McMurtrie, R. E. (1993) Modelling of canopy carbon and water balance. In: Hall, D. O., Scurlock, J. M. O., Bolhar-Nordenkampf, H. R., Leegood, R. C. and Long, S. P. (Eds) *Photosynthesis and Production in a Changing Environment*, pp. 220–231. Chapman & Hall, London.

McNaughton, S. J. (1983) Compensatory plant growth as a response to herbivory. *Oikos* **40**, 329–336.

Malaisse, F., Freson, R., Goffinet, G. and Malaisse-Mousset, M. (1975) Litterfall and litter breakdown in miombo. In: Golley, F. B. and Medina, E. (Eds) *Tropical Ecological Systems: Trends in Terrestrial and Aquatic Research*, pp. 137–152. Springer Verlag, New York.

Menaut, J.-C., Abbadie, L., Lavenu, F., Loudjani, P. and Podaire, A. (1991) Biomass burning in West African savannas. In: Levine, J. S. (Ed.) *Global Biomass Burning: Atmospheric, Climatic and Biospheric Implications*, pp. 133–142. MIT Press, Cambridge, Mass.

Menaut, J.-C. and César, J. (1979) Structure and primary production of Lamto savannas Ivory Coast. *Ecology* **60**, 1197–1210.

Montgomery, R. F. and Askew, G. P. (1983) Soils of tropical savannas. In: Bourliere, F. (Ed.) *Ecosystems of the World 13: Tropical Savannas*, pp. 63–78. Elsevier, Amsterdam.

Moore, A. W. (1960) The influence of annual burning on a soil in the derived savanna zone of Nigeria. *Trans. 7th Int. Congr. Soil Sci.* **4**, 257–264.

Moore, A. W., Russel, J. S. and Coaldrake, J. E. (1967) Dry matter and nutrient content of a subtropical semi-arid forest of *Acacia harpophylla* F. Muell. (Brigalow). *Austr. J. Bot.* **15**, 11–24.

Murphy, P. G. (1977) Rates of primary productivity in tropical grassland, savanna and forest. *Geo-Eco-Trop* **1**, 95–102.

Noy-Meir, I. (1973) Desert ecosystems: environment and producers. *Annu. Rev. Ecol. Syst.* **4**, 25–51.

Nye, P. H. and Greenland, D. (1960) *The Soil Under Shifting Cultivation.* Technical Communication 51, Commonwealth Agriculture Bureau, Wallingford, UK.

Olson, J. S., Watts, J. A. and Allison, L. J. (1983) *Carbon in Live Vegetation of Major World Ecosystems.* Report ONRL-5862, Oak Ridge National Laboratory, Oak Ridge, Tenn.

Palm, C. A. and Sanchez, P. A. (1990) Nitrogen release from the leaves of some tropical legumes as affected by their lignin and polyphenol contents. *Biotropica* **22**, 330–338.

Pandey, C. B. and Singh, J. S. (1992) Rainfall and grazing effects on net primary productivity in a tropical dry savanna, India. *Ecology* **73**, 2007–2021.

Parker, L. W., Fowler, H. G., Ettershank, G. and Whitford, W. G. (1982) The effects of subterranean termite removal on desert soil nitrogen and ephemeral fauna. *J. Arid Environ.* **5**, 53–59.

Raich, J. W. and Nadelhoffer, K. J. (1989) Belowground carbon allocation in forest ecosystems: global trends. *Ecology* **70**, 1346–1354.

Rajvanshi, R. and Gupta, S. R. (1985) Biomass, productivity and litterfall in a tropical *Dalbergia sissoo* Roxb. forest. *J. Tree Sci.* **4**, 73–78.

Rosenberg, N. J., Blad, B. L. and Verma, S. B. (1983) *Microclimate: The Biological Environment.* 2nd Edn. John Wiley, New York.

Rushworth, J. E. (1978) Kalahari sand scrub: something of value. *Rhod. Sci. News* **12**(8), 193–195.

Rutherford, M. C. (1982) Woody plant biomass distribution in *Burkea africana* savannas.

In: Huntley, B. J. and Walker, B. H. (Eds) *The Ecology of Tropical Savannas*, pp. 120–141. Ecological Studies 42, Springer-Verlag, Berlin.

Sanchez, P. A., Gichuru, M. P. and Katz, L. B. (1982) Organic matter in major soils of the tropical and temperate regions. *12th Int. Congr. Soil Sci.* 99–114.

San Jos, J. J. and Medina, E. (1975) Effects of fire on organic matter production in a tropical savanna. In: Golley, F.B. and Medina, E. (Eds) *Tropical Ecological Systems*, pp. 251–264. Springer-Verlag, Berlin.

Sarmiento, G. (1984) *The Ecology of Neotropical Savanna.* Harvard University Press., Cambridge, Mass.

Scanlan, J. C. and Burrows, W. H. (1990) Woody overstory impact on herbaceous understory in *Eucalyptus* spp. communities in central Queensland. *Austr. J. Ecol.* **15**, 191–197.

Scholes, M. C., Scholes, R. J. and McKean, S. K. (1992) The origin and stability of soil organic matter in a semi-arid savanna. *Proceedings of a Seminar on Stable Isotopes in Plant Nutrition, Soil Fertility and Environmental Studies*, pp. 57–60. Institute of Soil, Climate and Water, Private Bag X79, Arcadia, South Africa.

Scholes, R. J. (1988) Response of three semi-arid savannas on contrasting soils to removal of the woody component. Ph.D. thesis, University of the Witwatersrand, Johannesburg, South Africa.

Scholes, R. J. and Walker, B. H. (1993) *An African Savanna: Synthesis of the Nylsvley Study.* Cambridge University Press, Cambridge, UK.

Schütz, H., Seiler, W. and Rennenberg, H. (1990) Soil and land-use related sources and sinks of methane in the context of the global methane budget. In: Bouwman, A. F. (Ed.) *Soils and the Greenhouse Effect*, pp. 269–288. John Wiley & Sons, Chichester.

Seiler, W. and Crutzen, P. J. (1980) Estimates of gross and net fluxes of carbon between the biosphere and atmosphere from biomass burning. *Clim. Change* **2**, 207–247.

Smith, E. L. (1960) Effects of burning and clipping at various times during the wet season on tropical grass range in northern Australia. *J. Range Manage.* **13**, 197–203.

Snaydon, R. W. (1991) The productivity of C3 and C4 plants: a reassessment. *Funct. Ecol.* **5**, 321–330.

Stromgaard, P. (1985) Biomass, growth and burning of woodland in a shifting cultivation area of south-central Africa. *For. Ecol. Manage.* **12**, 163–178.

Swaminath, M. H. (1988) Studies on the response of fast-growing forestry species for biomass production under irrigation. *Myforest* **24**, 117–123.

Tietema, T. (1993) Possibilities for the management of indigenous woodland in southern Africa: a case study from Botswana. *Proceedings of the Symposium on Ecology and Management of Indigenous Forests in Southern Africa*, Victoria Falls, July, pp. 134–142. Zimbabwe Forestry Commission, Harare.

Trapnell, C. G. (1959) Ecological results of woodland burning in Northern Rhodesia. *J. Ecol.* **47**, 161–172.

Van Vegten, J. A. (1984) Thornbush invasion in a savanna ecosystem in eastern Botswana. *Vegetatio* **56**, 3–7.

Varshney, C. K. (1972) Productivity of Delhi grasslands. In: Golley, P. M. and Golley, F. B. (Eds) *Tropical Ecology with Emphasis on Organic Production*, pp. 27–42. Univ. of Georgia, Athens, Ga.

Vyas, L. N., Garg, R. K. and Agarwal, S. K. (1972) Net aboveground production in the monsoon vegetation at Udiapur. In: Golley, P. M. and Golley, F. B. (Eds) *Tropical Ecology with Emphasis on Organic Production*, pp. 95–99. Univ. of Georgia, Athens, Ga.

Walker, B. H., Ludwig, D., Holling, C. S. and Peterman, R. S. (1981) Stability of semi-arid savanna grazing systems. *J. Ecol.* **69**, 473–498.

Walker, J., Moore, R. M. and Robertson, J. A. (1972) Herbage response to tree and shrub thinning in *Eucalyptus populnea* shrub woodlands. *Austr. J. Agric. Res.* **23**, 405–410.

Ward, H. K. and Cleghorn, W. B. (1964) The effects of ringbarking trees in *Brachystegia* woodland on the yield of veld grasses. *Rhod. Agric. J.* **61**, 98–107.

Whittaker, R. H. and Likens, G. E. (1973) Carbon in the biota. In: Woodwell, G. M. and Pecan, E. V. (Eds) *Carbon and the Biosphere.* AEC Symposium Series 30, 281–302. US Dept of Commerce, Springfield, Va.

Zietsman, P. C., Grobbelaar, N. and van Rooyen, N. (1988) Soil nitrogenase activity of the Nylsvley Nature Reserve. *South Afr. J. Bot.* **54**, 21–27.

Zinke, P. J., Strangenberger, A. G., Post, W. M., Emanuel, W. R. and Olson, J. S. (1986) Worldwide organic soil carbon and nitrogen data. ORNL/CDIC-18 NDP-018.Carbon Dioxide Information Centre, Oak Ridge, Tenn.

5 Carbon Budgets of Temperate Grasslands and the Effects of Global Change

O. E. SALA[1], W. K. LAUENROTH[2] and I. C. BURKE[3]
[1]*Department of Ecology, Faculty of Agronomy, Argentina*
[2]*Departments of Range Land Ecosystem Science and Natural Resource Ecology Laboratory, Colorado State University, Fort Collins, USA*
[3]*Departments of Forest Sciences and Natural Resource Ecology Laboratory, Colorado State University, Fort Collins, USA*

5.1 INTRODUCTION

Temperate grasslands account for a large fraction of the vegetation of the earth (Coupland 1992). Large expanses of temperate grasslands and derivative croplands are located at mid latitudes in Asia, North, and South America (Singh *et al.* 1983) (Figure 5.1). In North America, the grasslands include the central grassland region, which ranges along a precipitation gradient from the semiarid shortgrass steppe to the subhumid tallgrass prairie, and a temperature gradient from subtropical to boreal. In South America, temperate grasslands encompass large units, such as the pampa grasslands on the mesic end, and the patagonian steppe on the xeric end of the gradient (Soriano *et al.* 1991). Large expanses of temperate grasslands also occur in Asia, stretching from the Ukraine to the Peoples' Republic of China. There are also smaller units in Europe, Australia, and Southern Africa.

Temperate grasslands are located in one of the regions where the impact of global climate change is predicted to be large (Mitchell *et al.* 1990). Most global circulation models (GCMs) predict for a double CO_2 scenario an increase in temperature between 4 and 6 °C for the regions of the world where temperate grasslands are now located. The climate change impact is predicted to be larger only in the region where boreal forests and arctic tundra are now located. The interaction between the large area encompassed and the expected magnitude of the predicted climate change suggests that changes in the carbon budgets of temperate grasslands may have an important impact upon the earth's atmosphere.

Climate change is only one of the components of a set of global-scale en-

Global Change: Effects on Coniferous Forests and Grasslands.
Edited by A. I. Breymeyer, D. O. Hall, J. M. Melillo and G. I. Ågren

Figure 5.1 Map of the temperate grasslands of the world, adapted from Coupland (1992)

vironmental changes affecting ecosystems. Global change includes changes in the composition of the atmosphere such as the observed increases in the concentration of CO_2, methane, and CFCs. These changes in atmospheric composition may have direct impacts upon vegetation, with or without climate change. Global change also includes changes in land use which are driven by economic, demographic, and social forces.

A major feature of global change is that climate, atmospheric composition and land use are not independent factors but they are strongly interactive. For example, changes in land use, such as from grassland to cropland, may be important contributors to changes in atmospheric composition (Burke *et al.* 1991). This change, in turn, may modify the climate which could then speed up or slow down the shift in land use. This, again, would likely affect the emissions of greenhouse gases.

The objective of this chapter is to evaluate the carbon budgets of present temperate grasslands and to assess the possible effects that global change may have. To accomplish our objective we will first discuss how carbon is distributed among different compartments and how the size of these carbon pools is associated with major environmental variables. This is a static view of the carbon budget. Second, we will assess the import inputs and outputs and their controls. Finally, we will evaluate how the three factors driving global change may affect the carbon budget of temperate grasslands.

5.2 A STATIC VIEW OF THE CARBON BUDGET

A simple scheme of the partitioning of carbon among compartments indicates the relative size of carbon pools in temperate grasslands, the major inputs and outputs and their controls (Figure 5.2). Carbon is fixed from the atmosphere by means of photosynthesis which, at the ecosystem level, is called gross primary production (GPP). We represented in the diagram just net primary production (NPP), because it is the variable for which data are available. Carbon is translocated from above- to belowground parts of the plants and in the opposite direction depending upon seasonality and grazing. Herbivory is the flow from plant parts to herbivores. This flow is particularly large in grasslands. Unlike forests, where loss of biomass to herbivores is usually less than 10%, grasslands are frequently characterized by herbivory rates of 25–50% or more of aboveground NPP (McNaughton 1976, 1985; Lauenroth and Milchunas 1992; Milchunas and Lauenroth 1993) and perhaps 25% of the belowground productivity (Coleman 1976; Scott *et al.* 1979; Ingham and Detling 1984; Lauenroth and Milchunas 1992).

Above- and belowground plant parts are transferred into the soil organic matter pool as they become senescent. There is also a flow from herbivores to soil organic matter. We followed the CENTURY model scheme (Parton *et al.* 1987)

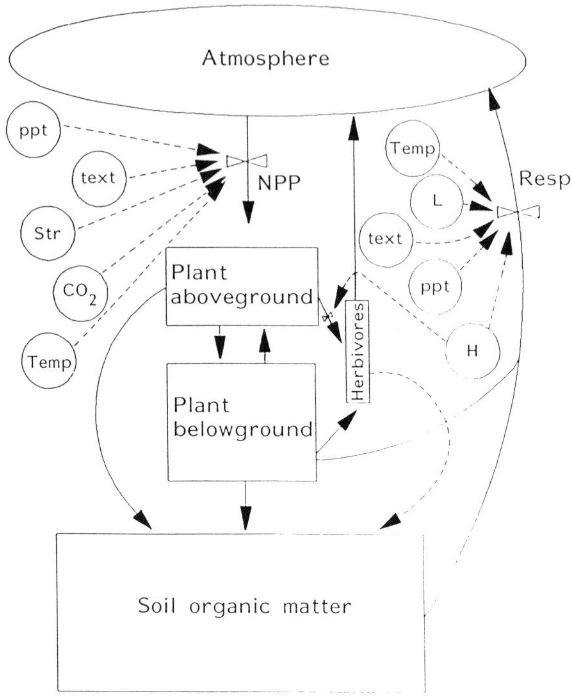

Figure 5.2 A scheme of the C budget of temperate grasslands. State variables are plant aboveground, plant belowground, herbivores, and soil organic matter. The major C flows represented in this diagram are net primary production (NPP), and belowground respiration (Resp). Control variables are precipitation (ppt), soil texture (text), community structure (Str), CO_2 concentration in the atmosphere (CO_2), air temperature (Temp), land use (l), and human intervention (H)

by including soil microorganisms in the soil organic matter compartment. Soil organic matter can be divided further into different compartments according to their turnover rates. We chose not to divide this compartment because of the scarcity of data with that level of detail, and the difficulty of physical characterization of such kinetic pools. Finally, major carbon losses in grasslands occur as a result of respiration from plants, animals and microorganisms (decomposition).

Some patterns are evident across temperate grasslands. Most of the carbon in grasslands is located belowground in the soil organic matter compartment and in the belowground plant compartment (Atjay *et al.* 1979; French *et al.* 1979; Parton *et al.* 1987). The location of the major carbon store belowground provides temperate grasslands with many of their unique characteristics.

During the International Biological Program in the late 1960s and early 1970s, a large data base was collected across temperate grasslands. For North American grasslands, the size of the aboveground plant compartment is correlated with

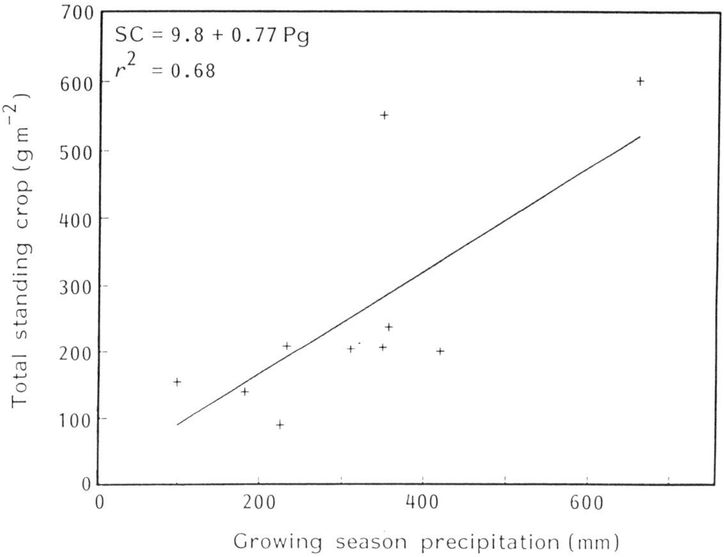

Figure 5.3 Relationship between aboveground total standing crop (SC) and growing season precipitation for North American grasslands (after Sims *et al.* 1978)

growing season precipitation (total standing crop = 9.78 + 0.77P, $R^2 = 0.68$) (Figure 5.3) (Sims *et al.* 1978).

Plant belowground biomass is most closely related to annual air temperature. We collated data from North and South America encompassing a range of precipitation from 150 to 1000 mm yr^{-1} and a range of mean annual temperature from 3 to 15 °C. We found that temperature accounted for a large fraction of the variability in root biomass among sites (Figure 5.4) (root biomass = 1996.7 − 94.7 t, $R^2 = 0.74$) (Sims *et al.* 1978; Doll and Deregidus 1986; Soriauo *et al.* 1987).

Burke et al. (1989) analyzed 500 grassland soil profiles and 300 cropland soil profiles distributed across the central grassland region of North America in an attempt to identify large-scale controls of soil organic carbon. They found that the amount of carbon in the soil was mainly explained by temperature, land use, and precipitation (Figure 5.5a). Temperature was the variable which had the strongest effect upon total soil carbon. Soil organic carbon decreased with increasing mean annual temperature. Moreover, the relationship between carbon in the soil and temperature has a quadratic term, and therefore the accumulation of carbon decreases sharply as a consequence of a small increase in temperature. The authors use this as evidence that decomposition is a major control over soil carbon. Precipitation also explained a significant fraction of the variance in soil carbon, with soil organic carbon increasing with precipitation up to 800 mm

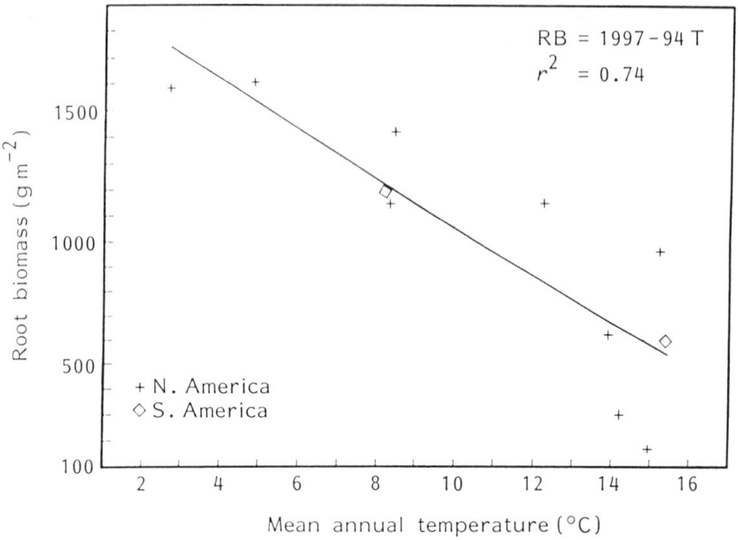

Figure 5.4 Relationship between root biomass (RB) and mean annual temperature for sites from North and South America

annual precipitation, then leveling off. This is likely due to the influence of precipitation in increasing the inputs into soil organic matter, NPP.

The authors fit similar regression models to grassland and cropland soils, and they assumed that the difference between the two models was the effect of tillage. In this way they were able to estimate the effect of cultivation upon carbon loss. In general, absolute organic carbon loss increased with total soil organic carbon content (Figure 5.5b).

The biomass of herbivores that an ecosystem can sustain depends directly upon aboveground net primary production (ANPP) (McNaughton *et al.* 1989). The relationship between herbivore carrying capacity and primary production is exponential. It indicates that, for example, arid grasslands support less animal biomass per unity primary production than mesic grasslands. This relationship encompasses a range of ecosystems which goes beyond that of grasslands, from deserts to forests (McNaughton *et al.* 1989).

5.3 A DYNAMIC VIEW OF THE CARBON BUDGET: INPUT AND OUTPUT FLOWS

To this point we have analyzed a snapshot of the carbon budget of temperate grasslands. We will next analyze the carbon inputs and outputs. This is an attempt to build our understanding of the carbon dynamics to provide a basis for suggesting responses to potential global change.

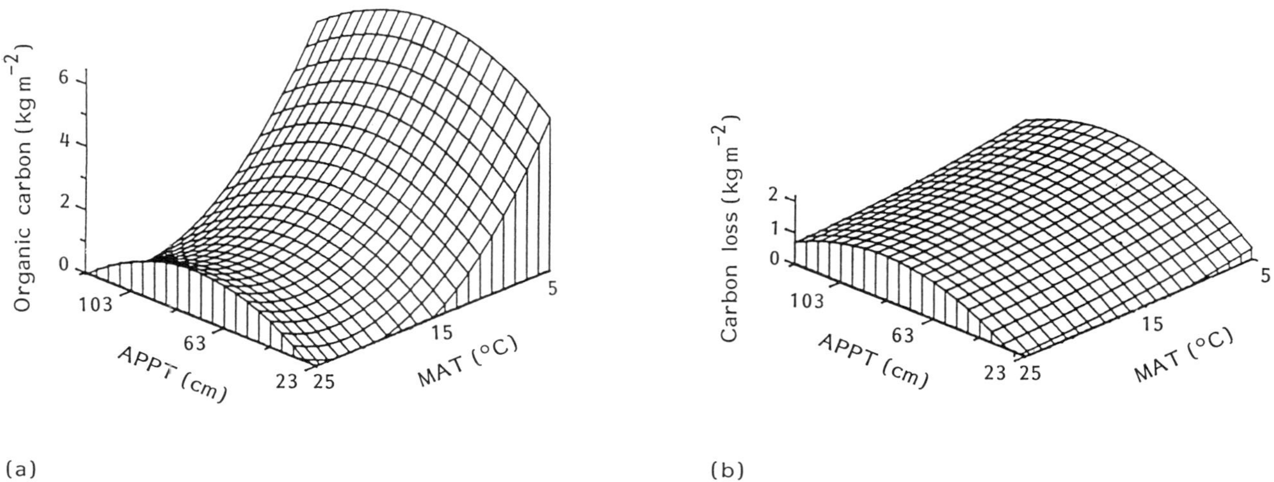

Figure 5.5 Predicted response surface of soil organic carbon (a) and carbon loss as a result of tillage (b) to variation in mean annual temperature (MAT) and annual precipitation (APPT) for loamy soils (after Burke *et al.* 1989)

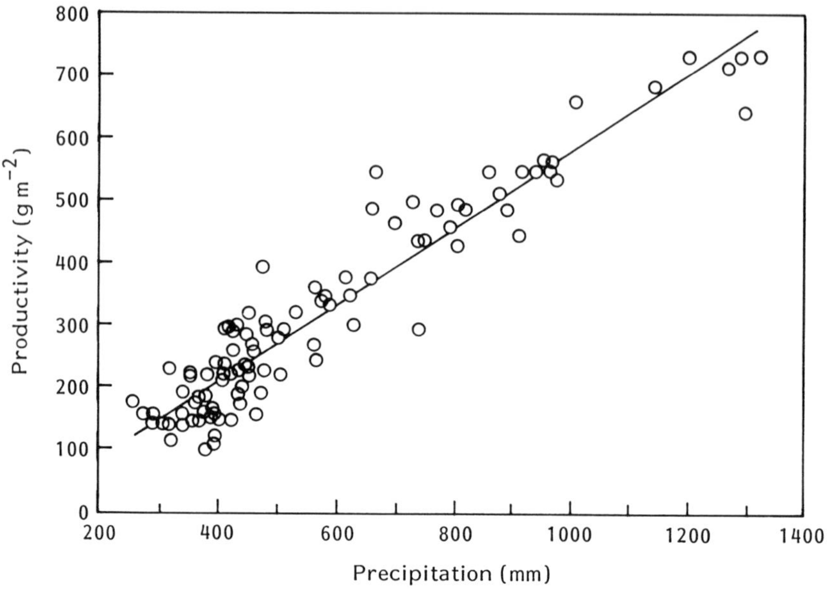

Figure 5.6 Relationship between average aboveground annual NPP and average annual precipitation for 100 ecological regions within the central grassland region of the USA (after Sala *et al.* 1988b)

5.3.1 Net primary production

There is a clear relationship between ANPP and annual precipitation. A linear model has proved to describe this relationship very well. For example, in the central grassland region of North America, 90% of the variance in production of 100 major ecological units was accounted for by annual precipitation (Sala *et al.* 1988b) (Figure 5.6). When the same model was independently fit to data from South America and Africa, there were very small differences in the parameters of the model (Figure 5.7). This gives us some confidence that this is likely to be a universal model for temperate grasslands. Others have found similar results over a wider range of semiarid and arid ecosystems (Walter 1979; Le Houérou 1984).

At a more detailed scale, soil information needs to be included to explain a similar fraction of the variability in ANPP. In central North America, water-holding capacity interacted with annual precipitation to explain ANPP (Figure 5.8). The statistical model was developed using data from a broad range of sites. The data provided an empirical test of the inverse texture hypothesis (Noy Meir 1973). This hypothesis states that production in dry regions should be greater on coarse-textured (low water-holding capacity) soils than on fine-textured (high water-holding capacity) soils. In contrast, in humid regions, production should be greater on fine-textured soils than on coarse-textured soils. Precipitation and

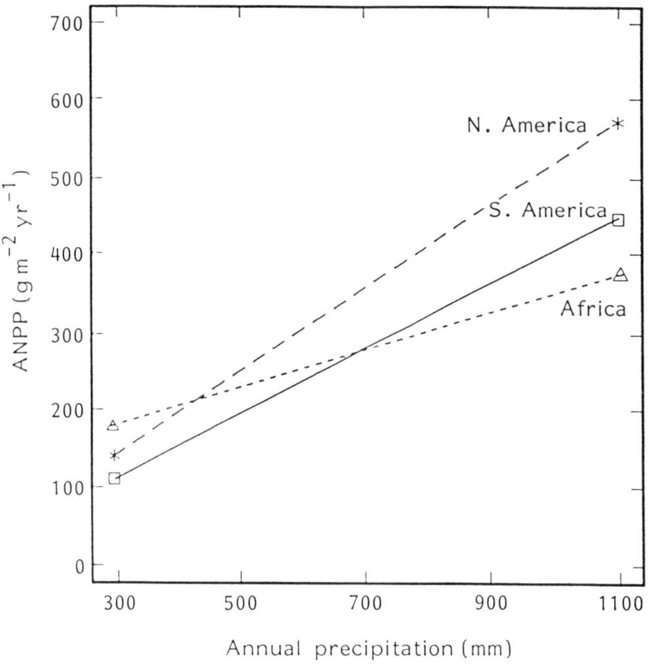

Figure 5.7 Relationship between average aboveground annual NPP and average annual precipitation for the central grassland region of the USA (Figure 5.6), South America and Africa

soil water-holding capacity interact through changes in soil water dynamics. In dry regions, important losses of soil water occur via bare soil evaporation. Bare soil evaporation is lower in sandy soils than loamy soils because water penetrates deeper. In humid regions, substantial water losses can occur via deep percolation, which is relatively low in soils with high water-holding capacity. Therefore, in arid regions, sandy soils with low water-holding capacity have more water available for plant growth than soils with higher water-holding capacity. The opposite pattern occurs in humid regions.

So far we have explored the relationship between primary production and environmental factors across sites. Based upon the assumption that space and time are interchangeable, we might speculate that the effect upon primary production of changes in precipitation resulting from climate change will be depicted by these same models. However, when we analyzed a 52-yr production time series for the shortgrass steppe in north-central Colorado, USA, we found that the relationship between production and precipitation across time was different from the relationship between production and precipitation across space (Figure 5.9) (Lauenroth and Sala 1992). The temporal model had a slope

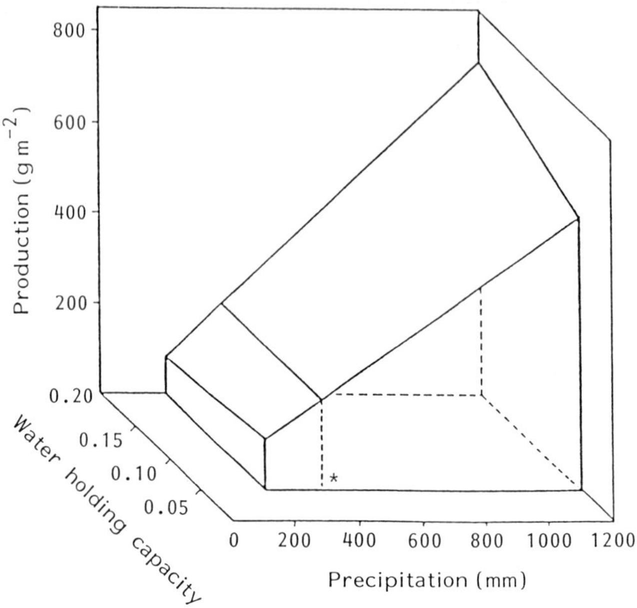

Figure 5.8 Relationship between aboveground NPP, soil water holding capacity, and mean annual precipitation for 9498 sites across the central grassland region of the USA (after Sala *et al.* 1988b)

much smaller than the spatial model. The spatial model compared the average production to average precipitation for different sites whereas the temporal model compared the production and precipitation for different years but always for the same site.

Structural vegetation constraints to primary production and time lags in responding to favorable conditions suggest an explanation for the difference in slope between the two models. In the spatial model each data point represents a site with a different vegetation structure; that is, a particular species composition, plant density, plant cover, and root pattern. By contrast, in the temporal model, it is the same community that responds to changes in water availability. Changes in structure as a result of increases or decreases in water availability involve different processes ranging from leaf expansion, tillering, germination and establishment of new individuals up to migration of new species or ecotypes. These processes have characteristic time constants, some of them exceeding the growing season. We suggest that availability of roots and leaves to capture energy, water, and nutrients can constrain the ecosystem response to increases in water availability.

The difference between spatial and temporal models indicates the difficulty of exchanging space for time. Predictions of the effects of climate change on ANPP, based upon regional models, may contain important sources of previously

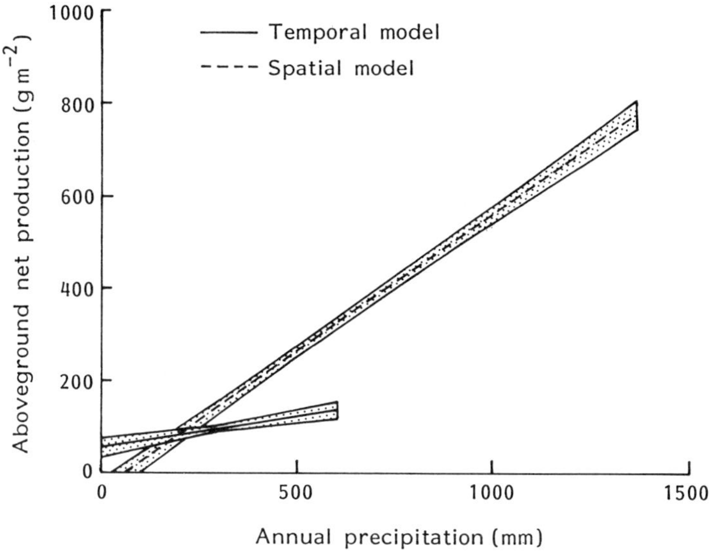

Figure 5.9 Relationship between aboveground NPP and annual precipitation for a regional model (Sala *et al.* 1988b) for the central grassland region of the USA and for the long-term model for the Central Plains Experimental Range in North central Colorado, U.S. (Lauenroth and Sala 1992). Shaded areas represent 95% confidence intervals

unrecognized error. Spatial models depict equilibrium, or at least slowly changing, conditions. The use of such models under scenarios of rapidly changing climatic conditions will result in large errors in estimates of ANPP.

As noted earlier, the regional and long-term analysis pointed out that precipitation was the major environmental correlate of ANPP. Temperature did not account for a significant fraction of the variability in primary production in either the spatial (Sala *et al.* 1988b) or the temporal (Lauenroth and Sala 1992) analysis. There are a number of possible explanations for this finding. There are photosynthetic responses to temperature at the plant level, which apparently are dampened at the ecosystem level. Precipitation and temperature tend to be correlated temporally. Wet days, months, or years tend to be cooler than their dry counterparts. Spatially, the situation is much more complex, and correlations can be positive or negative. The small variability among years in temperature and in potential evapotranspiration relative to the variability in precipitation also affect the predictive power of temperature in the temporal model. Precipitation was almost twice as variable (c.v. = 31%) as potential evapotranspiration (c.v. = 18%) (Sala *et al.* 1992). On the contrary, this explanation does not appear to be valid for the regional model which encompassed a broad range of temperature from the US–Mexico border to the US–Canada border, and in which temperature also was not significant. This analysis suggests that water availability is the most frequent limiting factor of primary production in temperate grasslands. Its availability in time and space is the single most important factor determining primary production in temperate grasslands.

The discussion to this point has dealt exclusively with aboveground production. The portion of total production that is accounted for by the belowground organs of grassland plants (BNPP) is substantial, equaling or exceeding aboveground production (Coleman 1976; Fogel 1985). A problem encountered by attempts to characterize estimates of BNPP for grasslands or any other ecosystem type is that estimates are few and subject to a large amount of uncertainty (Singh *et al.* 1984; Sala *et al.* 1988a; Milchunas and Lauenroth 1992). While estimates of BNPP for temperate grasslands range from two to three times ANPP (Sims and Singh 1978), estimates based upon the turnover of roots labeled with ^{14}C suggest a relationship between BNPP and ANPP of approximately 1:1 (Dahlman 1968; Milchunas and Lauenroth 1992).

5.3.2 Decomposition

Decomposition rates in temperate grasslands are strongly controlled by temperature, soil texture, and litter quality (Jenkinson 1977; Sorenson 1981; Schimel *et al.* 1985). Decomposition rates rapidly increase with temperature (Jenny 1980). Meentemeyer (1984) constructed maps of decomposition rates (k) for North America, and reported that decomposition rate decreased from the warm and wet southeastern corner of the continent to the cool and dry north-central part of Canada.

The control of decomposition by temperature suggests an explanation for the relationship between root biomass and temperature (Figure 5.4). Most studies of root biomass do not distinguish between dead and live roots. Consequently, dead roots will persist in the soil longer in cool sites where decomposition is slower than in warm sites. We hypothesize that the high root biomass of cool sites is the result of a high proportion of dead roots.

Fine-textured soils generally have slow decomposition rates (Sorenson 1981). Therefore, landscape and regional locations which are net deposition sites for fine soil materials tend to accumulate soil organic matter rapidly. This is attributed to the high surface tension of soils with high clay contents, and subsequent physical and chemical protection of organic matter from decomposers. This important control by soil texture causes strong small- and large-scale (spatial) variations in soil carbon associated with parent material and topography in temperate grasslands (Schimel *et al.* 1985; Yonker *et al.* 1988). Plant litter quality is also an important control over decomposition rates and nutrient availability in grasslands (Wedin and Tilman 1990). Meentemeyer (1978) found that lignin content accounted for part of the variability in decomposition. Although decomposition rates increase with increasing water availability (Meentemeyer 1978), they are much more responsive to changes in temperature, as demonstrated by regional patterns.

5.3.3 Differential controls of production and decomposition

In synthesis, we found that temperature, on balance, seems to influence decomposition more than NPP, and precipitation, on balance, seems to influence production more than decomposition. Regional assessments of soil organic matter suggest that soil organic carbon accumulation is controlled more by output (decomposition) than by input (primary production).

We suggest that the differential response of production and decomposition to temperature and precipitation may be related to the adaptations of decomposing organisms for low-moisture environments. A number of investigators have found that decomposition in desert and semi-desert environments is not limited by moisture availability (Santos *et al.* 1984; Whitford *et al.* 1986, 1988). These authors suggest two explanations for the continued decomposer activity at low water availability, with little or no response to increases in moisture. First, microflora and fauna associated with decomposition in dry areas are exceptionally well-adapted to low moisture conditions. Such adaptations include migration, mucilaginous compounds that protect microorganisms from desiccation, and transport of water across fungal hyphae. The second explanation for high decomposer activity in low-moisture environments is related to the upward transport and condensation of water vapor at the soil surface (Santos *et al.* 1984). Finally, it has been suggested that in environments dominated by pulse availability of moisture, the fluctuations in microclimate may be much more important controls over decomposition than average values of moisture (Whitford *et al.* 1981; Sala and Lauenroth 1982; Schimel and Parton 1986).

5.4 THE POSSIBLE EFFECTS OF GLOBAL CHANGE UPON THE CARBON BUDGET

5.4.1 Land use change

The greatest changes in the carbon budgets of the temperate grasslands as a result of management are related mainly to cultivation. Transformation of grasslands to croplands can result in a net carbon flux to the atmosphere. Historically, all cropland management practices resulted in carbon losses and depletion of soil organic matter (SOM). Tillage increases decomposition by breaking soil aggregates and allowing previously protected material to be exposed to decomposers, and by placing surface residues in contact with the soil (Tieszen *et al.* 1982; Elliott 1986; Doran and Werner 1990). In addition, the shift from perennial to annual plant species, and the removal of a portion of the aboveground biomass during harvest may decrease inputs into SOM. Losses of SOM are nearly exponential in the first 10–20 years of cultivation (Haas *et al.* 1957). A combination of modelling and historical data analysis suggested (Burke *et al.* 1991) that the emissions that occurred as a result of land use change in the central grasslands of North America

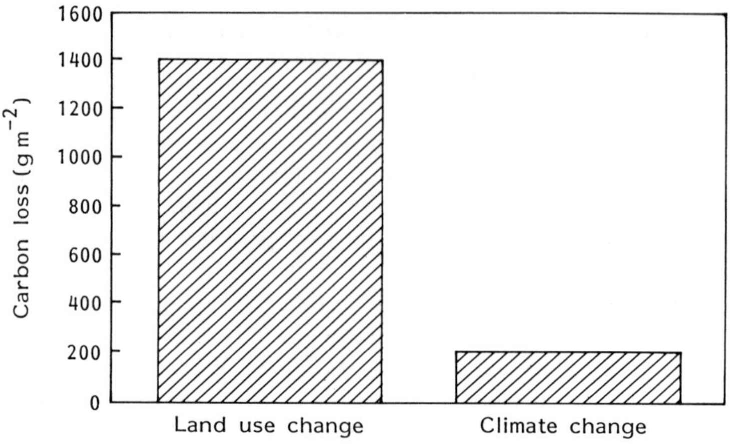

Figure 5.10 The effect of tillage and climate change on carbon losses from the central grassland region of theUSA (after Burke *et al.* 1991)

were much larger than those which may occur as a result of the increase in temperature as predicted by the GISS global circulation model (Figure 5.10). Direct effects of CO_2 and changes in precipitation were not considered. Using the CENTURY ecosystem model, Burke et al. (1991) estimated that after 50 years of climate change, there will be a net loss of carbon ranging from 50 to 350 g m^{-2}. In contrast, an historical analysis of field data indicated a carbon loss between 800 and 2000 g m^{-2} as a result of cultivation since settlement of the plains. The authors indicated that, on a regional basis, past effects of cultivation have been much more important than the expected climate change as a control upon carbon emissions. The simulated climate effect resulted in emissions of 1.4 Tg C yr^{-1} for the entire region, which is smaller than the 18 Tg C yr^{-1} associated with the estimated large-scale effects of historical cultivation.

Concern over losses of SOM and net carbon release to the atmosphere has resulted in new research on cultivation practices over the past several decades. This new research has had conservation of SOM as one of its objectives and has resulted in crop production methods that either minimize tillage, or eliminate it altogether. Preliminary results suggest that these new methods have the potential to conserve or, in some cases, increase soil carbon storage. These new methods decrease or eliminate several of the negative effects of cultivation on ecosystem carbon balance (Wood *et al.* 1990).

A recent trend in many developed countries of converting cropland back to the original ecosystem will create a new and potentially large category of recovering or successional ecosystems. In the temperate grassland regions in the northern hemisphere, such conversions will have important effects on regional carbon budgets. The long-term significance of the conversion of cropland to grassland will depend upon regional, national, and worldwide surpluses of agricultural products.

5.4.2 Atmospheric composition change

The response of temperate grasslands to an increase in the atmospheric concentration of CO_2 is predicted to have a strong interaction with the availability of nutrients and water (Mooney *et al.* 1991). There has been evidence of enhanced photosynthesis in C3 plants with elevated atmospheric CO_2. Elevated CO_2 has two effects: upon the photosynthesis mechanism itself and upon stomatal conductance. An increase in external CO_2 concentration results in an increase in internal CO_2 concentration of C_3 plants, which leads to a higher rate of sucrose formation. The duration of the enhancement of photosynthesis as a result of increased CO_2 concentrations is variable among species. Some species show long-term increases in photosynthesis while others show a short-term increase and then a return to levels associated with ambient CO_2 concentrations. Carbon dioxide enhancement also reduces stomatal conductance and consequently transpiration water loss. Since transpiration is reduced and photosynthesis is enhanced or at least maintained, an increase in CO_2 concentration results in an increase in the amount of carbon fixed per unit of water transpired. The improvement of water use efficiency is the physiological explanation for why the largest response to CO_2 enhancement may occur in water-limited systems. Most of the predictions for the differential response of ecosystems to CO_2 are based upon our current understanding of the response of individual plants to CO_2, water, and nutrient availability. However, ecosystem level experiments to identify these response patterns are absolutely necessary since in several cases plant level patterns are not translated into ecosystem level patterns (Snaydon 1991).

In nutrient-poor environments, the shortage of nutrients may mask the response in growth to CO_2 enhancement. Some experiments have shown a decrease in the nitrogen content of plants growing under elevated CO_2 conditions. On the map of relative response to CO_2, temperate grasslands are located in the central region, which suggests that we may expect a moderate response to CO_2 enhancement. Currently few data are available from field experiments on temperate grasslands with elevated levels of CO_2. Those that have been conducted report fewer responses than had been predicted from extrapolations from laboratory experiments (Knapp *et al.* in prep).

5.4.3 Climate change

There is agreement among climate change predictions from GCMs that an increase in greenhouse gases will result in a temperature increase (Mitchell *et al.*

1990). Discrepancies are associated with the timing, the magnitude, and the spatial distribution of the increase. Our analysis suggests that an increase in temperature will likely result in a reduction of the SOM pool in temperate grasslands as a result of an increase in decomposition rates. Two studies have conducted simulations of the effects of such climate change on North American temperate grasslands (Schimel *et al.* 1991; Burke *et al.* 1991). The studies used the same simulation model, but different climate change scenarios. Simulated decomposition rates increased in response to rising temperatures, and carbon emissions to the atmosphere thus increased.

The effect of an increase in temperature upon primary production is likely to be small. A larger impact may occur via increased potential evapotranspiration and a relative reduction of available water. In the simulations of Schimel et al. (1991), increased decomposition resulted in increased nitrogen mineralization which stimulated NPP in the more mesic grasslands. Such interactions are difficult to assess without experimental evidence.

Predictions regarding change in precipitation and soil moisture vary among regions and GCMs. Most models predict a decrease in precipitation for the temperate northern hemisphere grasslands and an increase for the southern hemisphere temperate grasslands. Increases in precipitation, depending upon magnitude, will have a high probability of increasing primary production and carbon input to soils. This may or may not compensate the increase in carbon losses resulting from the relatively more certain temperature increase. For those regions in which a reduction in precipitation is predicted, a decrease in inputs will occur simultaneously with an increase in the outputs. It will clearly drive these grasslands into a negative carbon balance.

The relative importance of the three factors driving global change may vary with time. We suggest that the maximum change in land use in temperate grasslands has already occurred. Most of the areas with some crop-production potential have been transformed into croplands. Although the increase in greenhouse gases has been documented, their importance may continue to increase for the next several decades. The effect of climate change will likely be observed last because of the time lags built in the climate system.

5.5 REFERENCES

Atjay, G. L., Ketner, P. and Duvigneaud, P. (1979) Terrestrial primary production and phytomass. In: Bolin, B., Degens, E., Kempe, S. and Ketner, P. (Eds) *The Global Carbon Cycle*, SCOPE 13, pp. 129–182. Wiley, Chichester.

Burke, I. C., Kittel, T. G. F., Lauenroth, W. K., Snook, P., Yonker, C. M. and Parton, W. J. (1991) Regional analysis of the central Great Plains. *BioScience* **41**, 685–692.

Burke, I .C., Yonker, C., Parton, W. J., Cole, C. V., Flach, K. and Schimel, D. S. (1989) Texture, climate, and cultivation effects on soil organic matter content in U.S. grassland soils. *Soil Sci. Soc. Am. J.* **53**, 800–805.

Coleman, D. C. (1976) A review of root production processes and their influence on soil biota in terrestrial ecosystems. In: Anderson, J. M. and Macfadyen, A. (Eds) *The Role of Terrestrial and Aquatic Organisms in Decomposition Processes*, pp. 417–434. Blackwell Scientific, Oxford, England.

Coupland, R. T. (1992) Approach and generalizations. In: Coupland, R.T. (Ed.) *Ecosystems of the World 8A, Natural Grassland, Introduction and Western Hemisphere*, pp. 1–6, Elsevier, New York.

Dahlman, R. C. (1968) Root production and turnover of carbon in the root–soil matrix of a grassland ecosystem. In: Ghilarov, M. S. (Ed.) *Methods of Productivity Studies in Root Systems and Rhizosphere Organisms*, pp. 11–21.

Doll, U. M., Deregibus, V. A. (1986) Efecto de la exclusion del pastoreo sobre il subsistuna susterraneo di un postizal templado hómedo. *Turrialba* **36**: 337–344.

Doran, J. W. and Werner, M. R. (1990) Management and soil biology. In: Francis, C. A., Flora, C. B. and King, L. D. (Eds) *Sustainable Agriculture in Temperate Zones*, pp. 205–230. John Wiley & Sons, New York.

Elliott, E. T. (1986) Aggregate structure and carbon, nitrogen, and phosphorus in native and cultivated soils. *Soil Sci. Soc. Am. J.* **50**, 627–632.

Fogel, R. (1985) Roots as primary producers in belowground ecosystems. In: Fitter, A. H., Atkinson, D., Read, D. J. and Usher, M. B. (Eds) *Ecological Interactions in Soil*. British Ecol. Soc. Publ. No. 4, pp. 23–35. Blackwell Scientific, Oxford, England.

French, N. R., Steinhorst, R. K. and Swift, D. M. (1979) Grassland biomass trophic pyramids. In: French, N. R. (Ed.) *Perspectives in Grassland Ecology*, Ecol. Studies 32, pp. 59–87. Springer-Verlag, New York.

Haas, H. J., Evans, C. E. and Miles, E. R. (1957) Nitrogen and carbon changes in soils as influenced by cropping and soil treatments. *USDA Tech. Bull. 1164*. US Govt Printing Office, Washington, DC.

Ingham, R. E. and Detling, J. K. (1984) Plant–herbivore interactions in a North American mixed-grass prairie. III. Soil nematode populations and root biomass on *Cynomic indovicianus* colonies and adjacent uncolonized areas. *Oecologia* **63**, 307–313.

Jenkinson, D. S. (1977) Studies on the decomposition of plant material in soil. V. The effects of plant cover and soil type on the loss of carbon from 14-C labeled rye grass decomposing under field conditions. *J. Soil Sci.* **28**, 424–494.

Jenny, H. (1980) *The Soil Resource*. 377 pp. Springer-Verlag, New York.

Lauenroth, W. K. and Milchunas, D. G. (1992) Short-grass steppe. In: Coupland, R. T. (Ed.) *Ecosystems of the World 8A, Natural Grasslands, Introduction and Western Hemisphere*. Chapter 11, pp. 183–226, Elsevier, New York.

Lauenroth, W. K. and Sala, O. E. (1992) Long-term forage production of North American shortgrass steppe. *Ecol. Appl.* **2**, 397–403.

Le Houérou, H. N. (1984) Rain use efficiency: a unifying concept in arid-land ecology. *J. Arid Environ.* **7**, 213–247.

McNaughton, S. J. (1976) Serengeti migratory wildebeest: facilitation of energy flow by grazing. *Science* **191**, 92–94.

McNaughton, S. J. (1985) Ecology of a grazing system: the Serengeti. *Ecol. Mono.* **55**, 259–294.

McNaughton, S. J., Oesterheld, M., Frank, D. A. and Williams, K. J. (1989) Ecosystem-level patterns of primary productivity and herbivory in terrestrial habitats. *Nature* **341**, 142–144.

Meentemeyer, V. (1978) Macroclimate and lignin control of litter decomposition rates. *Ecology* **59**, 465–472.

Meentemeyer, V. (1984) The geography of organic decomposition rates. *Ann. Assoc. Am. Geograph.* **74**, 551–560.

Milchunas, D. G. and Lauenroth, W. K. (1992) Carbon dynamics and estimates of primary production by harvest, ^{14}C dilution, and ^{14}C turnover. *Ecology* **73**, 593–607.

Milchunas, D. G. and Lauenroth, W. K. (1993) A quantitative assessment of the effects of grazing on vegetation and soils over a global range of environments. *Ecol. Monogr.* In press.

Mitchell, J. F. B., Manabe, S., Meleshko, V. and Tokioka, T. (1990) Equilibrium climate change–and its implications for the future. In: Houghton, J. T., Jenkins, G. J. and Ephraums, J. J. (Eds) *Climate Change, The IPCC Scientific Assessment*, Chapter 5, pp. 131–172. Cambridge University Press, New York.

Mooney, H. A., Luxmoore, B. G., Oechel, W. C. and Pitelka, L. F. (1991) Predicting ecosystem responses to elevated CO_2 concentrations. *BioScience* **41**, 96–104.

Noy Meir, I. (1973) Desert ecosystems: environment and producers. *Ann. Rev. Ecol. Syst.* **4**, 25–52.

Parton, W. J., Schimel, D. S., Cole, C. V. and Ojima, D. S. (1987) Analysis of factors controlling soil organic matter levels in Great Plains grasslands. *Soil Sci. Soc. Am. J.* **51**, 1173–1179.

Sala, O. E., Biondini, M. E. and Lauenroth, W. K. (1988a) Bias in estimates of primary production: an analytical solution. *Ecol. Model.* **44**, 43–55.

Sala, O. E. and Lauenroth, W. K. (1982) Small rainfall events: an ecological role in semiarid regions. *Oecologia* **53**, 301–304.

Sala, O. E., Lauenroth, W. K. and Parton, W. J. (1992) Long-term soil water dynamics in the shortgrass steppe. *Ecology* **73**, 1175–1181.

Sala, O. E., Parton, W. J., Joyce, L. A. and Lauenroth, W. K. (1988b) Primary production of the central grassland region of the United States. *Ecology* **69**, 40–45.

Santos, P. F., Elkins, N. Z., Steinberger, Y. and Whitford, W. G. (1984) A comparison of surface and buried *Larrea tridentata* leaf litter decomposition in North America hot deserts. *Ecology* **65**, 278–284.

Schimel, D. S., Coleman, D. C. and Horton, K. A. (1985) Soil organic matter dynamics in paired rangelands and cropland toposequences in North Dakota. *Geoderma* **36**, 201–214.

Schimel, D. S., Kittel, T. G. F. and Parton, W. J. (1991) Terrestrial biogeochemical cycles: global interactions with atmosphere and hydrology. *Tellus* **43**, 188–203.

Schimel, D. S. and Parton, W. J. (1986) Microclimatic controls of nitrogen mineralization and nitrification in shortgrass steppe soils. *Plant and Soil* **93**, 347–357.

Scott, J. A., French, N. R. and Leetham, J. W. (1979) Patterns of consumption in grasslands. In: French, N. R. (Ed.) *Perspectives in Grassland Ecology*, pp. 89–105. Springer-Verlag, New York.

Sims, P. L. and Singh, J. S. (1978) The structure and function of ten western North American grasslands. III. Net primary production, turnover, and efficiencies of energy capture and water use. *J. Ecol.* **66**, 573–597.

Sims, P. L., Singh, J. S. and Lauenroth, W. K. (1978) The structure and function of ten western North American grasslands. I. Abiotic and vegetation characteristics. *J. Ecol.* **66**, 251–285.

Singh, J. S., Lauenroth, W. K., Hunt, H. W. and Swift, D. M. (1984) Bias and random errors in estimators of net root production: a simulation approach. *Ecology* **65**, 1760–1764.

Singh, J. S., Lauenroth, W. K. and Milchunas, D. G. (1983) Geography of grassland ecosystems. *Progress in Physical Geography* **7**, 46–80.

Snaydon, R. W. (1991) The productivity of C3 and C4 plants: a reassessment. *Funct. Ecol.* **5**, 321–330.

Sorenson, L. H. (1981) Carbon–nitrogen relationships during the humification of cellulose in soils containing different amount of clay. *Soil Biol. Biochem.* **13**, 313–321.

Soriano, A., Eolluscio, R. A. and Satorre, E. (1987) Spatial heterogeneity at the root system of grasses in the Patagonian arid steppe. *Bulletin of the Torrey Botanical Club* **114**: 103–108.
Soriano, A., León, R. J. C., Sala, O. E., Lavado, R. S., Deregibus, V. A., Cahuépé, M. A., Scaglia, O. A., Velázquez, C. A. and Lemcoff, J. H. (1991) Ríode la Plata grasslands. In: Coupland, R.T. (Ed.) *Ecosystems of the World 8A, Natural Grasslands, Introduction and Western Hemisphere*, Chapter 19, pp. 367–407. Elsevier, New York.
Tieszen, H., Stewart, J. W. B. and Bettany, J. R. (1982) Cultivation effects on the amounts and concentration of carbon, nitrogen and phosphorus in grassland soils. *Agron. J.* **74**, 831–835.
Walter, H. (1979) *Vegetation of the Earth and Ecological Systems of the Geo-biosphere*, 274 pp. Heidelberg Science Library. Springer-Verlag, New York.
Wedin, D. A. and Tilman, D. (1990) Species effects on nitrogen cycling: a test with perennial grasses. *Oecologia* **84**, 433–441.
Whitford, W. G., Meentemeyer, V., Seastedt, T. R., Cromack, K., Crossley, D., Santos, P., Todd, R. L. and Waide, J.B. (1981) Exceptions to the AET model: deserts and clearcut forests. *Ecology* **62**, 275–277.
Whitford, W. G., Steinberger, Y., MacKay, W., Parker, L.W., Freckman, D., Wallwork, J. A. and Weems, D. (1986) Rainfall and decomposition in the Chihuahuan Desert. *Oecologia* **68**, 512–515.
Whitford, W. G., Stinnett, K. and Anderson, J. (1988) Decomposition of roots in a Chihuanhuan Desert ecosystem. *Oecologia* **75**, 8–11.
Wood, C. M., Westfall, D. G., Peterson, G. A. and Burke, I. C. (1990) Impacts of cropping intensity on carbon and nitrogen mineralization under no-till dryland agroecosystems. *Agron. J.* **82**, 1115–1120.
Yonker, C. M., Schimel, D. S., Paroussis, E. and Heil, R. D. (1988) Patterns of organic carbon accumulation in a semiarid shortgrass steppe, Colorado. *Soil Sci. Soc. Am. J.* **52**, 478–483.

6 Photosynthesis, Rising Atmospheric Carbon Dioxide Concentration and Climate Change

S. P. LONG, C. P. OSBORNE and S. W. HUMPHRIES
Department of Biology, University of Essex, Colchester, UK

6.1 INTRODUCTION

Photosynthesis provides the driving step to the biogeochemical global carbon cycle. Whilst much is known of the direct effects of temperature, water vapor pressure deficit and nutrient supply on this process (reviewed: Long 1985; Long and Woodward 1988), until recently little has been known about the long-term effects of growth in elevated CO_2 concentrations. Because CO_2 is commonly a limiting substrate for photosynthesis, the current and projected increases in atmospheric CO_2 concentration (c_a) are of direct significance to this step in the carbon cycle. The biophysics and biochemistry of photosynthetic carbon assimilation appear highly conserved across terrestrial vegetation, regardless of taxonomic affinity. This has allowed the construction of mechanistic models of photosynthetic response to increase in c_a which apply to the vast diversity of terrestrial plants, and provides a rare opportunity for scaling from biochemical to landscape levels. The dominant primary producers of all ecosystems except grasslands are C_3 species. These are species in which the first product of photosynthetic CO_2 assimilation is a 3-carbon compound, phosphoglycerate (PGA). CO_2-fixation is catalyzed by the enzyme ribulose-1,5-bisphosphate carboxylase/oxygenase (Rubisco). PGA is metabolized through the Calvin cycle to resynthesize ribulose-1,5-bisphosphate (RubP), the primary acceptor of CO_2. Carbohydrate for storage and post-photosynthetic metabolism is gained by the transfer of intermediates of the Calvin cycle into synthetic pathways, particularly that leading to sucrose, the major carbohydrate exported from photosynthetic cells. Virtually all grass species of tropical grasslands and a significant proportion of the grasses of the steppe/prairie biome are C_4 species, i.e. plants in which the first product of photosynthetic carbon metabolism is a 4-carbon compound. These species utilize an additional carboxylase, phosphoenol-pyruvate carboxylase (PEPc) in the C_4-phosynthetic cycle, to concentrate CO_2 at the site of

Global Change: Effects on Coniferous Forests and Grasslands.
Edited by A. I. Breymeyer, D. O. Hall, J. M. Melillo and G. I. Ågren

Rubisco within the bundle sheath cells of the leaf. Essentially, these additional steps provide a light-driven CO_2 pump. The result is that photosynthesis in C_4 species is saturated at a c_a below the current ambient concentration whilst in C_3 species saturation is only approached at a c_a of three times current atmospheric levels (Edwards and Walker 1983). This review is therefore primarily concerned with C_3 species, since no direct effect of elevated c_a on photosynthesis is anticipated for C_4 species. Whilst the primary concern of this book is with coniferous forest and grasslands, it is appropriate to consider the wider knowledge of the effects of elevated c_a on terrestrial plants, since the basic mechanisms are the same in all C_3 species regardless of biome. Atmospheric concentrations of CO_2 are predicted to rise from the current concentration of 355 μmol mol^{-1} of CO_2 in air to about 500 μmol mol^{-1} in 2050 and *ca.* 700 μmol mol^{-1} by 2100, following the Intergovernmental Panel of Climate Change (IPCC) IS92a scenario (Leggett *et al.* 1992). This chapter examines the implications of these changes for photosynthesis, by examining field and laboratory experiments, and model predictions. There is a good mechanistic understanding of the instantaneous effects of variation in temperature and c_a on C_3 photosynthesis. This, coupled with the conservative nature of the mechanism of C_3 photosynthesis, has allowed the development of generic models of cell, leaf and canopy photosynthesis in which the effects of both external changes, e.g. temperature increase, and internal changes, e.g. decrease in Rubisco activity, may be explored. Such a model, WIMOVAC (Humphries and Long 1995) is used here to examine the implications of changes in both climate and atmospheric composition on the process of photosynthesis and on carbon uptake, and how these may be modified by acclimation.

6.1.1 Photosynthesis as the primary site of response

Carbon dioxide is a substrate of photosynthesis and is limiting to the rate of photosynthesis in C_3 species at all light levels. In the short term, increase in c_a will, in C_3 plants, lead to increased net leaf photosynthetic CO_2 uptake (A) both by increasing the velocity of carboxylation of RubP and suppressing photorespiration. Because increase in c_a increases the net efficiency of photosynthesis, an increase in photosynthesis must therefore be expected regardless of whether other factors are limiting photosynthesis or not. For example, an increase occurs regardless of whether light is limiting or saturating (Long and Drake 1991). This results because an increase in c_a diverts ATP and NADPH generated on the photosynthetic membrane to the anabolic reactions of the Calvin cycle and away from the catabolic reactions of the photosynthetic C_2 pathway leading to photorespiration (Long and Drake 1992).

Photosynthesis is not only the major physiological process by which plants sense change in c_a (Mott 1990), but is also the process by which plant production is most likely to be affected by change in c_a. Other physiological processes influencing gaseous fluxes are also affected independently by change in c_a. In both C_3 and C_4 plants, stomatal aperture and hence conductance (g_s) and transpiration per unit leaf area decrease with increasing c_a (Mott 1990). This effect of rising c_a on stomata is independent of the effect on photorespiration. Increase in c_a has also been shown to depress night-time respiration (R_d) (Bunce 1990). Many other effects have been attributed to increased CO_2 (Bazzaz 1990). Relative stimulation of CO_2 uptake and dry matter production have frequently been shown to occur under conditions of environmental stress, suggesting that elevated CO_2 may ameliorate stress (reviewed: Long and Drake 1992). Amelioration of stress effects may be attributed to an increased capacity for photosynthesis, an increased supply of photosynthate and/or an increased water and nitrogen use efficiency (Idso 1989). There are also examples where elevated c_a has increased the negative effects of nitrogen and phosphorus deficiency on growth or crop yield (Conroy *et al.* 1988; Mitchell *et al.* 1993).

6.1.2 Effects of rising CO_2 on photosynthesis and predictive models for natural vegetation

Prediction of crop and natural ecosystem production in a changed climate has often reflected the view that there would be little stimulation of growth with an increase in c_a. The models of Parry and Carter (1988, 1990), used for predicting regional patterns of change in crop production, make no allowance for any direct effect of rising c_a. The model of Esser (1987) underlies the 'Osnabrück Biosphere Model', which has been used to assess future changes in global primary productivity, yet it ignores the direct effects of rising c_a on photosynthetic carbon uptake. The '1990 – Greenpeace Assessment' (Schimel 1990) used the CENTURY model to predict the likelihood of ecosystem feedbacks on rising atmospheric CO_2, and considered only the indirect effect of rising c_a on the water use efficiency of vegetation, not its direct effects on photosynthesis (Ågren *et al.* 1991). Failure to include a direct stimulation of carbon gain by rising c_a may have various roots:

(a) observations which show photosynthesis to decline following transfer to an elevated c_a (Bazzaz 1990). Downward adjustments of photosynthetic enhancement during growth in an elevated c_a are usually referred to as 'photosynthetic acclimation' (Gunderson and Wullschleger 1994);
(b) the view that production in natural ecosystems is limited by factors other than c_a (Melillo *et al.* 1990; Schimel 1990);
(c) the argument that decrease in photorespiratory losses resulting from rising c_a may be canceled by an increase in photorespiration resulting from the projected concurrent rise in mean temperature (Eamus 1991).

The empirical evidence that crops grown in elevated c_a show increased dry matter production is overwhelming, however. Kimball (1983) cited over 450 studies in which crops were grown at an elevated c_a. He concluded that average crop yield is increased by 34% and transpiration is decreased by 34%. Of course it may be argued that increased yield is the indirect result of decreased transpiration, rather than increased photosynthesis. Other predictions have considered that the effects of elevated c_a are adequately accounted for by multiplication with a constant or β-factor (reviewed: Hendrey, 1992). Multiplication with a single constant ignores the interactions of c_a with other key variables. Yet these interactions can dramatically modify or even reverse the response to increase in c_a – most are well understood at a mechanistic level (Long 1991).

Here it is demonstrated that the acclimation of photosynthesis to elevated c_a is often associated with artificial environments and less common in field experiments. Where acclimation does occur, it seems to be an adaptive response which benefits the plant, and the positive effect of CO_2 on photosynthesis tends to be sustained in the long term. It is further demonstrated that failure to take account of the interactions of c_a with other environmental parameters can lead to serious errors in the prediction of increase in photosynthesis with rising c_a.

6.2 A MECHANISTIC MODEL OF LEAF AND CANOPY PHOTOSYNTHESIS

In order to reliably predict changes in photosynthesis in response to elevated c_a, a mechanistic approach to modeling the effect of CO_2 on leaf biochemistry is required (Reynolds and Acock 1985). This has been achieved with a mechanistic model of photosynthetic gas exchange, WIMOVAC, which scales from C_3/C_4 biochemistry to canopy CO_2-uptake (Humphries and Long 1995). Figure 6.1 illustrates the submodels of WIMOVAC used, their interconnection and the original sources of the equations used. The mechanistic model of leaf photosynthesis developed by Farquhar *et al.* (1980) and Farquhar and von Caemmerer (1982) has been widely used and validated (Long 1985; Long & Drake 1992; Harley *et al.* 1992). Equations originally derived by Farquhar *et al.* (1980) and given in Long and Drake (1992) were modified in WIMOVAC to include a potential phosphate limitation arising from the sequestration of cytosolic phosphate in sugar phosphates and failure of inorganic phosphate supply to keep up with triose phosphate production in the Calvin cycle (Sharkey 1985b). WIMOVAC was used to predict leaf photosynthetic rates of CO_2-uptake and the significance of changes in temperature and c_a via their effects at the level of carboxylation and oxygenation of RubP. Parameters for the leaf biochemistry module, except where stated otherwise in individual figure legends, are as given by Farquhar *et al.* (1980) for a leaf temperature of 25 °C.

The biochemical model uses c_i (intercellular CO_2 concentration) rather than c_a as a driving variable. Assimilation of CO_2 and stomatal conductance is assumed to determine c_i. To be useful in predicting leaf response to varying environmental

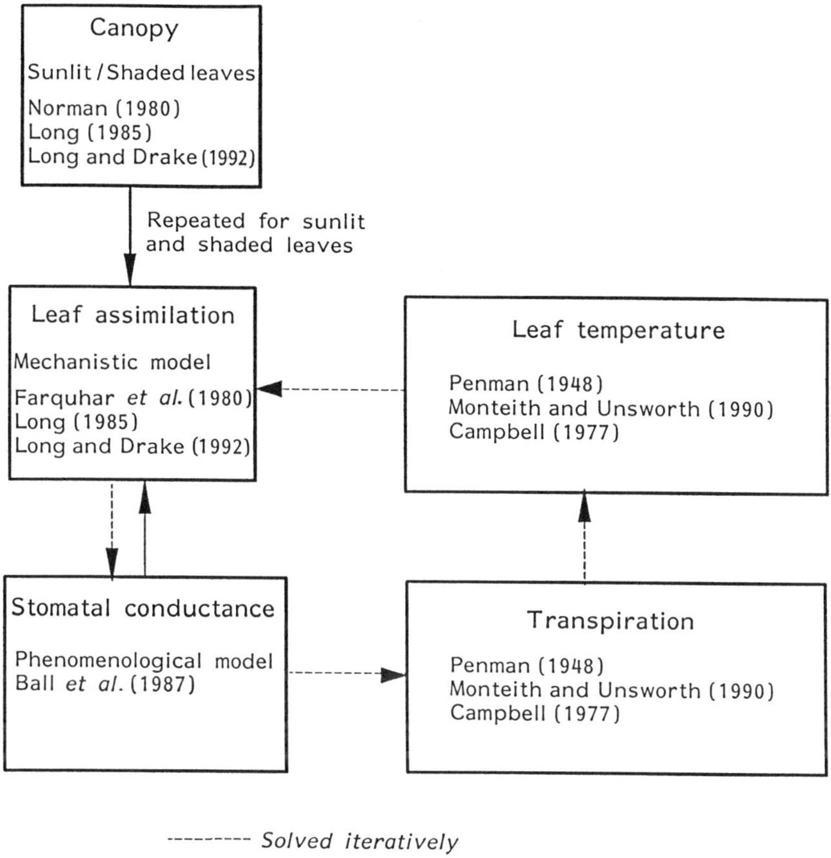

Figure 6.1 WIMOVAC mechanistic model for the prediction of leaf/canopy carbon and water vapor exchange from biochemical mechanisms and leaf microclimatic relations. Submodels, their interconnections and primary sources of equations

conditions, the biochemical model of CO_2 assimilation must be integrated with a model of stomatal behavior. A mechanistic understanding of the control processes involved in regulating g_s remains elusive, but a successful and robust phenomenological model of the response of g_s to c_i, A and humidity of the atmosphere is provided by Ball *et al.* (1987). Harley *et al.* (1992) adapted this model to use c_a rather than c_s (CO_2 concentration at the leaf surface) and relative humidity of the air outside the boundary layer. This revision is used in WIMOVAC. Because A and g_s are interdependent, the value of c_i is determined in WIMOVAC in an iterative fashion.

In order to examine the effects of variation in temperature, model equations for predicting diurnal change in air temperature were obtained from Spain and Keen

(1992). In addition, solubilities for O_2 and CO_2 were recalculated relative to their values at 25 °C using polynomial relationships fitted to tabular values of solubility at different temperatures (Linke 1965; Kaye and Laby 1973). Jordan and Ogren (1984) provided data on the response of the kinetic constants and specificity of Rubisco to temperature, which are used here.

Effects of varying leaf nitrogen content on the biochemistry of leaf processes have been introduced into WIMOVAC using data supplied by Field (1983). Field (1983) and Harley *et al.* (1992) both proposed a linear relationship between leaf nitrogen content and maximum RubP-saturated carboxylation velocity ($V_{c,max}$), light-saturated potential rate of electron transport (J_{max}) and R_d. All leaves, either singularly or as part of a canopy, were assumed here to have a leaf nitrogen concentration of 2 g m^{-2} unless otherwise specified.

Although leaf gas exchange provides many insights into plant adaptation to the environment, canopy processes are critical in determining crop and biome-level productivity (Norman 1980). To evaluate the significance of changes inferred at the leaf level to the canopy level, a simplified model of canopy photosynthesis has been used (Long 1991; Long and Drake 1992). Norman (1980) showed that by treating a canopy as two populations of leaves, sunlit and shaded, and by calculating the mean irradiance and in turn mean assimilation rate for each leaf population, the estimated canopy photosynthesis differed little from more complex canopy models. However, this division into sunlit and shaded leaves did provide a substantial improvement in prediction over models which simply assumed an exponential decline in light through homogeneously lit canopy layers. These equations, used in WIMOVAC, are given by Forseth and Norman (1993) and Long and Drake (1992). The parameters used here, unless otherwise stated in individual legends, are as given in Long and Drake (1992).

To facilitate investigation of elevated c_a and concomitant temperature effects on predicted canopy water usage, an expression has been introduced into the canopy submodel of WIMOVAC for the calculation of instantaneous transpiration of water vapor from the canopy, according to Penman (1948) and Monteith and Unsworth (1990). The expression has been combined with a model to describe the transfer of water vapor from the evaporating surface to the bulk air stream in terms of the aerodynamics of the turbulent air above the canopy. This is based upon a boundary layer conductance model proposed by Campbell (1977), and the theory behind this is discussed in some detail by Thornley and Johnson (1990). Transpiration rates for both the sunlit and shaded leaves within the canopy are considered according to the appropriate light and temperature microclimate conditions within the canopy. A derivation of the Penman (1948) equation in which transpiration is eliminated, is used to predict the difference between canopy leaf temperature and the ambient air temperature outside the canopy. Parameters for both the transpiration and leaf temperature modules were as given by Campbell (1977). However, the expression relating apparent sink momentum to canopy height, given by Campbell (1977), was corrected here and has the form, $d = 0.77h$ where d is the apparent sink momentum and h is the canopy height in meters. Leaf transpiration and leaf temperature are not independent quantities, and so an iterative procedure is used here to establish their respective equilibrium values in WIMOVAC.

6.3 ELEVATED c_a AND PHOTOSYNTHETIC ACCLIMATION

6.3.1 Instantaneous responses of photosynthesis to CO_2

Instantaneous responses of photosynthesis to an elevated c_a are obtained when plants grown under the current ambient c_a are transferred to an elevated c_a (Figure 6.2). The simulation shows that an increase in c_a would increase A at all light levels from below the light compensation point for photosynthesis (LCP) to the maximum photon fluxes experienced in the natural environment.

6.3.1.1 Light-limited photosynthesis

Rubisco catalyzes both the carboxylation and oxygenation of RubP; CO_2 is a competitive inhibitor of oxygenation and O_2 a competitive inhibitor of car-

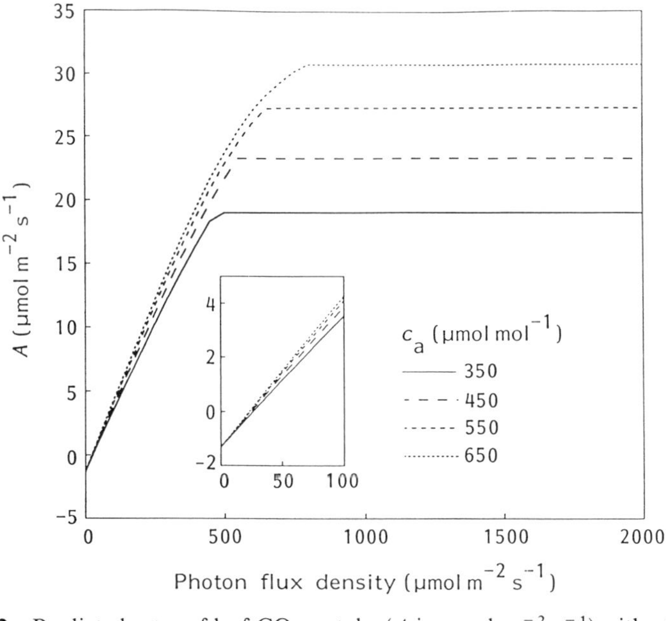

Figure 6. 2 Predicted rates of leaf CO_2-uptake (A in μmol m^{-2} s^{-1}) with photon flux density (μmol m^{-2} s^{-1}) for four atmospheric CO_2 concentrations (c_a in μmol mol^{-1} of CO_2 in air). Inset highlights early response of CO_2-uptake at low photon flux values

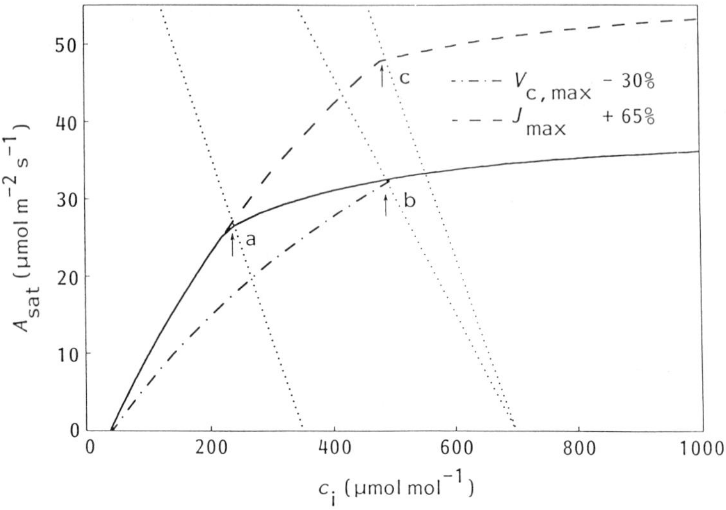

Figure 6.3 The simulated response of light-saturated rates of leaf CO_2 uptake (A_{sat} in μmol m^{-2} s^{-1}) with intercellular CO_2 concentration (c_i, μmol mol^{-1} of CO_2 in air). The solid line indicates the curve based on leaf biochemistry parameters given in Long and Drake (1992) and is typical of the response observed in many C_3 plants grown at current c_a. Arrows indicate the operating points, i.e. the c_i obtained for a given external c_a. The dotted lines joining c_i and c_a on the three curves indicate the supply function. The broken lines illustrate two potential patterns of acclimation to growth in elevated CO_2. The lower line indicates the result of a 30% decrease in $V_{c,max}$ and $V_{o,max}$, simulating a loss of Rubisco activity, and the upper line a 65% increase in J_{max}, simulating an increase in the maximum capacity for regeneration of RubP

boxylation. Uptake of CO_2 via carboxylation of RubP leads to carbohydrate synthesis through the Calvin cycle. Uptake of O_2 via oxygenation of RubP leads to carbohydrate anabolism via the photosynthetic carbon oxidation cycle, with the resultant release of CO_2 in photorespiration. Rising c_a will favor photosynthetic carbon accumulation at the expense of photorespiration through competitive inhibition of RubP-oxygenation. Maximum quantum yield (ϕ) is the initial and maximum slope of the hyperbolic response of A to absorbed photosynthetically active photon flux (I_{abs}). Light limitation restricts the supply of intermediates required in RubP-regeneration, and the rate of RubP supply limits CO_2-uptake as a consequence. ϕ increases after an instantaneous rise in c_a since suppression of photorespiration allows the use of a greater proportion of the available RubP in carboxylation (Figure 6.2).

6.3.1.2 Light-saturated photosynthesis

The response of light-saturated A (A_{sat}) to c_i reflects biochemical limitations to A_{sat}, independent of any stomatal effects. The A/c_i response shows two distinct phases predicted by the theory of Farquhar *et al.* (1980) (Figure 6.3): an initial slope at low c_i determined by Rubisco activity, which will be RubP-saturated, and a plateau or shallower slope at higher c_i where photosynthesis is limited by the rate of regeneration of the CO_2 acceptor, i.e. RubP. For a wide range of C_3 species it has been found that the c_i which they attain under current ambient c_a is often at the transition between RubP-saturated and RubP-limited photosynthesis, i.e. at the point where the quantity of active Rubisco and the capacity for regeneration of RubP are colimiting (Jarvis 1989) (Figure 6.3). An environmentally induced change in capacity for carboxylation of RubP appears to be commonly matched by a change in the capacity for RubP-regeneration, and *vice versa*, such that the two processes remain colimiting (Evans and Farquhar 1991).

6.3.2 Theoretical acclimation responses to CO_2

Decline in Rubisco activity represents one of the most consistent features of acclimation of the photosynthetic apparatus to elevated c_a so far identified (Idso 1989; Allen 1990). Why should such a decrease occur?

Growth in low nitrogen results in decreased concentrations of total Rubisco and corresponding decreases in the initial slope of the A/c_i response, sometimes termed the 'carboxylation efficiency'. However, in a range of species it has been found that capacity for regeneration of RubP changes in concert, such that the operating c_i remains at the point of inflection of the A/c_i response (Evans and Farquhar 1991). A similar pattern of acclimation in the A/c_i response is observed when plants are grown in low photon fluxes (Evans and Farquhar 1991; Sage 1994). These studies suggest an optimization of the distribution of resources within the chloroplast so that neither active Rubisco nor the apparatus for regeneration of RubP is in excess. If optimization of the distribution of resources between components of the photosynthetic apparatus is a ubiquitous phenomenon, then adjustment might also be expected in plants grown in elevated c_a. If c_i increases proportionately, then limitation would be shifted away from carboxylation to RubP-regeneration. Given the large investment of energy and nitrogen in Rubisco, up to 25% of leaf nitrogen, strong selective pressures for modulation of Rubisco carboxylase levels are likely.

Figure 6.3 illustrates how this might occur. Without any adjustment of the photosynthetic apparatus the A/c_i curve would be unchanged. Considering a typical A/c_i curve for a leaf developed at current c_a (based on the parameters of Long and Drake 1992), the operating c_i when $c_a = 350\ \mu\text{mol mol}^{-1}$ will be at the point of inflection of the curve (point 'a' on Figure 6.3), assuming $c_i/c_a \doteq 0.7$. If c_a is now doubled, c_i moves to 490 μmol mol^{-1}, and is now on the upper portion of the A/c_i response (point 'b' on Figure 6.3). Here A_{sat} would be limited by regeneration of RubP and Rubisco capacity would be in considerable excess. Acclimation of the A/c_i curve so that the inflection of the curve moves to 490 μmol mol^{-1} could be achieved in two ways (Figure 6.3). Either the quantity of active

Rubisco ($V_{c,max}$) is decreased, so that W_c is decreased to equal W_j at $c_i = 490$ μmol mol^{-1}, or J_{max} is increased, so that W_j is increased to equal W_c at $c_i = 490$ μmol mol^{-1} (point 'c', Figure 6.3). Using the model and parameters of Long and Drake (1992), this increase in the point of inflection of the A/c_i response is achieved either by simulating a 30% decrease in $V_{c,max}$, representing an equivalent decline in active Rubisco, or by increasing J_{max} by 65%, simulating an equivalent increase in capacity for regeneration of RubP.

Yelle *et al.* (1989) suggested that the decrease in photosynthetic capacity of plants growing in elevated CO_2 is caused by decrease in Rubisco activity. Figure 6.3 shows that a 30% decrease in Rubisco activity could occur without decreasing A_{sat} when c_a is elevated to 700 μmol mol^{-1}. Whilst these plants will show a reduced A_{sat} when measured at a c_a of 350 μmol mol^{-1} in comparison to non-acclimated plants because they are Rubisco-limited at this c_a, they should not show a reduced A_{sat} at 700 μmol mol^{-1}. This pattern of acclimation would explain why many studies have found that plants grown in elevated c_a show a decrease in photosynthesis relative to control plants when measured at 350 μmol mol^{-1}, but not when measured at the elevated growth concentration (500–800 μmol mol^{-1}).

To maintain $c_i = 0.7c_a$ as indicated in Figure 6.3, a 42% decrease in stomatal conductance would be required, which at a constant leaf-air vapor pressure deficit (VPD) would give a 42% decrease in transpiration. Thus acclimation of this nature would allow a substantial decrease of investment of nitrogen into Rubisco and provide a marked increase in water use efficiency, while supporting a 22% increase in A_{sat}. This would clearly be advantageous for plants growing with limited resources.

In summary, theoretical consideration of A/c_i responses suggests that acclimation of the response to elevated c_a could represent an optimization of Rubisco activity, capacity for regeneration of RubP and stomatal conductance according to other limitations, in particular nitrogen supply.

6.3.3 Is there empirical evidence for photosynthetic acclimation to an elevated c_a under natural conditions?

Acclimation is widely reported in studies of coniferous trees and grassland species. In a review covering 39 tree species, Gunderson and Wullschleger (1994) compared A_{sat} for plants grown at current ambient c_a and elevated c_a. Acclimation of photosynthesis measured at the current ambient c_a was apparent as an average 21% decrease in A_{sat} for plants grown at an elevated c_a compared with plants grown at the current ambient c_a (Figure 6.4). However, when A_{sat} was measured at the elevated c_a in which the plants were grown, an average enhancement of 44% was observed relative to plants grown and measured at the current ambient c_a (Figure 6.4). The observation is consistent with the theoretical prediction that relatively large decreases in Rubisco activity can occur without

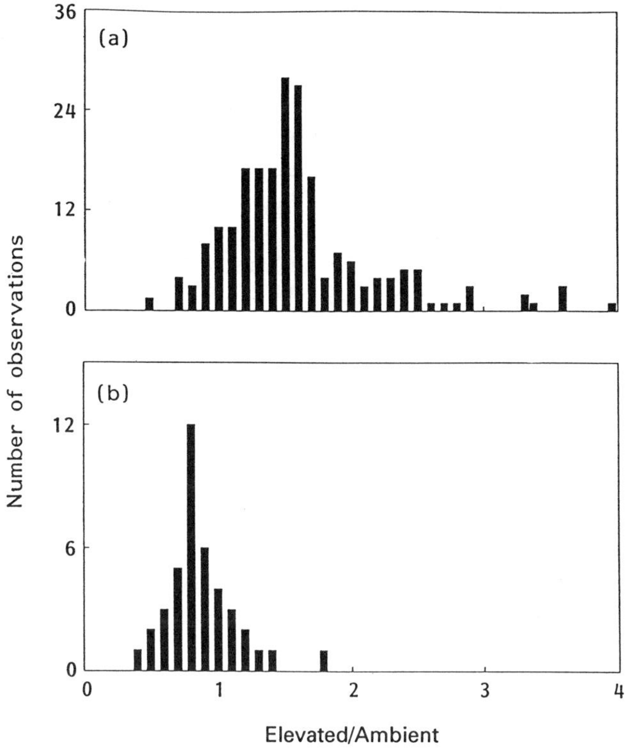

Figure 6.4(a) Frequency distribution of the ratios of leaf photosynthetic rate of CO_2 uptake (A in μmol m^{-2} s^{-1}) for trees grown and measured at elevated c_a relative to measurements on controls grown and measured at current ambient c_a. Each observation represents a single study; **(b)** frequency distribution of the ratios of A for trees grown in elevated c_a and measured at the current ambient c_a to controls grown and measured at the current ambient c_a. This figure is redrawn from Gunderson and Wullschleger (1994) who surveyed studies of the effects of growth in elevated c_a on photosynthesis in 39 tree species

affecting the enhancement effect of an elevated c_a on CO_2-uptake. Indeed this is illustrated by Figure 6.3 which shows that a ~30% decrease in A at a c_a of 350 μmol mol^{-1} would occur in leaves optimally acclimated to a c_a of 700 μmol mol^{-1} without any decrease in A when measured at 700 μmol mol^{-1}, relative to controls.

Eleven of the tree species covered by the review of Gunderson and Wullschleger (1994) were conifers. Average acclimation and enhancement of their photosynthesis after growth at elevated c_a were calculated as above, from the values of A_{sat} provided by Gunderson and Wullschleger (1994). Average photosynthetic enhancement at the growth c_a was 47% and average acclimatory reduction in photosynthesis measured at ambient c_a was 19%, i.e. these conifers showed no obvious difference from the combined values for the 39 species of trees. C_3 grasses also show evidence of photosynthetic acclimation, and enhancement of photosynthesis at elevated c_a similarly appears to persist. Photosynthesis was enhanced by 35–46% in *Lolium perenne* grown at an elevated c_a despite photosynthetic acclimation (Nijs *et al.* 1989; Ryle *et al.* 1992). These observations are consistent with the view that decreased photosynthetic capacity at current ambient c_a following growth at an elevated c_a simply reflects a remobilization or deactivation of excess Rubisco. This is also consistent with the view that Rubisco levels may be modulated to optimize nitrogen-use within the plant. Why then has earlier work suggested that acclimation reflects a decrease in the ability of plants to respond to increased c_a in photosynthesis?

6.3.4 Photosynthetic acclimation and pot size

Artificial limitation of sink size has probably confounded the summaries of effects of CO_2 offered in previous reviews where pot size was not considered (e.g. Cure and Acock 1986). Drawing upon the observations of Herold (1986) that pot size limits sink demand, Arp (1991) examined the relationship between rooting volume used in different experiments and the apparent stimulation of photosynthesis with growth in elevated c_a. Stimulation of photosynthesis in the long term was far more pronounced if the potential rooting volume was large for a wide range of plants. Thomas and Strain (1991) provided experimental evidence that restriction of root development by pot size could accentuate acclimation of photosynthesis to elevated c_a. Long and Drake (1992) accounted for this interaction with rooting volume in summarizing the effects of CO_2 on photosynthesis across a range of studies. Results were separated between those obtained using pots of size greater than 10 dm^3, and those with smaller rooting volumes. The survey of 112 studies suggested that when rooting volume is limited, a downward acclimation of A_{sat} at a given c_i results, but when there is no restriction of rooting volume, A_{sat} remains unchanged. CO_2-uptake from leaves which developed at an elevated c_a was compared to that of leaves developed at current c_a for common values of c_i, approximating c_i at current ambient c_a. Plants grown at elevated c_a in small pots showed statistically significant average reductions in photosynthesis of 21.4 and 12.5% when measured at the ambient and elevated c_a, respectively. Leaf photosynthetic rates of plants grown and measured in elevated CO_2 were on average higher than in plants grown and measured at current c_a. The effect was a 23.3% increase in photosynthesis for plants grown in small pots but a 50.5% increase for plants in large pots. These increases were statistically significant.

Sage (1994) conducted a detailed survey of the effects of growth at elevated c_a on the A/c_i response of photosynthesis. He examined the response of CO_2-uptake of ambient (A_{amb}) and elevated grown plants (A_{elev}) to c_i. To assess acclimation of an assimilation ratio independent of stomatal effects he plotted A_{elev}/A_{amb} against

c_i. He found that most plants (11 of 15) grown in relatively small pots (5 liters or less) had an assimilation ratio less than 1 across a range of c_i, indicating acclimation of both RubP-saturated and RubP-limited carboxylation. Many species (5 out of 13) grown in large pots (10 liters or more) exhibited A/c_i responses indicative of a shift in allocation from Rubisco toward factors limiting RubP-regeneration rate. However, most A/c_i responses indicated no acclimation, or an acclimation response involving proportionate responses in all components limiting to carboxylation. The latter pattern was observed in one field study, whilst two others showed no acclimation.

Results from pot studies may be relevant in natural habitats where rooting volume is limited. Such habitats may include shallow soils, such as those supporting chalk grassland, where the bedrock limits rooting depth; podsols, such as those supporting boreal forest, where an iron pan limits rooting depth; and cliffs and mountainous areas, where small volumes of soil are trapped in cracks in rock and on ledges. It may also have relevance to dense root mats where the physical presence of competitors' roots may limit potential rooting volume.

6.3.5 Photosynthetic components subject to acclimation responses

Analysis of A/c_i responses suggests that acclimation of the photosynthetic response to elevated c_a is uncommon in situations where rooting volume and nutrient availability are not limiting. Where acclimation does occur, it could represent an optimization of Rubisco activity and g_s, but less commonly the capacity for regeneration of RubP. The following sections examine the occurrence and significance of changes in these three potential limitations to CO_2-uptake.

6.3.5.1 Rubisco

The literature survey of Long and Drake (1992) showed that on average, elevated c_a produced a decrease of about 15% in both Rubisco content and Rubisco activity. Rubisco content was significantly decreased by 8% for plants grown with large rooting volumes, and there was also a significant reduction in the proportion of the enzyme which was active. This pattern of decrease in Rubisco content and activity has been clearly demonstrated for rice grown in a range of CO_2 concentrations. Here the activity of Rubisco per unit area and the amount of Rubisco protein were found to decline linearly with increase in c_a, such that the quantity of Rubisco in leaves developed at 900 μmol mol^{-1} was about one-third of that in leaves grown at about 150 μmol mol^{-1} (Rowland-Bamford *et al.* 1991). In contrast, growth of wheat in 550 μmol mol^{-1} under free-air CO_2 enrichment had no effect on the amount of Rubisco at any stage of plant development, except for the final stage of grain-filling (Nie *et al.* 1995).

Acclimation of carboxylation efficiency and hence $V_{c,max}$ may result from a decrease in the total quantity of Rubisco and/or a decrease in the activation state of the enzyme (Long and Drake 1992). In surveying a wide range of studies, Bowes (1991) noted marked between-species variation, but also evidence for and against acclimation within the same species. Whilst the majority of studies have shown a decline in Rubisco as a proportion of total soluble protein with growth in elevated c_a, others found no decrease, and a few an increase (see Long and Drake 1992; Gunderson and Wullschleger 1994). Leaf age also appears to affect acclimation, with young leaves apparently showing less acclimation in Rubisco levels than older ones (e.g. Yelle *et al.* 1989; Besford 1993), although in another study the effect was found to be greater in young leaves (Vu *et al.* 1983).

Given a common, although not complete, pattern of decrease in Rubisco, both as a proportion of total soluble protein and per unit leaf area in response to elevated c_a, the question arises as to what triggers the decrease. One obvious change indicated by almost all studies is an increase in carbohydrate within the leaf, indicating significant increases in both starch and sucrose, in plants grown in both large and small rooting volumes (Long and Drake 1992). Sheen (1990) and Sheen *et al.* (1992) have demonstrated that both sucrose and glucose can produce repression of *rbcs*, the nuclear gene which codes for the small subunit of Rubisco. The possibility that the carbohydrate status of the leaf affects Rubisco levels is attractive since it would explain the different patterns of acclimation observed in plants with different sink size and explain how acclimation of capacity for regeneration of RubP may be coordinated with Rubisco activity (Webber *et al.* 1994).

So far acclimation of Rubisco has only been considered in terms of quantity of protein and active enzyme, i.e. an effective decrease in maximum RubP-saturated velocity of carboxylation and oxygenation ($V_{c,max}$ and $V_{o,max}$). An acclimatory decrease in carboxylation efficiency would also be effected if the specificity of the enzyme for CO_2 relative to O_2 (τ) changed, occurring if the Michaelis constant for CO_2 (K_c) decreased. So far no evidence of decrease in τ as a phenotypic response to growth in elevated c_a has been reported, and the similarity of τ across contrasting species of C_3 terrestrial plants leads Bowes (1991) to suggest that there have been strong selective pressures for the maintenance of a high τ. However, forms of Rubisco with a lower τ are found where the enzyme is confined to environments with a low ratio of concentrations of O_2:CO_2 (Bowes 1993). Whilst τ in C_3 plants is about 100 at 25 °C, τ for Rubisco from seven C_4 plants ranged from 58 to 83 (Jordan and Ogren 1984). This lower specificity would have no effect on the rate of assimilation given the high c_i in the bundle sheath of C_4 plants (Long 1983). Similarly, τ for Rubisco from anaerobic photosynthetic bacteria can be as low as 9, and values for aquatic angiosperms can also be low (Bowes 1993). These results suggest that the selective pressure for maintenance of a high τ is removed in environments where potential for photorespiration is decreased or absent. C_4 plants have evolved from C_3 plants in relatively recent times (Long 1983); it may therefore be assumed that forms of Rubisco with lower τ

have evolved in these plants following the removal of the selective pressure for a high τ. Whilst there is no evidence of decreased τ as a phenotypic response to elevated CO_2, a decrease in τ within populations may result if elevated c_a decreases the selective pressure for maintenance of a high τ.

In summary, a decrease in the quantity of active Rubisco is common in plants grown in elevated c_a, even when the supply of nitrogen is high and rooting volumes large. Decrease in Rubisco quantity corresponds to a decrease in the initial slope of the A/c_i response, i.e. carboxylation efficiency, such that capacity for regeneration of RubP and carboxylation are colimiting. The resulting colimitation suggests that the decrease represents an acclimatory redistribution of resources, into other parts of the photosynthetic apparatus or other parts of the plant. Such reallocation has been observed in coniferous trees and grasses (Stulen and Hartog 1993). As noted in sections 6.3.2 and 6.3.3, enhancement of photosynthesis at growth c_a typically occurs despite acclimation in Rubisco at elevated c_a.

6.3.5.2 Stomata

In contrast to photosynthetic CO_2 assimilation within the mesophyll, the mechanism by which stomata respond to c_a remains unclear. Nevertheless, stomata of most species show predictable closure reactions with increase in c_a, and decrease in A and relative humidity (RH). Ball *et al.* (1987) provide a phenomenological model of the dependency of stomatal conductance on CO_2 concentration (c_s) and humidity (h_s) at the leaf surface, and A. This has proved valid for a wide range of C_3 species. Harley *et al.* (1992) developed a more practical approximation using the ambient RH and CO_2 concentration (c_a):

$$g_s = a + bA\frac{h_s}{c_s} \approx a' + b'A\frac{\text{RH}}{c_a}$$

where a, a', b, and b' are constants characteristic of the given plant material.

Stomatal aperture and conductance (g_s) generally decrease with increase in c_a, explaining the reduction in leaf transpiration commonly observed in plants grown in elevated c_a (Morison 1987). The stomata of conifers tend to respond less to elevated c_a than broad-leaved tree and herbaceous species (Morison 1985; Jarvis 1989). The mechanism by which stomata sense change in c_a is still a matter of debate (Mott 1990). Where stomata represent the major limitation to diffusion in the gas phase, leaf transpiration will decline linearly with decrease in g_s, if VPD remains constant. However, this does not automatically imply a parallel decrease in A. The significance of change in g_s to A is difficult to assess because of the nonlinearity of the A/c_i response. Farquhar and Sharkey (1982) suggested that the limitation (l) imposed on photosynthesis by stomata could be assessed by comparing the rate of CO_2 uptake (A) at a given c_a with that which would be obtained (A_0) if there was no diffusive barrier, i.e. $g_s = \infty$ and therefore $c_i = c_a$:

$$l = \frac{(A_0 - A)}{A_0}$$

Limitation (l) is calculated by determining A and A_0 from the A/c_i response. The theoretical A/c_i responses of Figure 6.3 suggest that leaves developed and measured in a c_a of 350 μmol mol^{-1} would have an l of 0.133. Increase in c_a to 700 μmol mol^{-1} would, by reference to Figure 6.3, decrease l by 64% and g_s by 42%, assuming no acclimation of the curve or assuming a decrease in carboxylation efficiency of up to 30%. If capacity for regeneration of RubP was increased by 65%, so raising the point of inflection of the curve to the operating point when $c_a = 700\,\mu$mol mol^{-1}, l would decrease by 53% and g_s by 8%. Thus, given all scenarios of acclimation expected from theoretical considerations, stomatal limitation would be expected to decline with rising c_a.

Because $\delta A/\delta c_i$ decreases with increased c_i, it follows that maintenance of a constant c_i/c_a will require a decrease in g_s as c_a rises. Typically, an increase in c_a from 350 to 700 μmol mol^{-1} will require about 40% decrease in g_s (Figure 6.3) and this agrees closely with values observed for a range of plants (Eamus and Jarvis 1989; Mott 1990). This is also indicated by decreased transpiration per unit leaf area in plants grown in elevated c_a.

Interpretation of A/c_i responses and predictions of the effects of elevated c_a on CO_2-uptake via models developed from that of Farquhar *et al.* (1980) have assumed that stomatal apertures are homogeneous within leaves. Water stress and treatment of leaves with abscissic acid (ABA) have been shown to induce stomatal heterogeneity (reviewed: Terashima 1992). In heterobaric leaves, i.e. leaves in which the intercellular air space system is divided into discrete patches by the pattern of leaf venation, differential closure of the stomata associated with these patches would result in heterogeneity of c_i and A. Cheeseman (1991) used a computer simulation to determine the effects of heterogeneous stomatal opening on the A/c_i response. His analysis suggested that variation in stomatal aperture between patches must exceed a coefficient of variation of 100% of the mean before the estimated A/c_i response differs significantly from the true response of the mesophyll. One effect of heterogeneity will be to broaden the point of transition between carboxylation and RubP-regeneration limitation, diminishing the advantage of an acclimatory change to optimize the balance between limitations. Decreased transpiration at elevated c_a is likely to improve plant water status and decrease ABA levels. Since declining water status and increased ABA levels are known to induce stomatal heterogeneity, reversal of these changes by increased c_a may tend to decrease the incidence of patchiness rather than *vice versa*. Further, the few field studies so far conducted suggest that stomatal heterogeneity may be rare under natural, as opposed to laboratory induced, drought conditions (Terashima 1992; Wise *et al.* 1992).

The above analyses have assumed that changes in g_s simply reflect changes in the A/c_i response and change in c_a. This ignores the possibility that the direct response of stomatal movement to c_a or the stomatal apparatus may be modified by development at elevated c_a. An acclimatory response of stomatal conductance could involve a change in sensitivity to c_a or a change in stomatal size or number.

ABA increases the sensitivity of the stomata to CO_2 concentration and is known to accumulate in leaves in response to drought. Since transpiration generally falls when c_a is elevated, a lower accumulation of ABA and decreased sensitivity to c_i might be expected. Plants of *Pinus radiata*, *P. menziesii* (Hollinger 1987) and cotton (Harley *et al.* 1992) grown in elevated c_a showed decreased sensitivity of g_s to CO_2. In these plants, decrease in g_s with increase in c_a is less in the plants grown in elevated c_a. Thus, slightly increased rates of photosynthesis and transpiration than those predicted are found. Evidence is contradictory for the observation that stomata acclimate to elevated c_a under well-watered conditions, independently of acclimation of photosynthesis (Eamus 1991; Long and Drake 1992; Sage 1994). Black spruce seedlings grown under elevated c_a had a lower g_s over a wide range of c_i than those grown under ambient c_a (Johnsen 1993).

Acclimation might also occur by change in stomatal numbers, which in the absence of variation in stomatal dimensions, will determine the maximum g_s that a portion of leaf could attain. One expectation might be that at increased c_a fewer stomata are required, since the rate of CO_2 diffusion into the leaf will be a decreasing limitation to photosynthesis as c_a rises. Studies have been conducted investigating responses of stomatal numbers to elevated c_a over experimental periods of weeks to the 1000s of years covered by paleoecological studies. Experimental evidence shows no clear developmental acclimation of stomata. Reported changes in stomatal density with growth at elevated c_a include increases, decreases and no change (Long and Drake 1992; Gunderson and Wullschleger 1994). Long-term studies drawing on herbarium material and paleoecological evidence are more conclusive, showing an inverse relation between variation in c_a and variation in stomatal numbers (Beerling *et al.* 1992; Beerling and Chaloner 1993a, b; Woodward 1993). Pearson *et al.* (1995) have shown an asymmetric response of stomatal conductance between the two surfaces of the leaf of *Rumex obtusifolius* with growth in elevated c_a, with the functional result that a leaf which is amphistomatous at the current c_a would become hypostomatous at elevated c_a. This could effect a substantial increase in instantaneous water use efficiency (ω).

In summary, stomata appear to become less limiting to photosynthesis with elevation of c_a, even though g_s is invariably decreased. Acclimation of stomatal aperture to elevated c_a may occur, allowing greater conductance at a particular c_a. The response could be driven by changes in leaf ABA resulting from reduced transpiration rates. Evidence for acclimation of stomatal development to elevated c_a is conflicting, but evidence showing evolutionary adaptation of stomatal numbers is clearer. Over a longer timescale, continued evolutionary adaptation to rising c_a might be expected in the form of decreased stomatal density.

6.3.5.3 Factors affecting RubP-regeneration rate

At the leaf level, optimization suggests that at elevated c_a, investment would be shifted away from Rubisco toward increased capacity for regeneration of RubP, such that the two remain colimiting, i.e. there is no over-investment in any one part of the photosynthetic apparatus. This is illustrated in Figure 6.3. However, when the plant as a whole is considered when nitrogen is limiting, it may be expected that optimization would redistribute nitrogen away from Rubisco to other parts of the plant, e.g. for root growth, rather than into photosynthetic apparatus for regeneration of RubP. Elevated c_a increases photosynthesis even when it is limited by the rate of regeneration of RubP. Thus, not only is less Rubisco required to maintain a given photosynthetic rate when c_a is elevated, but there is a decreased requirement for capacity for regeneration of RubP. It might therefore be predicted that when nitrogen is strongly limiting, optimization of nitrogen distribution would require a decrease in investment in Rubisco and, to a lesser extent, capacity for RubP-regeneration. If, however, resources are not limiting, then an increase in capacity for RubP-regeneration might be expected (upper line of Figure 6.3). The latter pattern has rarely been observed, whilst decreased capacity has (Sage *et al.* 1989; Sage 1994). Decreased capacity is influenced by the preponderance of studies in which pot sizes may have been limiting, however. Of the three field studies considered by Sage (1994), one showed an increase in capacity for regeneration of RubP, conflicting with the pattern suggested from pot studies.

The model of leaf photosynthesis used in the theoretical predictions of section 6.3.2 is formulated to assume that either Rubisco activity (Wc) or whole chain electron transport (W_j) limit the rate of carboxylation. Although some A/c_i responses (Sage 1994) indicate an increased capacity for regeneration of RubP following acclimation to elevated c_a, there is as yet very little evidence that reduced allocation of nitrogen to Rubisco is matched by any increased allocation to other enzymes of the Calvin cycle or to thylakoid proteins (Mott 1990). Limitation on regeneration of RubP may also be imposed by capacity to utilize the carbohydrate synthesized (Sage 1994), since increase in source strength relative to sink strength is well known to feedback onto photosynthetic rates (Stitt 1991). The sink is conceived as any pool of carbon outside the chloroplast envelope into which the products of photosynthesis are fed. The source strength is the potential of the chloroplasts to provide assimilate for the sinks; following the definition of Patrick (1993). Long and Drake (1992) showed that increased sucrose contents were a consistent feature of plants grown in elevated c_a. Accumulation of sucrose within the leaf has the potential to feed back on the capacity for regeneration of RubP in the short term via cytosolic inorganic phosphate (P_i) concentration and enzyme activation, and in the longer term via gene expression. Accumulation of phosphorylated intermediates of sucrose synthesis in the cytosol occurs when chloroplast formation of sugar phosphates exceeds the rate of sucrose synthesis and export. This may starve the chloroplast of the P_i required for the phosphorylation of Calvin cycle intermediates. In an adaptation of the Farquhar *et al.* (1980) model, Harley and Tenhunen (1991) suggested that P_i-limitation of regeneration of RubP should be considered as a third limitation to CO_2-uptake (W_p) where A will be determined by the minimum of W_j, W_c and W_p. If photosynthesis is limited by return of P_i to the chloroplast, then this will be apparent in C_3 species by a lack of O_2 sensitivity (Sharkey 1985a). The chloroplast envelope triose phosphate translocator exchanges P_i. Decreased return of P_i will inhibit translocation of triose phosphates and promote starch synthesis within the chloroplast. Decreased P_i could also decrease the chloroplast ATP/ADP ratio, which in turn would inhibit Rubisco activase (Bowes 1991), so leading to the decreased Rubisco activities commonly observed in the literature (Long and Drake 1992). P_i-regulation seems to explain many of the changes observed in plants grown in elevated c_a, but it does not provide a universally applicable solution. Empirical evidence suggests that changes in enzymes and metabolites expected with P_i-limitation are not always found when source activity exceeds sink capacity (reviewed: Stitt 1991). An alternative mechanism may be a decrease in activities or quantities of proteins involved in the regeneration of RubP, essentially matching the plant's photosynthetic capacity to its capacity to utilize end-products. Support for this hypothesis comes from the observations that soluble carbohydrates can suppress the expression of a number of photosynthetic genes (reviewed: Webber *et al.* 1994).

Increase in capacity for regeneration of RubP therefore requires an increase in sink strength relative to the increase in potential source strength within the plant. This is consistent with changes for the range of plants grown in elevated CO_2 and large rooting volumes observed in the literature survey by Long and Drake (1992). Below-ground parts are a major sink for carbon, and increases in the average root:shoot ratio suggested may be indicative of whole-plant physiological responses which lead to an increase in plant sink size. The change in source–sink balance and the potential increase in nitrogen uptake coming from an enlarged root system would both tend to offset acclimation of Rubisco and allow increased RubP-regeneration capacity. Efficient use of nitrogen resources in the plant may drive acclimatory changes in the photosynthetic apparatus (see section 6.3.2), and changes in plant nitrogen content may also drive changes in root:shoot ratio. Plant tissue nitrogen content tends to be diluted at an elevated c_a because of increases in carbon uptake not matched by nitrogen uptake (Long and Drake 1992). This may also be partly due to a decreased requirement for Rubisco and possibly apparatus for the regeneration of RubP with growth in elevated c_a. Decreases in plant tissue nitrogen content caused by growth in elevated c_a with a limiting nitrogen-supply commonly correspond with an increase in root:shoot ratio (Gutschick 1993). Increases in root:shoot ratio at elevated c_a may thus be a feedback response which attempts to rectify the relative decrease in nitrogen. If so, then total nitrogen in the vegetation, per unit ground area, should rise despite an increase in the C/N ratio of that tissue with growth at elevated c_a. This has been observed with growth of a C_3 tidal marsh community grown over two years at elevated c_a in the field (Curtis *et al.* 1989). Other morphological changes suggesting an increased sink size in elevated c_a are reported in the literature. These include increased tillering in a tundra sedge (Tissue and Oechel 1987), increased main shoot development in trees (Oberbauer *et al.* 1985), induction of flowering in trees (Mousseau and Saugier 1992), and increased root nodule activity in legumes (Idso 1989).

In summary, there is limited evidence that acclimation of photosynthesis to elevated c_a can involve an increase in capacity for the regeneration of RubP. Variation in response might be explained by variation in sink strength and the ability of sink strength to increase in elevated c_a. Whilst limitation on RubP-regeneration could theoretically be imposed through P_i-limitation, this is inconsistent with a number of biochemical and physiological observations. Other mechanisms which cause variation in RubP-regeneration should be sought, particularly those involving modulation of levels of proteins involved in the regeneration of RubP.

6.4 INTERACTIONS BETWEEN CO_2 AND OTHER ENVIRONMENTAL PARAMETERS

6.4.1 Temperature

Theoretical considerations lead to the view that simulation of instantaneous CO_2-uptake by elevated c_a will be relatively small at low temperatures, but large at high temperatures (Long 1991). The view is supported both by mechanistic modeling of leaf and canopy photosynthesis, and by empirical evidence.

6.4.1.1 Leaf light-saturated photosynthesis

According to the mechanistic model described in section 6.2, increase in c_a produces a progressive increase in A_{sat}, accentuated at higher leaf temperature (Figure 6.5). On elevation of c_a from 350 to 550 μmol mol^{-1}, an elevated c_a commonly used in experimental studies, A_{sat} is predicted to increase by 14, 32 and 57% at leaf temperatures of 10, 20 and 30 °C, respectively (Figure 6.5). It also follows from this interaction of c_a and temperature (T) that the temperature optimum (T_{opt}) of A_{sat} must increase with c_a. T_{opt} is predicted to increase by 2, 5 and 6 °C with increase in c_a to 450, 550 and 650 μmol mol^{-1}, respectively. The predicted upper temperature at which a positive A_{sat} may be maintained is

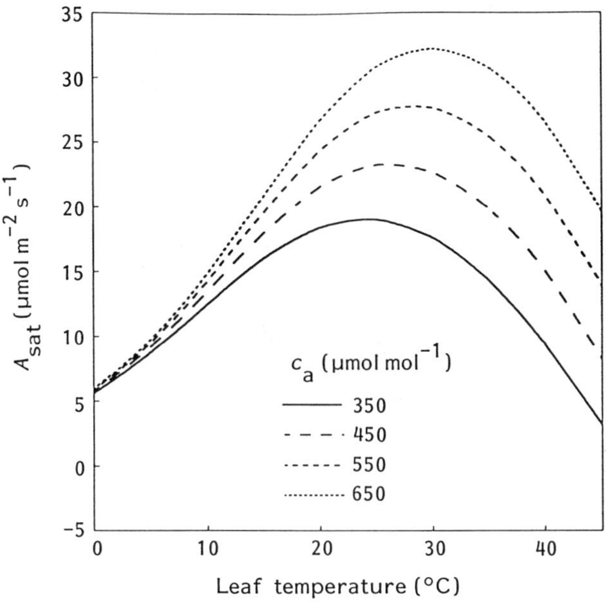

Figure 6.5 Predicted light-saturated rates of leaf CO_2-uptake (A_{sat} in μmol m^{-2} s^{-1}) with leaf temperature (°C) for four atmospheric CO_2 concentrations (c_a, μmol mol^{-1} of CO_2 in air)

similarly increased. The change in these characteristic temperatures underlies the importance of considering rise in c_a not just as a factor which increases photosynthetic rate, but also as one which strongly modifies the response to T. When the interactive effects of increased c_a and temperature are taken into account, a small increase in A_{sat} is indicated on increasing leaf temperature from 25 to 30 °C. If T was assumed to affect A_{sat} independently of c_a, i.e. the optimum remains unchanged at elevated c_a, then it would have been predicted that A_{sat} would fall rather than increase between 25 and 30 °C. Thus, failure to take account of the interaction would not only fail to predict the magnitude of change in photosynthesis with T but also the direction of change. Increase in T_{opt} with rising c_a is of particular significance to expected change in climate. The 'IPCC business-as-usual scenario' suggests an increase in mean temperatures of ~3 °C, averaged across latitudes, in the second half of the next century (Watson *et al.* 1990). In temperate climates this would increase the frequency of temperatures which are supraoptimal for A_{sat}. The results suggest that this inhibitory effect will be partially negated if the temperature increase coincides with an increase in c_a to 500 μmol mol^{-1}, as projected (Leggett *et al.* 1992), since this increase in c_a would apparently increase T_{opt} by 3 °C. The modeled interactive increase in A_{sat} and T_{opt} with elevated c_a and increasing temperature is very similar to that observed with

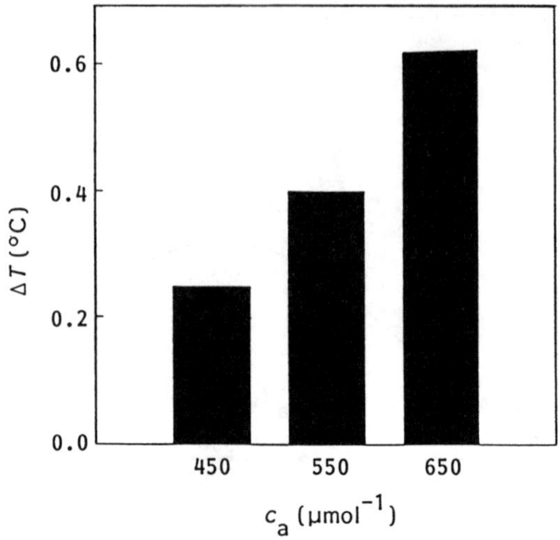

Figure 6.6 Predicted temperature difference (ΔT, °C) between leaves at an atmospheric CO_2 concentration (c_a, μmol mol^{-1} of CO_2 in air) of 350 μmol mol^{-1}, i.e. current ambient, and leaves in a c_a of 450, 550 or 650 μmol mol^{-1}. A total radiant energy flux of 900 W m^{-2}, an air temperature of 25 °C and a relative humidity of 0.7 are assumed. Leaf absorptance of 0.8 and emissivity of 0.9 are assumed. Stomatal conductance and energy balance are calculated by the procedures described in section 6.2

elevation of c_a and temperature for C_3 species (Berry and Björkman 1980; Pearcy and Björkman 1983; Sage and Sharkey 1987).

Predicted temporal patterns of temperature increase with global warming suggest that in the boreal forest biome, temperature increase will be large in autumn and spring (Mitchell *et al.* 1990), when T is suboptimal for A_{sat}. Thus, in this situation the interaction of rising c_a and T will be strongly synergistic. Growth of plants in different temperatures can result in an acclimatory shift in the response to A_{sat} to T. However, the relative stimulation of A resulting from increase in c_a would remain unchanged (Berry and Björkman 1980).

Will an increase in c_a, in the absence of any increase in air temperature, affect leaf temperature? As noted in section 6.1.1, transpiration per unit leaf area is decreased at elevated c_a, and thus latent heat loss will similarly be decreased. From energy budget considerations, increase in c_a from 350 to 550 μmol mol^{-1} with a photon flux of 1500 μmol m^{-2} s^{-1} would increase the temperature of a 'typical' C_3 leaf by 0.4 °C (Figure 6.6). Empirical evidence from three field-rooted species grown in open-top chambers suggests a strong positive correlation between reduction in g_s at elevated c_a, and concomitant leaf temperature rises (Idso *et al.* 1993). Leaf temperature increases have also been measured at elevated c_a under the natural micrometeorological conditions provided by free-air CO_2 enrichment (FACE) of a cotton crop. Here foliage temperature was on average 0.8 °C warmer at 550 μmol mol^{-1} c_a than at current ambient c_a (Kimball *et al.* 1992).

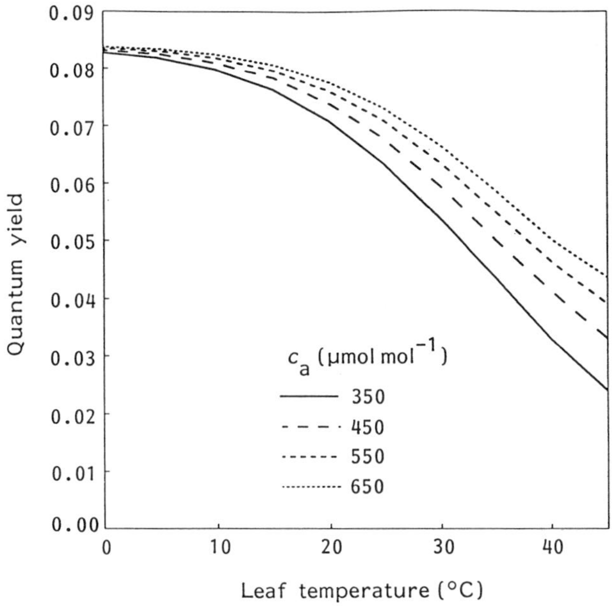

Figure 6.7 Predicted response of the maximum quantum yield of CO_2-uptake (ϕ) to leaf temperature (°C) at four atmospheric CO_2 concentrations (c_a, in μmol mol^{-1} of CO_2 in air)

Thus, even in the absence of any increase in mean air temperatures, some increase in mean leaf temperature is expected. For leaves which commonly experience suboptimal temperatures, e.g. boreal forest trees and temperate grassland species in the spring, this indirect effect of elevated c_a would produce a further stimulation of photosynthetic rate (Long 1991; Long and Drake 1992; see Figure 6.5).

6.4.1.2 Light-limited leaf photosynthesis

Experimental observation has established that ϕ in C_3 species decreases with increase in T (Ehleringer and Björkman 1977). The biochemical basis of this decrease is that increased temperature decreases the ratio of carboxylation to oxygenation rates (V_c/V_o). With a decrease in V_c/V_o an increasing proportion of the limiting supply of NADPH and ATP produced by electron transport is diverted into photorespiration. Because CO_2 is a competitive inhibitor of the oxygenation reaction, progressive increase in c_a decreases the rate of decline in V_c/V_o, and hence ϕ. Figure 6.7 illustrates from the mechanistic model WIMOVAC how various increases in c_a would affect the efficiency of light-limited photosynthesis at a range of temperatures. Long and Drake (1991) showed for plants grown for three years at elevated c_a, that ϕ in shoots grown and measured at a c_a of 700 μmol mol^{-1} at a leaf temperature of 25 °C was 0.080, 25% higher than the value of 0.064 for shoots grown and measured at the current ambient c_a. This increase agrees closely with the predictions modeled from consideration of the underlying biochemistry (Figure 6.7). Since ϕ declines with temperature, increased temperature will in part counteract the individual effect of increased c_a. For example, if the increase in c_a were accompanied by a 4 °C increase in leaf temperature, then ϕ would increase to only 0.077 with a doubling of c_a, i.e. a 20% as opposed to a 25% increase. Unlike light-saturated photosynthesis, light-limited photosynthetic rate by definition depends on the quantity of light received and the efficiency with which it is transformed, and becomes independent of both quantities of Rubisco and capacity for regeneration of RubP. The light-limited rate of CO_2-uptake increases with increase in c_a, simply because the efficiency of energy transformation is increased by the inhibition of photorespiration. At a constant c_a, quantum efficiency has been shown to be remarkably constant in C_3 terrestrial plants regardless of their taxonomic and ecological origins (Long *et al.* 1993). This experimental observation reflects the constancy of the photosynthetic mechanism across C_3 species. Thus, the rate of light-limited photosynthesis may be predicted mechanistically for C_3 vegetation without any need for parameterization, in contrast to light-saturated photosynthesis.

If R_d remains constant and ϕ rises with increase in c_a, it follows that the LCP must decrease. At all temperatures, LCP is depressed by increase in c_a; again the effect being greatest at high temperatures (e.g. Long and Drake 1992). Such decreases are of considerable significance in warm and shaded environments, where they would significantly alter the number of leaf layers that a canopy could maintain above the LCP for net carbon gain. They would greatly increase the photosynthetic productivity of forest ground-flora and understorey plants existing at photon fluxes which are little above the LCP in the current ambient c_a. Decreases in LCP in accordance with these predictions were observed in *Scirpus olneyi* growing in an elevated c_a (Long and Drake 1991).

6.4.1.3 Canopy photosynthesis

The mechanistic model WIMOVAC, described in section 6.2, uses predictions of A to calculate net canopy CO_2 accumulated over 24 hours ($A_{c,tot}$), where the quantity of an angular distribution of light is continually varying with change in sun angle. For a canopy of leaf area index 5, a substantial portion of the leaf area will be shaded even during the course of a sunny summer day, and thus the canopy response to T will not be dependent solely on the response of light-saturated leaf photosynthesis (A_{sat}) to temperature. Nevertheless, the pattern of response of $A_{c,tot}$ to temperature with increase in c_a is similar to that described for A_{sat} (Figure 6.8). For a rise in c_a from 350 to 550 μmol mol^{-1}, $A_{c,tot}$ is predicted to increase by 15, 30 and 74% at mean daily temperatures of 10, 20 and 30 °C, respectively for a summer day at latitude 52° N. The optimum mean daily temperature of $A_{c,tot}$ is predicted to rise by 10 °C and become more sensitive to temperature. This is a larger increase than predicted at the single leaf level (Figure

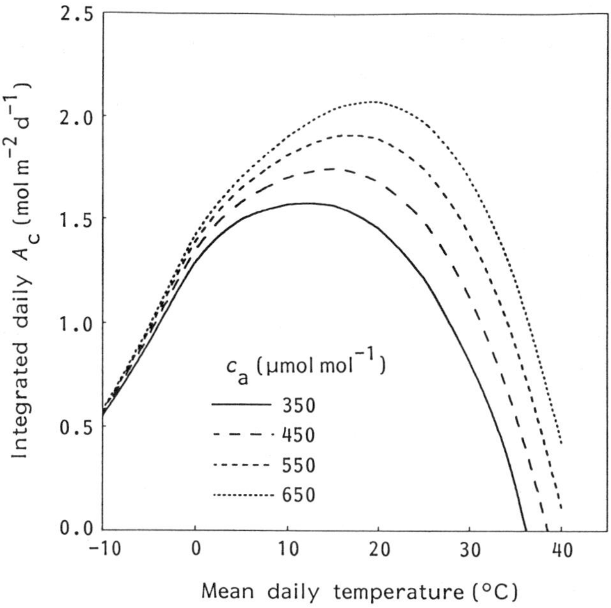

Figure 6.8 Predicted rates of canopy CO_2-uptake integrated over a 24 hr period ($A_{c,tot}$) with mean daily temperature (°C) and four atmospheric CO_2 concentrations (c_a, in μmol mol^{-1} of CO_2 in air). Figures are for a canopy of leaf area index of 5 and a ratio of horizontally to vertical projected leaves of 1, on Julian day 190, assuming clear sky conditions (atmospheric transmittance = 0.75) at a latitude of 0°. A diurnal temperature range of ± 5 °C is also assumed

6.5) and may be accounted for by the differing temperature relations of light-saturated and light-limited photosynthesis (cf. Figure 6.7), both of which affect the daily integral of canopy CO_2-uptake. However, as suggested at the leaf level, the predictions for canopy photosynthesis suggest that the enhancement of canopy photosynthesis induced by elevated c_a will be minimal in cold climates and maximal in the hottest climates (Figure 6.8). This expectation is consistent with experimental observations. Grulke *et al.* (1990) found little stimulation of net canopy CO_2-uptake (A_c) in Arctic tussock tundra, whilst model pasture ecosystems in N. Germany and a C_3 wetland vegetation in Maryland showed average increases of 25–45 and 88% respectively, with a doubling of c_a in the field (Overdieck 1990; Drake and Leadley 1991).

6.4.1.4 Conclusion

Theoretical considerations suggest that the relative stimulation of both light-limited and light-saturated C_3 photosynthesis resulting from elevated c_a will increase with temperature. These are consistent with field observations. The conserved nature of the mechanisms of C_3 photosynthesis and its temperature relations provides a strong basis for the development of canopy and vegetation scale models across biomes. The validity of this approach will depend largely on the significance and understanding of second-order effects determined by nutrient limitation, water availability and acclimation, which may alter light-saturated photosynthesis.

6.4.2 Nutrient limitation

Primary production in natural and crop ecosystems is commonly limited, at least for a part of the growing season, by nutrient supply; most commonly the supply of nitrogen or phosphorus. Elevated c_a over long periods could lead to secondary feedbacks at the ecosystem level, and impose further nutrient limitation (Bazzaz 1990). Indeed this has led to the suggestion that nutrient limitation would negate any stimulation of carbon gain, in the longer term, with rising c_a. However, two factors may act against this. First, photosynthesis is more resource-use efficient in elevated c_a because photorespiration is inhibited, so increasing the overall efficiency of net photosynthesis. Second, continued anthropogenic release of nitrogen oxides and ammonia to the atmosphere is increasing nitrogen in many ecosystems (Morris 1991). The interaction between nutrient limitation and elevated c_a on photosynthesis could be of great importance when predicting future photosynthetic production. What might be expected if there is a decrease in either the availability of nitrogen or the amount per unit leaf area?

6.4.2.1 Theoretical expectations

Field (1983) and Harley *et al.* (1992) proposed a linear relationship between leaf nitrogen content and both the maximum activity of Rubisco ($V_{c,max}$) and capacity for regeneration of RubP (J_{max}). Figure 6.9 couples these relationships with the mechanistic model of photosynthetic response to c_a. It should be noted that this is a worst case scenario, since it ignores the possibility that there might be a decline in $V_{c,max}$ without a decline in J_{max} (see section 6.3.2). Figure 6.9 suggests that whilst A_{sat} declines almost linearly with leaf nitrogen content, the proportionate increase in photosynthesis with increase in c_a is predicted to be similar at all nitrogen contents from 0.5 to 4 g [N] m^{-2}. However, if growth at elevated c_a leads to decreased leaf nitrogen contents over prolonged periods, then Figure 6.9 predicts that the gain resulting from increase in c_a will be partially offset by the decreased nitrogen content.

Will the gain through the direct effect of increased c_a on photosynthesis be fully offset by decrease in leaf nitrogen content? This will depend first on the extent of nitrogen decrease and secondly on temperature. Decrease in nitrogen will affect photosynthetic capacity at all temperatures, but the direct stimulation of photosynthesis by increased c_a will be greater at higher temperatures. Figure 6.10

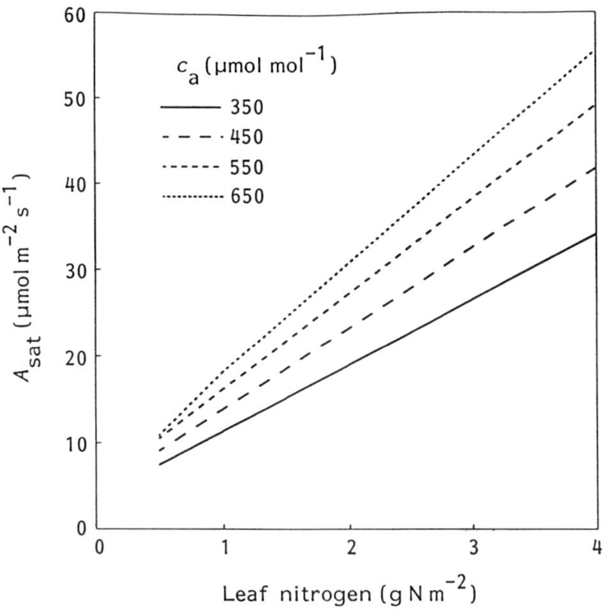

Figure 6.9 The predicted response of CO_2-uptake (A, μmol m^{-2} s^{-1}) to leaf nitrogen content (g N m^{-2}) at four leaf-N concentrations. A leaf temperature of 25 °C, relative humidity of 0.7 and photon flux of 1500 μmol m^{-2} s^{-1} are assumed

shows that if leaf nitrogen is decreased by 50% with growth at 500 μmol mol^{-1} c_a, then net photosynthetic CO_2-uptake would be lower in the elevated c_a than at the current ambient concentration at all temperatures below 30 °C. At 5 °C there would be a c_a. 40% loss of photosynthetic rate. However, the leaf could lose 40% of its nitrogen with growth in elevated CO_2 at temperatures above 14 °C, and still show a 50% increase in A at 35 °C.

6.4.2.2 Empirical evidence

Recent literature reviews suggest that photosynthetic acclimation to elevated c_a tends to be more pronounced under nutrient limitation (Bazzaz 1990; Gunderson and Wullschleger 1994; Sage 1994), although there are examples of the converse, i.e. acclimation is less pronounced or absent under nutrient limitation (Bunce 1992; Yuncong and Gupta 1993). Nutrient limitation has been suggested as the driving force for acclimation in general, since acclimation tends to increase nitrogen-use efficiency, but experimental evidence supporting the hypothesis is not conclusive (Bunce 1992; Gunderson and Wullschleger 1994).

Photosynthetic response was correlated with nitrogen-availability in *Pinus*

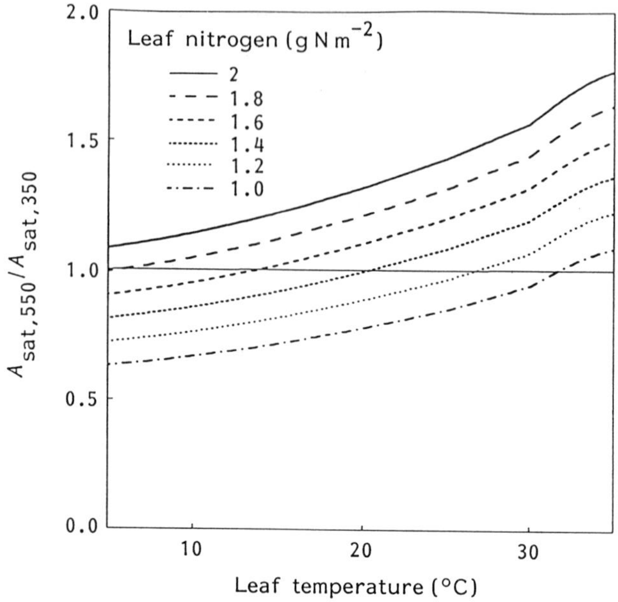

Figure 6.10 The upper line illustrates the ratio of CO_2-uptake (A, μmol m^{-1} s^{-1}) at the current ambient CO_2 concentration (c_a = 350 μmol mol^{-1}) relative to A at an elevated c_a of 550 μmol mol^{-1}, versus temperature, for leaves with a nitrogen content of 2 g m^{-2}. Growth in elevated c_a may lead to a dilution of leaf nitrogen. The lower lines compare A at c_a = 350 μmol mol^{-1} with A at c_a = 550 μmol mol^{-1}, but illustrate the effect of a progressive decrease in the nitrogen content of the leaves grown at the elevated c_a

taeda seedlings; acclimation being greatest where nitrogen-limitation was the greatest. In the same study, phosphorus-limited plants with an ample nitrogen-supply showed a greater enhancement of photosynthesis under elevated c_a than plants with an ample phosphorus supply (Tissue *et al.* 1993).

Studies of cotton suggest that the type of acclimation response occurring can depend on nutrient supply (Wong 1979). Plants grown at a c_a of 640 μmol mol^{-1} and a strongly limiting supply of nitrate showed a marked decrease in the initial slope of their A/c_i curve without any increase in the c_i-saturated rate, relative to plants grown in a c_a of 330 μmol mol^{-1}. Plants grown with a good supply of nitrate and in a c_a of 640 μmol mol^{-1} showed little decrease in the initial slope of the A/c_i curve relative to those grown in a c_a of 330 μmol mol^{-1}, but an increase in A_{sat} at high c_i, indicating an increased capacity for regeneration of RubP.

6.4.2.3 *Conclusion*

Although it is commonly suggested that nutrient limitation may negate the stimulation of carbon gain by elevated c_a (e.g. Melillo *et al.* 1990), theoretical

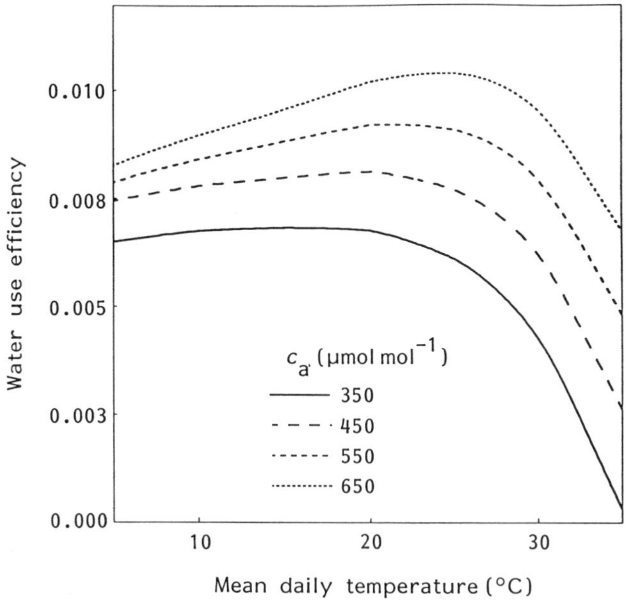

Figure 6.11 The predicted daily water use efficiency with respect to air temperature for a canopy of leaf area index 5 in a range of CO_2 concentrations (c_a, μmol mol^{-1}). A constant relative humidity of 0.7 is assumed. Other assumptions are as in Figure 6.9

considerations suggest that elevated CO_2 will allow C_3 plants to be more efficient in their use of nutrients. In a doubled c_a, an average C_3 leaf could lose 30–40% of its Rubisco activity before the light-saturated rate of photosynthesis is affected. Decreased Rubisco levels in leaves developed in elevated c_a, rather than reflecting a loss of photosynthetic capacity, may represent a redistribution of limiting resources, in particular nitrogen (Webber *et al.* 1994).

6.4.3 Interaction with water

6.4.3.1 Water use efficiency

Increased CO_2-uptake, and decreased transpiration due to reduced g_s, would both lead to improvements in water use efficiency with rising c_a. Instantaneous water-use efficiency (ω) is defined as the ratio of the rate of CO_2-uptake to rate of water loss. Figure 6.11 uses WIMOVAC to predict the daily ω for a canopy at four c_a with air temperatures from 5 to 35 °C, assuming a constant RH of 0.7 in the ambient air. Instantaneous water-use efficiency is predicted to rise with increase in c_a at all temperatures, but the relative rise due to elevated c_a is predicted to increase with temperature. The predicted rise in ω at a c_a of 550 relative to 350

μmol mol^{-1} is 83% at 35 °C, but only 21% at 10 °C. Figure 6.11 suggests a near linear increase in ω with increase in c_a at any given temperature, in agreement with the relationship described by Morison (1993) in a survey of studies covering 20 species.

Although there are no field data on the interactive effects of elevated c_a and temperature, there is much information showing an increase in ω. Elevated c_a leads to increased water use efficiency, especially in the short term, and this can result in improved drought-resistance through avoidance of water-stress. Increased CO_2-uptake and decreased transpiration due to reduced g_s at elevated c_a both lead to improvements in ω. Averaging over 29 observations of 20 C_3 species, Morison (1993) found that ω was approximately doubled in plants grown at elevated c_a. Improvements in ω can lead to the improved water status of plants grown at elevated c_a when subjected to water-stress. Black spruce seedlings were more drought-tolerant when grown at elevated c_a, as indicated by a higher xylem pressure potential after a drought episode (Johnsen 1993). Enhancement of net photosynthesis for plants growing in the elevated c_a increased from 33 to 333% in drought-treated plants. An interactive effect also occurred in g_s, where a 28% reduction at elevated c_a in well-watered plants compared to no difference between plants grown at elevated and control c_a under drought conditions. Similarly, Nie *et al.* (1992) examined the interactive effect of mild water-stress and elevated c_a on C_3 grasses in a natural prairie ecosystem. Again, leaf transpiration rate was not affected by elevated c_a in water-stressed plants, but was reduced in well-watered plants because of a reduction in g_s at elevated c_a. Both results might be explained by the initial decrease in transpiration at elevated c_a. This would decrease the rate of soil water depletion so that soil and plant water status will be better in the elevated c_a treatment as the drought develops. These plants might therefore be able to maintain a water status sufficient to attain the maximal stomatal aperture for the given c_a for a longer period into the drought. Indeed, Kimball *et al.* (1995) have recently shown under free-air CO_2 enrichment on a field scale, that for wheat grown under drought conditions (i.e. 50% of the water required for maximum yield) the enhancement of dry matter production and grain yield by elevated c_a, relative to plants grown at the current ambient c_a, was greater than in plants grown with 100% of the water required for maximum yield. On average, the increase in c_a to 550 μmol mol^{-1} decreased the rate of soil moisture depletion and crop water loss by ca. 10%. If however, increased c_a is accompanied by increased air temperature, then this gain may be offset by an increased water vapor pressure deficit (VPD).

The literature survey of Long and Drake (1992) suggested that where rooting volumes are large, acclimatory increases in water use efficiency are smaller than in plants grown with small rooting volumes. These results suggest a balance between increased water use efficiency and photosynthesis, where increase in water use efficiency will be least in elevated c_a when other resources allow the plant to make maximum use of the increased supply of CO_2. This is supported by

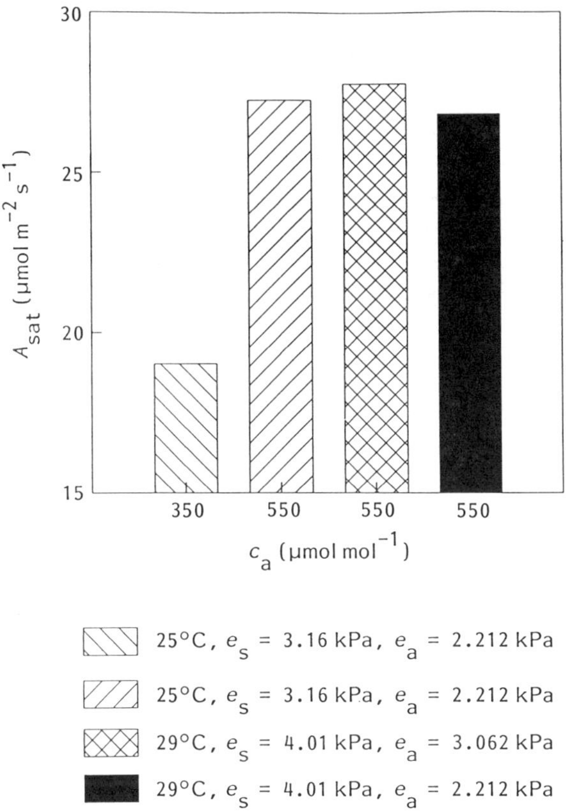

Figure 6.12 Predicted light-saturated rate of leaf CO_2-uptake (A_{sat}) at a photon flux of 1500 μmol m^{-2} s^{-l}, illustrating the effect of an increase in c_a from 350 to 550 μmol mol^{-1} with different assumptions about temperature and humidity change. First, no change in temperature or VPD, i.e. the saturation (e_s) and ambient humidity (e_a) difference (VPD = 0.9 kPa). Second, with a 4 °C rise in temperature and a concurrent increase in e_a to maintain a constant VPD. Third, with a 4 °C increase in temperature with no increase in e_a, which would increase the VPD to 1.8 kPa

the observations of Wong (1979) on cotton where stimulation of CO_2-uptake at the lowest nitrogen fertilization rate was least and increase in water-use efficiency greatest.

Decreased transpiration may add an important complication to attempts to identify stomatal acclimation to elevated c_a. If canopy transpiration is decreased, then soil moisture may remain higher, inducing a secondary difference in plant water status, which might be interpreted as stomatal acclimation to elevated c_a.

6.4.3.2 *Rising water-air vapor pressure deficit (VPD)*

Because the saturation humidity of the atmosphere (e_s) rises almost exponentially with temperature, it is likely that an increase in mean global air temperature will also increase VPD, i.e. the difference between actual and saturation humidity. This will be further increased if leaf-air temperature differences increase. An increase in VPD might also result if increased c_a can inhibit transpiration on a regional scale; on the interior of land masses transpiration is the major source of water vapor in the air. Stomatal conductance often declines with increase in VPD. If photosynthetic capacity remains unchanged, then this increase in VPD would force a decrease in c_i, and hence A. How significant would this decrease be and could it offset the increase expected from the direct effects of elevated c_a on photosynthesis? Figure 6.12 explores this question with WIMOVAC. If an increase of c_a to 550 μmol mol^{-1} is concurrent with an increase in mean air temperature of 4 °C, then in the worst case scenario there would be no increase in ambient humidity. As a result, VPD would rise. Figure 6.12 shows that if there was no increase in e_a, and assuming an RH of 0.7 at 25 °C, with an increase in temperature of +4 °C, VPD would rise from 0.9 to 1.8 kPa. However, this marked increase in VPD is predicted to remove only 3.5% of the gain in A that would result if c_a and T are simultaneously increased from 350 to 550 μmol mol^{-1} and 25 to 29 °C, respectively.

If acclimation of stomatal aperture or density to elevated c_a occurs, it could lead to even greater benefits under water stress. Another acclimatory response of stomatal conductance to elevated c_a could involve a change in sensitivity to humidity. Sage (1994) cites evidence from crop plants suggesting that stomata close to a greater degree in response to water stress in plants grown at elevated c_a.

6.5 CONCLUSION

In considering the effects of rising c_a on the carbon cycle in terrestrial biomes, photosynthesis occupies a unique position as the point of energy input into the cycle. Photosynthesis is also unique as the one process in which direct effects at the substrate level can be anticipated. The instantaneous effects of c_a on photosynthesis are well understood and the mechanisms highly conserved across terrestrial vegetation. This has provided a sound platform for the development of mechanistic models that are valid at scales from the cell to the landscape. Exploration of the mechanisms show that it is unreasonable to describe the effects of rising CO_2 on photosynthetic carbon uptake by use of a simple multiplier, such as the β-factor. Interactions with temperature, nitrogen and water can all markedly modify the degree and even direction of change. This is further complicated by the effects of growth at elevated c_a on the photosynthetic apparatus. Although much of the ‘down-regulation’ of photosynthesis may be explained as an artifact of the pot size used or the measurements made, there are species differences and clear interactions with nutrient supply. Understanding these species differences and nutrient effects will be critical to the development of sound models for the prediction of the long-term changes of photosynthetic carbon uptake by coniferous forests and grasslands.

6.6 REFERENCES

Ågren, G. I., McMurtrie, R. E., Parton, W. J., Pastor, J. and Shugart, H. H. (1991) State-of-the-art models of production–decomposition linkages in conifer and grassland ecosystems. *Ecol. Appl.* **1**, 118–138.

Allen, L. H. (1990) Plant responses to rising carbon dioxide and potential interactions with air pollutants. *J. Environ. Qual.* **19**, 15–34.

Arp, W. J. (1991) Effects of source–sink relations on photosynthetic acclimation to elevated CO_2. *Plant Cell Env.* **14**, 869–876.

Ball, J. T., Woodrow, I. E. and Berry, J. A. (1987) A model predicting stomatal conductance and its contribution to the control of photosynthesis under different environmental conditions. In: Biggins, J. (Ed.) *Progress in Photosynthesis Research*, pp. 221–224. Nijhoff, Dordrecht.

Bazzaz, F. A. (1990) The response of natural ecosystems to the rising global CO_2 levels. *Ann. Rev. Ecol. Syst.* **21**, 167–196.

Beerling, D. J. and Chaloner, W. G. (1993a) The impact of atmospheric CO_2 and temperature change on stomatal density: observations from *Quercus robur* lammas leaves. *Ann. Bot.* **71**, 231–235.

Beerling, D. J. and Chaloner, W. G. (1993b) Stomatal density responses of Egyptian *Olea europaea* L. leaves to CO_2 change since 1327 BC;. *Ann. Bot.* **71**, 431–435.

Beerling, D. J., Chaloner, W. G., Huntley, B., Pearson, J. A., Tooley, M. J. and Woodward, F. I. (1992) Variations in the stomatal density of *Salix herbacea* L. under the changing atmospheric CO_2 concentrations of late- and post-glacial time. *Phil. Trans. R. Soc. Lond. B* **336**, 215–224.

Berry, J. and Björkman, O. (1980) Photosynthetic response and adaptation to temperature in higher plants. *Ann. Res. Plant Physiol.* **31**, 491–543.

Besford, T. (1993) Photosynthetic acclimation in tomato plants grown in high CO_2. *Vegetatio* **104/105**, 441–448.

Bowes, G. (1991) Growth at elevated CO_2: photosynthetic responses mediated through Rubisco. *Plant Cell Env.* **14**, 795–806.

Bowes, G. (1993) Facing the inevitable: plants and increasing atmospheric CO_2. *Ann. Rev. Plant Phys. Mol. Biol.* **44**, 309–332.

Bunce, J. A. (1990) Short and long term inhibition of respiratory carbon dioxide efflux by elevated carbon dioxide. *Ann. Bot.* **65**, 637 642.

Bunce, J. A. (1992) Light, temperature and nutrients as factors in photosynthetic adjustment to an elevated concentration of carbon dioxide. *Phys. Plant.* **86**, 173–179.

Campbell, G. S. (1977) *An Introduction to Environmental Biophysics*. Springer, New York.

Cheeseman, J. (1991) PATCHY: simulating and visualizing the effects of stomatal patchiness on photosynthetic CO_2 exchange studies. *Plant Cell Env.* **14**, 593–600.

Conroy, J. P., Kuppers, M., Kuppers, B., Virgona, J. and Barlow, W. R. (1988) The influence of CO_2 enrichment, phosphorus deficiency and water stress on growth, conductance and water use of *Pinus radiata* D. Don. *Plant Phys.* **II**, 91–98.

Cure, J. D. and Acock, B. (1986) Crop responses to carbon dioxide doubling: a literature survey. *Agric. For. Met.* **38**, 127–145.

Curtis, P. S., Drake, B. G.and Whigham, D. F. (1989) Nitrogen and carbon dynamics in C3 and C4 estuarine marsh plants grown under elevated CO_2 *in situ. Oecologia* **78**, 297–301.

Drake, B. G. and Leadley, P. W. (1991) Canopy photosynthesis of a crops and native plant communities exposed to long-term elevated CO_2. *Plant Cell Env.* **14**, 853–860.

Eamus, D. (1991) The interaction of rising CO_2 and temperatures with water use efficiency: commissioned review. *Plant Cell Env.* **14**, 843–852.

Eamus, D. and Jarvis, P. G. (1989) The direct effects of increase in the global atmospheric CO_2 concentration on natural and commercial temperate trees and forests. *Adv. Ecol. Res.* **19**, 1–55.

Edwards, G. E. and Walker, D. A. (1983) *C3, C4 Mechanisms, Cellular and Environmental Regulation of Photosynthesis.* Blackwell, Oxford.

Ehleringer, J. and Björkman, O. (1977) Quantum yield for CO_2 uptake in C3 and C4 plants. *Plant Physiol.* **59**, 86–90.

Esser, G. (1987) Sensitivity of global carbon pools and fluxes to human and potential climatic impacts. *Tellus* **39B**, 245–260.

Evans, J. R. and Farquhar, G. D. (1991) Modeling canopy photosynthesis from the biochemistry of the C3 chloroplast. In: Boote, K. J. and Loomis, R. S. (Eds) *Modeling Crop Photosynthesis–from Biochemistry to Canopy*, pp. 1–16. Crop Science Society of America, Inc., Madison, Wis.

Farquhar, G, D. and Sharkey, T. D. (1982) Stomatal conductance and photosynthesis. *Ann. Rev. Plant Physiol.* **33**, 317–345.

Farquhar, G. D. and von Caemmerer, S. (1982) Modeling of photosynthetic responses to environmental conditions. In: Lange, O. L., Nobel, P. S., Osmond, C. B. and Ziegler, H. (Eds) Physiological Plant Ecology. 11. *Encyclopaedia of Plant Physiology*, New Series, pp. 548–577. Springer, Berlin.

Farquhar, G. D., von Caemmerer, S. and Berry, J. A. (1980) A biochemical model of photosynthetic CO_2 assimilation in leaves of C3 species. *Planta* **149**, 78–90.

Field, C. (1983) Allocating leaf nitrogen for the maximization of carbon gain: leaf age as a control on the allocation program. *Oecologia* **56**, 341–347.

Forseth, I. N. and Norman, J. M. (1993) Modeling of solar irradiance, leaf energy budget and canopy photosynthesis. In: Hall, D. O., Scurlock, J. M. O., Bolhár-Nordenkampf, H. R., Leegood, R. C. and Long, S.P. (Eds) *Photosynthesis and Productivity in a Changing Environment.* pp. 207–219. Chapman & Hall, London.

Grulke, N. E., Rieckhers, G. H., Oechel, W. C., Hjelm, U. and Jaeger, C. (1990) Carbon balance in tussock tundra under ambient and elevated atmospheric CO_2. *Oecologia* **83**, 485–494.

Gunderson, C. A. and Wullschleger, S. D. (1994) Photosynthetic acclimation in trees to rising atmospheric CO_2: a broader perspective. *Photosynth. Res.* **39**, 369–388.

Gutschick, V. P. (1993) Nutrient-limited growth rates: roles of nutrient-use efficiency and of adaptations to increase uptake rate. *J. Exp. Bot.* **44**, 41–51.

Harley, P. C. and Tenhunen, J. D. (1991) Modeling the photosynthetic response of C3 leaves to environmental factors. In: Boote, K. J. and Loomis, R. S. (Eds) *Modeling Crop Photosynthesis–from Biochemistry to Canopy*, pp. 17–39. Crop Science Society of America, Inc., Madison, Wis.

Harley, P. C., Thomas, R. B., Reynolds, J. F. and Strain, B. R. (1992) Modeling photosynthesis of cotton grown in elevated CO_2. *Plant Cell Env.* **15**, 271–282.

Hendrey, G. R. (1992) Global greenhouse studies: need for a new approach to ecosystem manipulation. *Critical Rev. Plant Sci.* **11**, 61–74.

Herold, A. (1986) Regulation of photosynthesis by sink activity – the missing link. *New Phytol.* **86**, 131–144.

Hollinger, D. Y. (1987) Gas exchange and dry matter allocation responses to elevation of atmospheric CO_2 concentration in seedlings of three species. *Tree Physiol.* **3**, 193–202.

Humphries, S.W. and Long, S.P. (1995) WIMOVAC: a software package for modelling the dynamics of plant leaf and canopy photosynthesis. *CABIOS* **11**, 361–371.

Idso, S. B. (1989) *Carbon Dioxide and Global Change: Earth in Transition.* IBR Press, Tempe.

Idso, S. B., Kimball, B. A., Akin, D. E. and Kridler, J. (1993) A general relationship between CO_2-induced reductions in stomatal conductance and concomitant increases in foliage temperature. *Env. Exp. Bot.* **33**, 443–446.

Jarvis, P. G. (1989) Atmospheric carbon dioxide and forests. *Phil. Trans. R. Soc. Lond. B* **324**, 369–392.

Johnsen, K. H. (1993) Growth and ecophysiological responses of black spruce seedlings to elevated CO_2 under varied water and nutrient additions. *Can. J. For. Res.* **23**, 1033–1042.

Jordan, D. and Ogren, W. L. (1984) The CO_2/O_2 specificity of ribulose 1,5 bisphosphate carboxylase/oxygenase dependence on ribulose bisphosphate concentration, pH and temperature. *Planta* **161**, 308–313.

Kaye, G. W. C. and Laby, T. H. (1973) *Tables of Physical and Chemical Constants.* 4th Edn. Longman, London.

Kimball, B. A. (1983) Carbon dioxide and agricultural yield: an assemblage and analysis of 430 prior observations. *Agron. J.* **75**, 779–788.

Kimball, B.A., Pinter, P.J., Garcia, R.L., LaMorte, R.L., Wall, G.W., Hunsaker, D.J., Wechsung, G., Wechsung, F. and Kartschall, T. (1995) Productivity and water use of wheat under free-air CO_2 enrichment. *Global Change Biol.* **1**, 425–445.

Kimball, B. A., Pinter, P. J. and Mauney, J. R. (1992) Cotton leaf and boll temperatures in the 1989 FACE experiment. *Critical Rev. Plant Sci.* **11**, 233–240.

Leggett, J., Pepper, W.J. and Swart, R.J. (1992) Emission scenarios for the IPCC: an update. In: Houghton, J.T., Callander, B.A. and Varney, S.K. (Eds) *Climate Change 1992: The Supplementary Report to the IPCC Scientific Assessment*, pp. 17–39. Cambridge University Press, Cambridge, UK.

Linke, W. F. (1965) *Solubilities: Inorganic and Metal-organic Compounds.* 14th Edn. American Chemical Society, Washington DC.

Long, S. P. (1983) C4 photosynthesis at low temperature. *Plant Cell Env.* **6**, 345–363.

Long, S. P. (1985) Leaf gas exchange. In: Barber, J. and Baker, N.R. (Eds) *Photosynthetic Mechanisms and the Environment*, pp. 453–499. Elsevier, Amsterdam.

Long, S. P. (1991) Modification of the response of photosynthetic productivity to rising temperature by atmospheric CO_2 concentrations: has its importance been underestimated? *Plant Cell Env.* **14**, 729–740.

Long, S. P. and Drake, B. G. (1991) Effects of the long-term elevation of CO_2 concentration in the field on the quantum yield of photosynthesis of the C3 sedge, *Scirpus olneyi*. *Plant Physiol.* **96**, 221–226.

Long, S. P. and Drake, B. G. (1992) Photosynthetic CO_2 assimilation and rising atmospheric CO_2 concentrations. In: Baker, N.R. and Thomas, H. (Eds) *Crop Photosynthesis: Spatial and Temporal Determinants*, pp. 69–103. Elsevier Science Publishers BV, Amsterdam.

Long, S. P., Postl, W. F. and Bolhár-Nordenkampf, H. R. (1993) Quantum yields for uptake of carbon dioxide in C3 vascular plants of contrasting habitats and taxonomic groupings. *Planta* **189**, 226–234.

Long, S. P., and Woodward, F. I. (Eds) (1988) *Plants and Temperature*. Company of Biologists, Cambridge.

Melillo, J., Callaghan, T. V., Woodward, F. I., Salati, E. and Sinha, S. K. (1990) Effects on ecosystems. In: Houghton, J. T., Jenkins, G. J. and Ephraums J. J. (Eds) *Climate Change: The IPCC Scientific Assessment*, pp. 283–310. Cambridge University Press, Cambridge.

Mitchell, J. F. B., Manabe, S., Meleshko, V. and Tokioka, T. (1990) Equilibrium climate change – and its implications for the future. In: Houghton, J. T., Jenkins, G.J. and Ephraums, J. J. (Eds) *Climate Change: The IPCC Scientific Assessment*, pp. 131–175. Cambridge University Press, Cambridge.

Mitchell, R. A. C., Mitchell, V. J., Driscoll, S. P., Franklin, J. and Lawlor, D. W. (1993) Effects of increasing CO_2 concentration and temperature on growth and yield of winter wheat at two levels of nitrogen application. *Plant Cell Env.* **16**, 521–529.

Monteith, J. L. and Unsworth, M. J. (1990) *Principles of Environmental Physics*. 2nd Edn. Edward Arnold, London.

Morison, J. I. L. (1985) Sensitivity of stomata and water use efficiency to high CO_2. *Plant Cell Env.* **8**, 467–474.

Morison, J. I. L. (1987) Intercellular CO_2 concentration and stomatal response to CO_2. In: Zeigler, E., Farquhar, G. D. and Cowan, I. R. (Eds) *Stomatal Function*, pp. 229–251. Stanford University Press, Stanford, Calif.

Morison, J. I. L. (1993) Responses of plants to CO_2 under water limited conditions. *Vegetatio* **104/105**, 193–209.

Morris, J. T. (1991) Effects of nitrogen loading on wetland ecosystems with particular reference to atmospheric deposition. *Ann. Rev. Ecol. Syst.* **22**, 257–279.

Mott, K. A. (1990) Sensing of atmospheric CO_2 by plants. *Plant Cell Env.* **13**, 731–737.

Mousseau, M. and Saugier, B. (1992) The direct effect of increased CO_2 on gas exchange and growth of forest tree species. *J. Exp. Bot.* **43**, 1121–1130.

Nie, D., He, H., Kirkham, M. B. and Kanemasu, E. T. (1992) Photosynthesis of a C3 grass and a C4 grass under elevated CO_2. *Photosynthetica* **26**, 189–198.

Nie, G.-Y., Long, S.P., Garcia, R.L., Kimball, B.A., LaMorte, R.L., Pinter, P.J., Wall, G.W. and Webber, A.N. (1995) Effects of free-air CO_2 enrichment on the development of the photosynthetic apparatus in wheat, as indicated by leaf proteins. *Plant Cell Env.* **18**, 855–864.

Nijs, I., Impens, I. and Behaeghe, T. (1989) Leaf and canopy responses of *Lolium perenne* to long-term elevated atmospheric carbon-dioxide concentration. *Planta* **177**, 312–320.

Norman, J. M. (1980) Interfacing leaf and canopy light interception models. In: Hesketh, J. D. and Jones, J. W. (Eds) *Predicting Photosynthesis for Ecosystem Models*, pp. 49–67. CRC Press, Boca Raton.

Oberbauer, S. F., Strain, R. and Fetcher, N. (1985) Effects of CO_2-enrichment on seedling physiology and growth of two tropical tree species. *Physiol. Plant.* **65**, 352–356.

Overdieck, D. (1990) Direct effects of elevated CO_2 concentration levels on grass and clover in 'model ecosystems'. In: Beukema, J.J., Wolff, W.J. and Brouns, J. J. W. M. (Eds) *Expected Effects of Climatic Change on Marine Coastal Ecosystems*, pp. 41–47. Kluwer, Dordrecht.

Parry, M. L. and Carter, T. R. (1988) The assessment of effects of climatic variations on agriculture: aims, methods and summary of results. In: Parry, M. L., Carter, T. R. and Konijn, N. T. (Eds) *The Impact of Climatic Variation on Agriculture*, pp. 11–95. Kluwer, Dordrecht.

Parry, M. L. and Carter, T. R. (1990) An assessment of the effects of climatic change on agriculture. In: Jackson, M., Ford-Lloyd, B. V. and Parry, M. L. (Eds) *Climatic Change and Plant Genetic Resources*, pp. 61–84. Belhaven, London.

Patrick, J. W. (1993) Sink strength: whole plant considerations. *Plant Cell Env.* **16**, 1019–1020.

Pearcy, R. W. and Björkman, O. (1983) Physiological effects. In: Lemon, E. R. (Ed.) *CO_2 and Plants: The Response of Plants to Rising Levels of Atmospheric Carbon Dioxide*, pp. 65–105. Westview Press, Boulder, Colo.

Pearson, M., Davies, W.J. and Mansfield, T.A. (1995) Asymmetric responses of adaxial and abaxial stomata to elevated CO_2: impacts on the control of gas exchange by leaves. *Plant Cell Env.* **18**, 837–843.

Penman, H. L. (1948) Natural evaporation from open water, bare soil and grass. *Proc. R. Soc. Lond. B* **193**, 120–145.

Reynolds, J. F. and Acock, B. (1985) Predicting the response of plants to increasing carbon dioxide: a critique of plant growth models. *Ecol. Modeling* **29**, 107–129.

Rowland-Bamford, A. J., Baker, J. T., Allen, H. J. and Bowes, G. (1991) Acclimation of rice to changing atmospheric carbon dioxide concentration. *Plant Cell Env.* **14**, 577–583.

Ryle, G. J. A., Powell, C. E. and Tewson, V. (1992) Effect of elevated CO_2 on the photosynthesis, respiration and growth of perennial ryegrass. *J. Exp. Bot.* **43**, 811–818.

Sage, R, F. (1994) Acclimation of photosynthesis to increasing atmospheric CO_2: the gas exchange perspective. *Photosynth. Res.* **39**, 351–368.

Sage, R. F. and Sharkey, T. D. (1987) The effect of temperature on the occurrence of O_2 and CO_2 insensitive photosynthesis in field grown plants. *Plant Physiol.* **84**, 658–664.

Sage, R. F., Sharkey, T. D. and Seemann, J. R. (1989) Acclimation of photosynthesis to elevated CO_2 in five C3 species. *Plant Physiol.* **89**, 590–596.

Schimel, D. (1990) Biogeochemical feedbacks in the Earth system. In: Leggett, J. (Ed.) *Global Warming. The Greenpeace Report*, pp. 68–82. Oxford University Press, Oxford.

Sharkey, T. D. (1985a) O_2-insensitive photosynthesis in C3 plants. *Plant Physiol.* **78**, 71–75.

Sharkey, T. D. (1985b) Photosynthesis in intact leaves of C3 plants: physics, physiology and rate limitations. *Bot. Rev.* **51**, 53–105.

Sheen, J. (1990) Metabolic repression of transcription in higher plants. *Plant Cell* **2**, 1027–1038.

Sheen, J., Huang, H., Schaeffner, A. R., Leon, P. and Jang, J.-C. (1992) Sugars, fatty acids, and photosynthetic gene expression. *Photosynth. Res.* **34**, 107.

Spain, J. D. and Keen, R. E. (1992) Temperature and biological activity. In: Keen, R.E. and Spain, J. D. (Eds) *Computer Simulation in Biology*, pp. 183–200. Wiley-Liss, New York.

Stitt, M. (1991) Rising CO_2 levels and their potential significance for carbon flow in photosynthetic cells. *Plant Cell Env.* **14**, 741–762.

Stulen, I. and Hartog, J. D. (1993) Root growth and functioning under atmospheric CO_2 enrichment. *Vegetatio* **104**, 99–115.

Terashima, I. (1992) Anatomy of non-uniform photosynthesis. *Photosynth. Res.* **31**, 195–212.

Thomas, R. B. and Strain, B. R. (1991) Source–sink balance as a factor in the photosynthetic acclimation of cotton to long-term elevated CO_2. *Plant Physiol.* **96**, 627–634.

Thornley, J. H. M. and Johnson, I. R. (1990) *Plant and Crop Modeling – a Mathematical Approach to Plant and Crop Physiology.* Oxford Univ. Press, Oxford.

Tissue, D. T. and Oechel, W. C. (1987) Response of *Eriophorum vaginatum* to elevated CO_2 and temperature in the Alaskan tussock tundra. *Ecology* **68**, 401–410.

Tissue, D. T., Thomas, R. B. and Strain, B. R. (1993) Long-term effects of elevated CO_2 and nutrients on photosynthesis and Rubisco in loblolly pine seedlings. *Plant Cell Env.* **16**, 859–865.

Vu, C. V., Allen, L. H., Jr and Bowes, G. (1983) Effects of light and elevated atmospheric CO_2 on the ribulose bisphosphate carboxylase activity and ribulose bisphosphate level of soybean leaves. *Plant Physiol.* **73**, 729–734.

Watson, R. T., Rodhe, H., Oescheger, H. and Siegenthaler, U. (1990) Greenhouse gases and aerosols. In: Houghton, J. T., Jenkins, G. J. and Ephraums, J. J. (Eds) *Climate Change: The IPCC Scientific Assessment*, pp. 1–40. Cambridge University Press, Cambridge.

Webber, A. N., Nie, G.-Y. and Long, S. P. (1994) Acclimation of photosynthetic proteins to rising atmospheric CO_2. *Photosynth. Res.* **39**, 413–426.

Wise, R. R., Ortiz-Lopez, A. and Ort, D. R. (1992) Spatial distribution of photosynthesis during drought in field-grown and acclimated and non-acclimated growth chamber-grown cotton. *Plant Physiol.* **100**, 26–32.

Wong, S. C. (1979) Elevated atmospheric partial pressure of CO_2 and plant growth. 1. Interactions of nitrogen nutrition and photosynthetic capacity in C3 and C4 plants. *Oecologia* **44**, 68–74.

Woodward, F. I. (1993) Plant responses to past concentrations of CO_2. *Vegetatio* **104/105**, 145–155.

Yelle, S., Beeson, R. C., Jr, Trudel, M. J. and Gosselin, A. (1989) Acclimation of two tomato species to high atmospheric CO_2. I. Sugar and starch concentrations. *Plant Physiol.* **90**, 1465–1472.

Yuncong, L. and Gupta, G. (1993) Photosynthetic changes in soybean with and without nitrogen and increased carbon dioxide. *Plant Sci.* **89**, 1–4.

6.7 ABBREVIATIONS

A	net rate of leaf CO_2-uptake per unit leaf area (μmol m^{-2} s^{-1})
A_c	net rate of canopy CO_2-uptake per unit ground area (μmol m^{-2} s^{-1})
$A_{c,tot}$	A_c integrated over the course of 1 day (mol m^{-2} d^{-1})
A_0	A at infinite stomatal conductance
A_{sat}	light-saturated A
ABA	abscissic acid
ADP	adenosine diphosphate
ATP	adenosine triphosphate
c_a	atmospheric CO_2 concentration (μmol mol^{-1})
c_i	intercellular concentration of CO_2 (μmol mol^{-1})
c_s	concentration of CO_2 at the leaf surface (μmol mol^{-1})
e_a	ambient atmospheric humidity (kPa)
e_s	saturation atmospheric humidity (kPa)
g_s	stomatal conductance to CO_2 diffusion (mol m^{-2} s^{-1})
I	photosynthetically active photon flux (μmol m^{-2} s^{-1})
I_{abs}	absorbed I
J	potential rate of coupled photosynthetic electron transport (μmol m^{-2} s^{-1})
J_{max}	light-saturated J
K_c	Michaelis constant for CO_2 recalculated for an equivalent gas concentration in air (μmol mol^{-1})

K_o	Michaelis constant for O_2 recalculated for an equivalent gas concentration in air (μmol mol^{-1})
l	stomatal limitation (dimensionless)
LCP	light compensation point of photosynthesis (μmol m^{-2} s^{-1})
NADPH	reduced form of nicotinamide adenine dinucleotide phosphate
P_i	inorganic phosphate ion
PEPc	phosphoenol-pyruvate carboxylase
PGA	phosphoglycerate
rbcs	nuclear gene coding for the small subunit polypeptide of Rubisco
R_d	'dark respiration rate' (μmol m^{-2} s^{-1})
RH	relative humidity (dimensionless)
RubP	ribulose-1,5-bisphosphate
Rubisco	ribulose-1,5-bisphosphate carboxylase/oxygenase
T	temperature (°C)
T_{opt}	optimum temperature for CO_2-uptake (°C)
V_c	carboxylation velocity expressed on a leaf area basis (μmol m^{-2} s^{-1})
$V_{c,max}$	maximum RubP-saturated V_c
V_o	oxygenation velocity expressed on a leaf area basis (μmol m^{-2} s^{-1})
$V_{o,max}$	maximum RubP-saturated V_o
VPD	leaf-air water vapor pressure deficit (kPa)
W_c	RubP-saturated rate of carboxylation (μmol m^{-2} s^{-1})
W_j	RubP-limited rate of carboxylation (μmol m^{-2} s^{-1})
W_p	cytosolic P_i-limited rate of carboxylation (μmol m^{-2} s^{-1})
τ	the specificity of Rubisco for CO_2 relative to O_2
ϕ	the quantum yield of CO_2-uptake determined from the initial slope of the response of A vs I_{abs} (dimensionless)
ω	the ratio of A to instantaneous leaf transpiration rate (dimensionless)

7 Rising Atmospheric Carbon Dioxide and Plant Respiration

M. A. GONZÀLEZ-MELER,[1] B. G. DRAKE[1] AND J. AZCÓN-BIETO[2]
[1]Smithsonian Environmental Research Center, Edgewater, USA
[2]Department Fisiologia Vegetal, Universitat de Barcelona, Spain

7.1 INTRODUCTION

Rising atmospheric CO_2 concentrations are expected to cause important responses in photosynthesis, photorespiration and stomatal conductance, and because of this, these processes have been the subject of biochemical, physiological, agronomic and ecological studies (Allen 1990; Bazzaz 1990; Hogan *et al.* 1991, Morison 1990; Mott 1990). The effects of rising CO_2 and climate change on respiration may have a significant impact on whole plant and ecosystem carbon budgets, but little attention has been given to the possible effects of rising atmospheric CO_2 on this process although there has been considerable emphasis placed on the potential increase of global respiration resulting from increased temperature (review in Ryan 1991). Recent experimental evidence suggests that elevated CO_2 can reduce respiration. It appears that this effect is sufficient to increase plant growth over and above what might be expected strictly from the stimulation of photosynthesis (Reuveni and Gale 1985), and it may be large enough to counteract the expected stimulation of increased temperature of 2–3 °C on respiration.

Respiration is essential for growth and maintenance of tissues. The solar energy conserved during photosynthesis and stored as chemical energy in organic molecules is released in a regulated manner for the production of adenosine triphosphate (ATP), the universal currency of biological energy transactions. Phytomass and carbohydrate content generally increase in C_3 plants grown in elevated CO_2 levels as a result of increased photosynthetic activity (Allen 1990; Mott 1990), and when attention was first directed to the global importance of vegetation in the mediation of atmospheric CO_2, it was thought that the stimulation of photosynthesis and growth should increase dark respiration rates due to increased carbohydrate concentration. In some species, particularly young, growing tissues, increased dark respiration has been shown to accompany the

Global Change: Effects on Coniferous Forests and Grasslands.
Edited by A. I. Breymeyer, D. O. Hall, J. M. Melillo and G. I. Ågren

UNEP

stimulation of photosynthesis by elevated CO_2 (Gifford *et al.* 1985; Poorter *et al.* 1988). But not all studies in which photosynthesis has been increased have produced the expected stimulation of respiration by increased carbohydrates. The effects of photosynthesis on subsequent respiration rates were less when photosynthetic rates were increased by high CO_2 than when the rate was similarly increased by light (Ludwig *et al.* 1975) and many studies clearly indicate that long-term adaptation of plants to elevated atmospheric CO_2 results in significantly lower dark respiration rated (Bunce 1994). An important point is the recent rediscovery that plant respiration responds reversibly to rapid changes in atmospheric CO_2 concentration (Amthor *et al.* 1992).

The effects of rising atmospheric CO_2 and climate change on respiration have been reviewed by Amthor (1991), Ryan (1991), and Farrar and Williams (1991), Poorter *et al.* (1992) and Bunce (1994). These reviews emphasize the potential for direct effects of CO_2 on different aspects of respiration including maintenance respiration (Ryan 1991; Bunce 1994), enzymatic rates (Amthor 1991), carbon partitioning (Farrar and Williams 1991), and growth (Amthor 1991; Bunce 1994; Poorter *et al.* 1992). All of them indicate the considerable potential for rising CO_2 and climate change to alter the respiration of vegetation.

In many species, a direct, immediate and reversible inhibition has been demonstrated (Amthor *et al.* 1992; Bunce 1994). In these cases, the rate of CO_2 emission from the plant tissue is decreased within minutes after CO_2 concentration was increased and is reversed upon returning the CO_2 concentration to a lower level. The inhibition of dark respiration resulting from increasing the concentration of CO_2 in the ambient air of a plant has been reported by Kidd (1916), Nilovskaya and Razoryonova (1968), Kaplan *et al.* (1977), Silsbury and Stephens (1984), Reuveni and Gale (1985), Gifford *et al.* (1985), Bunce (1990), El Kohen *et al.* (1991), and Amthor *et al.* (1992). In all of these reports, dark respiration was reduced when the concentration of CO_2 was increased. Long-term effects are determined by comparing the rate of CO_2 emission (or O_2 assimilation) from tissues at a common CO_2 background. This class of effects represents the acclimation of tissues to higher CO_2 and involves effects on tissue composition, especially carbohydrate accumulation and reduced protein content which is reflected in the reduction of nitrogen concentration, as well as the effects of growth, response to stress, and other factors that are a consequence of the higher efficiency of key plant processes often found in high CO_2 concentration.

There are conflicting views expressed in the literature regarding the effects of elevated CO_2 on respiration (Poorter *et al.* 1992; Thomas *et al.* 1993). These differences represent unresolved conceptual problems as well as different methods of measurement and different bases for expressing the results of measurements. One view suggests that all measurements should be expressed per unit of leaf area, which would make it very difficult to compare results of different tissue types, such as stems and roots, with foliage. In addition, this approach places the emphasis on the photosynthesizing organ and the consequences of photosynthesis. We suggest that the basis of comparison should be the specific rate of respiration, expressed as the loss of carbon or the assimilation of oxygen per gram carbon ($g\ g^{-1}$). In addition, it is important that the background concentration of CO_2 during growth and during measurement be made clear.

7.2 RESPONSE OF DARK RESPIRATION TO ATMOSPHERIC CO_2 CONCENTRATION

The inhibitory effect of elevated atmospheric CO_2 on dark CO_2 evolution and O_2 uptake rates has been reported for different tissues and organs: whole plant (Reuveni and Gale 1985; Bunce 1990), seedlings (Reuveni and Gale 1985), shoot (El Kohen *et al.* 1991), fruit (Kerbel *et al.* 1988), roots (Reuveni and Gale 1985; Nobel and Palta 1989), leaves (Amthor *et al.* 1992; Bunce 1990; Thomas and Griffin 1994), leaf epiderman strips (Shaish *et al.* 1989), tissue culture cells (Kerbel *et al.* 1990), plant mitochondria (Spalding *et al.* 1980; Palet *et al.* 1992), microorganisms (Koizumi *et al.* 1991) and even animal tissues (Palet *et al.* 1991). Respiration of mature leaves of *Rumex crispus* was reversibly inhibited by CO_2 in the range of 5–950 ppm (Amthor *et al.* 1992). The magnitude of the inhibition is quite variable (Poorter *et al.* 1992; Reuveni *et al.* 1993), but averages when the increase in CO_2 is from 350 to 700 $\mu l\ l^{-1}$.

At the most fundamental level, respiration can be considered to have a dual nature: it is the source of metabolic intermediates used in the synthesis of cellular constituents as well as the source of ATP and reduced nucleotide (particularly pyridine nucleotide), the driving force of these syntheses. A variety of substrate (carbohydrates, lipids, and proteins) can be respired through pathways occurring in several cell compartments (e.g. cytosol, mitochondria, chloroplast) with the products converging on the tricarboxylic acid cycle and hence on mitochondrial oxidative phosphorylation (see Dennis and Turpin 1990). This duality of respiration, as a source of substrate as well as a source of the energy to drive the reactions, is probably reflected in the regulation of the rate by respiratory substrate and by adenosine diphosphate (ADP) availability, although it remains a goal of plant physiologists to fully establish the mechanism for this control. One of the most confounding facts about the effects of elevated CO_2 on respiration is that elevated CO_2 appears to cause both inhibition and stimulation of respiration (Amthor 1991; Azcón-Bieto 1992; Poorter *et al.* 1992). Stimulation of respiration by immediate exposure to elevated CO_2 has been shown in lemon fruit (Young *et al.* 1962), potato tubers (Perez-Trejo *et al.* 1981), tissue cultures (Palet *et al.* 1991), and plant mitochondria (Gonzàlez-Meler 1995; Gonzàlez-Meler 1996). We propose that the effects of different CO_2 concentrations on the rate of CO_2 efflux are in the first instance caused by the effect of CO_2 on the activities of certain key respiratory enzymes, discussed in section 7.2.1, and we suggest that some of the apparently contradictory responses to elevated CO_2 might be explained by the

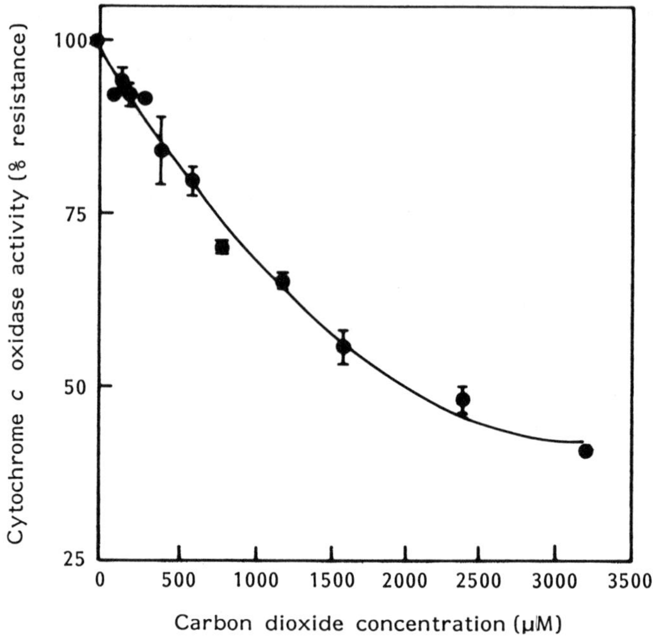

Figure 7.1 The effect of CO_2 concentration on the activity of cytochrome *c* oxidase *in vitro*

interaction of the alternative and cytochrome pathways as discussed in section 7.2.2 below.

7.2.1 Effects of CO_2 concentration on enzyme activity

Several authors have suggested that CO_2 might have direct inhibitory effects on some respiratory enzymes (Shipway and Bramlage, 1973; Kerbel *et al.* 1988, 1990) including cytochrome *c* oxidase (Palet *et al.* 1991). We will consider the effects reported for CO_2 inhibition of respiratory enzymes of the glycolytic pathway, Krebs cycle and mitochondrial electron transport. We conclude that cytochrome *c* oxidase could be the main target of the direct CO_2 inhibition of respiration.

There is evidence that atmospheric CO_2 concentration affects the activity of respiratory enzymes, although the CO_2 levels at which some of these enzymes were inhibited was much higher than projected atmospheric levels. Cytochrome *c* oxidase (cytox hereafter) is the final step in the mitochondrial electron transport chain and the rate of oxidative phosphorylation is tightly coupled to the activity of this enzyme. Cytox is inhibited by CO_2–bicarbonate (Miller and Evans 1956; Bendall *et al.* 1958; Palet *et al.* 1991, 1992) (Figure 7.1). Cytox activity of photomyxotrophic carnation callus homogenate was inhibited 38% by 10 mM of

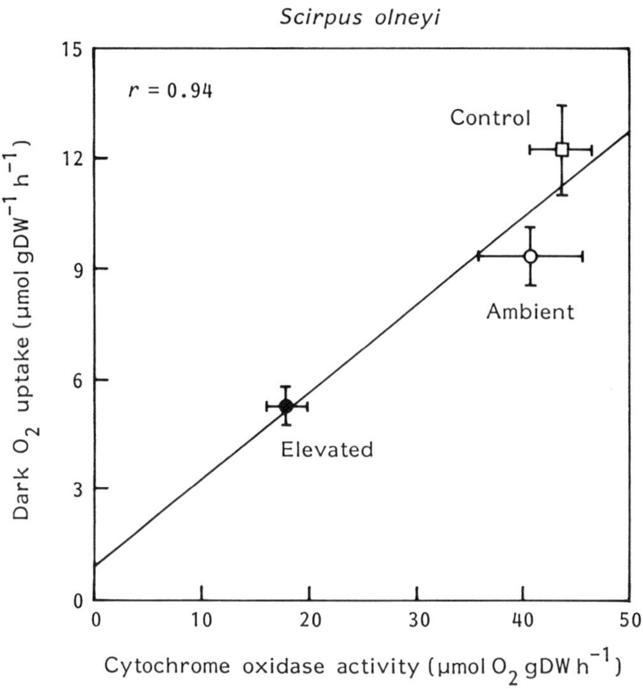

Figure 7.2 The relationship between respiration in whole tissues of the sedge, *Scirpus olneyi*, and the *in vitro* activity of cytochrome *c* oxidase from extracts of the sedge. Plants were grown in the field and in the field exposed to normal ambient CO_2 and in elevated CO_2 (680 ppm) in open top chambers (Azcón-Bieto *et al.* 1994)

potassium bicarbonate (about 2.5% CO_2 in air) at pH 7.2 (Palet *et al.* 1991). The activity of purified beef heart cytox was inhibited almost 50% by 20 mM of potassium bicarbonate (about 5% CO_2 in air) at the same pH (Palet *et al.* 1992). Palet et al. (1991) concluded that external free CO_2 is the species responsible for the effects on cytox and suggested a carbamate formation in proteins as possible mechanism. In contrast, cytox from roots of spinach and soybean were strongly inhibited by high sodium bicarbonate concentrations at pH 8.8 suggesting a competitive model as mechanism of inhibition (Miller and Evans 1956). Cytox from different plant species have different sensitivities to inhibitors: cytox from C_4 leaves was highly sensitive to CO compared with cytox from C_3 leaves (Naik *et al.* 1992), suggesting that there may be a range of sensitivity of inhibition of respiration to elevated CO_2 in the air.

The tight coupling between the activity of cytox and respiration has been demonstrated in plants having different levels of the enzyme. Figure 7.2 shows the linear relationship between respiration, measured by O_2 assimilation, in whole sections of green tissues from *Scirpus olneyi* grown in two different atmospheric

CO_2 concentrations, and the activity of cytox in extracts from these tissues (Azcón-Bieto *et al.* 1994). In *S. olneyi*, the effects of elevated CO_2 on the activity of cytox from this species was sufficient to explain the reduction of respiration measured by gas exchange in the field.

It appears that other enzymes are also inhibited by elevated CO_2 concentrations. Amthor (1991) suggested that the dehydrogenases are inhibited by increased CO_2 concentrations. Studies of apple tissues showed that a very high CO_2 concentration Krebs cycle activity was reduced (Monning 1983). In pear fruits, a concentration of 5% CO_2 in air inhibited the activity of succinate dehydrogenase by approximately 50% in the first month of CO_2 exposure and somewhat more after six months' exposure (Frenkel and Patterson 1973).

Other enzymes of the Krebs cycle and some glycolytic enzymes are also inhibited by high concentrations of CO_2 (Amthor 1991). Both ATP: phosphofructokinase (PFK) and PPI: phosphofructokinase (PFP), were strongly (up to seven times) inhibited by 10% CO_2 in air (Kerbel *et al.* 1988). Carbon dioxide (CO_2/bicarbonate) decreased the activity and the affinity of the NAD (nicotinamide adenine dinucleotide) malic enzyme from the CAM (reduced nicotinamide adenine dinucleotide) plant *Sedum praealtum* with a decrease in decarboxylation rate of malate and mitochondrial O_2 uptake (Spalding *et al.* 1980). Carbon dioxide CO_2–bicarbonate) stimulated malate oxidation, but reduced NADH (crassulacea acid metabolism) oxidation in apple mitochondria (Shipway and Bramlage 1973). Amthor (1991) reviews evidence suggesting that succinate dehydrogenase is particularly sensitive to elevated CO_2.

It is clear that some enzymes can be inhibited by increasing concentrations of CO_2 (Figure 7.3) and that some of these (e.g. PFP or PFK) are sensitive to very small changes in pH, which can be affected by very high atmospheric CO_2 (Bown 1985). We propose that the inhibition of respiratory enzymes, especially cytox, by increased atmospheric CO_2 is the basis for the inhibition of CO_2 efflux or O2 uptake by whole tissues. To apply this hypothesis to those cases in which increased CO_2 stimulates respiration it is necessary to consider the coordinated system of the mitochondrial electron transport chain.

7.2.2 CO_2 and the cytochrome and alternative pathways

A complication in relating the rate of respiration to the background CO_2 concentration is the partitioning of electron flow through the electron transport chain between the cytochrome oxidase pathway and the alternative, cyanide-resistant pathway (Figure 7.3). Both pathways are involved in the consumption of substrate and in CO_2 emission and O_2 uptake, but the alternative pathway is nonphosphorylating and therefore uncoupled from energy production.

Cyanide and other compounds inhibit aerobic respiration of most organisms by inactivation of cytox. In many plant tissues, however, cyanide has a small effect on respiration determined as CO_2 efflux or O_2 assimilation and the remaining

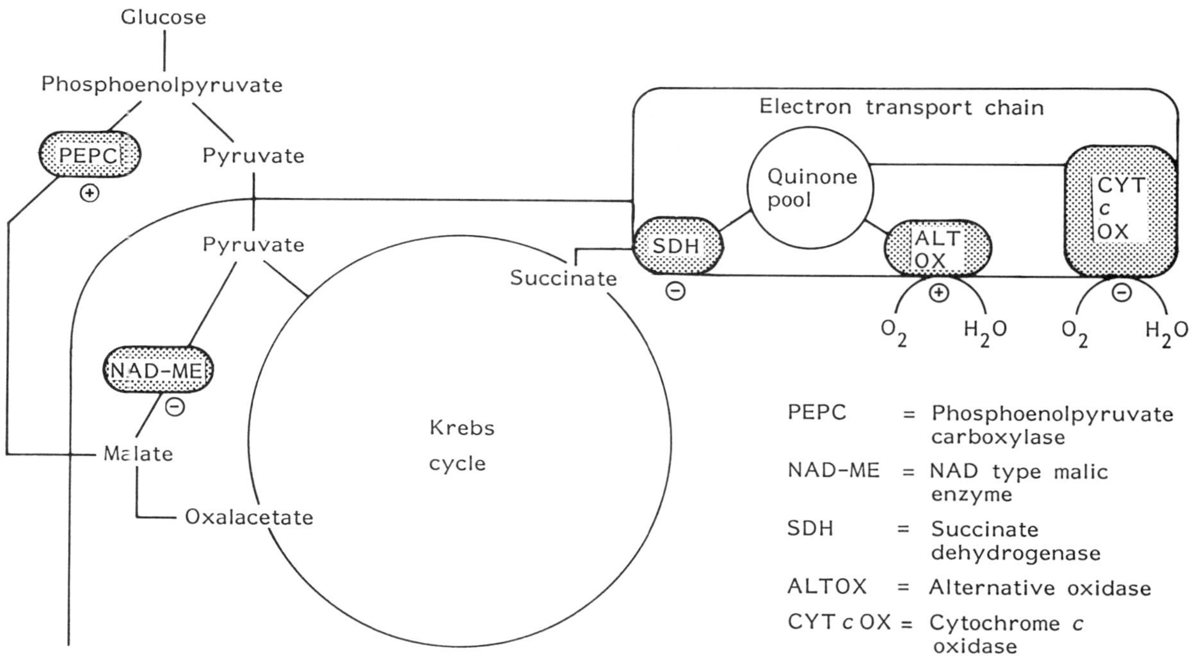

Figure 7.3 Points in metabolism and in the mitochondrial electron chain where CO_2 bicarbonate has an effect on the activity of enzymes, indicated as darkened background. The effect of CO_2 is given as + if a stimulation and − if inhibition. For other details see text

respiration in the presence of cyanide is said to be due to an alternative pathway from electron flow. Cyanide-resistant respiration is present in plants, bryophytes, several fungi, algae, a few bacteria and some animal tissues (Henry and Nyns 1975).

The basis of cyanide-resistant respiration is a short branch in the mitochondrial electron transport chain at a step prior to cytochrome *c* oxidase (Figure 7.3). This alternative branch takes electrons from the ubiquinone pool and passes them to O_2 through an oxidase (altox). Unlike the cytox pathway, there is no conservation of proton motive energy in the production of ATP by the alternative pathway and all free energy generated is lost as heat. For most plants, the physiological role of the alternative pathway is unknown, although the heat produced in this way is used in the spadix of arum lilies to volatilize attractants for insect pollination. High levels of carbohydrates can activate the alternative pathway (Azcón-Bieto *et al.* 1983) burning excess levels of photosynthate wastefully. The observation that excess levels of carbohydrate could be consumed through this pathway, led to the hypothesis proposed for its role in higher plants (Lambers 1982).

The system is believed to function as an electron overflow regulating the supply of energy within the cell (Lambers, 1982, 1985). The alternative pathway can act as an overflow when the cytochrome pathway has been saturated by rapid glycolysis or Krebs cycle activity (Lambers, 1985; Amthor 1989). Engagement of the alternative pathway in the consumption of carbohydrates seems to occur when the electron flow through the cytochrome phosphorylating pathway runs at 40% of its capacity (Dry *et al.* 1989).

Some reports cited earlier showed a stimulation of respiration in elevated CO_2. This stimulation of CO_2 emission or O_2 uptake can be due to consumption of excess carbohydrates through the alternative pathway. Stimulation of cyanide-resistant respiration by CO_2 has often been observed (Lange 1970; Day *et al.* 1978; Perez-Trejo *et al.* 1981; Palet *et al.* 1991 and 1992). Increase in the activity of the alternative pathway can be induced by inhibitors of components of the cytochrome pathway (Lambers *et al.* 1983; Møller *et al.* 1988). Blocking the cytochrome pathway with cyanide, sulfide, antimycin A or myxothiazol raises the activity of the alternative pathway towards its maximum capacity (Bahr and Bonner 1973). If the capacity of the alternative pathway is high enough, its stimulated activity can compensate or even increase the rate of CO_2 efflux or O_2 assimilation to a level higher than the initial rate from the uninhibited cytochrome pathway (Lambers 1985). The increase in the activity of the alternative oxidase was progressive with the partial inhibition of the saturated cytochrome pathway (Gonzàlez-Meler *et al.* 1992).

Carbon dioxide can stimulate activity of the alternative pathway through inhibition of components of the cytochrome pathway. In pea leaf mitochondria, CO_2–bicarbonate inhibited O_2 uptake and stimulated alternative pathway activity (Palet *et al.* 1992). In soybean cotyledon mitochondria, CO_2–bicarbonate stimulated O_2 assimilation via the alternative pathway, and this stimulation was inhibited by salycylhydroxamic acid (SHAM), a specific inhibitor of the alternative oxidase. Both the cytochrome pathway activity, determined as the rate of respiration in the presence of SHAM, and capacity, the rate of respiration uncoupled from the ubiquinone pool and fed by electron donors in the presence of SHAM, were also inhibited by CO_2 in the short term in culture cells (Palet *et al.* 1991). Thus, increased CO_2 may inhibit the activity of cytox and, simultaneously, in the presence of sufficient substrate and high capacity of altox, have the net effect of increasing CO_2 efflux.

The capacity of the alternative pathway appears to acclimate to different environmental conditions. Germination of wheat seeds under very high CO_2 concentrations stimulated the capacity of the alternative oxidase (McCaig and Hill 1977). In potato tubers, resistance to cyanide increased after incubation in 10% CO_2 in air for several hours (Day *et al.* 1978). Stimulation of CO_2 evolution or CO_2 uptake in potato tubers was increased by high concentrations of CO_2 in air, but this stimulation was not totally inhibited by cyanide, showing a stimulation of both pathways (Perez-Trejo *et al.* 1981).

It is clear that CO_2 inhibits components of the mitochondrial electron transport chain, especially cytox, because CO_2 can inhibit the rate of respiration when the enzyme is uncoupled from the supply of electrons via the ubiquinone pool. Carbon dioxide can indirectly increase the activity of the cyanide-resistant respiration. Thus, the compensating effects of CO_2 on these two pathways have to be accounted for in studies of the short-term effects of exposure to different concentrations of CO_2. When the exposure is to very high CO_2 concentrations or if it persists for several days, the capacity of both pathways has to be determined if the effects of CO_2 are to be correctly interpreted.

7.3 ACCLIMATION OF RESPIRATION TO LONG-TERM CO_2 ENRICHMENT

Long-term responses of respiration to elevated CO_2 measured by comparing the response of tissues grown in different CO_2 concentrations at a common CO_2 background are reported in Ludwig *et al.* (1975), Gifford *et al.* (1985), Reuveni and Gale (1985), Spencer and Bowes (1986), Allen (1989), Bunce (1990), Drake (1989), Wullschleger *et al.* (1991, 1992), Baker *et al.* (1992), Gary and Veyres (1992) and Mousseau and Saugier (1992).

Through stimulation of photosynthesis, high atmospheric CO_2 levels may modify adenylate energy charge and carbohydrate status, both of which are presumed to regulate respiration. Control of respiration by substrate and adenylate turnover is often superimposed as reflected in diurnal changes in respiration in response to light/dark cycles (Azcón-Bieto 1992). Increased substrate levels tend to raise the activity of the alternative oxidase path, rather than of the cytochrome path (Lambers 1985), which would result in increased CO_2 emission, but no increase in energy production. Plant respiration appears to respond to short-term changes in substrate availability (via photosynthesis) and in ATP demand, and also to long-term changes in substrate availability (see Azcón Bieto 1992). The details of the mechanism by which respiratory pathways are regulated by substrate and adenylate levels have not been established, but in principle, this dual system of regulation allows the plant to respond more efficiently to short-term and long-term environmental changes, including CO_2 changes.

7.3.1 Relationship between respiration and photosynthesis

There is a direct interdependence between photosynthesis and respiration: the greater the respiration and growth, the larger the photosynthetic organ size, and higher rates of photosynthesis result in greater availability of substrate for respiration (Azcón-Bieto and Osmond 1983). There are reductions in concentrations of the major enzymatic constituents of the photosynthetic apparatus of plants growing in elevated CO_2. The primary carboxylating enzyme, Rubisco, is often reduced in content in the foliage of plants grown in elevated CO_2 (Long and Drake 1992). Reduction of the content of this enzyme, which occupies a major fraction of the soluble protein content, may also be accompanied by the reduction in concentration of other soluble enzymes and this results in reduction of leaf level nitrogen concentration (Conroy 1992).

Reduced nitrogen concentration is commonly observed in plants grown in elevated CO_2 and it is to be expected that respiration rate would be lower in tissues having a lower nitrogen concentration because the respiratory cost of protein turnover and maintenance is relatively large. In studies of the salt marsh sedge, *Scirpus olneyi*, protein content was reduced 50% and the amount of Rubisco by a similar amount (Jacob *et al.* 1995). In this plant, Rubisco accounted for about 50% of the soluble protein. The metabolic cost for maintenance of tissues with high concentrations of protein is greater than the cost for the maintenance of tissues with low concentrations of nitrogen. Decrease in nitrogen concentration of the tissue seems to be an acclimation of plants growing at high CO_2 (Conroy 1992; Peñuelas and Matamala 1990). Thus, lower rates of respiration may result from the acclimation of the photosynthetic apparatus to elevated CO_2.

The higher level of respiration in preilluminated leaves is correlated with a high carbohydrate level and a higher CO_2 compensation point (Azcón-Bieto and Osmond 1983; Thomas *et al.* 1993). It has been suggested that the higher level of respiration of illuminated leaves may be associated with stimulated activity of the alternative pathway (Azcón-Bieto *et al.* 1983; Gardestrom and Edwards 1985; Thomas *et al.* 1993).

7.3.2 Long-term effects of high CO_2 on respiration in crops and native species

The general effect of increased photosynthesis by high CO_2 is to raise the levels of soluble carbohydrates in the cell (Bowes 1993). Although there are some exceptions (e.g. Ziska and Teramura 1992) growth in elevated CO_2 seems to cause increased respiration in crop species, which are typically fast growing. In such species, increasing CO_2 levels result in excess carbohydrate concentrations and stimulation of activity of the alternative pathway. The energy overflow hypothesis (Lambers 1982) suggests that the alternative oxidase could wastefully burn carbohydrates present in excess to the demands of growth, maintenance, ion transport and other physiologic requirements.

The alternative pathway can be engaged in leaves and roots when additional carbohydrates are supplied exogenously or by higher rates of photosynthesis. Addition of exogenous sugars increased respiration in wheat leaves and part of the increased respiration was due to increased engagement of the nonphosphorylating alternative pathway (Azcón-Bieto *et al.* 1983). Azcón-Bieto and Osmond (1983) showed that dark respiration rate of wheat leaves correlated with leaf carbohydrate levels determined by previous photosynthetic rate. The alternative oxidase was engaged by high CO_2 in sunflower roots (Gifford *et al.* 1985). Increased respiration rates in soybean leaves after 2–4 weeks in 1000 ppm of CO_2 correlated with increased respiration, but were not mediated by changes in the activity of cytochrome *c* oxidase, suggesting that the high CO_2 treatment had stimulated the alternative pathway (Hrubec *et al.* 1985). Similar results were found in cotton (Thomas *et al.* 1993), barley (Williams *et al.* 1992), sunflower (Gifford *et al.* 1985) pea (Musgrave *et al.* 1988) and soybean (Thomas and Griffin 1994). Respiration in detached maize root tips was increased without concomitant increase in the energy charge (Saglio and Pradet 1980) suggesting that the increased respiration was due to the stimulation of the alternative pathway.

Finally, some crops species showed decreased respiration. In the cases of alfalfa (Bunce and Caulfield 1991), tomato (Bunce 1990) and soybean (Bunce 1990) elevated CO_2 increased biomass production, but not growth rate.

Thus, in crop species, the general pattern seems to be that elevated CO_2 increases both carbohydrate levels and perhaps the alternative oxidase activity leading to greater efflux of CO_2. In the few cases where it has been examined, it does not appear that increased CO_2 efflux necessarily leads to increased energy production.

Native herbaceous species grown in elevated CO_2 usually show an increase in photosynthesis (Drake and Leadley 1991; Long and Drake 1992), an increase in biomass (Curtis *et al.* 1989; Hunt *et al.* 1991), a reduction in nitrogen concentration (Curtis *et al.* 1989; Peñuelas and Matamala 1990; Drake 1992a, b) and a decrease in dark respiration. We will now examine the effects of these changes on respiration.

Most reports of respiration in tree species show a decrease in respiration rates under elevated CO_2. Generally, experiments with trees were run for several months or even years and the magnitude of the response of respiration to high CO_2 often varies throughout the season. Chestnut seedlings grown for two years at 700 ppm CO_2 decreased respiration by about 50% of the controls at normal ambient CO_2 at the beginning and end of the growing season (El Kohen *et al.* 1991). Orange trees had decreased respiration by about 20% after four years of exposure to 700 ppm of CO_2 (Idso and Kimball 1992). White oak leaves and yellow-poplar leaves grown under 650 ppm CO_2 for two months had decreased respiration rate of 59 and 48% on a dry weight basis and 50 and 37% on an area basis and starch content was increased, but not sucrose levels (Wullschleger *et al.* 1991). After three years in high CO_2, white oak showed a decrease in young leaf respiration of 20% (Wullschleger and Norby 1992). Three years of exposure in 650 ppm of CO_2 decreased respiration rate (dry weight basis) by 45% without changes in specific growth rate (Wullschleger *et al.* 1992). Similar results were obtained in leaf respiration rates (area basis) in two species of maple (35 and 52% reduction), oak (45%) and apple (29%) during one month of growth in 700 ppm of CO_2 (Bunce 1992). Respiration was also decreased in the tropical trees *Acacia mangium* (30%), *Psychotria limonensis* (25%) and *Ficus obtusifolia* (20%) when grown at high CO_2 (Drake 1989). Seedling of loblolly pine increased photosynthesis, carbohydrates and biomass production, but there were no differences in dark CO_2 evolution after 100 days of growth at 650 ppm of CO_2 (Griffin *et al.* 1993).

The effects of elevated CO_2 on native herbaceous species is variable. Respiration increased with increased growth rate in *Plantago major* grown in elevated CO_2 (Poorter *et al.* 1988). In cases in which there was no significant change in tissue composition, as for example in *Lolium* and clover (Ryle *et al.* 1992a and b), and *Dactylus glomerata* (Bunce and Caulfield 1991), elevated CO_2 had no effect on whole plant respiration.

Reduced respiration is often observed in herbaceous plants grown in elevated CO_2. Respiration (biomass basis) of *Lolium* (Bunce and Caulfield 1991) and orchard grass were reduced in plants grown in elevated CO_2 (Ziska and Bunce 1993). Oxygen uptake decreased in the sedge, *Scirpus olneyi*, grown in high CO_2 for seven years. This effect occurred whether results were expressed on a dry weight or area basis and correlated with reduced activity and capacity of the cytochrome pathway and to the amount of the complex III and IV enzymes of the mitochondrial electron transport chain (Azcón-Bieto *et al.* 1994). These results also showed that respiration was tightly coupled to the activity of the cytochrome *c* oxidase and there was no change in the activity of the alternative pathway (Figure 7.2).

In summary, while there are exceptions in both cases, higher respiration rates of crop species and lower respiration rates of perennial native species grown in high CO_2 appear to be the pattern. Moreover, it appears that the stimulation of the alternative pathway accompanies increased respiration rates of annual, fast growing crops, but not slower growing, perennial species. In this survey, no effort was made to examine the influence of small rooting volume on these effects, but there may be an artifact of this effect in the results because many of the annual crop studies were done in controlled environments in which small pots may have been used. If there is an artifact in these data, the results may not apply to the vast majority of species growing in native environments, but it would emphasize our point that increased carbohydrates, especially soluble sugars, accompany stimulation of the alternative pathway reflecting the overflow of additional photosynthate.

7.4 INTERACTION BETWEEN GROWTH, RESPIRATION AND CO_2

McCree (1970) fitted an empirical equation in which a substantial portion of respiration of white clover plants was proportional to the daily net carbon gain by photosynthesis or growth. The remainder of the respiration was proportional to dry weight of living material in the plant, and was asssumed to reflect maintenance cost because it is apparently linked to the maintenance of plant life functions. This division of respiration has provided a useful construct within which to assess the role of various environmental and internal factors on the relationship between growth and respiration. Considerable experimental evidence agrees with the essential basis of this respiration model (see Amthor 1989), but from the strictly biochemical point of view, there is no distinction between growth and maintenance respiration. The relationship between these two components of respiration and the existence of nonphosphorylating pathways in plant mitochondria, which consume carbohydrate, but produce no useful energy, is also not clear at the moment.

High atmospheric CO_2 usually results in decreased nitrogen concentration and higher C/N ratios. It has been shown that the C/N ratio and the proportion of photosynthate invested in respiration increases with decreasing nitrogen supply (van der Werf *et al.* 1992). However, in plants growing in elevated CO_2 the ratio of respiration to photosynthesis decreases and it is therefore difficult to interpret the reduced nitrogen concentration in these plants as limiting plant growth. Higher C/N ratios can represent lower maintenance respiration (Wullschleger and Norby 1992; van der Werf *et al.* 1992; Bunce 1994) and lower costs in the growth of new tissue (Wullschleger *et al.* 1992; Wullschleger and Norby 1992; Griffin *et al.* 1993).

Wullschleger and Norby (1992) determined the effect of elevated CO_2 on growth and maintenance of leaves of white oak seedlings grown in open top chambers in the field in native, unfertilized soil. Specific respiration rates for leaves of this species were 56% lower in elevated CO_2 than in normal ambient CO_2 . The relative growth was reduced 31% and the maintenance coefficient 45%. These effects were attributed to the reduced cost of maintaining tissues having reduced nitrogen concentrations. Similar results were obtained in leaves of yellow poplar (Wullschleger *et al.* 1991). However, Ziska and Bunce (1993) report instances in which maintenance respiration was reduced for plants grown in elevated CO_2 even though these plants had no decrease in protein or nitrogen content. Thomas *et al.* (1993) found increased dark respiration resulting from enhanced maintenance respiration in cotton leaves growing at high CO_2. The authors related the increase in maintenance respiration to carbohydrate metabolism rather than to changes in leaf nitrogen concentration.

Plant respiration can decrease at high CO_2 even when there is no change in relative growth rate. A decrease of 30–40% in respiration at high CO_2 occurred even though growth rates were the same as for plants grown at normal CO_2 (Bunce and Caulfield 1991). Similar results were found in some C_3 crops (Bunce 1990). An increase in 50% in relative growth rate in plants grown under high CO_2 concentrations was supported without changes in CO_2 evolution (Bunce and Caulfield 1991). The lower respiration rates with constant relative growth rate seems to be related to changes in construction costs of the new tissue. High nitrogen availability increased the costs of leaf construction by 7%, but these costs were 3.5% lower in high CO_2 grown seedlings (Griffin *et al.* 1993).

Rising atmospheric CO_2 also appears to affect carbon use efficiency (CUE). CUE is the fraction of carbon taken up during the day (U) after accounting for nighttime emissions (E):

$$\text{CUE} = (U - E)/U$$

Gifford (1991) found that growth in elevated CO_2 either had no effect or increased CUE with an overall average of 4.4% for nine species. There was a significiant correlation between the increased CUE and decreased growth respiration (respiratory cost per unit of growth). On the other hand, the results for the effects of growth in elevated CO_2 on maintenance respiration were less clear with some species showing a decline in maintenance cost and others an increase (Gifford 1991). In short-term experiments, lasting up to 12 hours, Gifford found no significant effect of elevated CO_2 on whole plant respiration. Nevertheless, his data do show that when respiration of plants grown and tested in their respective CO_2 concentrations was compared, the plants in elevated CO_2 had lower rates of respiration than the controls in normal ambient CO_2.

7.5 CO_2 AND SOIL RESPIRATION

Energy requirements of the roots *vis-à-vis* the supply of assimilated carbon to the roots is an important factor determining the net carbon balance of whole plants. Some 30–60% of the daily photosynthetic carbon production is lost in respiration and a significant portion of this loss is from roots (Lambers 1985).

Respiration of roots of wheat, mung bean or sunflower was unchanged or slightly decreased by high CO_2, but the engagement of the alternative respiration was inconsistent (decrease, no effect, and increase, respectively) in plants grown at high CO_2 compared with plants grown at normal CO_2 (Gifford *et al.* 1985). Root respiration was higher in the first weeks of growth in roots of *Plantago major* growing at 700 ppm CO_2 (Poorter *et al.* 1988) and this did not correlate with shoot respiration.

Root respiration is rapidly affected by changes in aeration in the rhizosphere (Leshuk and Saltveit 1991; Drew 1992). The CO_2 tension in the soil can inhibit root (Palta and Nobel 1989) and soil microbial respiration (Koizumi *et al.* 1991).

There have been no studies of the effects of long-term exposure to elevated CO_2 on soil respiration. The microbial activity of the soil accounts for an average of 70% of the CO_2 efflux from the soil (Raich and Schlesinger 1992). Soil activity can produce a high CO_2 turnover: temperate forests have an estimated 680 g C m^{-2} yr^{-1} and tropical forests 1260 g C m^{-2} yr^{-1} soil respiration. Thus, soil microbial respiration is a major player in the global carbon budget and changes in the rate of respiration of these organisms, either through direct effects of CO_2 or indirectly through plant growth or climate change, can therefore affect atmospheric CO_2 (Koizumi *et al.* 1993; Wullschleger and Norby 1992).

In summary, not enough is known about the response of soil and root respiration to high atmospheric CO_2 in natural ecosystems where changes mediated by CO_2 in shoot growth requirements, biomass composition and decomposition, and changes in symbiotic associations are important components of the root + microbial CO_2 balance. Moreover, it is unclear how a change of 350 ppm CO_2 in air could influence the CO_2 in the soil, which can be of the order of 100 00 ppm. Priority should be given to the effects of elevated CO_2 on belowground processes, both of roots and microbes.

7.6 CONCLUSIONS

There is now abundant evidence that dark respiration in foliage and possibly in other plant tissues is reduced for a doubling of atmospheric CO_2. The primary effect is immediate and reversible and this effect appears to be altered by changes in tissue composition in some species. Elevated CO_2 inhibits *in vitro* activity of cytochrome *c* oxidase, succinate dehydrogenase and perhaps other enzymes of the mitochondrial electron transport system. Studies with cell-free extracts of enzymes, mitochondria, tissue cultures and of whole tissues all show that increased CO_2 reversibly inhibits both CO_2 efflux and O_2 assimilation consistent with the hypothesis that there is a direct effect of CO_2 on enzyme activity. In some cases the effect of changes in tissue composition caused by growth of plants in elevated CO_2 is to further decrease respiration and in others to cause a net increase of CO_2 efflux. In rapidly expanding whole leaves or in fast growing species exposed to elevated CO_2, high concentrations of carbohydrates stimulate activity of the alternate respiratory pathway. Stimulation of the alternative pathway is not accompanied by higher production of ATP.

In slow growing woody perennial species in which soluble carbohydrates do not tend to accumulate, respiration appears to be inhibited and there does not seem to be the level of engagement of the alternative oxidase that occurs in fast growing species including crops.

Although there is evidence that the inhibitory effect of elevated CO_2 on respiration extends to nonphotosynthetic microbes involved in decomposition, this would only have practical meaning for a high CO_2 world if these decomposers were exposed to near atmospheric levels of CO_2. It is hard to imagine that the high concentrations of CO_2 found in soil could be substantially altered by doubling of atmospheric CO_2 . Thus, the rates of soil respiration are not likely to be influenced by increasing atmospheric CO_2 except as related to changes in composition of decomposing plant tissues.

If increasing CO_2 concentrations reduce the rates of respiration in all plant tissues and in microbes and fungi as well, then it should also be expected that there would be a significant increase in the carbon accumulating in ecosystems as a result of the combined effects of stimulation of photosynthesis and reduction in respiration.

While the results of the effect of elevated CO_2 on respiration have been found to vary widely, even small effects to reduce respiration would have a significant impact on global carbon budgets. For this reason, additional study of the role of increasing atmospheric CO_2 in respiration should be given a high priority for research.

7.7 ACKNOWLEDGEMENTS

The authors are grateful to Drs Robert Gifford and James Bunce for useful comments in previous versions of the manuscript. This work was supported by US Department of Energy, the Smithsonian Institution and by fellowships programs of the Smithsonian Institution and Ministerio de Educación y Ciencia (FPI and FPU, Spanish Government) to M.A. G.-M.

7.8 REFERENCES

Allen, L. H. (1989) Responses of vegetation to carbon dioxide. In: *Temperature and CO_2 Effects on Rice*: Vol. 53. Carbon Dioxide Research Division, Office of Energy Research, United States Department of Energy, Washington, DC 20545.

Allen, L. H. (1990) Plant responses to rising carbon dioxide and potential interactions with air pollutants. *J. Env. Qual.* **19**, 15–34.

Amthor, J. S. (1989) *Respiration and Crop Productivity*. Springer-Verlag, New York, 215 pp.

Amthor, J. S. (1991) Respiration in a future, higher-CO_2 world. *Plant Cell Environ.* **14**: 13–20.

Amthor, J. S., Koch, G. W. and Bloom, A. J. (1992) CO_2 inhibits respiration in leaves of *Rumex crispus* L. *Plant Physiol.* **98**, 757–760.

Azcón-Bieto, J. (1983) Inhibition of photosynthesis by carbohydrates in wheat leaves. *Plant Physiol.* **73**, 681–686.

Azcón-Bieto, J. (1992) Relationships between photosynthesis and respiration in the dark in plants. In: Barber, J., Guerrero, M.G. and Medrano, H. (Eds) *Trends in Photosynthesis Research*, pp. 241–255. Intercept Ltd, Andover, UK.

Azcón-Bieto, J., Lambers, H. and Day, D. A. (1983) Effect of photosynthesis and carbohydrate status on respiratory rates and the involvement of the alternative pathway in leaf respiration. *Plant Physiol.* **72**, 598–603.

Azcón-Bieto, J. and Osmond, C.B. (1983) Relationship between photosynthesis and respiration. The effect of carbohydrate status on the rate of CO_2 production by respiration in darkened and illuminated wheat leaves. *Plant Physiol.* **71**, 574–581.

Azcón-Bieto, J., Gonzàlez-Meler, M. A., Dougherty, W. and Drake, B. G. (1994) Acclimation of respiratory O_2 uptake in green tissues of field grown native species after long-term exposure to elevated CO_2. *Plant Physiol.* **106**: 1163–1168.

Bahr, J. T. and Bonner, W. D. Jr (1973) Cyanide insensitive respiration. II. Control of the alternative pathway. *J. Biol. Chem.* **248**, 3446–3450.

Baker, J. T., Laugel, F., Boote, K. J. and Allen, L. H. Jr (1992) Effects of daytime carbon dioxide concentration on dark respiration in rice. *Plant, Cell Environ.* **15**, 231–239.

Bazzaz, F. A. (1990) The response of natural ecosystems to the rising global CO_2 levels. *Annu. Rev. Ecol. Syst.* **21**, 167–196.

Bendall, D. S., Ranson, S. L. and Walker, D. A. (1958) Some effects of carbon dioxide–bicarbonate mixtures on the oxidation and reduction of cytochrome c by Ricinus mitochondria. *Nature* **181**, 133–134.

Bowes, G. (1993) Facing the inevitable: plants and increasing atmospheric CO_2. *Annu. Rev. Plant Physiol. Plant Molecul. Biol.* **44**, 309–332.

Bown, A. W. (1985) CO_2 and intracellular pH. *Plant, Cell Environ.* **8**, 459–465.

Bunce, J. A. (1990) Short-term and long-term inhibition of respiratory carbon dioxide efflux by elevated carbon dioxide. *Ann. Bot.* **65**, 637–642.

Bunce, J. A. (1992) Stomatal conductance, photosynthesis and respiration of temperate deciduous tree seedlings grown outdoors at an elevated concentration of carbon dioxide. *Plant, Cell Environ.* **15**, 541–549.

Bunce, J. A. (1994) Response of respiration to increasing atmospheric carbon dioxide concentrations. *Physiol. Plant* **90**, 427–430.

Bunce, J. A. and Caulfield, F. (1991) Reduced respiratory carbon dioxide efflux during growth at elevated carbon dioxide in three herbaceous perennial species. *Ann. Bot.* **67**, 325–330.

Conroy, J. (1992) Influence of elevated atmospheric CO_2 on plant nutrition. *Austr. J. Bot.* **40**, 445–456.

Curtis, P. S., Drake, B. G and Whigham, D. F. (1989) Nitrogen and carbon dynamics in C_3 and C_4 estuarine marsh plants grown under elevated CO_2. *Oecologia* **78**, 297–301.

Day, D. A., Arron, G. P., Christoffersen, R. E. and Laties, G. G. (1978) Effect of ethylene and carbon dioxide on potato metabolism. Stimulation of tuber and mitochondrial respiration and inducement of the alternative path. *Plant Physiol.* **62**, 820–825.

Dennis, D. T. and Turpin, D. H. (1990) *Plant Physiology, Biochemistry and Molecular Biology.* Longman Scientific and Technical, Singapore.

Drake, B. G. (1989) *Effects of Elevated Carbon Dioxide on Chesapeake Bay Wetlands*, Vol. 051: IV *Ecosystem and Whole Plant Responses.* April-November 1988. Carbon Dioxide Research Division, Office of Energy Research, United States Department of Energy, Washington, DC 20545

Drake, B. G. (1992a) The impact of rising CO_2 on ecosystem production. *Water, Soil, Air Pollut.* **64**, 25–44.

Drake, B. G. (1992b) A field study of the effects of elevated CO_2 on ecosystem processes in a Chesapeake Bay wetland. *Aust. J. Bot.* **40**, 579–595.

Drake, B. G. and Leadley, P. W. (1991) Canopy photosynthesis of crops and native plant communities exposed to long-term elevated CO_2 treatment. *Plant, Cell Environ.* **14**, 853–860.

Drew, M. C. (1992) Soil aeration and plant root metabolism. *Soil Science* **154**, 259–268.

Dry, I. B., Moore, A. L., Day, D. A. and Wiskich, J. T. (1989) Regulation of alternative pathway activity in plant mitochondria: nonlinear relationship between electron flux and the redox poise of the quinone pool. *Arch. Biochem. Biophys.* **273**, 148–157.

El Kohen, A., Pontailler, J.-Y. and Mousseau, M. (1991) Effet d'un doublement du CO_2 atmosphérique sur la respiration à l'obscurité des parties aeriennes de jeunes chataigniers (*Castanea sativa* Mill). *C. R. Acad. Sci. Paris III*, **312**, 477–481.

Farrar, J. F. and Williams, M. L. (1991) The effects of increased atmospheric carbon dioxide and temperature on carbon partitioning, source–sink relations and respiration: commissioned review. *Plant, Cell Environ.* **14**, 819–830.

Frenkel, C. and Patterson, M. E. (1973) Effect of carbon dioxide on activity of succinic dehydrogenase in 'Bartlett' pears during cold storage. *Hort. Sci.* **8**, 395–396.

Gardestrom, P. and Edwards, G. E. (1985) Leaf mitochondria (C_3 + C_4 + CAM) In: Douce, R. and Day, D. A. (Eds) *Encyclopedia of Plant Physiology*, Vol. 18, pp. 314–346. Springer-Verlag, Berlin.

Gary, C. and Veyres, P. (1992) CO_2 concentration and the photosynthesis–respiration balance of vegetative tomato plants. In: Lambers, H. and van der Plas, L. H. W. (Eds) *Molecular, Biochemical and Physiological Aspects of Plant Respiration*, pp. 553–557. SPB Academic Publishing BV, The Hague.

Gifford, R. M. (1991) *Impact of Increasing Atmospheric Carbon Dioxide Concentration on the Carbon Balance of Vegetation.* Energy Research and Development Corporation (Australia). Canberra, Project Report No. ERDC 37.

Gifford, R. M., Lambers, H. and Morison, J. I. L. (1985) Respiration of crop species under CO_2 enrichment. *Plant Physiol.* **63**, 351–356.

Gonzàlez-Meler, M. A., Aranda, X., Ribas-Carbo, M., Peñuelas, J. and Azcón-Bieto, J. (1992) Titration of cytochrome pathway by azide, cyanide and sulfide in leaf slices of *Pisum sativum* L. In: Lambers, H. and van der Plas, L. H. W. (Eds) *Molecular, Biochemical and Physiological Aspects of Plant Respiration*, pp. 603–608. SPB Academic Publishing bv, The Hague.

Gonzàlez-Meler, M. A. (1995). Effects of increasing atmospheric concentration of carbon dioxide on plant respiration. PhD Thesis, Universitat de Barcelona, Barcelona.

Gonzàlez-Meler, M. A., Ribas-Carbo, M., Siedow, J. N. and Drake, B. G. (1996) The direct inhibition of plant mitochondrial respiration by elevated CO_2. *Plant Physiol.* in press.

Griffin, K. L. Thomas, R. B. and Strain, B. R. (1993) Effects of nitrogen supply and elevated carbon dioxide on construction cost in leaves of *Pinus taeda* (L) seedlings. *Oecologia* **95**, 575–580.

Henry, M. F. and Nyns, E. J. (1975) Cyanide-insensitive respiration. An alternative mitochondrial pathway. *Sub-Cellular Biochem.* **4**, 1–65.

Hogan, K., Smith, A. P. and Ziska, L. H. (1991) Potential effects of elevated CO_2 and changes in temperature on tropical plants: commissioned review. *Plant, Cell Environ.* **14**, 763–778.

Hrubec, T. C., Robinson, J. M. and Donaldson, R. P. (1985) Effects of CO_2 enrichment and carbohydrate content on the dark respiration of soybeans. *Plant Physiol.* **79**, 684–689.

Hunt, R., Hand, D. W., Hannah, M. A. and Neal, A. M. (1991) Response to CO_2

enrichment in 27 herbaceous species. *Funct. Ecol.* **5**, 410–421.

Idso, S. B. and Kimball, B. A. (1992) Effects of atmospheric CO_2 enrichment on photosynthesis, respiration, and growth of sour orange trees. *Plant Physiol.* **99**, 341–343.

Jacob, J., Greitner, C. and Drake, B. G. (1995) Acclimation of photosynthesis in relation to Rubisco and non-structural carbohydrate content and in situ carboxylase activity in *Scirpus olneyi* grown at elevated CO_2 in the field. *Plant Cell Environ.* **18**: 875–884.

Kaplan, A., Gale, J. and Poljakoff-Mayber, A. (1977) Effect of oxygen and carbon dioxide concentrations on gross dark CO_2 fixation and dark respiration in *Bryophyllum daigremontianum. Austr. J. Plant Physiol.* **4**, 745–752.

Kerbel, E. L. Kader, A. A. and Romani, R. J. (1988) Effects of elevated CO_2 concentrations on glycolysis in intact 'Bartlett' pear fruit. *Plant Physiol.* **86**, 1205–1209.

Kerbel, E. L. Kader, A. A. and Romani, R. J. (1990) Respiratory and glycolytic response of suspension-cultured 'Passe Crassane' pear fruit cells to elevated CO_2 concentrations. *J. Am. Soc. Hort. Sci.* **115**, 111–114.

Kidd, F. (1916) The controlling influence of carbon dioxide. Part III. The retarding effect of carbon dioxide on respiration. *Proc. R. Soc. L. Series B.* **89**, 136–156.

Koizumi, H., Nakadai, T., Usami, Y., Satoh, M. Shiyomi, M. and Oikawa, T. (1991) Effect of carbon dioxide concentration on microbial respiration in soil. *Ecol. Res.* **6**, 227–232.

Koizumi, H., Usami, Y. and Satoh, M. (1993) Carbon dynamics and budgets in three upland double-cropping agro-ecosystems in Japan. *Agric., Ecosyst., Environ.* **43**, 235–244.

Lambers, H. (1982) Cyanide-resistant respiration: a non-phosphorylating electron transport pathway acting as an overflow. *Plant Physiol.* **55**, 478–485.

Lambers, H. (1985) Respiration in intact plants and tissues: its regulation and dependence on environmental factors, metabolism and invaded organisms. In: Douce, R. and Day, D.A. (Eds) *Encyclopedia of Plant Physiology*, Vol. 18, pp. 418–473. Springer-Verlag, Berlin.

Lambers, H. Day, D. A. and Azcón-Bieto, J. (1983) Cyanide-resistant respiration in roots and leaves. Measurements with intact tissues and isolated mitochondria. *Physiol. Plant.* **58**, 148–154.

Lange, H. (1970) Atmungswege bei vernarbenden und proliferienden Gewebefragmenten der Kartoffelknolle. *Planta* **90**, 119–132.

Leshuk, J. A. and Salveit M. E., Jr (1991) Effects of rapid changes in oxygen concentration on the respiration of carrot roots. *Physiol. Plant.* **82**, 559–568.

Long, S. P. and Drake, B. G. (1992) Photosynthetic CO_2 assimilation and rising atmospheric CO_2 concentrations. Commissioned review. In: Baker, N. R. and Thomas, H. (Eds) *Topics in Photosynthesis*, Vol. 11, pp. 69–103. Elsevier Science Publishers BV, The Hague, Amsterdam.

Ludwig, R., Charles-Edwards, D. A. and Withers, A. C. (1975) Tomato leaf photosynthesis and respiration in various light and carbon dioxide environments. In: Marcelle, R. (Ed.) *Environmental and Biological Controls of Photosynthesis*: Proceedings of a Conference held at Limburgs Universitat, Diepenbeck, Belgium, pp. 29–36. W. Junk, The Hague, Amsterdam.

McCaig, T. N. and Hill, R. D. (1977) Cyanide-insensitive respiration in wheat: cultivar differences and effects of temperateature, carbon dioxide, and oxygen. *Can. J. Bot.* **55**, 549–555.

McCree, K. J. (1970) An equation for the rate of respiration in white clover grown under controlled conditions. In: Setlik, I. (Ed.) *Prediction and Measurement of Photosynthetic Productivity*, pp. 221–229. Centre for Agricultural Publishing and Documentation (PUDOC), Wageningen.

Miller, G. W. and Evans, H. J. (1956) Inhibition of plant cytochrome oxidase by bicarbonate. *Nature* **178**, 974–976.

Møller, I. A., Bérczi, A., van der Plas, L. H. W. and Lambers, H. (1988) Measurement of the activity and capacity of the alternative pathway in intact plant tissues: identification of problems and possible solutions. *Plant Physiol.* **72**, 642–649.

Monning, A. (1983) Studies on the reaction of Krebs cycle enzymes from apple tissue (cv. Cox orange) to increased levels of CO_2. *Acta Hortic.* **138**, 113–119.

Morison, J. I. L. (1990) Plant and ecosystem responses to increasing atmospheric CO_2. *Trends Ecol. Evol.* **5**, 69–70.

Mott, K. A. (1990) Sensing of atmospheric CO_2 by plants. *Plant, Cell Environ.* **13**, 731–737.

Mousseau, M. and Saugier, B. (1992) The direct effect of increased CO_2 on gas exchange and growth of forest tree species. *J. Exp. Bot.* **43**, 1121–1130.

Musgrave, M. E., Strain, B. Y. and Siedow, J. N. (1986) Response of two pea hybrids to CO_2 enrichment: A test of the energy overflow hypothesis for alternative respiration. *Proc. Natl. Acad. Sci.* **83**: 8157–8161.

Naik, R. M., Dhage, A. R., Munjal, S. V., Singh, P., Desai, B. B., Mehta, S. L. and Naik, M. S. (1992) Differential carbon monoxide sensitivity of cytochrome c oxidase in the leaves of C_3 and C_4 plants. *Plant Physiol.* **98**, 984–987.

Nilovskaya, N. T. and Razoryonova, T. A. (1968) Respiration rate of vegetable plants at various partial pressures of carbon dioxide. *Fiziol. Rast.* **5**, 876–873.

Nobel, P. S. and Palta, J. A. (1989) Soil O_2 and CO_2 effects on root respiration of cacti. *Plant and Soil* **120**: 263–271.

Palet, A., Ribas-Carbó, M., Argilés, J. M. and Azcón-Bieto, J. (1991) Short-term effects of carbon dioxide on carnation callus cell respiration. *Plant Physiol.* **96**, 467–472.

Palet, A., Ribas-Carbó, M., Gonzàlez-Meler, M. A., Aranda, X. and Azcón-Bieto, J. (1992) Short-term effects of CO_2/bicarbonate on plant cell respiration. In: Lambers, H. and van der Plas, L.H. W. (Eds) *Molecular, Biochemical and Physiological Aspects of Plant Respiration*, pp. 597–601. SPB Academic Publishing BV, The Hague, Amsterdam.

Palta, P. S. and Nobel, P. S. (1985) Influence of soil O_2 and CO_2 on root respiration for *Agave deserti. Physiol. Plant.* **76**, 187–192.

Peñuelas, J. and Matamala, R. (1990) Changes in N and S leaf content, stomatal density and specific leaf area of 14 species during the last three centuries. *J. Exp. Bot.* **41**, 1119–1124.

Perez-Trejo, M. S. (1981) Mobilization of respiratory metabolism in potato tubers by carbon dioxide. *Plant Physiol.* **67**, 514–517.

Poorter, H., Gifford, R. M., Kriedemann, P. E. and Wong, S. C. (1992) A quantitative analysis of dark respiration and carbon content as factors in the growth response of plants to elevated CO_2. *Austr. J. Bot.* **40**, 501–513.

Poorter, H. Pot, S. and Lambers, H. (1988) The effect of an elevated atmospheric CO_2 concentration on growth, photosynthesis and respiration of *Plantago major. Physiol. Plant.* **73**, 553–559.

Raich, J. W. and Schlesinger, W. H. (1992) The global carbon dioxide flux in soil respiration and its relationship to vegetation and climate. *Tellus*, **44B**, 81–89.

Reuveni, J. and Gale, J. (1985) The effect of high levels of carbon dioxide on dark respiration and growth of plants. *Plant Cell Environ.* **8**, 623–628.

Reuveni, J., Gale, J. and Mayer, A. M. (1993) Reduction of respiration by high ambient CO_2 and the resulting error in measurements of respiration made with O2 electrodes. *Ann. Bot.* **72**, 129–131.

Ryan, M. G. (1991) Effects of climate change on plant respiration. *Ecol. Appl.* **1**, 157–167.

Ryle, G. J. A., Powell, C. E. and Tewson, V. (1992a) Effect of elevated CO_2 on the photosynthesis, respiration and growth of perennial ryegrass. *J. Exp. Bot.* **43**, 811–818.

Ryle, G. J. A., Powell, C. E. and Tewson, V. (1992b) Influence of elevated CO_2 and temperature on the photosynthesis and respiration of white clover dependent on N_2 fixation. *Ann. Bot.* **70**: 213–220.

Saglio, P. H. and Pradet, A. (1980) Soluble sugars, respiration, and energy charge during aging of excised maize root tips. *Plant Physiol.* **66**, 516–519.

Shaish, A., Roth Bejarano, N. and Itai, C. (1989) The response of stomata to CO_2 related to its effect on respiration and ATP level. *Physiol. Plant.* **76**, 107–111.

Shipway, M. R. and Bramlage, W. J. (1973) Effects of carbon dioxide on activity of apple mitochondria. *Plant Physiol.* **51**, 1095–1098.

Silsbury, J. H. and Stephens, R. (1984) Growth efficiency of *Trifolium subterraneum* at high carbon dioxide concentration. In: *Advances in Photosynthesis Research*, Vol. IV. pp. 133–136. Marinus Nijhoff/Dr W. Junk, The Hague, Amsterdam.

Spalding, M. H., Goffrey, P. A. and Edwards, G. E. (1980) Malate decarboxylation in isolated mitochondria from the crassulaceum acid metabolism plant *Sedum praealtum*. *Arch. Biochem. Biophys.* **199**, 448–456.

Spencer, W. and Bowes, G. (1986) Photosynthesis and growth of water hyacinth under CO_2 enrichment. *Plant Physiol.* **82**, 528–533.

Thomas, R. B. and Griffin, K. L. (1994) Direct and indirect effects of atmospheric carbon dioxide enrichment on leaf respiration of *Glycine max*. (L.) Merr. *Plant Physiol.* **104**: 351–361.

Thomas, R. B., Reid, C. D., Ybema, R. and Strain, B. R. (1993) Growth and maintenance components of leaf respiration of cotton grown in elevated carbon dioxide partial pressure. *Plant, Cell Environ.* **16**, 539–546.

Van der Werf, A., Welschen, R. and Lambers, H. (1992) Respiratory losses increase with decreasing inherent growth rate of a species and with decreasing nitrate supply: a search for explanations for these observations. In: Lambers, H. and van der Plas, L.H. W. (Eds) *Molecular, Biochemical and Physiological Aspects of Plant Respiration*, pp. 421–432. SPB Academic Publishing BV, The Hague, Amsterdam.

Williams, M. L., Jones, D. G., Baxter, R. and Farrar, J. F. (1992) The effect of enhanced concentrations of atmospheric CO_2 on leaf respiration. In: Lambers, H. and van der Plas, L. H. W. (Eds) *Molecular, Biochemical and Physiological Aspects of Plant Respiration*, pp. 547–551. SPB Academic Publishing BV, The Hague, Amsterdam.

Wullschleger, S. D.and Norby, R. J. (1992) Respiratory cost of leaf growth and maintenance in white oak saplings exposed to atmospheric CO_2 enrichment. *Can. J. For. Res.* **22**, 1717–1721.

Wullschleger, S. D., Norby, R. J. and Gunderson, C. A. (1992) Growth and maintenance respiration in leaves of *Liriodendron tulipifera* L. saplings exposed to long-term carbon dioxide enrichment in the field. *New Phytol.* **121**, 515–523.

Wullschleger, S. D., Norby, R. J. and Hendrix, D. L. (1991) Carbon exchange rates, chlorophyll contents, and carbohydrate status on two forest species exposed to carbon dioxide enrichment. *Tree Physiol.* **10**, 21–31.

Young, R. E., Romani, R. J. and Biale, J. B. (1962) Carbon dioxide effects on fruit respiration II. Response of avocados, bananas and lemons. *Plant Physiol.* **37**: 416–422.

Ziska, L. H. and Bunce, J. A. (1993) Inhibition of whole plant respiration by elevated CO_2 as modified by growth temperature. *Physiol. Plant.* **87**, 459–456.

Ziska, L. H. and Teramura, A. H. (1992) CO_2 enhancement of growth and photosynthesis in rice (*Oryza sativa*). Modification by increased ultraviolet-B radiation. *Plant Physiol.* **99**, 473–481.

8 Carbon and Nutrient Allocation in Terrestrial Ecosystems

G. R. SHAVER[1] and J. D. ABER[2]
[1]*The Ecosystems Center, Marine Biological Laboratory, Woods Hole, USA*
[2]*Complex Systems Research Center, Institute for the Study of Earth, Oceans, and Space, University of New Hampshire, Durham, USA*

8.1 INTRODUCTION

The term 'allocation' refers to the ways in which plants distribute elemental and energy resources among the various plant functions including growth, reproduction, maintenance, storage, defense, and uptake of additional resources. The relative distribution of plant resources among these functions (i.e. the allocation pattern) varies greatly among species, and it is also responsive to environmental change and developmental changes within individual plants. This chapter provides a brief review of the variation in plant allocation patterns and their controls in terrestrial ecosystems, along with a discussion of the implications of variation in plant allocation for vegetation and ecosystem responses to climate change.

From a plant perspective, allocation is important because it is closely related to the plant's long- and short-term survival, productivity, and reproductive output. In this view, controls on allocation are a major component of a plant's overall adaptive 'strategy' (e.g. Harper and Ogden 1970; Grime 1979; Tilman 1988). Plant allocation is thus viewed as an optimization process in which survival, productivity, or reproduction are maximized at some 'optimum' allocation pattern. However, the 'optimum' allocation pattern may differ depending on the environment, the time scale of interest, and the variable being maximized. For example, short-term productivity might be maximized when a plant allocates most of its resources into growth of organs, such as leaves and roots, that allow further uptake of resources and thus further growth. Long-term productivity, on the other hand, might be maximized by greater allocation to storage or to defense against herbivores, which would increase long-term plant survival and decrease losses of resource-acquiring organs like leaves. An entirely different pattern of allocation might be considered optimal if reproductive output were the variable to be maximized.

From an ecosystem perspective, plant allocation patterns have profound im-

Global Change: Effects on Coniferous Forests and Grasslands.
Edited by A. I. Breymeyer, D. O. Hall, J. M. Melillo and G. I. Ågren

UNEP

plications for overall element cycling and especially for the accumulation and storage of organic matter in soils (Field *et al.* 1992). Because plant allocation pattern influences total primary productivity, it also influences total litter inputs to the soil. Because allocation determines the relative amounts of leaves, stems, roots, and other plant parts in vegetation, it also determines the relative amounts of different litter types and the location where the litter is added to the soil. Finally, allocation controls the chemical composition of litter as well and thus has a major influence on its rate of decomposition.

Effects of plant allocation pattern on soil organic matter dynamics may have important feedbacks on future plant allocation patterns (Field *et al.* 1992). The feedbacks result from the effects of variation in litter quality on soil resource recycling and future resource availability to plants. If resource availability changes as a result of plant allocation patterns, then the 'optimum' plant allocation pattern to resource uptake must also change.

8.2 ALLOCATION WITHIN PLANTS

Most of our understanding of allocation within plants is phenomenological, not mechanistic. That is, the broad patterns of allocation in relation to environment have been described well enough to allow useful but rough predictions about the types of allocation patterns one should expect in a given environment and how those patterns might change in response to environmental change (e.g. Grime 1979). One particularly useful approach has been to apply the principles of cost–benefit analysis and other economic analogies to develop such predictions (Bloom *et al.* 1985; Chapin *et al.* 1990). However, the underlying physiological mechanisms that drive allocation are still largely unknown (Wilson 1988; Ingestad and Ågren 1991); many of these mechanisms appear to be hormonally mediated (Chapin 1991) but are rarely studied by ecologists. It is largely because our mechanistic understanding is lacking that predictions often differ depending on assumptions about the 'goal' of allocation.

8.2.1 Allocation among tissues

One of the oldest and still most commonly used assumptions about within-plant allocation is that its 'goal' is to help achieve a more optimal functional balance among the organs responsible for resource uptake, particularly roots and shoots (Brouwer 1966). For example, Davidson (1969) proposed that the allocation of photosynthate to production of new roots and shoots should be controlled so as to maintain a constant proportional relationship between whole-plant photosynthesis and root absorption, i.e.

$$\text{Root mass} \times \text{rate}_{(\text{absorption})} \propto \text{Leaf mass} \times \text{rate}_{(\text{photosynthesis})}$$

where the rates of absorption or photosynthesis are expressed per unit root or leaf mass and the constant of proportionality represents some 'optimal' nutritional balance. Thus, plants compensate for changes in carbon or nutrient availability by increasing allocation to either tissue mass or biochemical uptake machinery in the tissue responsible for the resource with lower availability.

Functional balance models like Davidson's (1969), and more detailed analyses based on economic theory (Bloom *et al.* 1985) are often useful in making predictions about and interpreting variation in root:shoot ratio among sites, species, and vegetation types. They are certainly more general than simple allometric models that do not allow for plant responses to variation in resource availability (Wilson 1988). However, functional balance models are not mechanistic and they typically do not consider allocation to defense, storage, or reproduction.

Mechanistic models of root–shoot allocation are most often based on assumptions about the supply, internal concentration gradients, and consumption of labile carbon (derived from leaf photosynthesis) and labile nitrogen (N) or 'nutrients' (derived from root uptake). Many of these models are based on an original formulation by Thornley (1972; reviewed by Wilson 1988). In these models, transport of labile carbon to roots and labile nutrients to shoots is a function of the concentration differences between roots and shoots. The concentration gradients are determined by the rate of photosynthesis in shoots or nutrient uptake by roots, followed by consumption of labile carbon and nutrients at both ends of the gradient due to growth of roots and shoots. The growth rates of these organs are determined by the available concentrations of both substrates. In effect, roots and shoots compete for labile carbon and nutrient substrates to support growth, with roots having the advantage with respect to nutrient supply and shoots with respect to carbon supply. The result is that root growth is favored over shoot growth when nutrient supply is low, and shoot growth is favored over root growth when carbon supply is low.

Thornley-type models of root–shoot allocation often do a good job at predicting changes in allocation in response to a wide range of environmental conditions, including changes in light, water, CO_2 concentration, temperature, and pruning or defoliation (Wilson 1988). However, Ingestad and Ågren (1991; Ågren and Ingestad 1987) have recently argued that root–shoot allocation patterns are in fact less sensitive to external factors affecting carbon supply than predicted by Thornley-type models (e.g. Johnson and Thornley 1987) and by 'optimal allocation' theory (e.g. Bloom *et al.* 1985). At least in part, the differences between these models may be explained by their differing assumptions about external controls on nutrient supply (Rastetter and Shaver 1992). Thornley and others usually assume that nutrients are available at constant concentrations that are unaffected by root uptake, and thus, additional nutrients can always be taken up if allocation to root growth is increased. Ingestad and Ågren, on the other hand, assume that nutrient supply rate and uptake rate are equal, and thus additional allocation to roots has no effect on uptake unless the supply rate is increased. Reconciliation of these two approaches depends on testing them under conditions of both constant external nutrient concentrations and constant nutrient supply rates (Rastetter and Shaver 1992).

8.2.2 Structural and maintenance costs

Plant organs vary widely in their biochemical composition, i.e. in the relative abundance of lignin, cellulose, tannin, protein, sugar, starch, nucleic acids, and other essential molecules (Mooney 1972; Penning de Vries *et al.* 1974; Chapin 1989). This biochemical composition is dependent not only on the organ's principal function (e.g. photosynthesis, support, nutrient uptake) but also on other factors such as its expected longevity, exposure to environmental extremes, and exposure to herbivory. Because of this wide variation in composition among tissues, and because the costs of synthesis of specific molecules vary widely, one might expect the overall resource requirements for allocation to growth to be similarly variable. In other words, the same amount of plant resources allocated to growth might end up producing very different amounts of growth.

In fact, the overall carbon and energy costs of organ growth are fairly similar both within and among tissues, so that organ mass alone is a reasonable indicator of these allocation costs (Abrahamson and Caswell 1982). The reason for this similarity in carbon and energy costs is that almost all plant parts have fairly high concentrations of some energetically expensive compounds like lignin, protein, or tannin, even though the relative abundances of these compounds is variable (Chapin 1989; Ryan 1991b).

Nitrogen and phosphorus costs of growth, on the other hand, are much more variable among plant species and organs, and organ mass alone is not a good indicator of these costs. Variation in N and P costs is in general related to variation in N and P concentration, with leaves typically being more costly than other tissues. Leaves of deciduous tree and shrub species and forbs have higher nutrient costs that those of graminoids or (especially) evergreens (Chapin 1989; Reich *et al.*1992).

Once a tissue or organ is produced, at least some plant resources must be allocated to its maintenance (Penning de Vries 1975; Ryan 1991a). Maintenance allocation is related to the overall metabolic activity of an organ and especially to the need for replacement and resynthesis of its protein or N content. The respiratory costs of maintenance are also strongly affected by temperature, in contrast to respiration associated with growth which is more related to the specific biochemical products being synthesized than to temperature. The temperature dependence of allocation to maintenance respiration is particularly important in that it is a major component of any plant's carbon budget that is not subject to control by the plant.

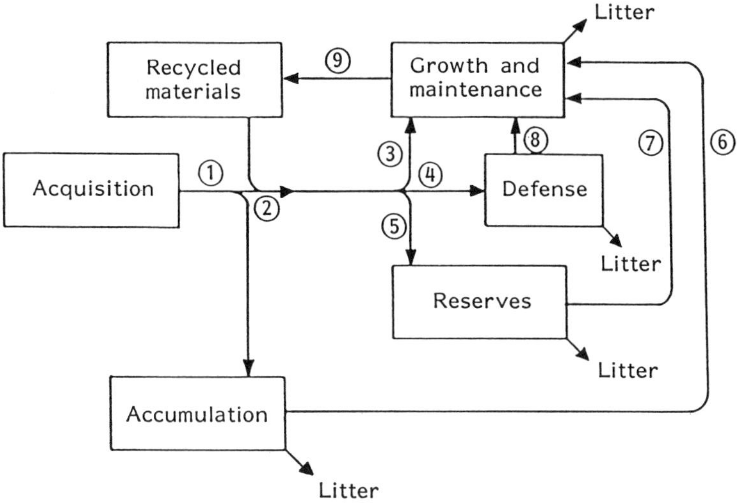

Figure 8.1 Allocation of resources to the three major classes of storage, and to defense, growth, and maintenance involves a number of interrelated resource pools and fluxes (from Chapin *et al.* 1990). *Accumulation* occurs when resource acquisition exceeds resource utilization in reserve formation, defense, and growth and maintenance (i.e. when flux 1 > flux 2). *Reserve formation* (flux 5), on the other hand, is more strongly metabolically regulated and competes directly with defense and growth (fluxes 3 and 4) for resources. Accumulation, reserves and defense (flux 8) may be used in future growth (fluxes 6 and 7), as can *recycled materials* (flux 9)

8.2.3 Allocation to storage and defense

Storage and defense against herbivory are often major components of a plant's overall allocation pattern (Figure 8.1). Storage includes a wide range of chemical compounds such as sugars, starches and other polysaccharides, some lipids, proteins or free amino acids, and both soluble organic and inorganic forms of P (Chapin *et al.* 1990). Many of these compounds also perform other, nonstorage functions as well, making it difficult to estimate the total allocation to storage. However, at times (especially during dormant periods) these storage compounds may account for more than half of the mass of perennial, nonwoody tissues in many plants, and even in woody tissues storage compounds may form the major portion of the mobile, high-turnover chemical fractions.

Chapin et al. (1990) defined three major classes of storage: accumulation, reserve formation, and recycling (Figure 8.1). Accumulation is simply storage that results from excess resource supply relative to immediate demands for growth or maintenance. For example, accumulation of one resource commonly occurs when its availability relative to other resources increases, and the plant cannot instantaneously reallocate its uptake effort to maintain a perfectly balanced nutrition. This form of storage may or may not be used for future growth and maintenance; if it is not it may be lost in litterfall (with consequent effects on litterfall chemistry and decomposition). Reserve formation, on the other hand, is the metabolically regulated production of storage products that might otherwise be used in growth or other plant functions. An example is the autumnal buildup of starch in roots and other perennial tissues of many plants, presumably to meet winter maintenance needs and to support growth the following spring. Recycling involves the breakdown of compounds already functioning to support growth, defense, or other plant processes, and storage of the breakdown products for reuse at a later time. Because most of the materials recycled come from senescent tissues, recycling is a major control over litter chemistry.

The broad patterns of storage and its variation among plants and ecosystems have been reasonably well-described although again the mechanisms controlling storage are poorly understood (Chapin *et al.* 1990). Accumulation is most common in plants with conservative growth patterns that are relatively unresponsive to short-term environmental fluctuation; these are often slow-growing plants found in more stressful environments. Recycling may be most important in species from infertile environments although within-species variation in recycling across fertility gradients does not show a consistent pattern. Recycling is generally more important for nutrients such as N or P than for carbon. Reserve formation is difficult to distinguish from accumulation or recycling although it clearly is important in many plants with long dormant periods or in plants with a predictable asynchrony in the timing of resource uptake and demand for resources in growth. Time scale is also important, with accumulation generally more important on daily or weekly periods, and storage or recycling more important on annual cycles (Chapin *et al.* 1990).

Defense includes structures such as thorns or spines, but more importantly it includes the wide array of 'secondary' plant metabolites such as tannins, alkaloids, phenolics, many resins, and more esoteric compounds that serve to reduce or inhibit herbivory. These defense compounds may account for 20–25% (or more) of the mass of many plant organs, such as leaves and young stems, that would otherwise be especially vulnerable to herbivory (Chapin 1989). Current theories of control over allocation to defense emphasize the importance of carbon and nutrient resource availability to plants and the relative degree of limitation by different resources (Bryant *et al.* 1983; Coley *et al.* 1985; Bazzaz *et al.* 1987). Allocation to defense is also influenced by a host of additional factors, most importantly including the coevolutionary history of specific plant–herbivore interactions (Ehrlich and Raven 1964; Feeney 1975; Rhoads and Cates 1976). The theory is continually evolving (Stamp 1992). However, from the ecosystem perspective of this book, a resource-based approach (Bazzaz *et al.* 1987; Field *et al.* 1992) to allocation to defense probably has the greatest utility.

In the resource-based approach, plants growing in low-resource environments are expected to allocate more to defense than plants in resource-rich environments because (1) carbon and nutrient resources lost to herbivory are more costly to replace than they are to defend in low-resource environments, and (2) plants growing in low-resource environments tend to grow slowly and to produce tissues that are relatively costly to build and that have slow turnover rates, making allocation to their defense even more advantageous relative to replacement (Coley *et al.* 1985). Plants from high-resource environments, on the other hand, allocate less to defense and more to resource acquisition, grow rapidly, and are much more responsive to variation in resource availability.

Not only the amount but also the chemical form of defense is expected to vary with resource availability, with carbon-based defenses (e.g. tannins) being favored when nutrients are strongly limiting and N-based defenses (e.g. alkaloids) when carbon or light are strongly limiting (Bryant *et al.* 1983; Coley *et al.* 1985). Both the amount allocated to defense and its chemical form may change over a plant's life history. For example, juvenile trees are often more heavily defended than adult plants, probably because juveniles are at greater risk to herbivory than adults (Bryant *et al.* 1983).

8.2.4 Allocation to reproduction

The literature on allocation to reproduction is particularly vast and varied (Harper 1977; Willson 1983; Bazzaz and Reekie 1985; Bazzaz *et al.* 1987). From a long-term, population/evolutionary perspective, reproductive allocation and its controls are clearly key questions. From an ecosystem/biogeochemical perspective, however, allocation to reproduction where the vegetation is dominated by long-lived, perennial plants usually amounts to only a small portion of annual carbon and nutrient budgets (the major exceptions being vegetation such as bamboo forests, with occasional, extreme mast-seeding). In many cases reproductive allocation is driven by storage from previous years (Chapin *et al.* 1990), thus effectively buffering its impact on allocation of currently acquired resources. For lack of space, we will not consider reproductive allocation further in this chapter.

8.3 ECOSYSTEM LEVEL EFFECTS

8.3.1 The need for a general theory

To develop a truly global, physiologically based model of terrestrial ecosystem response to global change, we must search for general patterns which emerge from the physiological literature, including the literature of allocation. Allocation presents a particular problem in this regard in that the basic patterns of allocation in relation to environment have been described, but the physiological mechanisms controlling allocation are generally unknown.

In nearly all of the models presented in this book, vegetation is described not in terms of species composition, but rather in terms of physiological responses to climate drivers. This is in part because there are no global data bases of species composition to provide data at this level of detail. Models applied globally will have to generalize to the level of detail available as input, including generalizations about allocation. In many cases, this is only to the level of biome or physiognomic group. To match this coarse level of resolution on descriptors of vegetation, we must generalize input parameters of vegetative response to the same level. Either physiognomic groups (e.g. broad-leaved deciduous forests versus annual herbaceous grasslands) or physiological groups (e.g. C_3 versus C_4 plants, symbiotic N fixers versus nonfixers) may be useful. Fortunately most of the major generalizations about allocation are at this level.

Such generalizations seem crude at the physiological level, given the depth of understanding of critical processes at the cellular and biochemical level that has emerged over the last two decades. It will also result in less accurate predictions of ecosystem dynamics at any one location than could be obtained using site- and species-specific data for any one area. However, given that only a very small fraction of the earth will ever be intensively studied, the intensity of model input demands must be relaxed to a level that is realizable for the entire globe.

8.3.2 An ecosystem-level theory of plant allocation

8.3.2.1 Above ground patterns

The last 10 years have seen tremendous progress in the development of precisely the kind of generalized theory of allocation required for global-scale modeling. Since 1972, Mooney and colleagues have published a series of papers linking several previously distinct physiological, morphological and life-history characteristics into a unified theory based on the fitness of plants in sites of different resource 'richness' (e.g. Mooney 1972; Mooney and Gulmon 1982; Coley *et al.* 1985; Bloom *et al.* 1985; Field and Mooney 1986; Bazzaz *et al.* 1987; Chapin *et al.* 1990; Chapin 1991; Field *et al.* 1992). This theory can be represented as a circular set of interactions resulting in positive feedbacks between site quality, plant response, and future site quality (Figure 8.2). Using the example of N as the growth limiting resource, this theory can be summarized as follows (see citations above for references). Sites with higher N availability will support foliage with higher N concentrations. There is a positive relationship between the N content of foliage and its maximum rate of photosynthesis. A higher N content in foliage therefore means that less time will be required for a unit of leaf mass to 'repay' the cost of production of that leaf (replacing the carbon that was required for structure and growth respiration). This shorter repay period, along with the high cost of producing antiherbivore compounds, creates selective pressures for low defense allocation and a shorter foliar retention time. All of the interactions above

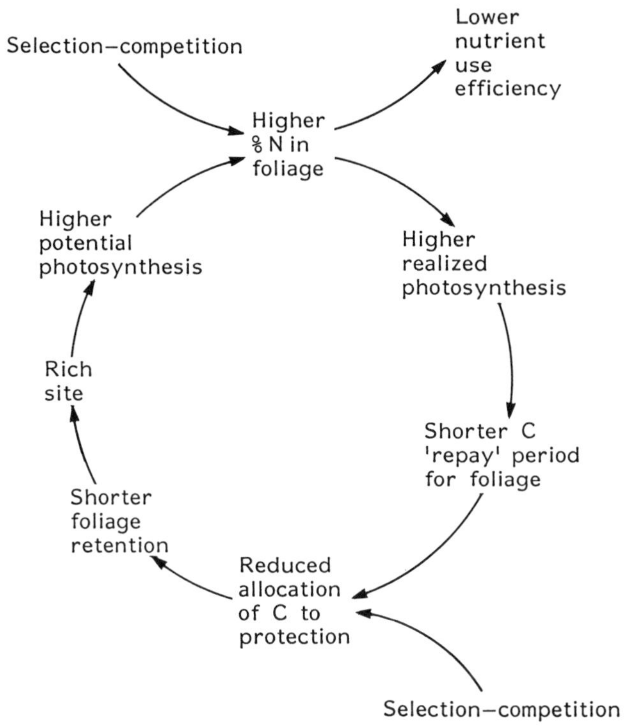

Figure 8.2 A generalized theory of interactions between resource availability (nitrogen in this case), maximum rates of photosynthesis, levels of investment in herbivore defense, and foliar longevity (see text for description and references)

combine to produce high inputs of relatively rich litter materials which decay rapidly and reinforce high N cycling, or a 'richer' site.

The power of this theory is that it unifies the processes of photosynthesis (and transpiration, see Field and Mooney 1986; Aber and Federer 1993), allocation, foliar longevity, herbivory and element cycling in a single interactive whole (Field *et al.* 1992). It is also well supported by a wealth of field data.

Recently, Reich, Gower and colleagues have elaborated on this theory and provided a rich and broad, cross-biome data set in support. Building on within-biome patterns shown by Reich *et al.* (1991) and Gower *et al.* (1993), Reich et al. (1992) show the following set of patterns (R^2 values are for log-log plots): Maximum net photosynthesis is strongly correlated with N concentration (mg g^{-1}, $R^2 = 0.74$) but not with N per unit leaf area. Leaf N concentration is also correlated with leaf lifespan ($R^2 = 0.52$), as is maximum leaf conductance ($R^2 = 0.59$). Other morphological and life-history interactions include strong relationships between leaf N concentration and leaf specific area (cm^2 g^{-1},

Table 8.1 Characteristics of plants with long and short leaf lifespans (adapted from Reich *et al.* 1992)

Characteristic	Leaf lifespan	
	Short	Long
Leaf level		
N concentration (mg g^{-1})	High	Low
Maximum photosynthetic rate	High	Low
Maximum leaf conductance	High	Low
Specific leaf area (cm^2 g^{-1})	High	Low
Plant level		
Relative growth rate	High	Low
Height growth	High	Low
Foliar N use efficiency	Low	High
Canopy level		
Total leaf biomass	Low	High
Net production/leaf mass	High	Low

$R^2 = 0.54$) and also between leaf specific area and maximum photosynthetic rate. Not surprisingly, these are all then related in turn to relative growth rate and height growth.

At the whole stand level, additional relationships emerge between leaf lifespan and total stand foliar biomass ($R^2 = 0.74$) and also aboveground net primary production normalized for leaf mass (inverse relationship, $R^2 = 0.78$), while there is no relationship between leaf lifespans and *total* above ground net primary production. Finally, there are significant relationships between leaf lifespan and both litterfall N use efficiency (1/[N in litterfall], $R^2 = 0.27$) and total N retranslocated from foliage during senescence (inverse, $R^2 = 0.57$), but not with the *fraction* of N retranslocated. The differences between N-poor and N-rich end points can be described, according to these results, as in Table 8.1.

It is important to realize that none of these relationships would necessarily be expected to hold within a single community. At that level, interspecific and intergeneric variation might be expected to swamp the very broad and general patterns shown here. Natural selection and evolution cannot be expected to have the precision (or monotony) of human engineered systems, and there are other selective pressures besides resource acquisition and allocation which drive the evolutionary process.

Still, taken together, these relationships describe a global pattern of carbon and N allocation, and carbon assimilation, that reinforces the modelling utility of the theories of Mooney, Field, Chapin *et al.* with some important system-level additions. First, the higher leaf specific area in N-poor foliage, along with the higher total foliar mass in N-poor systems, suggest that these result from the dilution of a relatively fixed amount of canopy N by different amounts of carbon, in order to provide protection against herbivory and as a result of differences in allocation to structural components. The notion that *total* canopy N should be held fairly constant across sites of differing N availability is an intriguing one, and worth further measurement. This would be consistent with optimization of canopies for full utilization of incoming radiation, with the amount of foliage required to accomplish this varying with the amount of carbon dilution required for foliar protection, in turn a function of site richness, foliar N concentration and leaf lifespan.

8.3.2.2 Below ground patterns

Far less is known about plant allocation and metabolism below ground. Studies of fine root dynamics are far fewer than for foliage, and methods are less precise and more subject to interpretation. The greatest wealth of information available on fine root biomass under different types of vegetation and site conditions comes from direct measurements of root mass in soil cores. This is a tremendously tedious and time-consuming task, which explains, in part, the lack of data. Estimating production and turnover of root biomass from sequential cores has proven controversial (Vogt *et al.* 1986; Lauenroth *et al.* 1986; McClaugherty *et al.* 1982), and in conflict with estimates derived from system-level measurements of N and carbon cycling (Raich and Nadelhoffer 1989; Nadelhoffer and Raich 1992). Two conflicting theories regarding relationships between foliar and root production have arisen from the existing data. Both agree that fine root biomass tends to increase with declining site quality. Although developed mainly for forest systems, this would extend to semiarid and arid systems as well, where the depth of root penetration is related to exploitation of deep soil horizons for water. The controversy arises over the rates of turnover of roots. The first theory suggests that root turnover is fairly constant and thus fine root production is higher in relation to foliar production on poor sites (e.g. Keyes and Grier 1981). The second theory suggests that fine root turnover is higher on richer sites and, thus, fine root production is relatively constant in proportion to foliar production across sites of different qualities (e.g. Aber *et al.* 1985).

In a recent paper, Hendricks *et al.* (1993) have reviewed the current literature and presented arguments in favor of the second theory, that fine root production is relatively constant in relation to foliar production. A key data set in this area is that of Raich and Nadelhoffer (1989), who did not use any direct fine root data at all! Instead, they summarized all the available studies in which aboveground carbon input to soils and total annual soil carbon efflux (as CO_2) were measured. They assumed, if soils were near steady-state for total carbon storage, that the difference between these two was due either to root respiration or the decomposition of root litter, the sum of which is total plant carbon allocation below ground. What they found was a very significant linear relationship between the two, suggesting relatively constant allocation and higher turnover rates on richer sites.

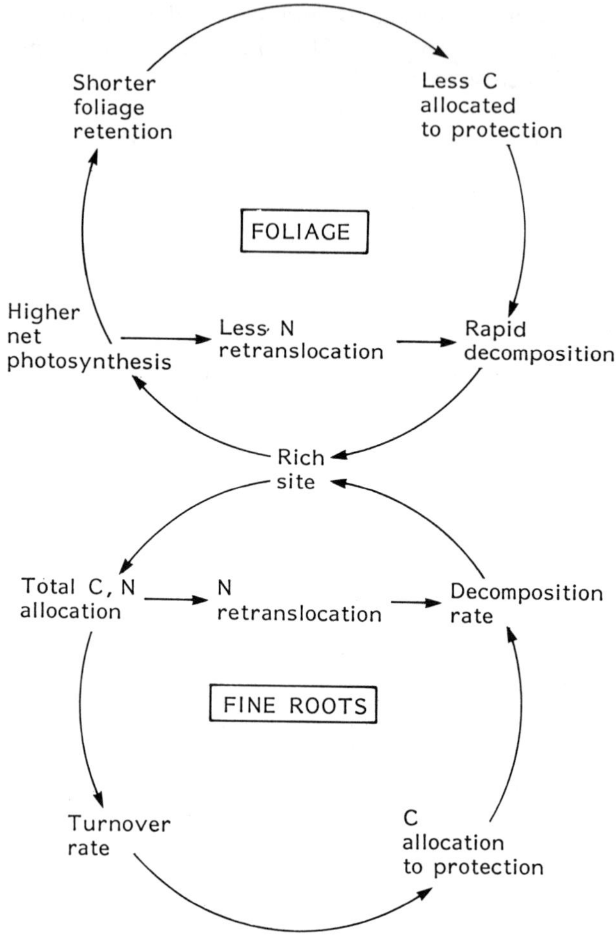

Figure 8.3 Extension of the generalized theory in Figure 8.1 to belowground tissues (after Hendricks *et al.* 1993)

A similar analysis of N cycling limitations on fine root production yielded similar conclusions (Nadelhoffer and Raich 1992; Hendricks *et al.* 1993).

Hendricks *et al.* (1993) have incorporated these results into the general theory in Figure 8.2 by extending it to the belowground part of the system (Figure 8.3). Here, analogous positive feedback loops operate in both halves of the system, again reinforcing the rich or poor status of a site. It should be stressed that this second half of the theory has not yet been supported by extensive field data. Studies are currently underway to test the patterns proposed here.

8.3.2.3 *The cost of producing and maintaining biomass*

Accurate predictions of carbon allocation in ecosystems must include losses due to respiration. While the carbon cost of growth respiration is generally agreed to be about 25% of the mass of carbon incorporated into tissues (Penning de Vries *et al.* 1974; Ryan 1991a), maintenance respiration can vary widely depending on the amount of live biomass maintained, the level of physiological activity in that biomass, and the temperature regime of the site. System-level measurements of maintenance respiration are even fewer than for fine root production, and generalizations have only recently begun to emerge.

Ryan (e.g. 1991a, b) has pioneered the development of a generalized method for estimating rates of maintenance respiration at several levels of organization. At the tissue level, he summarizes previous studies on the relationship between N content and respiration rates and presents a new data set covering tissues between 0.04 and 6% N by mass and showing a highly significant relationship between the two. He then develops a methodology for applying this to whole systems using stem biomass (heartwood and sapwood), N content, seasonal changes in temperature and a Q_{10} factor (Ryan 1991a). These methods are incorporated into a full carbon balance algorithm which is validated against whole ecosystem carbon balances as predicted by an ecosystem model (Ryan 1991b). An additional step to an even coarser resolution is the development of a relationship between mean annual temperature and the fraction of total gross photosynthesis expended in maintenance respiration for whole forest ecosystems. Preliminary estimates (M. Ryan, pers. comm.) suggest that maintenance costs vary between 7 and 15% of gross photosynthesis as mean annual temperature increases from 5 to 21 °C. At this level, the question is whether there is a higher order set of plant-level carbon allocation optimization 'rules' that constrain maintenance respiration to these levels, or if this is just a chance correlation resulting from similarity in structure and function between the forests studied.

8.3.2.4 *Modelling whole ecosystem carbon budgets*

The generalized theories of carbon gain and allocation described here offer, to the extent that they are validated by future research, tremendous power for modelling vegetation responses within *global*-scale models of terrestrial ecosystems. As these relationships cross biome boundaries, they overcome the limitations that occur in applying biome-specific parameters at or near ecotones that might be altered by climate change. They would predict smooth transitions between end-states in response to chronic atmospheric deposition. They could offer a very interesting set of predictions on the expected distribution of major physiognomic groups (e.g. grasslands, deciduous forests, evergreen forests) which could be validated using existing vegetation distribution data.

However, it should be remembered that this combination of algorithms is not likely to produce an accurate carbon balance at any particular site. The significance of the relationships described here results entirely from the wide range of conditions over which the measurements were accumulated. At any one site, greater precision could easily be achieved using more detailed models requiring more specific input data and taking into account that site's recent disturbance and vegetation history. The tradeoff between generality and accuracy is very clear in this case.

8.3.3 Conclusions

The major implication of the set of theories presented here is that the interaction between vegetation and site quality should be that of a positive feedback (accentuating either the richness or poorness of a site) rather than a negative feedback. The other factors which eventually limit the extent to which this positive feedback operates (e.g. limitations by other factors [water, radiation, climate] on rich sites, and system 'resetting' by fire on poor sites) should be better understood. The direct interactions with processes such as decomposition are explored in other chapters in this volume.

8.4 REFERENCES

Aber, J. D. and Federer, C. A. (1993) A generalized, lumped-parameter model of photosynthesis, evapotranspiration and net primary production in temperate and boreal forest ecosystems. *Oecologia* **92**, 463–474.

Aber, J. D., Melillo, J. M., Nadelhoffer, K. J., McClaugherty, C. A. and Pastor, J. (1985) Fine root turnover in forest ecosystems in relation to quantity and form of nitrogen availability: a comparison of two methods. *Oecologia* (Berlin) **66**, 317–321.

Abrahamson, W. G. and Caswell, H. (1982) On the comparative allocation of biomass, energy, and nutrients in plants. *Ecology* **63**, 982–991.

Ågren, G. and Ingestad, T. (1987) Root/shoot ratios as a balance between nitrogen productivity and photosynthesis. *Plant Cell Environ.* **10**, 579–586.

Bazzaz, F. A., Chiariello, N. R., Coley, P. D. and Pitelka, L. F. (1987) Allocating resources to reproduction and defense. *BioScience* **37**, 58–67.

Bazzaz, F. A. and Reekie, E. G. (1985) The meaning and measurement of reproductive effort in plants. In: White, J. (Ed.) *Studies in Plant Demography: A Festschrift for John L. Harper*, pp. 259–275. Academic Press, London.

Bloom, A. J., Chapin III, F. S. and Mooney, H. A. (1985) Resource limitation in plants – an economic analogy. *Ann. Rev. Ecol. Syst.* **16**, 363–392.

Brouwer, R. (1966) Root growth of grasses and cereals. In: Milthorpe, F. L. and Ivins, D. J. (Eds) *The Growth of Cereals and Grasses*, pp. 153–166. Butterworths, London. 358 pp.

Bryant, J. P., Chapin F. S. III, and Klein, D. R. (1983) Carbon/nutrient balance of boreal plants in relation to vertebrate herbivory. *Oikos* **40**, 357–368.

Chapin F. S. III (1989) The costs of tundra plant structures: evaluation of concepts and currencies. *Am. Nat.* **133**, 1–19.

Chapin F. S. III (1991) Integrated response of plants to stress. *BioScience* **41**, 29–36.

Chapin F. S. III, Schulze, E.-D. and Mooney, H. A. (1990) The ecology and economics of storage in plants. *Annu. Rev. Ecol. Syst.* **21**, 423–427.

Coley, P. D., Bryant, J. P. and Chapin III, F. S. (1985) Resource availability and plant antiherbivore defense. *Science* **230**, 895–899.

Davidson, R. L. (1969) Effect of root/leaf temperature differentials on root/shoot ratios in some pasture grasses and clover. *Ann. Bot.* **33**, 561–569.

Ehrlich, P. R. and Raven, P. H. (1964) Butterflies and plants: a study in coevolution. *Evolution* **18**, 586–608.

Feeney, P. (1975) Biochemical coevolution between plants and their insect herbivores. In: Gilbert, L.E. and Raven, P.H. (Eds) *Coevolution of Animals and Plants*, pp. 3–19. University of Texas, Austin.

Field, C. B., Chapin F. S. III, Matson, P. A. and Mooney, H. A. (1992) Responses of terrestrial ecosystems to the changing atmosphere. *Ann. Rev. Ecol. Syst.* **23**, 201–235.

Field, C. J. and Mooney, H. A. (1986) The photosynthesis–nitrogen relationship in wild plants. In: Givinish, T. J. (Ed.) *On the Economy of Plant Form and Function*, pp. 25–55. Cambridge University Press, New York.

Gower, S. T., Reich, P. B. and Son, Y. (1993) Canopy dynamics and aboveground production for five tree species with different leaf longevities. *Tree Physiol.* (in press).

Grime, J. P. (1979) *Plant Strategies and Vegetation Processes.* John Wiley & Sons, New York. 222 pp.

Harper, J. L. (1977) *Population Biology of Plants.* Academic Press, London. 892 pp.

Harper, J. L. and Ogden, J. (1970) The reproductive strategy of higher plants: 1. The concept of strategy with special reference to *Senecio vulgaris*. *J. Ecol.* **58**, 681–698.

Hendricks, J. J., Nadelhoffer, K. J. and Aber, J. D. (1993) The role of fine roots in energy and nutrient cycling. *Trends Ecol. Evol.* **8**,174–178.

Ingestad, T. and Ågren, G. (1991) The influence of plant nutrition on biomass allocation. *Ecol. Appl.* **1**, 168–174.

Johnson, I. R. and Thornley, J. H. M. (1987) A model of root:shoot partitioning with optimal growth. *Ann. Bot.* **60**, 133–142.

Keyes, M. R. and Grier, C. C. (1981) Above- and below-ground net production in 40 year old Douglas-fir stands on low and higher productivity sites. *Can. J. For. Res.* **11**, 599–605.

Lauenroth, W. K., Hunt, H. W., Swift, D. M. and Singh, J. S. (1986) Reply to Vogt *et al.* *Ecology* **67**, 580–582.

McClaugherty, C. A., Aber, J. D. and Melillo, J. M. (1982) The role of fine roots in the organic matter and nitrogen budgets of two forested ecosystems. *Ecology* **63**, 1481–1490.

Mooney, H. A. (1972) The carbon balance of plants. *Ann. Rev. Ecol. Syst.* **3**, 315–346.

Mooney, H. A. and Gulmon, S. L. (1982) Constraints on leaf structure and function in relation to herbivory. *BioScience* **32**, 198–206.

Nadelhoffer, K. J. and Raich, J. W. (1992) Fine root production estimates and belowground carbon allocation in forest ecosystems. *Ecology* **73**, 1139–1147.

Penning de Vries, F. W. T. (1975) The cost of maintenance processes in plant cells. *Ann. Bot.* **39**, 77–92.

Penning de Vries, F. W. T., Brunsting, A. H. M. and van Laar, H. M. (1974) Products, requirements and efficiency of biosynthesis: a quantitative approach. *J. Theor. Biol.* **45**, 358–377.

Raich, J. W. and Nadelhoffer, K. J. (1989) Belowground carbon allocation in forest ecosystems: global trends. *Ecology* **70**, 1346–1354.

Rastetter, E. B. and Shaver, G. R. (1992) A model of multiple element limitation for an acclimating plant. *Ecology* **73**, 1157–1174.

Reich, P. B., Uhl, C., Walters, M. B. and Ellsworth, D. S. (1991) Leaf lifespan as a determinant of leaf structure and function among 23 Amazonian tree species. *Oecologia* **86**, 16–24.

Reich, P. B., Walters, M. B. and Ellsworth, D. S. (1992) Leaf lifespan in relation to leaf, plant and stand characteristics among diverse ecosystems. *Ecol. Monogr.* **62**, 365–392.

Rhoads, D. F. and Cates, R. G. (1976) Toward a general theory of plant antiherbivore chemistry. *Rec. Adv. Phytochem.* **19**, 168–213.

Ryan, M. G. (1991a) Effects of climate change on plant respiration. *Ecol. Appl.* **1**, 157–167.

Ryan, M. G. (1991b) A simple method for estimating gross carbon budgets for vegetation in forest ecosystems. *Tree Physiol.* **9**, 255–266.

Stamp, N.E. (1992) The theory of plant–insect herbivore interactions on the inevitable brink of re-synthesis. *Bull. Ecol. Soc. Am.* **73**, 28–34.

Thornley, J. H. M. (1972) A balanced quantitative model for root:shoot ratios in vegetative plants. *Ann. Bot.* **36**, 431–441.

Tilman, D. (1988) *Plant Strategies and the Dynamics and Structure of Plant Communities.* Monographs in Population Biology 26. Princeton University Press, Princeton, NJ. 360 pp.

Vogt, K. A., Grier, C. C., Gower, S. T., Sprugel, D. G. and Vogt, D. J. (1986) Overestimation of net root production: a real or imaginary problem? *Ecology* **67**, 577–579.

Willson, M. F. (1983) *Plant Reproductive Ecology.* John Wiley & Sons, New York.

Wilson, J. B. (1988) A review of evidence on the control of shoot:root ratio, in relation to models. *Ann. Bot.* **61**, 433–449.

9 Elevated Carbon Dioxide, Litter Quality and Decomposition

J. M. MELILLO
The Ecosystems Center, Marine Biological Laboratory, Woods Hole, USA

9.1 INTRODUCTION

Plant litter decomposition is one of the important biogeochemical processes that could be affected by these aspects of global change. One long-standing hypothesis in the global-change literature is that elevated CO_2 will result in the production of plant litter with wide carbon to nitrogen (C/N) ratios and perhaps elevated concentrations of structural compounds such as cellulose and lignin (Melillo 1983; Strain and Bazzaz 1983; Melillo *et al.* 1990). In addition, this 'nitrogen-poor', litter that is enriched in structural compounds will decompose slowly. The slowed decomposition will, in turn, reduce nutrient cycling rates and, in the long term, will diminish the plants' ability to respond to elevated CO_2 in a classic negative feedback.

9.2 FACTORS THAT CONTROL LITTER DECOMPOSITION RATE

More than a half century ago, Tenney and Waksman (1929) developed a 'short list' of factors they considered primarily responsible for controlling the rate of plant litter decay. Included in the list were substrate quality and climate. Research on decomposition processes since then has supported their choices.

Indexes of substrate quality include element concentrations and concentrations of various classes of organic compounds. A high correlation between initial nitrogen concentration and decomposition rate has been demonstrated by a number of researchers (e.g. Bal 1922; Hill 1926; Waksman and Tenney 1928; Waksman and Gerretsen 1931; Monnier and Jeanson 1964; Satchell and Lowe 1966; Witkamp 1966; Cowling and Merrill 1966; Kaushik and Hynes 1971; Harrison and Mann 1975; and Taylor *et al.* 1989). Litter mass loss has also been correlated with initial lignin concentration (e.g. Cromack 1973; Cromack and Monk 1975; Fogel and Cromack 1977; Berg and McClaugherty 1987; Aber *et al.* 1991). In some studies, the initial lignin to nitrogen ratio has been shown to be

Global Change: Effects on Coniferous Forests and Grasslands.
Edited by A. I. Breymeyer, D. O. Hall, J. M. Melillo and G. I. Ågren

UNEP

the best predictor of plant litter decay rate (e.g. Melillo *et al.* 1982; Aber and Melillo 1982; Melillo *et al.* 1984). In an attempt to unify these various interpretations of what component of initial chemistry of litter controls its decay rate, Taylor *et al.* (1989) have proposed a two-phase exponential model, in which initial nitrogen content controls decomposition rate early in the decay sequence, while lignin is the controlling factor later in decay.

Climate regime, including both temperature and moisture, is also an important controller of decay rate. In well-drained grasslands and coniferous forests, the general rule is the warmer and wetter the site, the more rapid the litter decay rate (e.g. Meentemeyer 1978). Many models of plant litter decomposition have integrated climate and litter quality controls on litter decay rate (Bunnell *et al.* 1977a, b; Meentemeyer 1978; Parton *et al.* 1987; Rastetter *et al.* 1992).

9.3 CARBON DIOXIDE ENRICHMENT AND PLANT LITTER QUALITY

Changes in litter quality may be induced by an increased level of atmospheric CO_2. Observed litter quality changes include a reduction in nitrogen concentration and, in a few cases, increases in lignin concentration. A reduction in nitrogen concentration in leaf litter has often been found in CO_2 enrichment experiments with trees (e.g. Melillo 1983; Norby *et al.* 1986; O'Neil *et al.* 1987; Luxmoore *et al.* 1986; Brown 1991; Kohen *et al.* 1992; Boerner and Rebbeck 1993; Johnson 1993; Cotrufo *et al.* 1994) and with grasses (e.g. Overdieck and Reining 1986; Larigauderie *et al.* 1988). Most of the experiments involving trees are with deciduous species because the deciduous tree experiments require much less time to conduct than experiments with coniferous species. In CO_2-enrichment research designed to study the litter-quality issue, it is important that leaves or needles be permitted to senesce naturally so that the process of nitrogen reabsorption can occur prior to abscission. In conifers as much as 50% of the nitrogen in green needles may be reabsorbed (e.g. Prescott *et al.* 1989). Time to natural abscission is several years for conifers, while it is less than a year for deciduous species. Carbon dioxide experiments in which needle senescence and abscission are induced by drying or some other technique, limit nitrogen reabsorption and so mask any reduction in nitrogen concentration in litter. For example, in their CO_2-enrichment study Cotrufo *et al.* (1994) observed that the nitrogen concentration in spruce needles litter was very high, and unlike the deciduous species studied. They suggested that the reason for this difference was related to the fact that the deciduous species were allowed to abscise naturally, while the spruce needle litter was obtained by forced drying.

Even when the experimental artifact of induced senescence and abscission is avoided, the nitrogen concentration in the litter of nonwoody plants does not always appear to decrease under elevated CO_2 or decreases only slightly. Results

Table 9.1 Initial litter chemistry (%)

Nitrogen level (relative):		High		Low	
CO_2 level (ppmv):		350	700	350	700
Species	Element/ compound				
Ash	N	1.01	0.64	0.43	0.45
	Lignin	10.67	10.63	7.09	6.83
G. birch	N	0.51	0.39	0.34	0.40
	Lignin	13.24	12.60	10.91	11.49
Oak	N	1.12	0.50	0.44	0.46
	Lignin	17.54	16.58	14.50	14.16
R. maple	N	0.59	0.41	0.34	0.31
	Lignin	8.12	11.47	6.35	9.33
S. maple	N	0.94	0.56	0.90	0.60
	Lignin	9.34	11.17	11.13	8.43
Y. birch	N	1.03	0.63	0.58	0.51
	Lignin	12.94	12.62	10.37	10.97

indicating no increases or slight decreases have been reported for a salt marsh (Curtis *et al.* 1989) and a tall grass prairie (Kemp *et al.* 1994).

The circumstances under which plants produce nitrogen-depleted litter upon exposure to elevated CO_2 are not well defined. One argument is that in nitrogen-deficient soils, plants exposed to elevated CO_2 will change their nitrogen-use efficiencies (plants will incorporate more carbon into plant tissue per unit nitrogen incorporated) and that this change will persist in the litter (Norby *et al.* 1986; O'Neil *et al.* 1987). There is evidence, however, that decreased nitrogen concentration can occur in plant material exposed to elevation CO_2 at both low and high levels of soil nutrition (Cotrufo *et al.* 1994).

There are several reports of an increase in lignin concentration in litter from plants grown in high CO_2 atmospheres. These increases are usually only a few percent (Melillo 1983; Boerner and Rebbeck 1993; and Cotrufo *et al.* 1994). In contrast, Norby *et al.* (1986) found a decrease in leaf-litter lignin from oak seedlings grown in elevated CO_2. In this experiment, the lignin concentrations of both the 'treated' and 'control' plants were unusually low for leaf litter of an oak species, perhaps because the leaves were from very small seedlings. In a CO_2-enrichment experiment I conducted with Bazzaz and his students at Harvard, we found small changes (positive and negative) in leaf litter lignin concentrations depending on species and treatment (Table 9.1).

In summary, there is widespread experimental evidence for a reduction in nitrogen concentration of litter from plants grown in an enriched CO_2 environment. The evidence for changes in lignin content is not as clear-cut. The nitrogen changes imply reductions in litter decay rates. There is a small amount of experimental evidence from laboratory and field decomposition studies that supports the notion of reduced decay rates following CO_2-induced litter quality changes. This evidence is reviewed next.

9.4 DECAY OF LITTER FROM PLANTS GROWN IN A CO_2-ENRICHED ENVIRONMENT

I know of five decomposition studies that have followed the decay rate of litter from plants grown in a CO_2-enriched environment. Three are laboratory studies and two are field studies. One of the laboratory studies (Boerner and Rebbeck 1993) combined effects of elevated ozone twice and CO_2 on the litter quality of Tulip poplar (*Lircodendron tulipifer* L.). The litter was then decayed in laboratory tanks that were continuously stirred. The effect of CO_2 enrichment on litter quality was about a 30% reduction in nitrogen concentration and a large ($\sim 2.5\times$) increase in the lignin to nitrogen ratio. Over the course of the first year of the study, decay rate was much lower in the treated litter than in the control one.

Another laboratory decay study was done in soil-free microcosms, where Cotrufo *et al.* (1994) used three deciduous litter types (ash, *Froximas excelsior* L.; birch, *Betula pubescens* Ehrb.; and *Acer pseudoplatanus* L.) and one coniferous species (spruce, *Picea sitchensis* (Beng.) Carr.). The plants were grown under both treated (600 ppmv CO_2) and control (300 ppmv CO_2) conditions. In the decomposition part of the study, Cotrufo *et al.* found a significant decrease in respiration and decomposition rates in the litter materials from the treated (CO_2-enriched) plants.

The above two sets of laboratory results are in agreement with those found by Coûteaux *et al.* (1991) for litter from chestnut seedlings grown in elevated CO_2. When the litter was incubated in the laboratory with only microflora and protozoa, the litter from plants grown at elevated CO_2 decayed slower than did the litter from plants grown at ambient CO_2. When Coûteaux *et al.* added a more complex fauna, total carbon loss was actually increased for CO_2-enriched litter above the litter produced at ambient CO_2. These latter results are difficult to interpret, but they do highlight the need for experiments investigating litter decomposition under field conditions.

In a field decomposition study in tallgrass prairie, Kemp *et al.* (1994) found that litter from plants grown at elevated CO_2 decomposed at about the same rate as litter grown at ambient CO_2. The grass species studied were *Andropogon gerardii*, *Sorghostum mutans*, and *Poa pratensis*.

I have worked with Bazzaz and his students on a field decomposition study at the Harvard Forest. We used litter from six species of deciduous plants grown in full light, at either high or low nitrogen availability, and at either ambient (350 ppmv) or doubled (700 ppmv) CO_2. Litter from each of the various combinations of treatments was incubated in mesh bags placed on the floor of a deciduous forest site. After one year of decay in the field, we found that within a nutrient

Table 9.2 Observed decay rate at 1 year

Nitrogen level (relative):	High		Low	
CO_2 level (ppmv):	350	700	350	700
Species	Decay rate ($-k$)			
Ash	0.77	0.35	0.93	0.80
Grey birch	0.73	0.54	0.90	1.03
Oak	0.41	0.41	0.35	0.24
Red maple	0.78	0.47	0.60	0.61
Stripped maple	0.72	0.45	0.94	0.74
Yellow birch	0.87	0.41	0.95	0.65

treatment, litter materials from plants grown at elevated CO_2 generally had lower decay rates than those grown at ambient CO_2. The magnitude of the effect varied among species and there were a few exceptions to the general trend (Table 9.2).

9.5 WARMING AND DECOMPOSITION RATE

If warming accompanies increases an atmospheric CO_2, the temperature increase, which is likely to increase decay rate, may offset the negative feedbacks to plants associated with the CO_2-induced decrease in litter quality. At the Harvard Forest, my colleagues and I have conducted a soil warming experiment to determine the potential effects of this aspect of climate change on soil organic matter decay and on nitrogen availability.

Results from the first year of study (July 1991 through June 1992) at the Harvard Forest soil warming experiment indicated that heating increased emission of CO_2 and nitrogen mineralization. We estimated CO_2 fluxes to be approximately 7 100, 7 900, and 11 100 kg C ha^{-1} yr^{-1} in the control, disturbance control, and heated plots, respectively; this represents an increase due to warming of approximately 3200 kg C ha^{-1} yr^{-1}. Nitrogen mineralization was doubled in the forest floor plus the top 10 cm of the mineral soil. During the first growing season, we observed that in the forest floor, heating increased the average net mineralization rates from 1.02 to 2.47 mg N kg^{-1} d^{-1}. In the mineral soil, heating increased the average net mineralization rates from 0.09 to 0.19 mg N kg^{-1} d^{-1}. No nitrification was observed in either forest floor or the mineral soil in any of the treatments.

Results from the second year of study (July 1992 through June 1993) indicated that warming had a much less dramatic effect on CO_2 flux, but a sustained dramatic effect on nitrogen mineralization. We estimated CO_2 fluxes in the second year to be approximately 6800, 7600, and 8700 kg C ha^{-1} for control, disturbed control and heated treatments, respectively. Nitrogen mineralization rates remained doubled. Again, no nitrification was observed in either the forest floor or the mineral soil.

One interpretation of the differences in the CO_2 response (heated versus disturbed control) between the first and second years versus the constant response in nitrogen mineralization is as follows:

(a) there are two major soil organic matter (SOM) pools – a fast and a slow pool;
(b) the fast pool contains high C:N ratio material (e.g. recent litter) – the decay of fast pool material results in large CO_2 losses per unit of material processed and small net nitrogen release in the mineralization process;
(c) the slow pool contains low C:N ratio material (e.g. metastable humus) – the decay of the slow pool material results in a smaller loss of CO_2 per unit of material processed and large net nitrogen release in the mineralization process. As elevated soil temperature increases the rate of decay in the slow pool, both carbon (as CO_2) and nitrogen (as inorganic N) are released. The nitrogen then becomes available to be taken up by plants. Since the C:N ratio of plant material is substantially larger than the C:N ratio of SOM in the slow pool, warming may lead to increased carbon storage in the ecosystem. The magnitude of such an increase would, of course, depend on how plant carbon balance is affected by other components of climate change including the availability of water and the effect of increased temperature on both photosynthesis and respiration.

9.6 REFERENCES

Aber, J. D. and Melillo, J. M. (1982) Nitrogen immobilization in decaying hardwood litter as a function of initial nitrogen and lignin content. *Can. J. Bot.* **60**, 2263–2269.

Aber, J. D., Melillo, J. M. and McClaugherty, C. A. (1991) Predicting long-term patterns of mass-loss, nitrogen dynamics and soil organic matter formation from initial litter chemistry in forest ecosystems. *Can. J. Bot.* **68**, 2201–2208.

Bal, D. V. (1922) Studies on the decomposition of some common green manuring plants at different stages of growth in the black cotton soil of Central Provinces. *Agr. J. India* **17**, 133–155.

Berg, B. and McClaugherty, C. (1987) Nitrogen release from litter in relation to the disappearance of lignin. *Biogeochemistry* **4**, 219–224.

Boerner, R. E. J. and Rebbeck, J. (1993) Decomposition of hardwood leaves growth under elevated O_3 and/or CO_2. *Bull. Ecol. Soc. Am.* (Suppl.) **74**, 166.

Brown, K. R. (1991) Carbon dioxide enrichment accelerates the decline in nutrients relative growth rate of *Populus tremuloides* Michx. seedlings. *Tree Physiol.* **8**, 161–173.

Bunnell, F. L., Tart, D. E. N., Flanagan, P. W. and van Cleve, K. (1977a) Microbial respiration and substrate weight loss. I. *Soil Biol. Biochem.* **9**, 33–40.

Bunnell, F. L., Tart, D. E. N., Flanagan, P. W. and van Cleve, K. (1977b) Microbial respiration and substrate weight loss. II. *Soil Biol. Biochem.* **9**, 41–47.

Cotrufo, M. F., Ineson, P. and Rowland, A. P. (1994) Decomposition of tree leaf litters grown under elevated CO_2: effect of litter quality. *Plant and Soil* **163**, 121–130.

Coûteaux, M.-M., Mousseau, M., Celerier, M. -L. and Bottner, P. (1991) Increased atmospheric CO_2 and litter quality: decomposition of sweet chestnut leaf litter with animal food webs of different complexities. *Oikos* **61**, 54–64.

Cowling, E. B. and Merrill, W. (1966) Nitrogen in wood and its role in wood deterioration. *Can. J. Bot.* **44**, 1539–1554.

Cromack, K., Jr (1973) Litter production and litter decomposition in a mixed hardwood watershed and in a white pine watershed at Coweeta Hydrologic Station, North Carolina. Ph.D. thesis, Univ. Georgia, Athens.

Cromack, K., Jr and Monk, C. D. (1975) Litter production, decomposition, and nutrient cycling in a mixed hardwood watershed and white pine watershed. In: Howell, F. G., Gentry, J. B. and. Smith, M. H. (Eds) *Mineral Cycling in Southeastern Ecosystems*, pp. 609–624. US Energy Research and Development Admin. Symposium Series, CONF-740613, Washington, DC.

Curtis, P. S., Drake, B. G. and Whigham, D. F. (1989) Nitrogen and carbon dynamics in C_3 and C_4 estuarine marsh plants growth under elevated CO_2 *in situ*. *Oecologia* **78**, 297–301.

Fogel, R., and Cromack, K. Jr (1977) Effects of habitat and substrate quality on Douglas fir litter decomposition in western Oregon. *Can. J. Bot.* **55**, 1632–1640.

Harrison, P. G. and Mann, K. H. (1975) Detritus formation from eel grass (*Zostera marina* L.): the relative effects of fragmentation, leaching and decay. *Limnol. Oceanogr.* **20**, 924–934.

Hill, H. H. (1926) Decomposition of organic matter in soil. *J. Agr. Res.* **33**, 77–99.

IPCC Climatic Change (1992) *The Supplementary Report to the IPCC Scientific Assessment. 1992.* Houghton, J. T., Callander, B. A. and Varney, S. K. (Eds) Cambridge University Press, New York.

Johnsen, K. H. (1993) Growth and ecophysiological responses of black spruce seedlings to elevated CO_2 under varied water and nutrient additions. *Can. J. For. Res.* **23**, 1033–1042.

Kaushik, N. K. and Hynes, H. B. N. (1971) The fate of dead leaves that fall into streams. *Arch. Hydrobiol.* **68**, 465–515.

Kemp, P. R., Waldecker, D., Reynolds, J. F., Virginia, R. A. and Owensby, C. E. (1994) Effects of elevated CO_2 and nitrogen fertilization pretreatments on decomposition of tallgrass prairie leaf litter. *Plant and Soil.* **165**: 115–127.

Kohen, E. I., Rouhier, H. and Mousseau, M. (1992) Changes in dry weight and nitrogen partiitioning induced by elevated CO_2 depend on soil nutrient availability in sweet chestnut (*Castanea sativa* Mill.). *Ann. Sci. For.* **49**, 83–90.

Larigauderie, A., Hilbert, D. W. and Oechel, W. C. (1988) Effect of CO_2 enrichment and nitrogen availability on resource allocation in a grass, *Bromus mollis*. *Oecologia* **77**, 544–549.

Luxmoore, R. J., O'Neil, E. G. Ells, J. M. and Rogers, H. H. (1986) Nutrient uptake and growth responses of Virginia pine to elevated atmospheric carbon dioxide. *J. Environ. Qual.* **15**, 244–251.

Meentemeyer, V. (1978) Macroclimate and lignin control of decomposition rates. *Ecology* **59**, 465–472.

Melillo, J. M. (1983) Will increases in atmospheric concentrations affect litter decay rates?. In: *The Ecosystems Center Annual Report*, pp. 10–11, Marine Biological Laboratory, Woods Hole, Mass.

Melillo, J. M., Aber, J. D. and Muratore, J. F. (1982) Nitrogen and lignin control of hardwood leaf litter decomposition dynamics. *Ecology* **63**, 621–626.

Melillo, J. M., Naiman, R. J., Aber, J. D. and Linkins, A. E. (1984) Factors controlling mass loss and nitrogen dynamics of plant litter decaying in northern streams. *Bulletin of Marine Science* **35**(3): 341–355.

Melillo, J. M., Callaghan, T. V., Woodward, F. I., Salati, E. and Sinha, S. K. (1990) Climate change-effects on ecosystems. In: Houghton, J T., Jenkins, G. J. and Ephraums, J. J. (Eds) *Climate Change – The IPCC Scientific Assessment*, pp. 282–310. Cambridge Univ. Press.

Monnier, G. and Jeanson, C. (1964) Studies on the stability of soil structure: influence of moulds and soil fauna. In: Hallsworth, H. G. and Crawford, D. V. (Eds) *Experimental Pedology*, pp. 244–254. Butterworths, London.

Norby, R. J., Pastor, J. and Melillo, J. M. (1986) Carbon–nitrogen interactions in CO_2-enriched white oak: physiological and long-term perspectives. *Tree Physiol.* **2**, 233–241.

O'Neil, E. G., Luxmoore, R. J. and Norby, R. J. (1987) Elevated atmospheric CO_2 effects on seedling growth, nutrient uptake, and rhizosphere bacterial population of *Lirodendron tulipifera. Plant and Soil* **104**, 3–11.

Overdieck, D. and Reining, E. (1986) Effect of atmospheric CO_2 enrichment on perennial ryegrass (*Lolium perenne* L.) and white clover (*Trifolium repens* L.) competing in managed model-ecosystems. II. Nutrient uptake. *Acta Oecol. Plant.* **7**, 367–378.

Parton, W. J., Schimel, D. S., Cole C. V. and Ojima, D. S. (1987) Analysis of factors controlling soil organic matter on Great Plains Grasslands. *Soil Sci. Soc. Am. J.* **51**: 1173–1179.

Peterjohn, W. T., Melillo, J. M., Steudler, P. A., Newkirk, K. M. Bowles, F. P. and Aber, J. D. (1994) Responses of trace gas fluxes and N availability to experimentally elevated soil temperatures. *Ecol. Appl.* **4**, 617–625.

Prescott, C. E., Corbin, J. P. and Parkinson, D. (1989) Biomass, productivity, and nutrient-use efficiency of aboveground vegetation in four Rocky Mountain coniferous forests. *Can. J. For. Res.* **19**, 309–317.

Rastetter, E. B., McKane, R. B., Shaver, G. R., Melillo, J. M., Nadelhoffer, K. J., Hobbie, J. E. and Aber, J. D. (1992) Changes in C storage by terrestrial ecosystems: how C-N interactions restrict responses to CO_2 and temperature. *Water, Air and Soil Pollution* **64**: 327–344.

Strain, B. R. and Bazzaz, F. A. (1983) Terrestrial Plant Communities. In: E. R. Lemon (Ed.) *CO_2 and Plants*, pp. 177–222. AAAS Selected Symposium. Westview Press, Boulder, Colo.

Taylor, B. R., Parkinson, D. and Parsons, W. F. J. (1989) Nitrogen and lignin content as predictor of litter decay rates: a microcosm test. *Ecology* **70**, 97–104.

Tenney, F. G. and Waksman, S. A. (1929) Composition of natural organic materials and their decomposition in the soil: IV. The nature and rapidity of decomposition of the various organic complexes in different plant materials, under aerobic conditions. *Soil. Sci.* **28**, 55–84.

Waksman, S. A. and Gerretsen, F. C. (1931) Influences of temperature and moisture upon the nature and extent of decomposition of plant residues by micro-organisms. *Ecology* **12**, 33–60.

Waksman, S. A. and Tenney, F. G. (1928) Composition of natural organic materials and their decomposition in the soil: III. The influence of nature of plant upon the rapidity of its decomposition. *Soil Sci.* **26**, 155–171.

Witkamp, M. (1966) Decomposition of leaf litter in relation to environment, microflora and microbial respiration. *Ecology* **47**, 194–201.

10 Ecosystem Physiology – Soil Organic atter

**G. I. ÅGREN,[1] M. U. F. KIRSCHBAUM,[2]
D. W. JOHNSON,[3] E. BOSATTA[1]**

[1]*Swedish University of Agricultural Sciences, Department of Ecology and Environmental Research, Uppsala, Sweden*
[2]*Division of Forestry, CSIRO, Canberra, Australia*
[3]*Desert Research Institute, Reno*

10.1 INTRODUCTION

10.1.1 Soil organic matter as a C reservoir

Detrital inputs to world soils are estimated at about 60×10^{15} g yr^{-1} and decomposition at 50–60 $\times 10^{15}$ g yr^{-1}. There is no a priori reason to suspect that litter and soil organic matter (SOM) are in a steady-state condition at present, given the large changes in land use patterns that continue to occur at a global scale and the continuous input of C to the global cycle from fossil fuel burning. Schlesinger (1990) analysed chronosequence studies for soil C accumulation rate, and concluded that, on a world-wide basis, only approximately 0.4×10^{15} g C yr^{-1}, or less than 1% of primary production, escapes decomposition and is added to a soil C store of 1500×10^{15} g. However, erosion balances this influx of soil C. Given the uncertainties associated with C balances on a global scale, let alone on a plot scale, an input–output imbalance of 5–10% could easily escape detection, and imbalances of this magnitude would make soils either a sizeable source or sink for C. It is therefore important to determine first of all if soils are a net source or a net sink of C, and, if possible, to make some estimate as to the magnitude of the imbalance between inputs and outputs of C to the soil.

10.1.2 Soil organic matter as a nutrient reservoir

The major part of the N in biological circulation is bound to living or dead organic matter. Söderlund and Svensson (1976) estimate that the soil contains 300×10^{15} g of N in organic form versus 16×10^{15} g as insoluble inorganic N. They do not provide any estimates of soluble inorganic N but it should be

Global Change: Effects on Coniferous Forests and Grasslands.
Edited by A. I. Breymeyer, D. O. Hall, J. M. Melillo and G. I. Ågren

UNEP

negligible; Burns and Hardy (1975) assume it to be 1% of the insoluble. Post *et al.* (1985), based on direct soil samples, give a lower value, 95×10^{15} g, for the total soil N store. However, this change in total N should not affect the relative distribution between N forms. There is also a considerable variability in N storage between life zones with a range from 100 to 2800 g m^{-2} to a depth of 1 m. However, much of this variability is associated with differences in SOM storage such that the range of C/N values is from 10 to 30.

Dalal (1977) has reviewed the occurrence of soil organic P and concludes that the organic P constitutes 20–80% of the total P in the surface layer of the soil, although both higher and lower values have been observed. The importance of the organic P to the turnover of P can be greater than indicated by just its relative contribution to the total soil P because in the soil water solution around 80% of the P is organic but with some variation depending on soil type (Pierre and Parker 1927).

As with N and P, S in soils is mainly in organic form. Biederbeck (1978) states that 'well over 90% of the total S in most noncalcareous surface soils is present in organic form'. Scott (1985) goes even further: 'It is well established that 95% or more of the total S in most soils from humid and semi-humid areas is organic.' The C:S (as the C:N:S) ratio is also bounded to rather narrow limits, between 70 and 300 (Biederbeck 1978).

Soil organic matter also has a major influence upon site productivity because of its effects upon physical (bulk density, water holding capacity), biological (microbial populations) and chemical (cations exchange capacity) properties of soils (Chen and Aviad 1990). The cation exchange capacity (CEC) of organic matter arises primarily from H+ dissociation from those carboxyl and phenolic hydroxyl groups, which are acidic enough to dissociate at normal soil pH (Coleman and Thomas 1967). The CEC of organic matter is therefore strongly pH-dependent, in contrast to the CEC of clay minerals, which is derived from cationic substitution (e.g. Al for Si) within the lattice structure of the clays. Thus, organic matter represents a potential cation exchanger for the whole soil, especially in view of the fact that the CEC of organic matter (100–900 meq g^{-1} soil) is generally much greater than that of clays (5-150 meq g^{-1} soil).

10.2 ENVIRONMENTAL CONTROLS ON SOM

The environment controls the levels of SOM by regulating both the inputs of organic matter to the soil and the rate of losses. An analysis of observed levels of SOM is therefore, in contrast to a theoretical analysis (see section 10.5) difficult to do in isolation from the plant and animal communities that provide the source of the organic matter.

10.2.1 Temperature and soil moisture controls

There are strong correlations between soil C pools and climate. An extensive analysis of the global variability was made by Post *et al.* (1982), who concluded that 'Soil carbon density generally increases with increasing precipitation, and there is an increase in soil carbon with decreasing temperature for any particular level of precipitation.' Temperature increases over most normal ranges could increase both inputs to and outputs from the soil C. The analysis by Post *et al.* (1982) shows that increasing temperatures increase the rate of soil C output more than the input. This implies that the temperature response function for decomposition must be steeper than for production. As the response of SOM to increasing temperature is one of the most important likely changes with climate change, it is important to attempt to understand how SOM is likely to respond.

The effects of climate on decomposition are often expressed as a temperature factor and a moisture factor. The temperature response is generally expressed as a Q_{10} function where

$$k_2 = k_1 Q_{10}{}^{(T_2 - T_1)/10} \tag{10}$$

where k_2 and k_1 are the rate constants at two observed temperatures T_2 and T_1. While this relationship is easy to use and provides a ready indication of the temperature sensitivity of any system, it lacks a theoretical justification. Other relationships, such as the Arrhenius function, have been used (e.g. Ellert and Betany 1992) to provide a better theoretical framework. It is unclear, however, how this should be applied to a system such as a population of soil organisms, where the total activity is determined by the combined activity of a whole range of different organisms with probably quite different individual responses to temperature.

In the absence of a useful theoretical model, the use of Q_{10} values is a convenient tool to summarize observed responses. It must be recognized, however, that Q_{10} values are not constant for any system, but change with temperature (Kirschbaum 1994). This is illustrated in Figure 10.1, which summarises Q_{10} values for SOM or N mineralization for a range of soils as reported by different workers. Figure 10.1 indicates that Q_{10}'s are generally around 2 for temperatures above 20 °C, but increase at lower temperatures to some very high values. There is a large amount of scatter in the data, part of which may be due to experimental error. Researchers who have compared the temperature responses in different systems, such as different soil types, different soil horizons or different litter types, also generally report differences in the temperature sensitivities of different systems. The figure also shows a widely used temperature dependence of net primary productivity (Lieth 1973). The Q_{10} for net primary production is generally lower than the Q_{10} for decomposition. A comparison of the temperature dependencies of decomposition and net primary production therefore suggests that with increasing temperature, decomposition

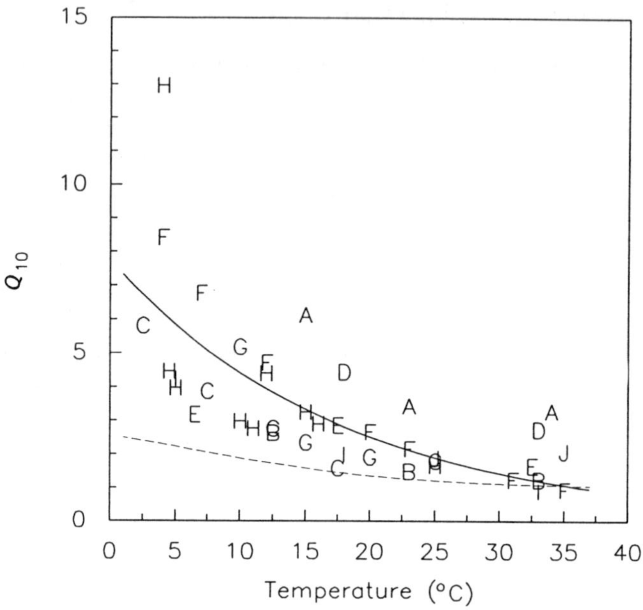

Figure 10.1 The Q_{10} of soil or litter respiration or net mineralization as a function of temperature (after Kirschbaum 1994). Shown are experimental data, indicated by different letters, and a least squares fit (solid line). The figure also shows the Q_{10} of net primary productivity according to Lieth (1973) (broken line). Letters refer to: (A) Bunt and Rovira (1955); (B) Drobnik (1962); (C) Flanagan and Veum (1974); (D) Moureaux (1967); (E) Nyhan (1976); (F) O'Connell (1990); (G) Ross and Cairns (1978); (H) Schleser (1982); (I) Waksman and Gerretsen (1931); (J) Wiant (1967)

should be stimulated much more than primary production leading to a considerable loss in SOM (Kirschbaum 1994).

The effects of soil moisture are also expressed in diverse ways. Various modifications of the precipitation to potential evapotranspiration ratio (P/PET) used in the Holdridgelife zone classification system provide easily accessible variables. In the CENTURY model this approach is used and parameterized such that for P/PET above 1, soil moisture has little effect on decomposition rates. Other alternatives are to calculate directly the soil moisture deficit (e.g. Jenkinson *et al.* 1987).

If we take the data by Post *et al.* (1982) and plot the cumulative fraction of soil C as a function of soil C density (Figure 10.2) we find that approximately 65% of the world's soil C is found in soils with a C density above 10 kg m^{-2}. In the CENTURY model, the C density of 10 kg m^{-2} corresponds approximately to a potential evapotranspiration ratio of 1. If we accept this parameterization, then, soil moisture has only a small influence on decomposition rates for a majority of soils. We can thus conclude that, for SOM, precipitation is more important in

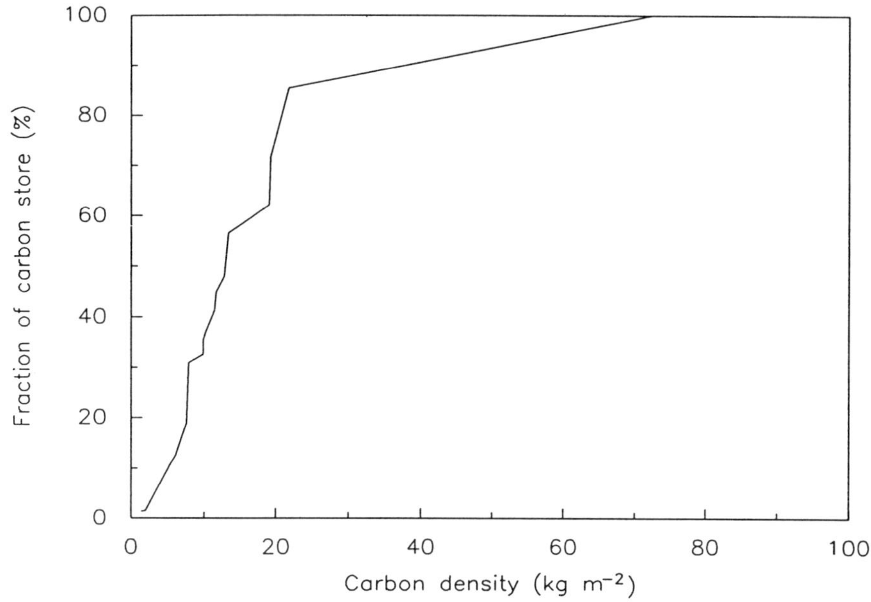

Figure 10.2 Cumulative distribution of soil carbon over carbon densities for the world's different biomes. Based on data in Post *et al.* (1982)

regulating the inputs to (plant production) than in regulating outputs from (decomposition) the soil.

10.2.2 Biological controls

10.2.2.1 Productivity

The discussion in the previous section shows that the understanding of SOM is inseparable from understanding production. Lashof (1989) compiled data on production and C storage in the world's major biomes. We have plotted these data in Figure 10.3. If the tundras are excluded, net primary production explains over 60% of the variation in soil C storage. A primary determinant of soil C storage appears therefore to be the production of the system. However, it is difficult to isolate productivity from decomposition as these two factors often depend in the same fashion on underlying environmental controls. There is also a large amount of small-scale variability within any biome (cf. section 10.2.2.2).

10.2.2.2 Litter quality

Differences in decomposition rates of litters of different origins have been

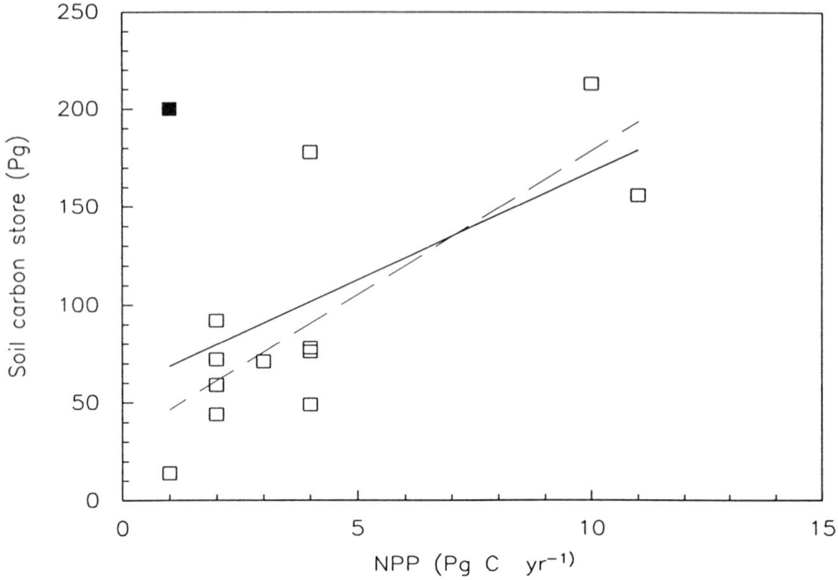

Figure 10.3 Relation between soil carbon store and net primary production for different biomes. Solid line with all biomes included ($r^2=0.30$), broken line with tundra (solid symbol) excluded ($r^2=0.62$). Based on data in Lashof (1989)

Table 10.1 Total amounts of C added and remaining with different types of added substrates in some of the treatments of the Ultuna long-term SOM experiment (from Hyvönen *et al.* 1996)

Added material	Amount of material added ($g\ C\ m^{-2}$)	Amount of material remaining ($g\ C\ m^{-2}$)	Per cent material remaining
Straw	6700	1160	17
Green manure	6710	1120	17
Sewage sludge	6480	1960	30
Farmyard manure	6610	1960	30
Sawdust	6760	2340	35
Peat	6790	5020	74

extensively studied (e.g. Aber *et al.* 1984; Berg *et al.* 1986, 1991a, b; Harmon *et al.* 1986). The studies of the long-term consequences of additions of different litters are fewer. In one of these, six different types of organic materials were added to an agricultural soil in central Sweden in equal amounts every second year from 1956 to 1991 (Kirchmann *et al.* 1994). Depending on the type of material added, between 17 and 74% of the added material was estimated to remain in 1991 (Table 10.1) (Hyvönen *et al.* 1996).

The same variability is found in e.g. the studies by McClaugherty *et al.* (1985) on litter dynamics and SOM formation in different forest types on Blackhawk Island, Wisconsin. Although aboveground litter production was approximately equal in three deciduous stands and a white pine stand, the forest floor of the latter contained twice as much organic matter. Likewise, a hemlock stand in which the litterfall was 40% of the litterfall in the deciduous stands had 60% more organic matter in the forest floor.

10.2.2.3 Decomposer community

It is generally recognized that the composition of the decomposer community influences decomposition rates. In particular, soil animals can accelerate the decomposition rate either directly by modifying the physicochemical environment or indirectly by stimulating microbial activity through grazing (Anderson and Ineson 1983, 1984). Quantitatively, less is known about the effects and variation with substrate but such effects are likely. For example, Coûteaux *et al.* (1991b) found in decomposition experiments with chestnut leaves from plants grown in elevated CO_2 (up to 700 ppm) that different combinations of soil fauna could produce a fourfold difference in evolved CO_2. With litter from plants grown in ambient CO_2 only a 50% difference was observed. However, Setälä and Huhta (1990) found much smaller faunal effects with birch and spruce litters, and during the later stage of their experiment, decomposition rates were lower with fauna present than without. This might be the result of a decline in litter quality due to the more rapid initial decomposition with fauna present. The decomposition studies on Blackhawk Island by Aber *et al.* (1984) show that litter quality dominates over differences in soil community, (Figure 10.4) (see also McClaugherty *et al.* 1985). There are no striking obvious differences in decomposition rate due to litter bag type (exclusion of fauna of different sizes) or placement in the soil. It is possible that there is a difference in decomposition rate between the different ecosystems, with the sugar maple stand having a higher rate than the other ones, but, again, the differences are small compared to the differences between the litter types. One could therefore expect that the diversity of the soil microbial community in natural systems is large enough to be able to cope with different litters almost similarly in any system.

10.2.3 Fire

The effects of fire on SOM are highly complex and differ greatly for different sites, with different soils, climates, species composition, time scales and for different nutrients. Here, we can only deal with fire in a brief and most general manner. For more detailed accounts, see Viro (1974), Walker *et al.* (1986), and Raison *et al.* (1989).

Fire potentially has three effects on SOM, and we will deal with them in turn:

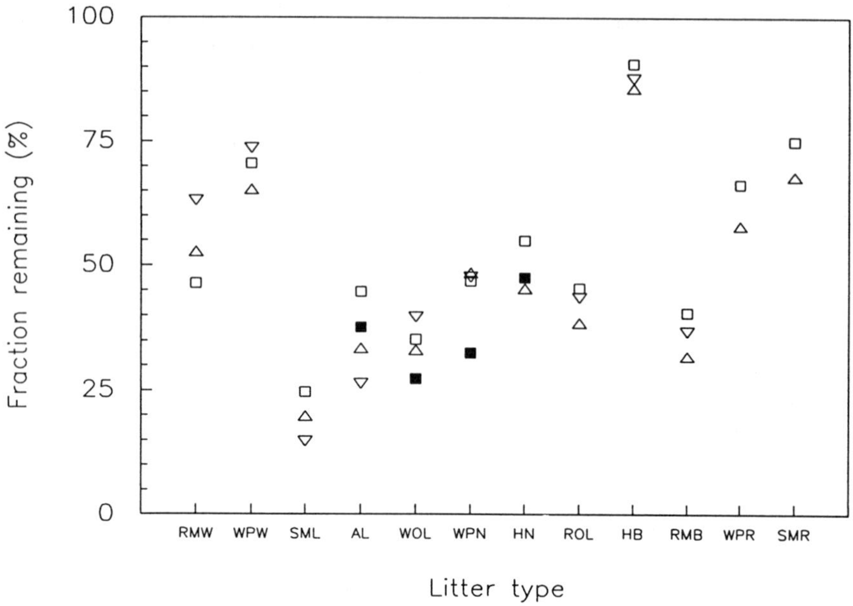

Figure 10.4 Fraction of different litters remaining after 732 d of decomposition in different environments. Open symbols: in sugar maple stand. Filled symbols: in the stand of the origin of the litter. (□): fine-meshed litter-bags on the forest floor; (△): fine-meshed litter-bags in the mineral soil; (▽): coarse-meshed litter-bags on the forest floor; (RMW) red maple wood; (WPW) white pine wood; (SML) sugar maple leaves; (AL) aspen leaves; (WOL) white oak leaves; (WPN) white pine needles; (HN) hemlock needles; (ROL) red oak leaves; (HB) hemlock bark; (RMB) red maple bark; (WPR) white pine roots; (SMR) sugar maple roots (from Aber *et al.* 1984)

(a) it causes a short-term loss of C due to direct combustion and indirectly due to changes in microclimate that enhance decomposition rate;
(b) it causes a short- to medium-term loss of N and a long-term loss of P and other nutrients;
(c) it may change species composition of a site that may restore lost N.

Fire leads to the direct combustion of large amounts of C above the soil surface. The combustion may be almost complete for dry grasslands or consume only a small fraction of potentially available C, such as in a low-intensity burn in an established and fire resistant forest (Walker *et al.* 1986). Intense fires may also burn part of the organic matter that is already in the soil (Walker *et al.* 1986), although soils tend to heat only very slowly, and even very intense fires are unlikely to directly affect the SOM to a depth greater than a few centimetres (Aston and Gill 1976; Humphreys and Craig 1981).

Of great importance are also post-fire changes to the soil microclimate. After the canopy and understory have been burned off, the soil is no longer protected by shading and much more sunlight may reach the soil surface. This can raise both maximum and minimum soil temperatures by several degrees for one or more years (Armson 1979). Viro (1974) reports an increase in mean summer temperature in the year after the fire in the humus layer from 18 °C under the unburned tree canopy to 31 °C in the open. With a loss of leaves from vegetation, the water transpiring capacity is also reduced; consequently, more water is retained in the soil profile. Hence, for one to several years after a fire, soils may be warmer and wetter than in the unburned state, which may greatly increase the rate of organic matter decomposition. However, the decrease in water-holding capacity due to loss of organic matter together with the increased soil temperature will increase evaporation from the soil (Pritchett 1979).

While fire clearly leads to a loss of soil C in the short term, the longer-term C dynamics are principally controlled by the soil's nutrient economy. During fires, N compounds readily volatilize so that large quantities may be lost (Raison *et al.* 1985). Viro (1974) estimates the average loss of N due to burning after clear-felling to be 320 kg ha^{-1} with about 180 kg coming from the slash and 140 kg from the humus layer. Phosphorus compounds, on the other hand, are generally less volatile and a greater proportion of P is likely to be retained within the system (Raison *et al.* 1985), although ash deposited on the soil surface may readily be lost due to erosion (Raison *et al.* 1989). Another consequence of fires can be a substantial increase in soil pH on acid soils, up to 2–3 pH units (Viro, 1974). This increase in soil pH can enhance N fixation and mineralization (Johansson 1984) and may increase the amount of available nutrients (Nykvist 1977). If the nutrients associated with the lost C become available to plants and promote greater C gain, then, after a number of years the soil C to nutrient ratios would return to the values characteristic for the site and the same soil C content as before the occurrence of fire would be re-established. The direct loss of C due to fires would then have few long-term consequences. In the long run, fire affects the amount of SOM primarily through the loss of nutrients that cannot be replenished by the ecosystem. Hence, the loss of P, or other nutrients if they are the primary limiting element, looms large in determining the long-term soil C dynamics of the system.

If fire frequency changes due to climate change it may have some short-term consequences for soil C storage. In the medium term, soil C storage is primarily controlled by the system's N economy. If the system is given sufficient time to recover after each fire episode, its overall N content can recover to pre-fire values, but very frequent fire intervals could prevent that. In the longer term, the movement of nutrients other than N are likely to be the most rate-limiting. In regions where fire frequency and intensity may increase, more nutrients may be lost and overall productivity and soil C storage may eventually decrease. The converse may be true for regions with fewer fires.

10.2.4 Edaphic and physical factors

In addition to temperature, moisture, and litter quality, there are several important edaphic and physical factors that strongly influence the accumulation of organic matter in soils. Chief among these are clay content, base status, and Fe and Al hydrous oxide content (Oades 1988; Anderson 1992). Intercalation of organic matter in smectite clays and carbohydrate adsorption onto clay surfaces can result in substantial SOM stabilization (Oades 1988; Anderson 1992). However, the often noted positive correlation between clay and organic matter content in soils may not represent a true cause–effect relationship because clay content is usually highly correlated with other factors that influence SOM accumulation such as polyvalent cations (Oades 1988). These cations stabilize SOM by a number of mechanisms, including electrostatic cation bridging of organic colloids (Ca and Mg in alkaline and circumneutral soils; Al in acidic soils) and specific adsorption onto Fe and Al hydrous oxide surfaces (acidic soils) (Oades 1988). Because Ca is seldom limiting to terrestrial plant productivity, the often noted long-term positive effects of liming on SOM (e.g. Gilmore 1980; Jenkinson 1991; see also review by Johnson 1992a) are likely due to these reactions rather than a direct effect upon plant primary productivity.

In addition to polyvalent cations, the effects of non-biological reactions between N and organic matter on SOM stabilization deserve mention here. Non-biological condensation reactions of phenols with ammonium and amino compounds are important in the production of humus (Mortland and Wolcott 1965; Paul and Clark 1989; see also review by Johnson 1992b). These reactions are enhanced by high pH (because NH_3 is the reactive form of N) and high NH_3 and/or NH_4^+ concentrations. The reaction can occur slowly at pH below neutrality, also (Mortland and Wolcott 1965), but the importance of non-biological reactions in forest soil or litter N retention under ambient pH conditions is unclear. Nömmik (1970) found little non-biological NH_4^+ retention in acid Norway spruce humus unless pH was raised to neutrality or higher. However, Schimel and Firestone (1989) found that non-biological reactions accounted for 20% of the N retained in an acid (pH 4.3–4.5) forest soil.

10.3 OBSERVED PATTERNS OF SOM

Long-term studies of SOM have mostly been done on agricultural soils. The most well-known of these is the 150-year-long Rothamsted experiments (e.g. Jenkinson 1991). A similar study, but of only a 35-year duration and on a smaller spatial scale, is the Ultuna long-term SOM experiment (Kirchmann *et al.* 1994). These studies show that the loss of C from old agricultural sites can be very slow; plots that were left bare lost annually on average 0.4% at Rothamsted over a 120-year period (Jenkinson and Rayner 1977) and 1.1% at Ultuna over a 35-year period, respectively. The slower rate at Rothamsted might be a result of the longer observation period and a decreasing rate with time. When allowed to return to deciduous woodlands, two Rothamsted areas increased their soil C content by 65 and 165%, respectively, in less than 100 years (Jenkinson 1991). Depending on type, the amounts of litter required to maintain these soils at a steady-state level vary between 1 and 3 Mg C ha^{-1} (Jenkinson *et al.* 1992).

There seem to be no similar studies of forest soils. The effects of forest management on soil C storage was recently reviewed by Johnson (1992c) showing that C storage does not decrease due to forest harvesting unless it is followed by intense burning or cultivation.

10.4 MODELLING SOM DYNAMICS

There exists at present a number of different approaches to summarize and explain the observations of SOM dynamics discussed above. They differ mainly in the degree to which they resolve the SOM into components of different properties but also in the way they handle external factors, e.g. climate and soil texture.

10.4.1 Single pool models

The simplest approach is found in Olson's (1963) classic paper which combines a rate of litterfall, I_0 with a loss rate constant, k, to calculate the change in stored C:

$$\frac{\mathrm{d}C}{\mathrm{d}t} = I_0 - kC \tag{2}$$

or in integrated form

$$C(t) = \frac{I_0}{k}(1 - \mathrm{e}^{-kt}) \tag{3}$$

In combination with estimates of litterfall and decomposition constants (k's) under different climatic conditions (e.g. Bray and Gorham 1964; Olson 1963), it is simple to calculate changes in the detrital pool under given climatic scenarios. This approach is extremely simple, but at the expense of lumping all information into a single parameter, which limits the possibilities of generalization. At steady state, equation (3) is formally equal to equation (11) (see section 10.4.3) but the term $1/k$ is replaced by a richer expression allowing more insights into the mechanisms exerting control over the decomposition rate. More complicated versions of the single pool model that take into account changes in litter quality with time have also been developed (e.g. Janssen 1984; Godshalk and Wetzel 1978)

10.4.2 Multiple pool models

As an alternative to letting all SOM be combined into one pool, other authors have divided it into several fractions characterized by different decomposition rates. The rationale for this is that there are differences in chemical structure and physical protection.

Using a matrix notation, all these models can be written in the format

$$\mathbf{C}(t+1) = \mathbf{TC}(t) + \mathbf{I} \tag{4}$$

where **C** is a vector representing the amounts of C in the different pools, **T** is a transition matrix describing the fluxes of C between the pools, and **I** a vector containing the litter inputs. To simplify the comparisons and stress the qualitative differences among the models as they appear in the topology of the models, we will only indicate by $-$, 0, or $+$ in the transition matrix whether a certain matrix element causes a decrease in a pool, has no effect or increases the pool. We will contrast three different approaches of this class (many of these models undergo continuous revisions and exist therefore in several versions with more or less substantial differences), all used to describe long-term carbon accumulation under grasslands. The three are:

(a) CENTURY (Parton *et al.* 1987; see also Chapter 11):

$$\mathbf{C} = \begin{pmatrix} \text{Structural C} \\ \text{Metabolic C} \\ \text{Active soil C} \\ \text{Slow soil C} \\ \text{Passive Soil C} \end{pmatrix} = \begin{pmatrix} - & 0 & 0 & 0 & 0 \\ 0 & - & 0 & 0 & 0 \\ + & + & - & + & + \\ + & 0 & + & - & 0 \\ 0 & 0 & + & + & - \end{pmatrix} \mathbf{C} + \begin{pmatrix} + \\ + \\ 0 \\ 0 \\ 0 \end{pmatrix} \tag{5}$$

(b) Rothamsted model (Jenkinson and Rayner 1977; Jenkinson *et al.* 1987; see also Jenkinson 1990; Jenkinson *et al.* 1992, for a newer version that is close to CENTURY):

$$\mathbf{C} = \begin{pmatrix} \text{Decomposable plant material} \\ \text{Resistant plant material} \\ \text{Biomass} \\ \text{Physically stabilized OM} \\ \text{Chemically stabilized OM} \end{pmatrix} = \begin{pmatrix} - & 0 & 0 & 0 & 0 \\ 0 & - & 0 & 0 & 0 \\ + & + & - & + & + \\ + & + & + & - & + \\ + & + & + & + & - \end{pmatrix} \mathbf{C} + \begin{pmatrix} + \\ + \\ 0 \\ 0 \\ 0 \end{pmatrix} \tag{6}$$

(c) V&P (van Veen and Paul 1981):

$$\mathbf{C} = \begin{pmatrix} \text{Plant material} \\ \text{Decomposable, not protected} \\ \text{Biomass} \\ \text{Decomposable, protected} \\ \text{Recalcitrant, not protected} \\ \text{Recalcitrant, protected} \end{pmatrix} = \begin{pmatrix} - & 0 & 0 & 0 & 0 & 0 \\ 0 & - & 0 & 0 & 0 & 0 \\ + & + & - & + & + & + \\ 0 & + & 0 & - & 0 & 0 \\ 0 & + & 0 & + & - & 0 \\ 0 & + & 0 & + & 0 & - \end{pmatrix} \mathbf{C} + \begin{pmatrix} + \\ + \\ 0 \\ 0 \\ 0 \\ 0 \end{pmatrix} \quad (7)$$

Although the models are structurally quite similar with entering litter divided into two classes of decomposability and the microbial biomass (active soil C in the terminology of CENTURY) as the wheel around which everything turns, there are also clear differences. On the one hand, the reason for differentiating the soil C into the pools differs. In CENTURY, there is no explicit physical protection of SOM but the division into slow and passive soil C is based on both soil texture and SOM chemical properties. The Rothamsted and the V&P model both make the corresponding division by explicitly distinguishing between physical and chemical stabilization. The differences are also apparent in the turnover rates associated with the different pools (Table 10.2).

10.4.3 Continuous quality models

As an alternative to the over-simplification in the Olson model and a way of avoiding the ambiguities arising when dividing SOM into a small number of pools, Bosatta and Ågren (1985, 1991a, b, 1994; Ågren and Bosatta 1987, 1988) have suggested that one could assume that there exists an infinite number of 'pools'. Thus, rather than assigning a certain C compound to a discrete C fraction, the compound is described by a continuous variable, q, named 'quality' (the CQ-model). Quality is a measure of the degradability of the compound as measured by the microbial growth rate for microbes feeding on that particular compound. Carbon in the soil is then described by a continuous

Table 10.2 Turnover times for the different pools of soil C. The values are only approximate as in all models climatic variables regulate the decomposition rates

	Turnover time (yr)		
	CENTURY	Rothamsted	V&P
Fraction 1	3	4	0.03
Fraction 2	0.5	1	0.03
Fraction 3	1.5	3	0.1
Fraction 4	25	70	3
Fraction 5	1000	3000	300
Fraction 6	—	—	30 000

distribution, such that $\rho c(q, t)dq$ is the amount of C in the quality interval $[q,q+dq]$ at time t.

To formulate the interaction between the soil C and the decomposing microbial community, the latter is described with three functions: $e(q)$, the microbial efficiency or production-to-assimilation ratio which tells how much soil C is lost in producing new microbial biomass; $u(q)$, the microbial growth rate; and $D(q, q')$, the dispersion function, which determines the fraction of C assimilated as quality q' that is returned to the soil C as quality q. The dispersion function plays the role of the off-diagonal elements in the matrix **T** in equation (4). In addition, the C concentration in the microbial biomass, f_C, is needed. The requirement of mass balance for C then leads to the following equation:

$$\frac{\partial \rho_C(q,t)}{\partial t} = -f_C\frac{u(q)}{e(q)}\rho_C(q,t) + \int f_C D(q,q')u(q')\rho_C(q',t)\mathrm{d}q' + I(q,t) \tag{8}$$

where $I(q, t)$ is the input of C.

This equation can be analysed at different levels of approximation depending upon the detail of information used to specify the three functions $e(q)$, $u(q)$, and $D(q,q')$. Equation (8) has also been modified to explicitly describe the microbial biomass (Bosatta and Ågren 1994). Decomposition of marine sediments has also been analysed by a similar approach (Boudreau and Ruddick 1991).

10.4.4 Comparison of the modelling approaches

Using the CQ-approach, it is easy to see how the various modelling approaches differ. The introduction of discrete classes of C is then simply a way of approximating the continuous spectrum with a set of discrete variables; the extreme case is Olson's model with a single variable. The effects of physical soil structure is another feature differentiating the models. In the CQ-model, the microbial growth rate on different qualities is differentially affected. In the CENTURY model, the way the C quality spectrum is discretized depends on soil texture. The Rothamsted model and the V&P model, on the other hand, both assume the existence of two C quality spectra, one for physically protected soil C and one for soil C that is not physically unprotected. The discretization of these spectra is fixed, but soil texture shifts C between them. The latest version of CENTURY (Chapter 11) has added a spatial dimension by separating surface litter from SOM.

10.5 EXPECTED EFFECTS OF CLIMATE CHANGE

The expected changes in SOM as a result of a climate change can be direct through some of the mechanisms discussed above or indirect through the changes in litter inputs. Anderson (1992) has written an extensive review of the effects of climate change on soils. In addition to what has already been discussed, he notes that organic mätter turnover becomes more superficial in the profile and divorced from SOM as soils become colder. Under cold conditions, SOM accumulation is a function of litter quality, there is little stabilization of SOM (e.g. by reactions with clays and polyvalent cations), and SOM is generally younger. Presumably the reverse is true as soils warm. Changes in soil physical structure through the suppression of freezing–thawing cycles could also alter the decomposition rates.

With respect to changes in plant properties, the most difficult to predict are those associated with changes in plant chemical properties. Some investigators have noted substantially decreased tissue N concentrations with elevated CO_2 concentrations (e.g. Strain 1985; Norby *et al.* 1986a, b; see also review by Johnson and Ball 1991). This litter quality–decomposition feedback is also known to occur as stands age due to increasing returns of a very high C:N ratio in woody litterfall following crown closure (Switzer and Nelson 1972; Turner 1981; Miller 1981; Johnson and Ball 1991) and could therefore occur at an earlier stand age if climate change were to increase production rates.

At the moment, the picture with respect to CO_2–litter quality–decomposition is somewhat confused. Based upon lignin:N and lignin:P ratios in white oak (*Querqus alba*) leaves, Norby *et al.* (1986b) conclude that rates of litter decomposition will not be greatly affected by CO_2. Coûteaux *et al.* (1991a) found that CO_2 enrichment caused increased litter C/N ratios but no differences in lignin, cellulose + hemicellulose, or hydrosoluble compounds in chestnut leaves. During subsequent decomposition tests of the litter, the authors noted two phases of CO_2 effects. Initially, litter quality effects dominate in that microbial respiration from untreated (ambient CO_2) leaves was greater than in treated (elevated CO_2) leaves. During the second period, the pattern reversed because of differences in organism composition (increased presence of lignin decomposers in the treated litter).

We can now elucidate some possible consequences of climate change for SOM stores by using the continuous quality approach (section 10.4.3). Although the other theories and models could have been used (e.g. Schimel *et al.* 1990; Jenkinson *et al.* 1991; Kohlmaier *et al.* 1991; Thornley *et al.* 1991; Kirschbaum 1993), we believe that this theory is the most convenient in this context as it summarizes in a very compact form a range of the features that have been discussed above. We will restrict the analysis to the soil only, neglecting feedbacks through the plant system.

For simplicity we will consider a system at steady state. Hence, there is a constant flux of C, I_0, of some organic matter fraction to the soil. The organic matter will start at some initial quality, q_0 but as it is used by the microbes as an energy source, it will decrease in quality on average for each cycle through the microbial community by an amount η_1 (η_1 is an approximation for $D(q, q')$). At each cycle a fraction, $1 - e(q)$, of the C is also lost as respiration, where $e(q)$ is the microbial efficiency (production-to-assimilation ratio). We assume that this efficiency can be described as

$$e(q) = e_1 q \tag{9}$$

where e_1 is a parameter. Similarly, the microbial growth rate, $u(q)$, is assumed to depend on the substrate quality as

$$u(q) = u_0 q^{\beta} \tag{10}$$

where u_0 is a base growth rate and β a shape parameter which measures the degree of physical protection of the substrate. We can then show that we will have a steady-state C store, C_{ss} of (Ågren and Bosatta 1987)

$$C_{ss} = \frac{I_0 e^{q_0 \cdot \eta_1}}{\eta_1 f_C u_0 q_0^{\beta - 1}} \int_0^1 x^{-\beta + 1/(e_1 \eta_1)} e^{-x q_0/\eta_1} \, dx \tag{11}$$

The steady-state store of C derived from one type of original organic matter can thus be expressed in terms of the seven parameters required to define equation (11), all of which have well-defined biological meanings. We can measure I_0 and f_C directly, whereas the other parameters at the moment need to be estimated from decomposition curves for each particular type of litter.

We can now use equation (11) to calculate, by changing the parameters, which category of processes can have the largest impact on soil C storage. To compare the importance of the different parameters, we define a relative sensitivity, s_p to a parameter p as

$$s_p = d(\ln C_{ss})/d(\ln p). \tag{12}$$

For three of the parameters (I_0, f_C, and u_0) the outcome of the sensitivity analysis is trivial because the steady-state C store is proportional (or inversely proportional) to the parameter. The effects of the other four parameters will depend upon where in the parameter space the analysis is performed. As an example, we will choose a set of parameters for Scots pine needles decomposing in a forest stand in central Sweden, $q_0 = 1$, $\eta_1 = 0.893$, $e_1 = 0.187$, $\beta = 3$ (Ågren and Bosatta 1987). We then get the results shown in Table 10.3.

For this type of system, the analyses show that, for equal relative changes in the parameters, litter quality (q_0) and microbial efficiency (e_1) are most important. Changes in microbial growth rate (climatic factors operating through u_0) rank equal to changes in litter production rate and changes in the growth rate–quality curve (β) come next. The average change in quality for each decomposition cycle (η_1) has least importance, which is fortunate as this parameter is the most difficult to estimate.

To see how representative a Scots pine forest in central Sweden is for other types of soils, we have performed a Monte Carlo simulation where we have created 1000 artificial soils by randomly assigning parameter values to q_0 η_1, e_1

Table 10.3 Relative sensitivity of steady-state soil C stores to the parameters defining the steady-state value

Parameter	I_0	I_C	u_0	q_0	η_1	e_1	β
s_p	1	−1	−1	−1.73	0.549	1.81	0.922

Table 10.4 Results from a Monte Carlo simulation of the relative sensitivity of the C store to the parameters defining the steady-state value. One thousand different parameter values have been used in the ranges given in the table. Mean, maximum, minimum values, and standard deviations of the relative sensitivity are shown

Parameter	q_0	η_1	e_1	β
Range	0.1–2	0.3–1.5	0.1–0.6	2–4
Sensitivities				
Mean	−1.45	0.494	1.95	1.33
SD	0.67	0.325	0.64	2.02
Minimum	−2.91	0.033	1.10	−2.23
Maximum	0.88	1.98	5.10	9.80

and β from a uniform distribution and then calculated the relative sensitivities. Ranges for the parameters were selected on the basis of the discussions in section 10.2 to be what we consider ecologically relevant. We have also taken into account restrictions on parameter combinations, e.g. $e_1 q_0 < 1$. The results are given in Table 10.4.

The mean sensitivities in this Monte Carlo simulation are close to what we found for the pine forest. This might, however, only reflect the ranges chosen for the parameters. The ranges and variability in the sensitivities are, however, large. In some cases we can even have a change in the sign of the sensitivity. We have also looked for patterns in the sensitivities, but have not detected any. We do not know to what extent the parameters we are using should be correlated in real systems, e.g. a high litter quality implies efficient microbes, which implies additional constraints on parameter combinations. Some of the extreme sensitivities can therefore be the consequence of unreal parameter combinations, but we should be aware that there might be large differences, both quantitatively and qualitatively, between different ecosystems. These differences are probably more important to understand than deviations from the presumed steady state.

In conclusion, it does not seem possible at the moment to predict the balance between the forces that increase SOM accumulation under climate change and those that decrease accumulation. The comparisons of forest models (Chapter 13) show that different ways of integrating our current understanding lead to predictions that even differ in the signs of the predictions, let alone the magnitudes. At the same time, the model analyses of the different grassland sites (Chapter 13) show that there are large differences between sites with different current climates.

10.6 REFERENCES

Aber, J. D., McClaugherty, C. A. and Melillo, J. M. (1984) Litter decomposition in Wisconsin forests – mass loss, organic-chemical constituents and nitrogen. *School of Natural Resources, University of Wisconsin-Madison. Report R3284.*

Ågren, G. I. and Bosatta, E. (1987) Theoretical analysis of the long-term dynamics of carbon and nitrogen in soils. *Ecology* **68**, 1181–1189.

Ågren, G. I. and Bosatta, E. (1988) Nitrogen saturation of terrestrial ecosystems. *Environ. Pollut.* **54**, 185–197.

Anderson, J. M. (1992) Responses of soils to climate change. *Adv. Ecol. Res.* **22**, 163–210.

Anderson, J. M. and Ineson, P. (1983) Interaction between soil arthropods and microorganisms in carbon, nitrogen and mineral element fluxes from decomposing leaf litter. In: Lee, J.A., McNeill, S. and Robinson, I. H. (Eds) *Nitrogen as an Ecological Factor*, pp. 413–432. Blackwell, Oxford.

Anderson, J. M. and Ineson, P. (1984) Interaction between microorganisms and soil invertebrates in nutrient flux pathways of forest ecosystems. In: Anderson, J. M., Rayner, A. D. and Walton, D. W. H. (Eds) *Invertebrate–Microbial Interactions*, pp. 59–88. Cambridge University Press, Cambridge.

Armson, K. A. (1979) *Forest Soils: Properties and Processes.* University of Toronto Press, Toronto.

Aston, A. R. and Gill, A. M. (1976) Coupled soil moisture, heat and water vapour transfer under simulated fire conditions. *Aust. J. Soil Res.* **14**, 55–66.

Berg, B., Booltink, H., Breymeyer, A., Ewertsson, A., Gallardo, A., Holm, B., Johansson, M.-B., Koivuoja, S., Meentemeyer, V., Nyman, P., Olofsson, J., Reurslag, A., Staaf, H., Staaf, I. and Uba, L. (1991a) Data on needle litter decomposition and soil climate as well as site characteristics for some coniferous forest sites. Part 1. Site characteristics. *Dep. Ecol. Environ. Res., Swed. Univ. Agric. Sci. Rep. 41.* Uppsala.

Berg, B., Booltink, H., Breymeyer, A., Ewertsson, A., Gallardo, A., Holm, B., Johansson, M.-B., Koivuoja, S., Meentemeyer, V., Nyman, P., Olofsson, J., Reurslag, A., Staaf, H., Staaf, I. and Uba, L. (1991b) Data on needle litter decomposition and soil climate as well as site characteristics for some coniferous forest sites. Part 2. Decomposition data. *Dep. Ecol. Environ. Res., Swed Univ. Agric. Sci. Rep. 42.* Uppsala.

Berg, B., Ewertsson, A., Ferenczi, J., Holm, B., Jansson, P.-E., Kuivuoja, S., Meentemeyer, V., Nyman, P., Staaf, H. and Staaf, I. (1986) Data on needle litter decomposition and soil climate as well as site characteristics for some coniferous forest sites. *Dep. Ecol. Environ. Res., Swed. Univ. Agric. Sci. Rep. 22.* Uppsala.

Biederbeck, V. O. (1978) Soil organic sulfur and fertility. In: Schnitzer, M. and Khan, S. U. (Eds) *Soil Organic Matter.* Developments in Soil Science 8, pp. 273–310. Elsevier, Amsterdam.

Bosatta, E. and Ågren, G. I. (1985) Theoretical analysis of decomposition of heterogeneous substrates, *Soil Biol. Biochem.* **17**, 601–610.

Bosatta, E. and Ågren, G. I. (1991a) Dynamics of carbon and nitrogen in the organic matter of the soil: a generic theory. *Am. Nat.* **138**, 227–245.

Bosatta, E. and Ågren, G. I. (1991b) Theoretical analysis of carbon and nutrient interactions in soils under energy-limited conditions. *Soil Sci. Soc. Am. J.* **55**, 728–733.

Bosatta, E. and Ågren, G.I. (1994) Theoretical analysis of microbial biomass dynamics in soils. *Soil Biol. Biochem.* **26**, 143–148.

Boudreau, B. P. and Ruddick, B. R. (1991) On a reactive continuum representation of organic matter diagenesis. *Am. J. Sci.* **291**, 507–538.

Bray, L. R. and Gorham, E. (1964) Litter production in forests of the world. *Adv. Ecol. Res.* **2**, 101–157.

Bunt, J. S. and Rovira, A. D. (1955) The effect of temperature and heat treatment on soil metabolism. *J. Soil Sci.* **6**, 129–136.

Burns, R. C. and Hardy, R. W. F. (1975) *Nitrogen Fixation in Bacteria and Higher Plants.* Springer-Verlag, Berlin, Heidelberg, New York.

Chen, Y. and Aviad, T. (1990) Effects of humic substances on plant growth. In: McCarthy, P., Clapp, C. E., Malcolm, R. L. and Bloom, P. R. (Eds) *Humic Substances in Soil and Crop Sciences: Selected Readings*, pp. 161–186. Am. Soc. Agron., Madison, Wis.

Coleman, N. T. and Thomas, G. W. (1967) The basic chemistry of soil acidity. In: Pearson, R. W. and Adams, F. (Eds) *Soil Acidity and Liming*, pp. 1–41. Am. Soc. Agron., Madison, Wis.

Coûteaux, M.-M., Bottner, P., Rouhier, H. and Billès, G. (1991a) Atmospheric CO_2 increase and plant material quality: production, nitrogen allocation and litter decomposition of sweet chestnut. In: Teller, A., Mathy, P. and Jeffers, J. N. R. (Eds) *Responses of Forest Ecosystems to Environmental Changes*, pp. 429–436. Elsevier Applied Science, London and New York.

Coûteaux, M.-M., Mousseau, M., Célérier, M.-L. and Bottner, P. (1991b) Increased atmospheric CO_2 and litter quality: decomposition of sweet chestnut leaf litter with animal food webs of different complexities. *Oikos* **61**, 54–64.

Dalal, R. C. (1977) Soil organic phosphorus. *Adv. Agron.* **29**, 83–117.

Drobnik, J. (1962) The effect of temperature on soil respiration. *Folia Microbiol.* **7**, 132–140.

Ellert, B. H. and Bettany, J. R. (1992). Temperature dependence of net nitrogen and sulfur mineralization. *Soil Sci. Soc. Am. J.* **56**, 1133–1141.

Flanagan, P. W. and Veum, A. K. (1974) Relationship between respiration, weight loss temperature and moisture in organic residues in tundra. In: Holding, A. J., Heal, O. W., MacLean, S. F. and Flanagan, P. W. (Eds) *Soil Organisms and Decomposition in Tundra*, pp. 249–277. Tundra Biome Steering Committee, Stockholm, Sweden.

Gilmore, A. R. (1980) Changes in a reforested soil associated with tree species and time. IV. Soil organic content and pH in pine plantations after 24 years. *Univ. Illinois Agr. Exp. Sta. For. Res. Rep. No. 80-3.* Urbana-Champaign.

Godshalk, G. L. and Wetzel, R. G. (1978) Decomposition of aquatic angiosperms. II. Particulate components. *Aquat. Bot.* **5**, 301–327.

Harmon, M. E., Franklin, J. F., Swanson, F. J., Sollins, P., Gregory, S. V., Lattin, J. D., Anderson, N. H., Cline, S. P., Aumen, N. G., Sedell, J. R., Lienkaemper, G. W., Cromack, K. Jr and Cummins, K. W. (1986) Ecology of coarse woody debris in temperate ecosystems. *Adv. Ecol. Res.* **15**, 133–302.

Humphreys, F. R. and Craig, F. G. (1981) Effects of fire on soil chemical, structural and hydrological properties. In: Gill, A. M., Groves, R. H. and Noble, I. R. (Eds) *Fire and the Australian Biota*, pp. 177–200. Aust. Acad. Science, Canberra, Australia.

Hyvönen, R., Ågren, G. I. and Andrén, O. (1996) Modelling long-term carbon and nitrogen dynamics in an arable soil receiving organic matter amendments. *Ecol. Appl.* (in press).

Janssen, B. H. (1984) A simple model for calculating decomposition and accumulation of 'young' soil organic carbon. *Plant Soil* **76**, 297–304.

Jenkinson, D. S. (1990) The turnover of organic carbon and nitrogen in soil. *Philos. Trans. R. Soc. Lond.* **B329**, 361–368.

Jenkinson, D. S. (1991) The Rothamsted long-term experiments: are they still of use? *Agron. J.* **83**, 2–10.

Jenkinson, D. S., Adams, D. E. and Wild, A. (1991) Model estimates of CO_2 emissions from soil in response to global warming. *Nature* **351**, 304–306.

Jenkinson, D. S., Harkness, D. D., Vance, E. D., Adams, D. E. and Harrison, A. F. (1992) Calculating net primary production and annual input of organic matter to soil from the amount and radiocarbon content of soil organic matter. *Soil Biol. Biochem.* **24**, 295–308.

Jenkinson, D. S., Hart, P. B. S., Rayner, J. H. and Parry, L. C. (1987) Modelling the turnover of organic matter in long-term experiments at Rothamsted. *INTECOL Bull.* **15**, 1–8.

Jenkinson, D. S. and Rayner, J. H. (1977) The turnover of soil organic matter in some of the Rothamsted classical experiments. *Soil Sci.* **123**, 298–305.

Johansson, M.-B. (1984) Litter decomposition rate at burned and clear-felled areas compared to closed stands. *Dep. For. Soils, Swed. Univ. Agric. Sci., Report 49*. Uppsala.

Johnson, D. W. (1992a) Effects of forest management on soil carbon storage. *Water Air Soil Pollut.* **64**, 83–120.

Johnson, D. W. (1992b) Nitrogen retention in forest soils. *J. Environ. Qual.* **21**, 1–12.

Johnson, D. W. (1992c) Effects of forest management on soil carbon storage. *NCASI Tech. Bull. No. 628*. New York.

Johnson, D. W. and Ball, J. T. (1991) Environmental pollution and impacts on soils and forests of North America. *Water Air Soil Pollut.* **54**, 3–20.

Kirchmann, H., Persson, J. and Carlgren, K. (1994) The Ultuna long-term soil organic matter experiment, 1956–1991. *Dep. Soil Sci., Swed. Univ. Agric. Sci., Rep. Diss. 17*, 55 pp. Uppsala.

Kirschbaum, M. U. F. (1993) A modelling study of the effects of changes in atmospheric CO_2 concentration, temperature and atmospheric nitrogen input on soil organic carbon storage. *Tellus* **45B**, 321–334.

Kirschbaum, M. U. F. (1994) The temperature dependence of soil organic matter decomposition and the effect of global warming on soil organic carbon storage. *Soil Biol. Biochem.* (in press).

Kohlmaier, G. H., Lüdeke, M., Janecek, A., Benderoth, G., Kindermann, J. and Klaudius, A. (1991) Land biota, source or sink of atmospheric carbon dioxide: positive and negative feedbacks within a changing climate and land use development. In: Schneider, S.H. and Boston, P.J. (Eds) *Scientists on Gaia*, pp. 223–239. MIT Press, Cambridge, Mass.

Lashof, D. A. (1989) The dynamic greenhouse: feedback processes that may influence future concentrations of atmospheric trace gases and climatic change. *Clim. Change* **14**, 213–242.

Lieth, H. (1973) Primary production: terrestrial ecosystems. *Hum. Ecol.* **1**, 303–332.

McClaugherty, C. A., Pastor, J., Aber, J. D. and Melillo, J. M. (1985) Forest litter decomposition in relation to soil nitrogen dynamics and litter quality. *Ecology* **66**, 266–275.

Miller, H. G. (1981) Forest fertilization: some guiding concepts. *Forestry* **54**, 157–167.

Mortland, M. M. and Wolcott, A. R. (1965) Sorption of inorganic nitrogen compounds by soil materials. In: Bartholomew, W. V. and Clark, F. E. (Eds) *Soil Nitrogen*, pp. 150–197. Agron. 10. Amer. Soc. Agron., Madison, Wis.

Moureaux, C. (1967) Influence de la température et de l'humidité sur les activités biologiques de quelques sols ouest-africains. *Cah. Pédol. Orstom* **5**, 393–420.

Nömmik, H. (1970) Non-exchangeable binding of ammonium and amino nitrogen by Norway spruce raw humus. *Plant Soil* **33**, 581–595.

Norby, R. J., O'Neill, E. G. and Luxmoore, R. J. (1986a) Effects of atmospheric CO_2 enrichment on the growth and mineral nutrition of *Querqus alba* seedlings in nutrient-poor soil. *Plant Physiol.* **82**, 83–89.

Norby, R. J., Pastor, J. and Melillo, J. M. (1986b) Carbon–nitrogen interactions in CO_2-enriched white oak: physiological and long-term perspectives. *Tree Physiol.* **2**, 233–241.

Nyhan, J. W. (1976) Influence of soil temperature and water tension on the decomposition rate of carbon-14 labelled herbage. *Soil Sci.* **121**, 288–293.

Nykvist, N. (1977) Skoliga åtgärders inverkan på storlek och tillgänglighet av ekosystemets näringsförråd. *Sver. skogsvårdsförb. tidskr.* **2–3**, 167–178.

Oades, J. M. (1988) The retention of organic matter in soils. *Biogeochemistry* **5**, 35–70.

O'Connell, A. M. (1990) Microbial decomposition (respiration) of litter in eucalypt forests of south-western Australia: an empirical model based on laboratory incubations. *Soil Biol. Biochem.* **22**, 153–160.

Olson, J. S. (1963) Energy storage and the balance of producers and decomposers in ecological systems. *Ecology* **44**, 322–331.

Parton, W. J., Schimel, D. S., Cole, C. V. and Ojima, D. S. (1987) Analysis of factors controlling soil organic matter levels in Great Plains grasslands. *Soil Sci. Soc. Am. J.* **51**, 1173–1179.

Paul, E. A. and Clark, F. E. (1989) *Soil Microbiology and Biochemistry*. Academic Press, New York.

Pierre, W. H. and Parker, F. W. (1927) Soil phosphorus studies: II. The concentration of organic and inorganic phosphorus in the soil solution and soil extracts and the availability of organic phosphorus to plants. *Soil Sci.* **24**, 119–128.

Post, W. M., Emanuel, W. R., Zinke, P. J. and Stangenberger, A. G. (1982) Soil carbon pools and world life zones. *Nature* **298**, 156–159.

Post, W. M., Pastor, J., Zinke, P. J. and Stangenberger, A. G. (1985) Global patterns of soil nitrogen storage. *Nature* **317**, 613–616.

Pritchett, W. L. (1979) *Properties and Management of Forest Soils*. John Wiley & Sons, New York.

Raison, R. J., Keith, H. and Khanna, P. K. (1989) Effects of fire on the nutrient-supplying capacity of forest soils. In: Dyck, W. J. and Mees, C. A. (Eds) *Impact of Intensive Harvesting on Forest Site Productivity*, pp. 39–54. IEA/BE T6/A6 Report No. 2, Forest Research Institute, Rotorua, New Zealand.

Raison, R. J., Khanna, P.K. and Woods, P.V. (1985) Mechanisms of element transfer to the atmosphere during vegetation fires. *Can. J. For. Res.* **15**, 132–140.

Ross, D. J. and Cairns, A. (1978) Influence of temperature on biochemical processes in some soils from tussock grasslands. *New Zeal. J. Sci.* **21**, 581–589.

Schimel, D. S., Parton, W. J., Kittel, T.G.F., Ojima, D. S. and Cole, C. V. (1990) Grassland biogeochemistry: *Clim. Change* **17**, 13–25.

Schimel, J. P. and Firestone, M. K. (1989) Inorganic N incorporation by coniferous forest floor material. *Soil Biol. Biochem.* **21**, 41–46.

Schleser, G. H. (1982) The response of CO_2 evolution from soils to global temperature changes. *Z. Naturfor.* **37a**, 287–291.

Schlesinger, W. H. (1990) Evidence from chronosequence studies for a low carbon-storage potential of soils. *Nature* **348**, 232–234.

Scott, N.M. (1985) Sulphur in soils and plants. In: Vaughan, D. and Malcolm, R. E. (Eds) *Soil Organic Matter and Biological Activity*. Developments in Plant and Soil Sciences Vol. 16, pp. 379–401. Martinus Nijhoff/Dr W. Junk Publishers, Dordrecht.

Setälä, H. and Huhta, V. (1990) Evaluation of the soil fauna impact on decomposition in a simulated coniferous forest soil. *Biol. Fertil. Soils* **10**, 163–169.

Söderlund, R. and Svensson, B. H. (1976) The global nitrogen cycle. In: Svensson, B. H. and Söderlund, R. (Eds) *Nitrogen, Phosphorus and Sulphur – Global Cycles. Scope Report 7. Ecol. Bull.* **22**, 23–73.

Strain, B. R. (1985) Physiological and ecological controls on carbon sequestering in terrestrial ecosystems. *Biogeochemistry* **1**, 219–232.

Switzer, G. L. and Nelson, L. E. (1972) Nutrient accumulation and cycling in loblolly pine

(*Pinus taeda* L.) plantation ecosystems: the first twenty years. *Soil Sci. Soc. Am. Proc.* **36**, 143–147.

Thornley, J. H. M., Fowler, D. and Cannell, M. G. R. (1991) Terrestrial carbon storage resulting from CO_2 and nitrogen fertilization in temperate grasslands. *Plant Cell Environ.* **14**, 1007–1011.

Turner, J. (1981) Nutrient cycling in an age sequence of Western Washington Douglas-fir stands. *Ann. Bot.* **48**, 159–169.

Van Veen, J. A. and Paul, E. A. (1981) Organic carbon dynamics in grassland soils. 1. Background information and computer simulation. *Can. J. Soil Sci.* **61**, 185–201.

Viro, P. J. (1974) Effects of forest fire on soil. In: Kozlowski, T. T. and Ahlgren, C. E. (Eds) *Fire and Ecosystems*, pp. 7–45. Academic Press, New York.

Waksman, S. A. and Gerretsen, F. C. (1931) Influence of temperature and moisture upon the nature and extent of decomposition of plant residues by microorganisms. *Ecology* **12**, 33–60.

Walker, J., Raison, R. J. and Khanna, P. K. (1986) Fire. In: Russell, J. S. and Isbell, R. F. (Eds) *Australian Soils: The Human Impact*, pp. 185–216. University of Queensland Press, St Lucia, Australia.

Wiant, H. V. (1967) Influence of temperature on the rate of soil respiration. *J. For.* **65**, 489–490.

11 Global Grassland Ecosystem Modelling: Development and Test of Ecosystem Models for Grassland Systems

W. J. PARTON,[1] M. B. COUGHENOUR,[1] J. M. O. SCURLOCK,[2] D. S. OJIMA,[1] T. G. GILMANOV,[3] R. J. SCHOLES,[4] D. S. SCHIMEL,[5] T. B. KIRCHNER,[1], J.-C. MENAUT,[6] T. R. SEASTEDT,[7] E. GARCIA MOYA,[8] A. KAMNALRUT,[9] J. I. KINYAMARIO[10] and D. O. HALL[2]

[1]*Natural Resource Ecology Laboratory, Colorado State University, Fort Collins, USA*
[2]*Division of Life Sciences, King's College London, UK*
[3]*Department of Vertebrate Zoology and Ecology, Moscow State University, Russia*
[4]*Division of Water, Environment and Forest Technology, CSIR, Pretoria, South Africa*
[5]*Climate System Modelling Program, NCAR, Boulder, USA*
[6]*Laboratoire d'Ecologie, Ecole Normale Supérieure, Paris, France*
[7]*Institute of Arctic and Alpine Research, University of Colorado, Boulder, USA*
[8]*Centro de Botanica, Colegio de Postgraduados, Chapingo, Mexico*
[9]*Faculty of Natural Resources, Prince of Songkla University, Hatyai, Thailand*
[10]*Department of Botany, University of Nairobi, Kenya*

11.1 INTRODUCTION

Natural grasslands, both in the tropics and in temperate regions, play a major but poorly defined role in the global carbon (C) cycle. Grasslands are one of the most widespread vegetation types worldwide, covering nearly one-fifth of the earth's land surface (24×10^6 km^2, Lieth 1972; Hall and Scurlock 1991), and containing approximately 30% of global soil C stocks (Anderson 1991). Ranging from the savannas of Africa to the North American prairies and the converted grasslands of Latin America and South-East Asia, these ecosystems are geographically extensive. The dynamics of these ecosystems are often jointly controlled by climate and biogeochemical factors. The interaction of these factors control fluxes of C through plant and soil organic matter (SOM) in grasslands. The

Global Change: Effects on Coniferous Forests and Grasslands.
Edited by A. I. Breymeyer, D. O. Hall, J. M. Melillo and G. I. Ågren

magnitude of the fluxes and storages of C in these ecosystems have been little studied, especially in the tropics, and may have been underestimated in the past (Long *et al.* 1989).

Grassland ecosystems are globally important as meat and grain producing regions. Human utilization of these ecosystems has occurred for thousands of years. These ecosystems are noted by their high climate variability and high frequency of drought events. The climate variability coupled with high human use makes these ecosystems more susceptible to rapid degeneration of ecosystem properties and soil C levels. An improved understanding of the factors affecting the long-term productivity of the ecosystems will be beneficial to those individuals living in those ecosystems, as well as to developing a better understanding of the role these ecosystems play in global environmental change.

The amount of C in terrestrial vegetation and soils worldwide is greater than the amount found in the atmosphere and the ocean surface layers, which cycle C on annual to millennial time scales. The role of grasslands in global biogeochemical cycles cannot be overlooked, especially when the contribution of worldwide grassland burning is considered (Hao *et al.* 1990; Hall and Scurlock 1991). Tropical grasslands occupy 15 million km^2, and in terms of both land area and productivity are nearly equal to tropical forests. Carbon stored in grassland soils, temperate and tropical, are the major C component in these ecosystems and are a significant C reservoir globally (Anderson 1991).

Grasslands are susceptible to changes in climate and land use. These semiarid and arid ecosystems tend to develop under highly variable rainfall regimes which are conducive to rapid changes in ecosystem structure given modifications in fire and grazing patterns (Archer *et al.* 1994; Ojima *et al.* 1994), and in altered climate regimes (OIES 1991). In order to better understand what factors control grassland dynamics across a broad environmental gradient, we have employed two types of ecosystem models. These models are able to evaluate the influence of climate, edaphic, and land use factors that control grassland dynamics.

Under the aegis of the SCOPE Project, the CENTURY and GRASS models for plant–soil ecosystems were used. This chapter provides general information of the different grassland sites used for model development and testing, methodology of parameterization, description of models, and finally the analysis of model behavior across a suite of grassland sites.

11.2 DESCRIPTION OF SITES

The grassland study sites used for the modelling work (Figure 11.1) reported here cover most of the world's major grassland types, ranging from the continental steppes and plains of the USA and Russia (22–81 cm yr^{-1} precipitation) to both natural and converted grasslands in wet and dry regions of the tropics (59–154 cm yr^{-1} precipitation). Soil types cover a wide range of textures: 20–85% sand,

Figure 11.1 Location of the sites used to test the CENTURY model

10–40% silt, and 2–70% clay. Six of the important biogeographic regions identified by the International Geosphere-Biosphere Programme (IGBP) for global change research are included: temperate Northern Hemisphere, central arid Asia, Caribbean, northern Africa (including West Africa), southern and eastern Africa, and the tropical Asian monsoon region (Eddy *et al.* 1991). Site descriptions are summarized in Table 11.1. The diversity of soils, climatic conditions and plant growth patterns of these sites provides a robust test of CENTURY for global grasslands. While the sites permit a general global intercomparison, each site has made a unique contribution to the analysis presented here.

The large grassland region of the US Central Plains represents a natural west-to-east moisture gradient. The Central Plains Experimental Range (CPER) at the arid end of the gradient is a shortgrass steppe experiencing wide diurnal, seasonal and annual variations in temperature and precipitation. Vegetation at the CPER is dominated by C_4 shortgrasses. The Konza Prairie Research Natural Area, at the more mesic end of the moisture gradient, is characterized by warm-season (C4) tallgrass prairie species. Data available at CPER include a five-year time series (1971–75) of aboveground live biomass and plant production, for control, irrigated, fertilized and both irrigated and fertilized plots. Konza data include estimates of aboveground plant production (1976–90) and seasonality of aboveground live biomass (1984–90) for annually burned and unburned plots. None of the data from Konza was used to develop the model. Soil data from the CPER were used in model development; CPER live biomass data were not used in model development.

The five sites in the Commonwealth of Independent States (CIS; formerly USSR) represent the 'continental' climate-gradient from the European temperate meadows of Kursk and Otradnoye in western Russia, through the typical steppes of Khomutov in Ukraine, to the ultracontinental steppes of Tuva, at the geographical centre of the Asian continent. The long time series of aboveground biomass data (1954–83) for the Kursk site covers a wide range of year-to-year variation in weather conditions. At Otradnoye, five-year monthly time series (1967–72) are available on aboveground biomass for a loamy and a sandy soil. The Khomutov site has four years of aboveground biomass data (1967–70). More detailed data on vegetation dynamics are available for the Shortandy site, with monthly observations of live and dead matter, both above- and belowground, over the period 1975–79. The cold, dry Tuva site (aboveground biomass, 1977–84) is a particularly interesting test for the CENTURY model, since its climatic and physiological characteristics differ markedly from the grasslands of the US Central Plains which were used originally to formulate CENTURY (Parton *et al.* 1987).

In view of the large areal extent of tropical grass-dominated ecosystems and savannas (Hall and Scurlock 1991), it was an important objective of this SCOPE Project to develop an ecosystem model to simulate these tropical grasslands.

Table 11.1 Grassland site descriptions

	Nairobi National Park Kenya	Klong HK, Hat Yai, Thailand	Montecillos, Chapingo, Mexico	Lamto, Ivory Coast	CPER, Pawnee, Colo., USA	Konza Prairie, Kansas USA	Khomutov, Ukraine	Kursk, Russia	Otradnoye 1, Russia	Shortandy, Russia	Tuva, Kazakhstan
Latitude/longitude (approx.)	1° S 36° E	6° N 101° E	19° N 99° W	6° N 5° W	40° N 105° W	39° N 97° W	47° N 38° E	52° N 37° E	61° N 30° E	52° N 71° E	52° N 94° E
Elevation (m)	1500	100	2220	100	1625	400	75	250	50	367	800
Annual precipitation (mm)	680	1540	590	1170	300	818	441	560	543	351	214
Mean monthly minimum temperature (°C) (month)	10.8 (Jul.)	21.5 (Jan.)	1.4 (Jan.)	21.5 (Jan.)	−9.8 (Feb.)	−10.2 (Feb.)	−4.1 (Dec.)	−12.6 (Jan.)	−9.2 (Feb.)	−24.6 (Jan.)	−39.2 (Jan.)
Mean monthly maximum temperature (°) (month)	28.5 (Mar.)	31.2 (Mar.)	26.9 (May)	36.0 (Feb.)	30.7 (Jul.)	36.5 (Aug.)	30.0 (Jul.)	24.7 (Jul.)	26.3 (Jul.)	27.2 (Jul.)	26.7 (Jul.)
Soil type	Black clay vertisol	Humic gley	Solonet	Sandy	Sandy loam	Clay loam	Chernozem/vermiboroll	Chernozem/haplaboroll	Soddy-podzolic I. Loamy II. Sandy	Calciboroll/calciustosol	Haplustosol/calciustosol
Grassland type	Dry grassland	Humid savanna	Saline	Savanna	Shortgrass prairie	Tallgrass prairie	Typical steppe	Meadow steppe	Meadow I. Loamy II. Sandy	Dry steppe	Semidesert ultra-continental steppe
Dominant species (C_3/C_4)	*Themeda triandra*? (C_4)	*Eulalia trispicata* (C_4)	*Distichlis spicata* (C_4)	*Louditia* (C_4)	*Bouteloua gracilis* (C_4)	*Andropogon scoparius/gerardii* (C_4)	*Stipa lessingiana* (C_3)	*Bromus riparius* (C_3)	I.*Alchemilla monticola*(C_3) II.*Alopecurus pratensis*(C_3)	*Stipa zalesskyi* (C_3)	*Agropyron crystatum* (C_3)
Maximum aboveground live biomass (g m^{-2}) (typical month)	337 (Jul.)	442 (Nov.)	430 (Aug.)	450 (Nov.)	40 (Jun.)	380 (Jul.)	340 (Jun.)	362 (Jul.)	I. 244 II. 277 (Jun.)	142 (Jul.)	103 (Jul.)
Historical long-term management regime (estimated)	Grazing with fire every 5 years	N/A (converted grassland)	N/A (alluvial grassland)	Annual burning	Annual grazing	Grazing with fire every 3 years	Annual mowing	Annual mowing	Annual mowing	Annual grazing	Annual grazing

Three of the four tropical grassland sites (Kenya, Thailand and Mexico) were studied under a UN Environment Programme Project, and have used a common methodology to collect data on live and dead matter, above- and belowground, over the period 1984–90 (Long *et al.* 1992). The fourth tropical site was located at Lamto, Ivory Coast (Menaut and César 1979).

The Kenyan grassland is a dry grassland, established over a long period on an ancient soil. By contrast, the Thailand site was probably converted from forest within the last 50–100 years, characteristic of many grasslands in South-East Asia. The Mexican grassland is derived from the bed of a saline lake, drained in 1911. The continuous record of above- and belowground data from this site is one of the most detailed for any grassland worldwide. The humid grass savanna site at Lamto, Ivory Coast, has been intensively studied by Dr J.-C. Menaut and co-workers for the past 30 years, and a discontinuous record kept of monthly aboveground live and dead matter, and total root biomass (Menaut and César 1979). Like most tropical grasslands, all four sites were subjected to burning and/or mowing removal during the period of study. The CENTURY model's ability to simulate the effects of burning has thus been tested under treatments ranging from annual fires to burning every six years. We selected two well-characterized grassland sites: the CPER site, a temperate grassland; and the Kenya site, a tropical grass-dominated savanna, for the comparison between CENTURY and GRASS.

11.3 MODELLING PROTOCOL

Two types of ecosystem models have been employed in this SCOPE study, a simple ecosystem model, CENTURY (Parton *et al.* 1987, 1988, 1992) and a revised version of the more detailed ecophysiologically based model, GRASS. The original description of the GRASS model, is provided by Coughenour (1984). The CENTURY model is a 'lumped' parameter model that aggregates plant production processes with SOM dynamics. It uses a monthly time step and simulates aboveground and belowground processes, soil water budgets at various depths, SOM C (Figure 11.2) and N dynamics in the top 20 cm of the soil, and incorporates various land use practices that include grazing, fire, and cropping systems. The GRASS model is a mechanistic physiologically based grassland model which simulates shoot number, C and N budgets, energy balance, and water budget (Coughenour 1984). The model is different from the CENTURY model in that it uses a daily time step, mechanistic photosynthetic and plant growth functions, and dynamic representations of plant C and N allocation. The GRASS model was originally designed to represent the impact of grazing on grassland systems in Africa.

Simulations of ecosystem dynamics (i.e. annual net plant productivity, seasonal patterns of live aboveground biomass, peak live aboveground biomass, soil

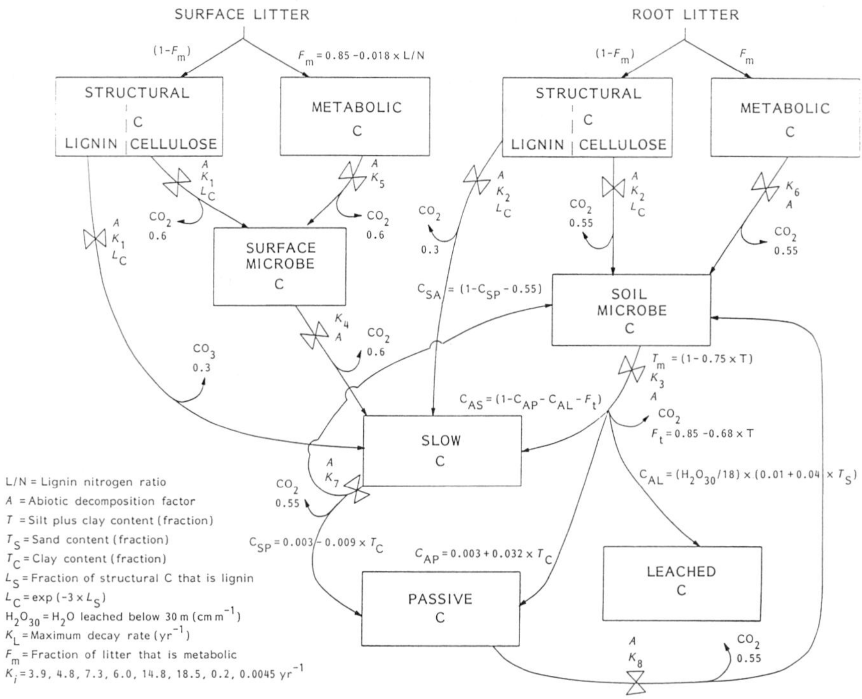

Figure 11.2 Flow diagram for the CENTURY soil organic C model

organic matter C and N dynamics, N mineralization, evapotranspiration) at all 11 sites were conducted using CENTURY. The ability of the model to respond to burning, irrigation and N fertilization was tested using observed data from these sites.

The CENTURY and GRASS models were both run for a shortgrass steppe site in Colorado, USA (CPER) and a site near Nairobi, Kenya and the results of the different models were compared. The comparison is intended to evaluate the dynamic responses of a more mechanistic model and the use of a finer time step (day vs month) on simulated grassland dynamics. Assessment of the relative merits of the two classes of models in determining various ecosystem processes was evaluated.

The sites used to evaluate CENTURY and GRASS all have at least 3 years (several sites have more than 10 years of data) of seasonal live biomass data, observed soil C and N levels and climate data. The statistical comparisons used to test the model include linear regression of observed and simulated results, comparison of model predictions with frequently used regression models (i.e. production vs annual precipitation), and an evaluation of the number of times that the difference between the model outputs and the observed data was less than 25% of the observed data.

While there are more parameters affecting plant growth in GRASS than in CENTURY, the number of parameters in GRASS that vary among sites has been minimized to facilitate cross-site parameterization and application over large spatial scales.

11.4 CENTURY ECOSYSTEM MODEL DESCRIPTION

The CENTURY model is a general ecosystem type which simulates plant–soil dynamics of grasslands, forests, crops and savannas (Parton *et al.* 1987, 1988; Sanford *et al.* 1991). It was developed during this SCOPE project (Version 3.0; Parton *et al.*, 1992) by incorporating routines derived from more mechanistic models or suggested by detailed site data. The model has three major submodels. These include:

(a) biophysical submodel which calculates hydrological and temperature drivers;
(b) a plant production submodel which calculates above- and belowground plant processes;
(c) a soil organic matter submodel that calculates changes in soil C and N and plant detritus.

Additional functions are used to simulate various management or disturbance regimes, such as fire and grazing intensity and frequency. The model has now been parameterized and successfully validated for many different grassland sites in temperate and tropical regions (Parton *et al.* 1987, 1989a, b). It incorporates simplified representations of key processes relating to C assimilation and turnover, based on existing models.

Modifications to the original version of CENTURY are described in Parton *et al.* (1992; 1994). These changes include adding the effect of clay on the formation of passive SOM; improving the surface litter decomposition model; including the effect of anaerobiosis on decomposition; simulating leaching of soluble organic material; improving the inorganic leaching equations; and developing a robust grassland production submodel.

11.4.1 Biophysical sybmodels

The CENTURY model includes a simplified water budget model which calculates monthly evaporation and transpiration water loss, water content of the soil layers, snow water content and saturated flow of water between soil layers. The potential evapotranspiration rate (PET) is calculated as a function of the average monthly maximum and minimum air temperature using the equations developed

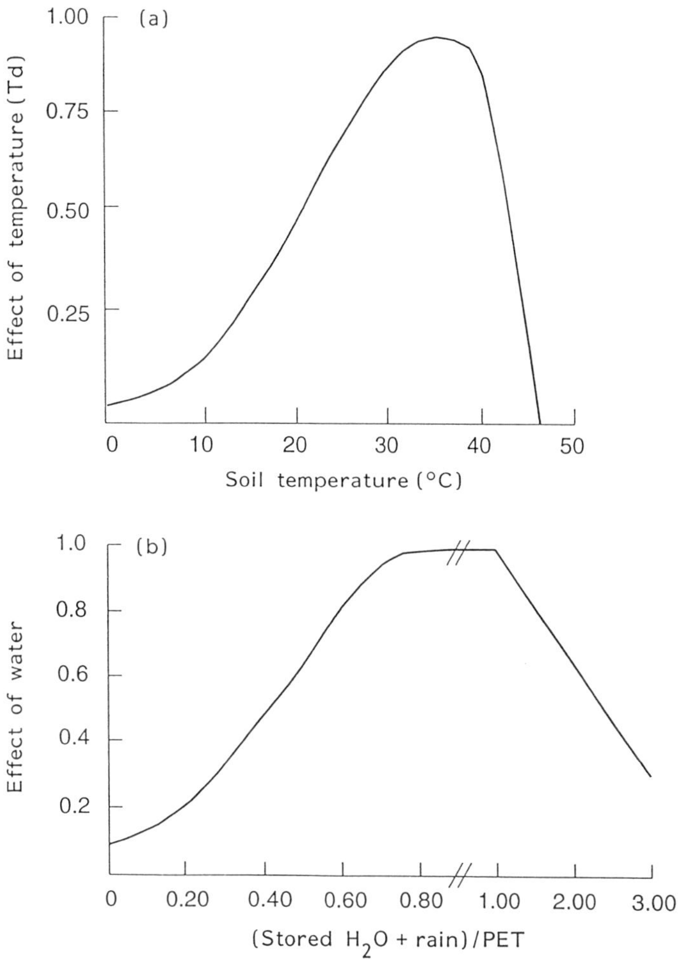

Figure 11.3 The abiotic effect of soil temperature (a) and soil moisture (b) on decomposition

by Linacre (1977). Bare soil and interception water loss is a function of aboveground biomass, rainfall, and PET. Transpiration water loss is a function of the live leaf biomass (exponential function of leaf biomass), rainfall, and PET. The field capacity and wilting point for the different soil layers are calculated as a function of the bulk density, soil texture, and organic matter content using an equation developed by Gupta and Larson (1979).

Average monthly soil temperature near the soil surface is calculated using equations developed by Parton (1984). These equations calculate near surface maximum soil temperature as a function of maximum air temperature (2 m height) and canopy biomass (lower for high biomass) while the minimum soil temperature is a function of minimum air temperature (2 m height) and canopy biomass (higher for higher biomass). The actual soil temperature used for decomposition and plant growth rate functions is the average of the minimum and maximum monthly soil temperatures (Figure 11.3).

11.4.2 Plant production submodel

The grassland submodel (Figure 11.4) simulates grass growth and includes the impact of grazing and fire on plant production. Potential plant production is calculated as a function of soil temperature, available water, and a self-shading factor using the following equation:

$$P_p = P_{max} \times f(T_p) \times f(M_p) \times f(S_p) \quad (1)$$

where P_p is the aboveground potential plant production rate (g m^{-2} month^{-1}), P_{max} is the maximum potential aboveground plant production rate (250 g m^{-2} month^{-1}), T_p (Figure 11.5b) is the effect of soil temperature on growth, $M_p$is the effect of moisture on production (Figure 11.5a), and S_p is the effect of plant shading on plant growth (Figure 11.5c). The effect of moisture on production M_p is a function of the ratio of current monthly precipitation plus the previous month's stored soil water (0–60 cm) to the potential evapotranspiration rate. The soil water-holding capacity also influences M_p by modifying the amount of stored soil water, so that lower water-holding soils (e.g. sandy soils) have a higher growth rate under dry conditions (see Figure 11.5a). This function was added to the model in order to represent the observation that sandy soils are more productive in dry environments (Sala *et al.* 1988). The temperature response curves for plant growth are based on data from Christie and Detling (1982) for C_3 and C_4 plants. The shading factor reduces plant production in the presence of large amounts of standing dead matter (Figure 11.5c).

The potential plant production rate (P_p) is reduced if there is insufficient N. A maximum and minimum C to N (C:N) ratio is specified for new production of live roots and shoots. The minimum C:N ratio sets the limit for maximum N concentration of new plant production. The live shoot C:N ratio is a linear function of the live plant biomass. The fraction of the labile nutrient pools that are available for plant growth (F_g) are a function of the live root biomass (ROOT; g m^{-2}) according to the following equation:

$$F_g = 1.0 - 0.8 \exp(-0.015\ \text{ROOT}) \quad (2)$$

This equation is based on data presented by Wedin and Tilman (1990). Live root and shoot death are calculated as a function of the available soil water in the 0–60 cm layer ($H_2O_{(a)}$) using the following equations:

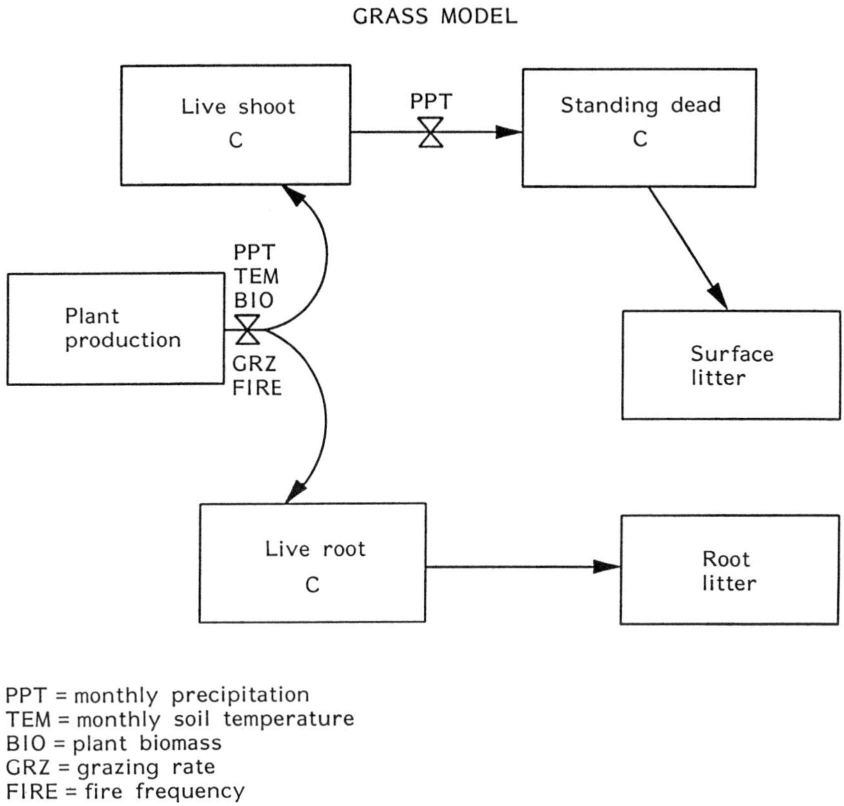

Figure 11.4 Flow diagram for the grassland production submodel

$$R_d = 0.12\ [\exp(-5.0\ H_2O_{(a)})] \quad (3)$$
$$S_d = 0.20\ [\exp(-5.0\ H_2O_{(a)})] \quad (4)$$

where R_d and S_d are the fraction of roots and shoots, respectively, which die per month. Shoot death also occurs when live plant biomass exceeds 400 g m^{-2} at a rate of 10% per month. This death rate is designed to simulate death due to shading of lower leaves in the canopy.

The model calculates root/shoot ratios as a function of the annual rainfall using equations presented in the original description of CENTURY (Parton *et al.* 1987). The effects of fire and grazing on plant production are based on data from Ojima *et al.* (1990, 1994) and Holland *et al.* (1992), respectively. The simulated impacts of fire are to increase the root/shoot ratio, and to increase the C:N ratio of live shoots (+10 for temperate sites; 0 for tropical sites) and roots (+30 for all sites), removing vegetation and returning inorganic nutrients (Ojima *et al.* 1990). Grazing removes vegetation, returns nutrients to the soil (by urination and

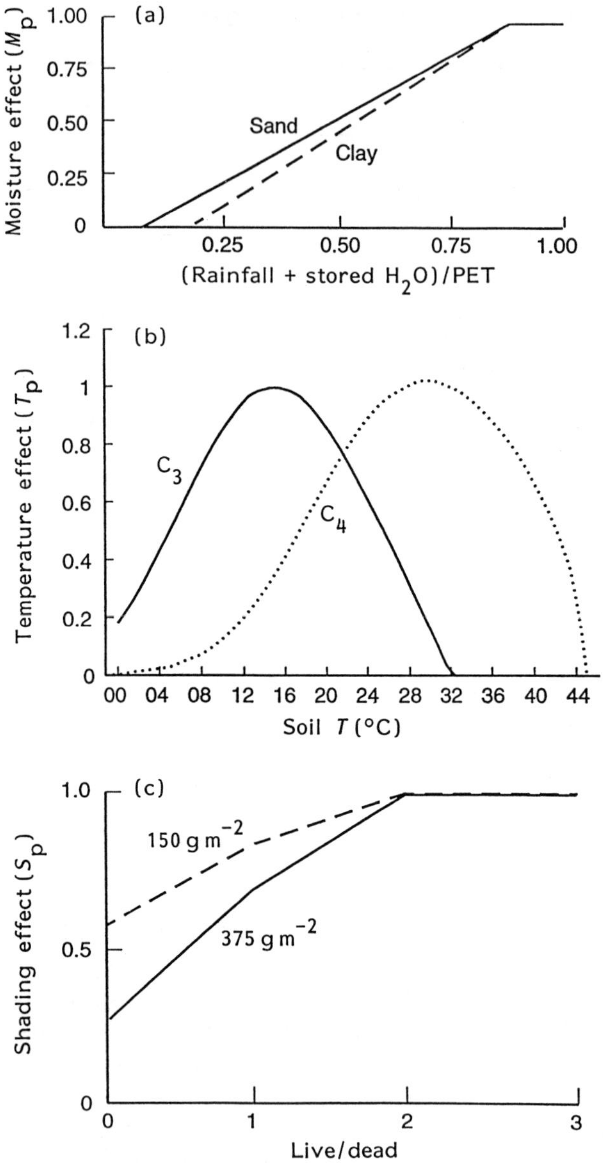

Figure 11.5 The effect of moisture (a), soil temperature (b) and shading (c) on potential grass growth

defecation), alters the root/shoot ratio and increases the N content of live shoot and roots (Holland *et al.* 1992).

11.4.3 Decomposition and soil organic matter submodel

The soil organic matter (SOM) submodel simulates the dynamics of C and N in the organic and inorganic parts of the soil system. The flow diagram for soil C (Figure 11.2) shows that soil C is divided into three components which include active, slow, and passive soil C. Active SOM includes live soil microbes plus microbial products (the total active pool is approximately two to three times the live soil microbial biomass), the slow pool includes resistant plant material (for instance, lignin-like components) and soil-stabilized plant and microbial material, while the passive material is very resistant to decomposition and includes physically and chemically stabilized SOM. The flows of C are controlled by the inherent maximum decomposition rate of the different pools and the water and temperature-controlled decomposition factor (Figure 11.3). Average monthly soil temperature at the soil surface controls the temperature function, and the ratio of stored water (0–30 cm depth) plus current monthly precipitation to potential evapotranspiration is the input for the moisture function. Microbial respiration occurs for each of the decomposition flows. The partitioning of decomposition between stabilized SOM and CO_2 flux is a function of soil texture for the stabilization of active C into slow C (increasing CO_2 flux for sandy soils and less soil C storage). Justification for these assumptions is presented in Parton *et al.* (1987).

Plant residues (shoots and roots) are partitioned into structural (resistant to decomposition) and metabolic (readily decomposable) plant material as a function of the initial residue lignin (L)-to-nitrogen (N) ratio. The lignin fraction is assumed to be part of the structural material, and (i.e. L:N ratio) controls the decomposition rate of structural material.

The N submodel has the same general structure as the soil C model (Figure 11.6). The organic-N flows follow the C flows and are equal to the product of the C flow and the N:C ratio of the state variable that receives the C. The C:N ratios of the soil state variables receiving the flow of C are a function of the mineral N pool (NO_3^- plus NH_4^+) and vary within the ranges 3–15, 12–20, and 7–10, respectively, for active, slow and passive SOM (Figure 11.4). The C:N ratio of newly formed surface microbial biomass is a linear function of the N content of the material being decomposed, and increases from 10 to 20 as the N content decreases from 2.0 to 0.01% (see Figure 11.6). The C:N ratio of slow SOM material formed from surface microbes is equal to the C:N ratio of the microbes plus 5.0. The N associated with C lost in respiration is assumed to be mineralized. Given the C:N ratio of the state variables and the microbial respiration loss, decomposition of metabolic residue, active, slow, and passive pools generally results in net mineralization of N, while decomposition of structural material

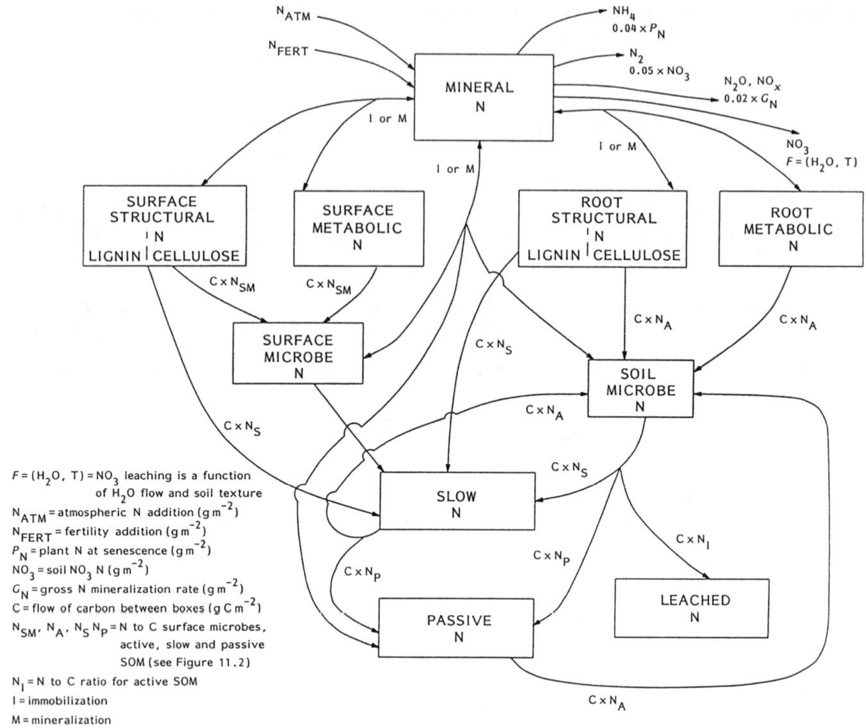

Figure 11.6 Flow diagram for the CENTURY soil N submodel

immobilizes N. The model also uses simple equations to represent N inputs due to atmospheric deposition and N fixation and calculates N losses due to N_2, NO, N_2O, and NH_3 gas fluxes (see Figure 11.6) and NO_3 leaching. A more complete description and justification for the N submodel is presented by Parton *et al.* (1987, 1988, 1994).

11.5 GRASS MODEL DESCRIPTION

The GRASS model is more physiologically oriented relative to CENTURY (Figure 11.7), as it links physiological processes and morphometric traits (Coughenour 1984). Plant growth submodels represent tillering and C and N budgets. Additional modules simulate light penetration and soil water and heat budgets (Coughenour 1984). The model uses a time step of two hours for diurnal processes such as leaf and soil energy balance and photosynthesis, and two days for other plant growth processes. The model is driven by daily weather data, although daily data can also be generated stochastically from monthly weather

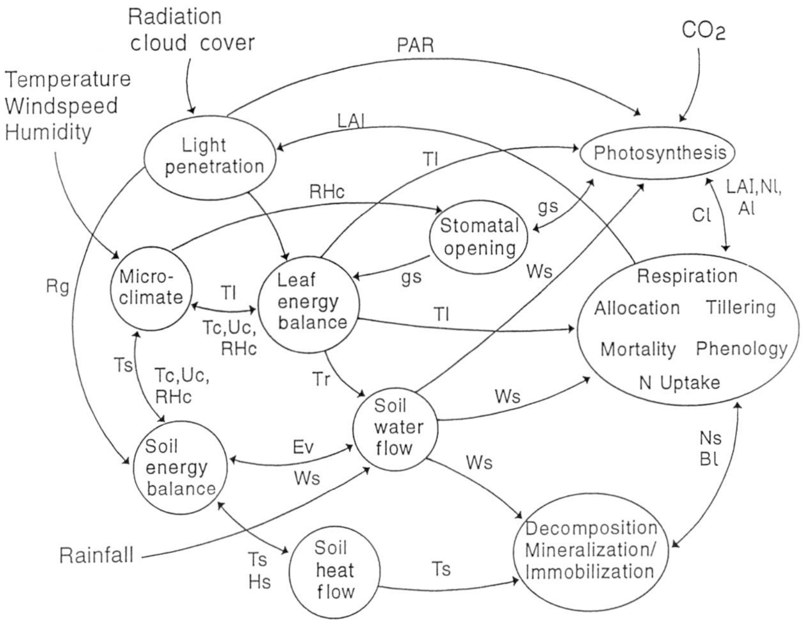

Figure 11.7 Major submodules in the GRASS model and flows of information among submodules. (LAI)-leaf area index; (Nl)-leaf nitrogen; (Al)-leaf age; (Cl)-labile carbon; (Ns)-soil N; (Bl)-litter biomass; (Ws)-soil water; (Ts)-soil temperature; (PAR)-photosynthetically active radiation; (g_s)-stomatal conductance; (Ev)-soil evaporation; (Ws)-soil water; (Tc)-canopy temperature; (Uc)-windspeed; (RHc)-humidity; (Tr)-transpiration; (Rg)-global radiation; (Tl)-leaf temperature

data. Long-term model runs are possible on current microcomputers (e.g. 100 years in 30 minutes). The original version of the model was developed to examine plant responses to herbivory in Africa (Coughenour 1984). The model has been modified and updated over the last decade into a new version (GRASS) that is linked to daily soil and water flow submodels (Parton 1978) and to a daily time step implementation of the decomposition and nutrient cycling (CSOM) routines from CENTURY. The daily CSOM routines use the same CENTURY model structure and nearly identical parameters, but decomposition responds to actual daily soil moisture content.

The submodel for C_3 photosynthesis used in GRASS is based on the model of Farquhar *et al.* (1980), which considers the relative limitations of rates of ribulose 1,5-biphosphate (RubP) carboxylase fixation of internal leaf CO_2 and RubP regeneration, and RubP oxygenation. Reaction rates respond to temperature according to Arrhenius functions, which generally exhibit temperature optima. The CO_2 fixation of RubP is limited by mesophyll CO_2 concentration in C_3 species and by bundle sheath CO_2 concentration in C_4 species. The maximum rate of RubP regeneration is asymptotically light limited. The C_4 cycle, present only in C_4 species, is limited by phosphoenylpyruvate (PEP) carboxylase fixation, which responds to mesophyll CO_2 concentration (Chen and Coughenour 1994). Assimilation rate is reduced by low soil water content and leaf N. Dark respiration responds to temperature with a Q_{10} function.

The empirical stomatal conductance submodel of Ball *et al.* (1987) is used to effect diffusion of CO_2 from the atmosphere into the mesophyll. Stomatal conductance (g_s) responds to leaf surface CO_2 concentration, relative humidity and net photosynthesis (A_n). The model calculates a simultaneous solution for mesophyll and bundle sheath CO_2, due to A_n and diffusion through stomates. The g_s value influences latent heat loss in the leaf energy balance.

Net fixed C can be stored as labile C within the plant. Respiration costs associated with tissue biosynthesis and maintenance are calculated after Ryan (1991). Maintenance respiration is a function of N content, as well as temperature according to a Q_{10} function. Carbon is allocated to structural root vs shoot tissues dependent upon water and N stress. Tissue mortality rates respond to water stress, and to tissue age in the case of leaves. Nitrogen is taken up by roots and allocated in relationship to leaf age. Nitrogen is retranslocated during tissue mortality to tissues that are still living.

Biomass production models interact with tillering and phenology submodels. Tillering depends on water, temperature, N and labile C. Phenology advances in response to growing degree day sums and daylength, but soil temperature and moisture may also trigger phenological change.

11.6 MODEL VERIFICATION OF CENTURY AND GRASS

11.6.1 CENTURY

The parameterization of CENTURY for the different grasslands was conducted to minimize the modifications of only a small number of site-specific model parameters (Appendix 11.1). Many of the parameters used by CENTURY are intended to remain constant in the majority of applications and are referred to as the fixed parameters. Other parameters which are particular to an individual site or group of sites are contained within a site-specific file. Simulated output of aboveground live biomass, aboveground net primary production and levels of SOM were used to validate the model relative to observations made at each of the sites. In evaluating the robust behavior of the model across the range of environmental factors, it was desirable for the majority of the model parameters to comprise a universal 'world grassland' set, with a relatively small number of site-specific parameters changing according to data available from each study site.

Some of these changes in parameters consist of site-specific data required to run the CENTURY model, including latitude and longitude, soil texture (sand, silt and clay), soil pH, soil depth for modelling water budget, plant type (C_3, C_4)

Table 11.2 Site-specific differences between functions controlling plant growth

	Kenya	Thailand	Mexico	Lamto	CPER	Konza	Khomutov	Kursk	Otadnoye	Shortandy	Tuva
Root/shoot ratio as a function of total precipitation	–	–	○	–	○	○	○	○	○	○	○
Root turnover as function of soil water	○	○	○	○	–	○	○	○	○	○	○
Atmospheric and soil N inputs	○	–	○	○	○	○	+	+	+	+	+
Number of soil layers available for plant growth	–	–	○	○	○	○	○	○	–	○	–
Shoot turnover as function of soil water	+	+	○	○	○	○	○	○	○	○	○
Fall rate of standing dead matter to litter	–	–	–	–	○	–	○	○	○	○	○
C/N ratio of shoot matter as a function of biomass	○	○	○	+	○	+	○	○	○	○	○

○ = standard function.

+/– = increase or decrease in value of function (see text for details).

and growing season, and weather data (monthly precipitation, monthly maximum and monthly minimum temperatures). The remaining site-specific parameters relate mainly to plant physiological functions controlling growth, death, turnover and N inputs. These differences were kept to a minimum (Table 11.2). Land management variables (frequency of burning and grazing) also differed among sites; CENTURY allows burning and grazing events to take place in specified months over a repeated pattern of up to 20 years.

For each site modelled, a site file was first created by entering basic site identification and 'minimum input' data into the 'universal' grassland file. Phenological differences (length of growing season) and plant growth responses to temperature were also entered at this stage. The CENTURY model was then run for a period of 500–5000 years, using repeated weather sequences based on long-term weather data from the sites. Appropriate patterns of grassland management (grazing and burning) were incorporated in these long-term runs, based on the known history of the sites (see Table 11.1).

Simulated levels of SOM were then evaluated for long-term stability, and compared with actual field measurements. This check for long-term stability was necessary to appropriately partition soil C to the slow and passive SOM pools of CENTURY, which have characteristic turnover times of 50 and 1000 years, respectively. The long-term values of SOM were then entered in the site file to provide stable initial conditions. The model simulations were made over periods of 20–30 years, using actual weather data recorded at the site, and a comparison of simulated and observed values of monthly biomass and dead matter, above- and belowground was made. For these short-term model runs, the management regime was based on recent grazing/burning practice at each site. For example, the CIS sites are all within Biosphere Reserves and not subject to either grazing or burning, although grazing and mowing for forage would have been common practice in the past and were included in the long-term initializations.

Differences in physiological parameters between the 11 grassland sites are summarized in Table 11.2. At the Kenyan, Lamto and Thailand sites, the relatively low belowground live biomass compared with aboveground live biomass required a change in the equations determining root/shoot ratio. Root death rate, determined by available soil water, was satisfactorily modelled by a common function at all the sites except for CPER, where a lower rate was required to achieve the observed balance between live and dead root biomass.

Estimating long-term N balance proved to be one of the most difficult aspects of the modelling exercise. Atmospheric deposition and N-fixation inputs are normally determined as a simple linear function of precipitation based on data from the US Great Plains, but increased inputs were required to predict the observed production at the (relatively dry) CIS sites, which is consistent with the higher rates of N deposition in continental Europe compared with the US Central Plains (Bónis *et al.* 1980). Reduced N inputs were necessary at the Thailand site in order to limit production under conditions of high rainfall. At

Konza, Lamto and Kenya, the function determining the C:N ratio was modified to allow wider ratios based on detailed information available on the N content of plant matter. This modification was important because these sites have frequent burning which would otherwise result in too much loss of N.

The death rate of aboveground biomass is determined in CENTURY by the number of soil layers contributing water to plant growth (and how quickly they dry), as well as by a direct soil water function. In order to produce sufficiently rapid turnover of aboveground matter, the number of soil layers was reduced at the Kenya, Thailand, Otradnoye and Tuva sites. This modification reflected differences in soil depth and structure. The death rate function was further increased at the Kenya and Thailand sites. The transfer of standing dead to litter was highest for continental sites with snowfall (0.20 per month) and much lower for the tropical sites (0.10 per month).

Models, like CENTURY, which simulate NPP are an important complement to more physiologically oriented models which simulate photosynthesis and respiration. Physiological models require detailed climatic information, such as incoming solar radiation and cloudiness, that are not widely available for retrospective analyses such as are presented here. Physiological models may also be sensitive to species-dependent traits, though this sensitivity may be tractable for large-scale calculations (Schimel *et al.* 1991; Sellers *et al.* 1992; McGuire *et al.* 1992). The extensive body of NPP data over time, reflecting both geographic variability and the effects of climate variability, allows the development of powerful calibrated models like CENTURY. Such data and models are a crucial benchmark test for predicted time-integral quantities (e.g. NPP, biomass levels) from models based on calculation of instantaneous exchanges of CO_2.

11.6.1.1 Model testing – CENTURY

In this section we show a comparison of observed data and simulated model output variables. The model output variables that we will test include live shoot biomass dynamics, aboveground net primary production, peak standing live shoot biomass, and soil C and N levels. These variables were selected because they characterize general grassland ecosystem dynamics and can be evaluated at a variety of sites. Soil C level and plant production are important for the global C cycle, live plant biomass has a big impact on the energy and water budgets, and peak live biomass is frequently used as an estimate of aboveground plant biomass. Live plant biomass influences the energy budget by altering the canopy albedo and soil temperature latent heat fluxes. The importance of these latter considerations is related to evaluating potential changes in the terrestrial ecosystem resulting from different climate change scenarios.

We employed three techniques for testing the CENTURY model:

(a) linear regression of observed data vs simulated model results;
(b) comparison of the CENTURY model predictions with empirical regression models predicted from climate variables (Lauenroth 1978);
(c) calculating the number of times the difference between CENTURY model predictions and observed data differed by less than a threshold proportion (25%) of the observed data.

The results show that in general the observed vs simulated r^2 for the CENTURY model and the empirical regression models are fairly similar. The CENTURY model is able to predict the grassland ecosystem dynamics in more mechanistic fashion which allows for more reliable scenarios of grassland dynamics to climate changes. The advantage of the CENTURY model is that it can predict total system behavior (e.g. nutrient cycling, N gas fluxes, water fluxes, leaching inorganic and organic compounds) and responses to manipulations such as fertilization, irrigation and land use changes, while regression models can only predict what will happen for the particular observed data set used to generate the regression model coefficients. As a minimum criterion, in general, we felt CENTURY should predict observed data at least as well as empirical climate regression models, which have limited extrapolation potential without including nutrient interactions (Esser 1986).

11.6.1.2 Aboveground NPP

Aboveground plant production data were available from the CPER shortgrass prairie site and the Konza tallgrass site. The CPER site has plant production from 1971 to 1975 for a control, fertilized, irrigated, and irrigated/fertilized site (Lauenroth 1978). The Konza site has plant production data from 1976 to 1990 for annual burned and unburned sites (Abrams *et al.* 1986; Briggs *et al.* 1989). The comparison of observed and simulated production data (Figure 11.8) shows that the model simulates these different treatments with an r^2 of 0.70. A linear regression where the independent factors were annual precipitation, fertilization and burning has an $r^2 = 0.67$. While CENTURY may not greatly improve predictions under current conditions, compared to a simple regression model, its concurrent ability to predict the seasonality of foliage production has substantial significance to vegetation–atmosphere interactions studies and models of ecosystem dynamics, including examination of the effects of a changed climate regime. Of CENTURY's predicted aboveground plant production values, 60% had errors less than $\pm 25\%$ of the observed plant production. In comparison, standard errors of observed plant productivities across replicates generally ranged from 10 to 20% of the observed means.

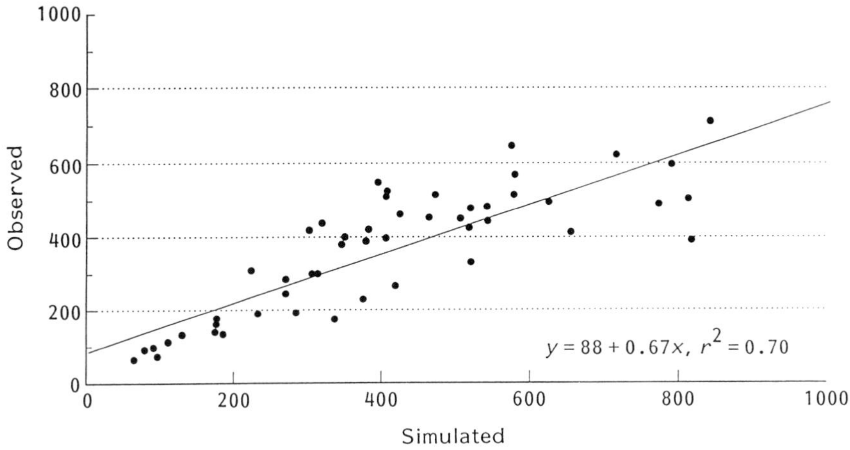

Figure 11.8 Comparison of observed and simulated aboveground plant production for all the CPER treatment and burned and unburned sites at Konza

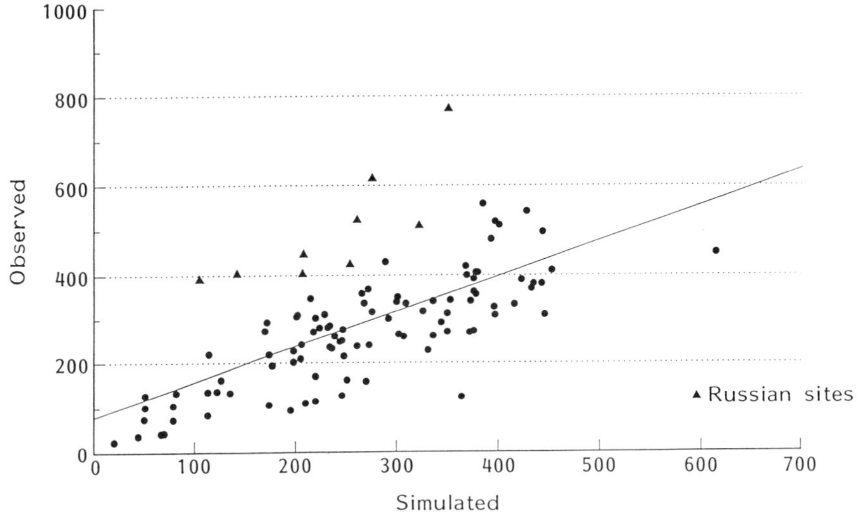

Figure 11.9 Comparison of observed and simulated peak live biomass from all of the sites. The Kursk and Khomutov site data are shown with a ''

11.6.1.3 Peak live biomass

The simulated peak live biomass from all of the sites and treatments were compared to the observed peak live biomass (Figure 11.9). (The observed peak live biomass is the maximum live biomass observed in a particular calendar year

(January to December).) The results from CENTURY have an $r^2 = 0.45$. The r^2 for the best regression model (independent variables include annual precipitation, fertilization and fire) was 0.40. Peak live biomass is frequently used to estimate annual plant production in ungrazed grasslands, with annual production being approximately 50% greater than the peak live biomass (Lauenroth and Sala 1992).

An analysis of the residual error showed that all of the large underestimates of peak live biomass by CENTURY came from the Khomutov and Kursk sites. Removing some of these points from the comparison increased the r^2 from 0.45 to 0.65. Of the CENTURY predictions, 57% had errors that were less than $\pm 25\%$ of the observed data. In general, the model simulated peak live biomass reasonably well for all of the sites, with the exception of peak live biomass for the observed high plant production years (9 years out of 32 years) at the Kursk and Khomutov sites. The reason for the Russian site errors is unclear. We suspect that changes in species composition occurred during the high production years. It is also important to note that these two sites recorded some of the highest peak live biomass observed at any site in spite of the fact that precipitation at the Russian sites is one-half to one-third of the values for the tropical sites.

11.6.1.4 Live plant biomass dynamics

The comparison of observed and simulated live biomass for some of the temperate (Figure 11.10) and tropical (Figure 11.11) sites shows that general seasonal patterns were well simulated by the model. Differences between very dry and wet years (1989 vs 1990 at Konza, and 1981 vs 1980 at Tuva) were well simulated; however, more subtle differences between years with similar precipitation (Konza 1984–87, and Kursk 1975–79) were not well simulated. These subtle interannual differences may result from factors including:

(a) interannual variations in incident PAR, not included in the model;
(b) effects of timing of rainfall not captured with a monthly time step, or
(c) lag effects of nutrient or photosynthate storage in plants, although this is not an inclusive list.

Investigation of the above effects is ongoing. For the tropical sites (Figure 11.11) the response to fire (1986 and 1989 at Thailand, 1986 at Lamto) was fairly well simulated. The death of live biomass at Mexico was underestimated during the dry season.

Figure 11.12 shows a comparison of observed vs simulated live plant biomass for all of the sites and different treatments ($r^2 = 0.39$). Again the model tended to underestimate live biomass for high production years at the Kursk and Khomutov sites. Removing the data from the Kursk and Khomutov sites increased the r^2 to 0.56, suggesting that the errors for the Russian high plant production years greatly reduce the r^2. Of the CENTURY predictions, 31% had

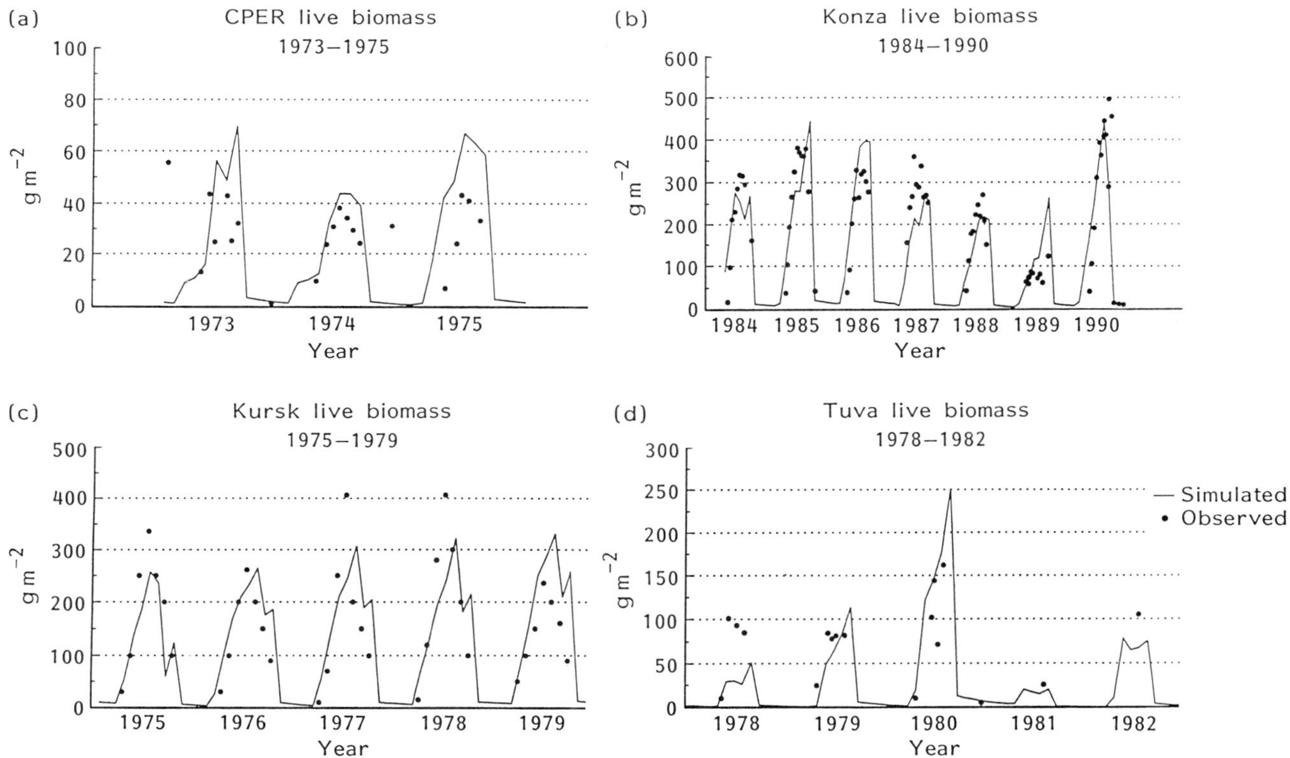

Figure 11.10 Comparison of simulated and observed live biomass for (a) CPER, (b) Konza, (c) Kursk and (d) Tuva sites

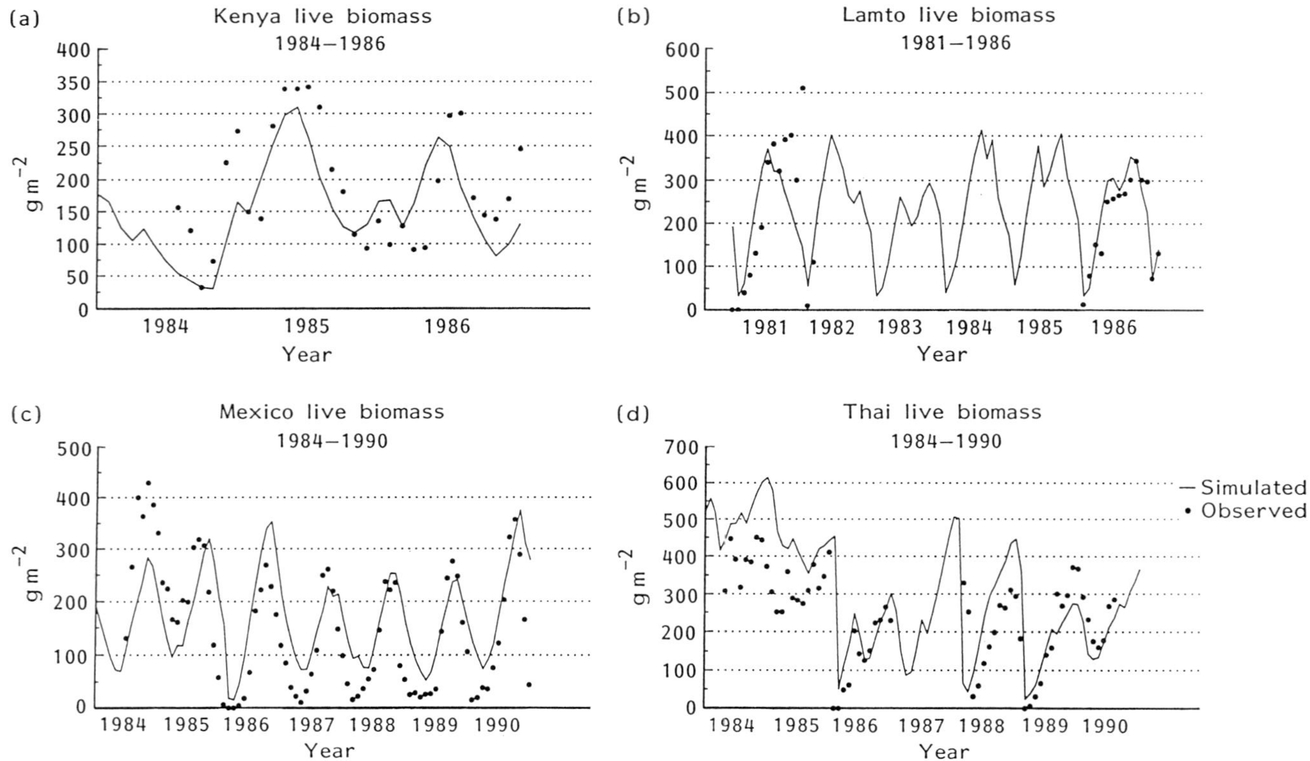

Figure 11.11 Comparison of simulated and observed live biomass for (a) Konza, (b) Lamto, (c) Mexico, and (d) Thailand sites

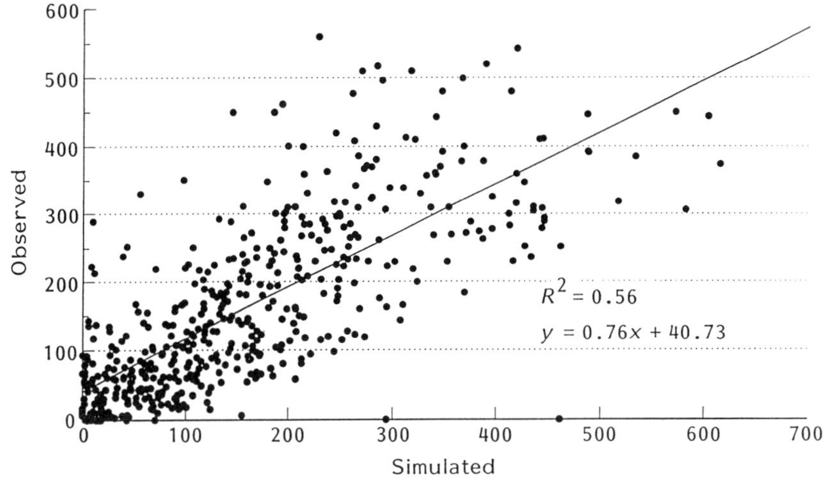

Figure 11.12 Comparison of simulated and observed live biomass for all of the sites and treatments (except Khom and Kursk)

errors less than ±25% of the observed data, while 57% of CENTURY predictions had errors less than ±50% of the observed data. These results show that the model simulates peak live biomass and total annual aboveground primary production more precisely than it simulates the monthly dynamics of live shoot biomass.

11.6.1.5 Soil C and N

Soil C and N levels are well simulated ($r^2 = 0.93$ and 0.89), respectively for C and N across a range of soil C levels ranging from less than 2000 g C m^{-2} at Lamto to greater than the 10000 g C m^{-2} at the Kursk site (Figure 11.13a, b). The assumptions used to control soil C stability and decomposition (clay impact on passive SOM and silt plus clay on slow SOM) seem to work across a diverse set of soil textures and soil mineralogies. All of the soil C simulated values and 75% of the soil N simulated values had errors less than ±25% of the observed values. Soil C data from the CPER and Konza site were used to fit some of the model parameters in the SOM model. We had insufficient data to develop empirical models for soil N and C storage in this study; however, other studies support the conclusion that climate and soil texture are required as predictors (Burke *et al.* 1989).

11.6.2 GRASS

The photosynthesis submodel was tested against measurements of photosynthetic responses to leaf internal (mesophyll) CO_2 concentration (A/C_i curves),

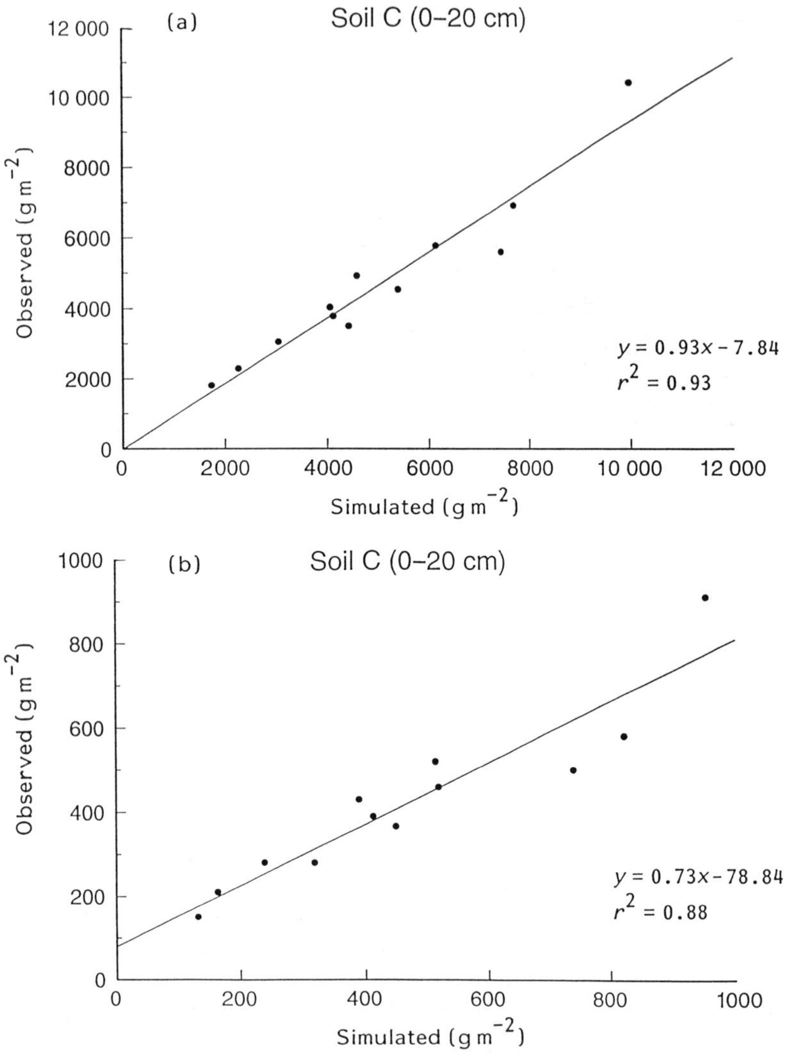

Figure 11.13 Comparison of simulated and observed steady-state soil C (a) and N (b) for 9 of the 11 sites. Mexico and Thailand sites were not used because they are not at equilibrium with their present land uses

incident photosynthetically active radiation (PAR), and leaf temperature (Monson *et al.* 1982; Knapp *et al.* 1993; Polley *et al.* 1992). The measurements provided data for a tall C_4 grass (*Andropogon gerardii*), a short C_4 grass (*Bouteloua gracilis*) and mid-height C_3 grass (*Agropyron smithii*), all from North America. To support our predictions of ecosystem responses to climatic change, it is important to demonstrate that simulated plants respond realistically to temperature and CO_2. Since C_i depends upon stomatal conductance, A/C_i curves are also affected by the stomatal submodel (Ball *et al.* 1987; Chen and Coughenour 1994).

Simulated A/C_i curves compared favorably with data (Figure 11.14a–c). Ambient CO_2 concentrations of 350 ppm produced C_i of 200–250 ppm. Assimilation rates of the C_4 plants were close to being saturated at 350 ppm CO_2 ambient, due to concentration of CO_2 by PEP carboxylase into bundle sheath cells. The ratio of A_n at 700 ppm to that at 350 ppm CO_2 was 1.04 in *Andropogon gerardii*, 1.03 in *B. gracilis*, and 1.31 in *Agropyron smithii*. The CO_2 compensation points (x-intercepts) were near 0 ppm for C_4 species, and 50 ppm for the C_3 species.

Light responses were accurately simulated (Figure 11.14d–f). The quantum efficiencies (initial slopes of the light response curves) were correct, as were light compensation points (x-intercepts). The C_3 species was nearly saturated at 1500 mmol m^{-2} s^{-1} while the C_4 plants were not. However, it was necessary to change the maximal electron transport rate J_m to simulate the light response curve of the C_3 plant at 800 ppm CO_2 and 2% O_2.

Simulated and measured temperature response curves compared favorably, despite a wide range of shapes (Figure 11.14g–i). The C_4 plants exhibited higher temperature optima than the C_3 plants (37–38 °C vs 25 °C) as expected. While A_n in the C_3 grass was still active at 5 °C, in the short C_4 grass *B. gracilis* it was much restricted and in the tall C_4 grass *Andropogon gerardii* it was insignificant. Conversely, the C_4 plants were still active at 50 °C while the C_3 plant was not. The temperature optima for photosynthesis were shifted higher under elevated CO_2 concentrations due to differential temperature sensitivities of solubilities and RubP affinities for CO_2 and O_2, in accordance with observations (Long *et al.* 1989).

Photosynthetic light responses of the Kenyan C_4 grass Themeda triandra (Kinyamario and Imbamba 1992) at different canopy heights were compared with model simulations at different leaf N concentrations (Figure 11.15). Leaves higher in the canopy tend to have higher N concentrations. Top canopy leaves were assumed to have maximal N concentrations. The favorable comparisons supported model predictions of maximal photosynthetic rates of this species, as well as responses to light and N.

The complete GRASS-CSOM model was tested against biomass dynamics and primary productivity data from the CPER and the Nairobi sites (Figure 11.16a, b). Both live and dead aboveground biomass dynamics were well simulated, with a few exceptions. Highly unusual data patterns such as at CPER in 1973 could not and possibly should not have been simulated by the model. Correlations between observed and predicted values for green shoots (Figure 11.17a, b) were highly significant, although the r^2 value was low for the CPER data, due to several outliers. For CPER, 46% of the model predictions fell within $\pm 25\%$ of the observed data value, while for Nairobi 46% fell within the $\pm 25\%$ interval. However for Nairobi, many predicted values were only just outside the

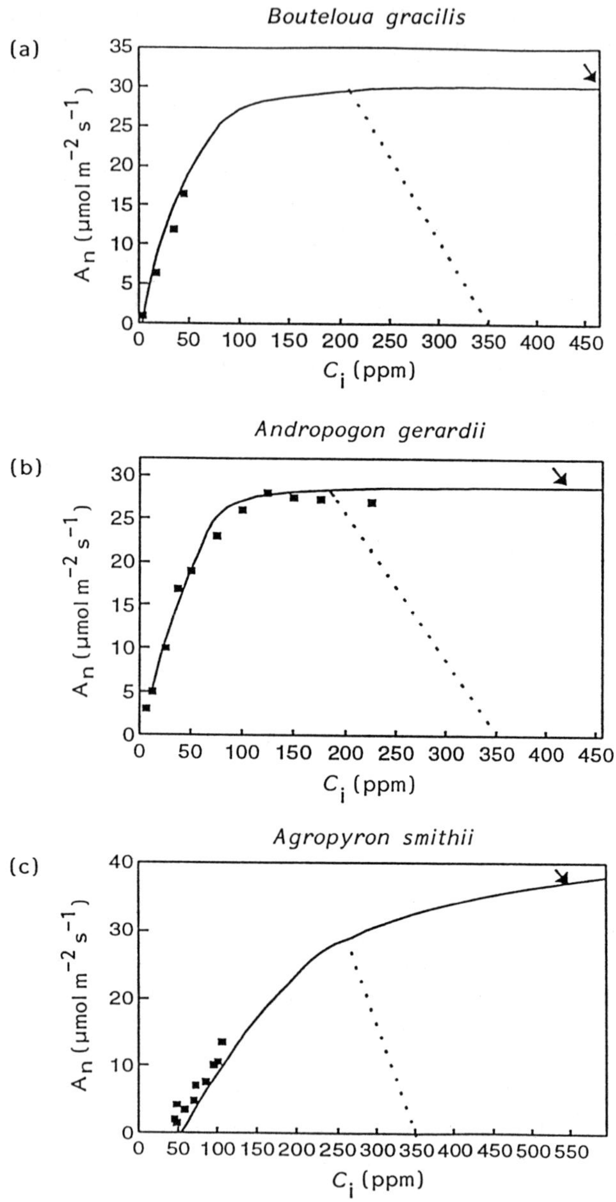

Figure 11.14 Comparison of observed (points) and simulated (lines) responses of net assimilation (A_n) to leaf internal CO_2, light (PAR) and temperature (*T*). Three grass species are compared. Observed data for *B. gracilis* (C_4) are from Monson *et al.* (1983) (A/C_i from warmgrown plants, temperature and light responses mean of cool- and warmgrown plants). Data for *Andropogon gerardii* (C_4) are from Polley *et al.* (1992) (A/C_i, PAR curves) and from A. Knapp (*T* curve, pers. comm., see Chen *et al.* 1993). Data for

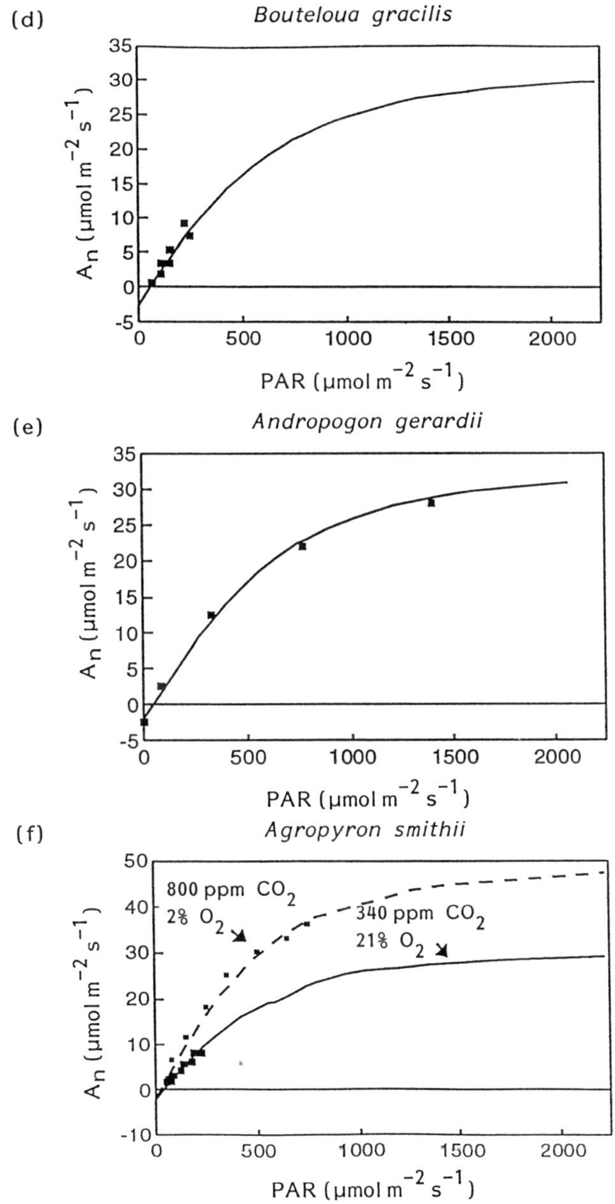

Figure 11.14 *(continued)* *Agropyron smithii* (C_3) are from Monson *et al.* (1983) (A/C_i curve and 340 ppm CO_2 PAR curve), and Monson *et al.* (1982) (temperature response, and 800 ppm CO_2 PAR curve). *B. gracilis* measured at PAR 2000, *T* 30°, RH 80%, and no water or temperature stress. *Agropyron smithii* measured at 25°, PAR 2000, and no water or N stress. *Andropogon gerardii* A/C_i and PAR curves determined at RH 40%, *T* 25°, PAR 1400, *T* 37°, soil water 1 MPa, N 1.8%, and *T* curve at RH 40%, PAR 1000, soil water 1 MPa, N 2%

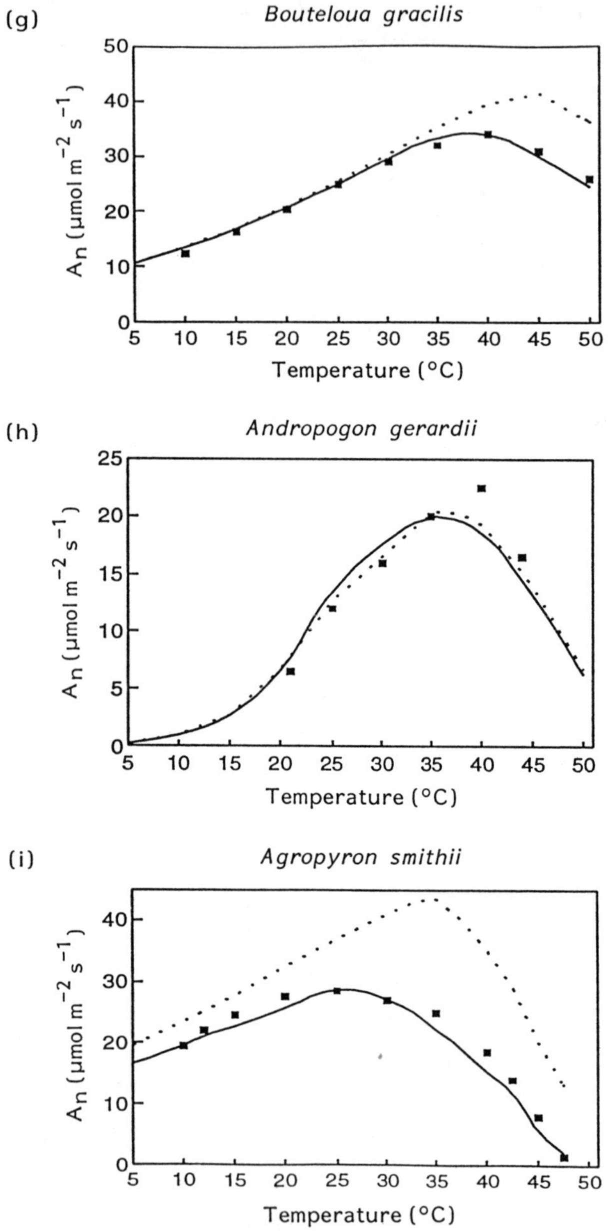

Figure 11.14 *(continued)*

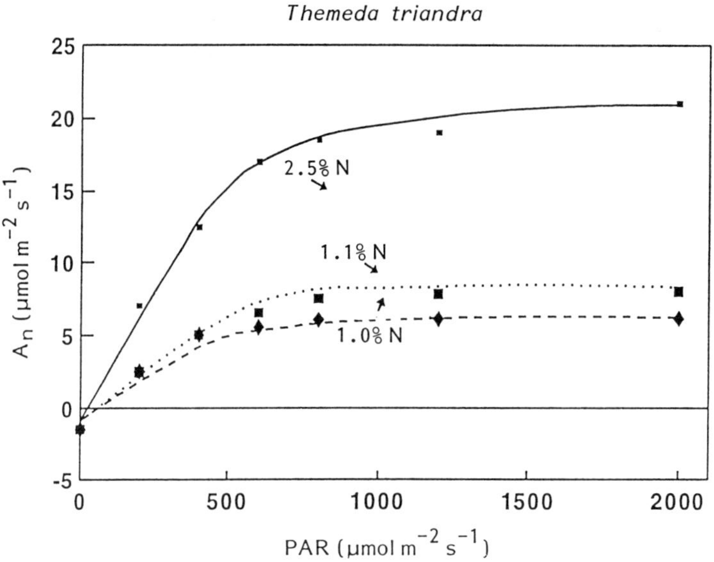

Figure 11.15 Light response curves of the African grass *T. triandra* at different leaf N levels, meant to represent leaves at different canopy positions. Observed data (points) from Kinyamario and Imbamba (1992)

$\pm$25% limits. Predictions of aboveground net primary production (ANPP) and peak live shoot mass were more precise (Figure 11.18a, b). Over 90% of the variance in ANPP and 56% of the variance in peak biomass were explained by the model. There was a slight tendency to overpredict ANPP at low values, as reflected by the positive *y*-intercept.

Simulated plant responses between GRASS and CENTURY were compared for a semiarid temperate steppe (CPER) and a tropical savanna (Kenya) across several years of simulated and observed study. We found that the two models behaved in similar fashion across a number of properties, including prediction of seasonal monthly live biomass, annual ANPP, and peak live biomass. For each of these properties, the differences between the two models in comparing the regression analysis between simulated and observed data from the two sites were found to be nonsignificant in correlation coefficient, slope, or intercept. The simulated monthly results of CENTURY indicate a bimodal bias in predicting live green plant biomass for the CPER site. The CENTURY model underpredicts live green biomass at the low levels of observed live green biomass and overpredicts live green biomass at higher levels. The Kenya simulations for CENTURY did not show this bias, nor did GRASS for either site.

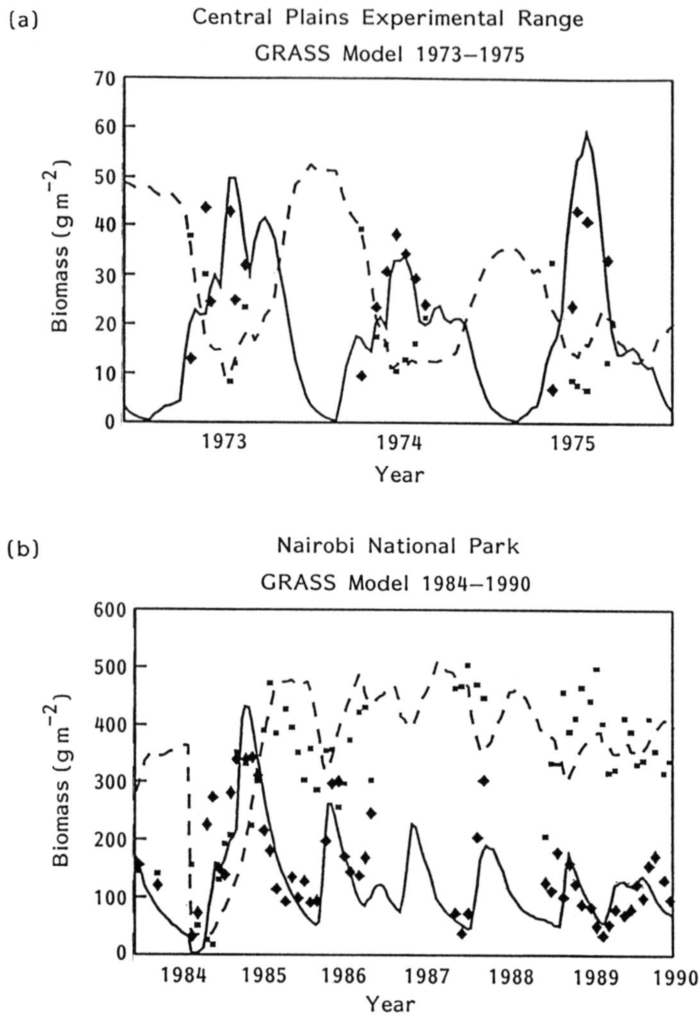

Figure 11.16 Comparison of observed and GRASS-simulated aboveground biomass dynamics at two sites. Large points, solid line = observed and simulated green shoots. Small points, dashed line = standing dead shoots. Observed CPER data from IBP/CPER-LTER data bank. Observed Nairobi data from D. Hall *et al.* and J. Kinyamario (pers. comm.)

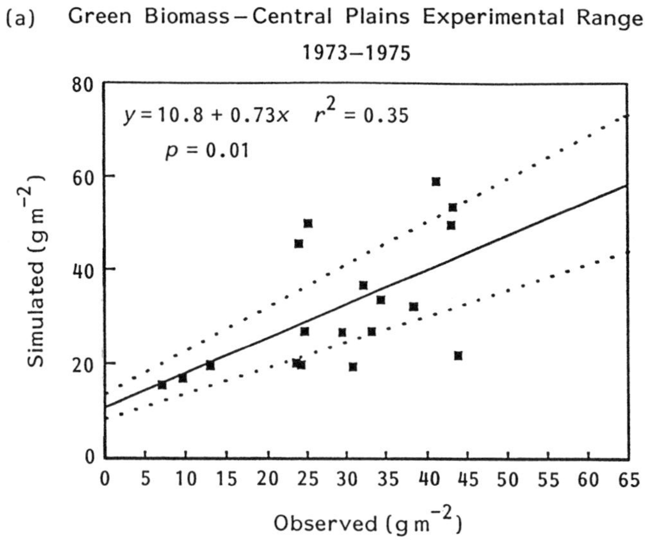

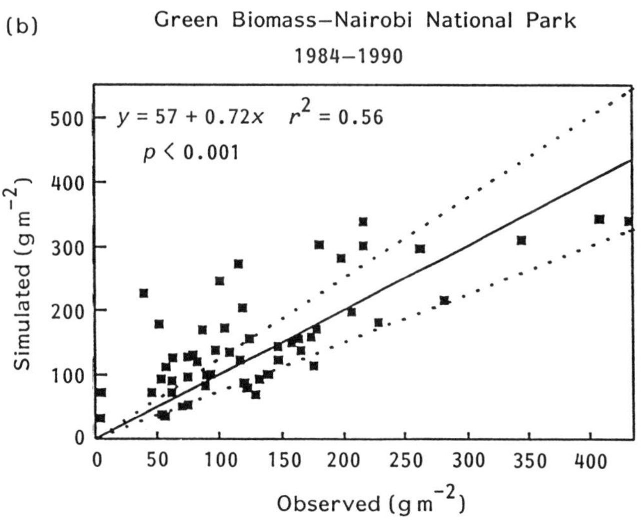

Figure 11.17 Correlations between observed and GRASS-simulated green biomass observations. Lines represent equality of simulated and observed data. Dashed lines indicate ±25% of observed data. Data sources as in Figure 11.4

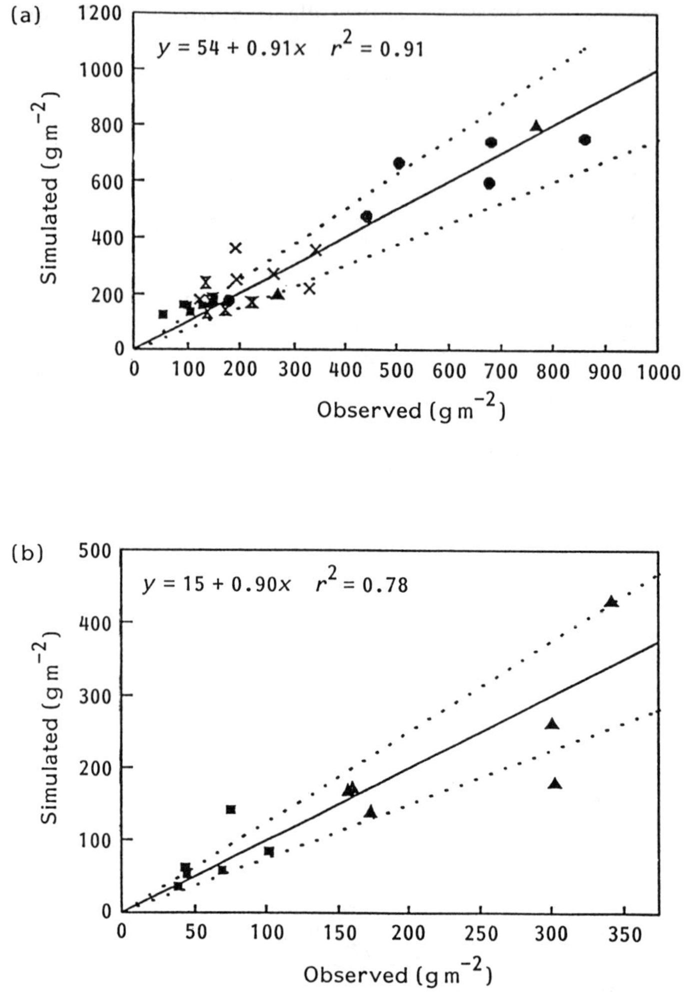

Figure 11.18 Correlations between observed and GRASS-simulated aboveground production (a) and peak live aboveground biomass (b) ● CPER nitrogen + irrigated; (X) CPER irrigated ⌛ CPER nitrogen; ■ CPER control; ▲ Nairobi. CPER data from 1970–75, Nairobi from 1988–89. Data sources as in Figure 11.4

11.7 SUMMARY

The comparison of observed and simulated live peak biomass and plant production had r^2 values of 0.45 and 0.70. The CENTURY error terms were generally less than $\pm 25\%$ of the observed data. Seasonal live biomass was not as well represented by the model ($r^2 = 0.39$) and had errors less than $\pm 50\%$ of the observed data. The model simulated differences between wet and dry years well, but was unable to simulate more subtle differences between years with similar precipitation. The model substantially underestimated live biomass for unusually high production years at the Kursk and Khomutov sites.

A comparison of the CENTURY model predictions for plant production and peak live biomass with empirical regression equations fitted to observed data showed that the CENTURY model was more successful than the regression equations. Steady-state soil C and N levels were well simulated by the model ($r^2 = 0.93$ and 0.89) for a set of sites with differing climate and soil textures. The model was able to predict soil C and N levels to within $\pm 25\%$ of observed data. Globally, key controls over C storage, NPP and biomass include climate, N inputs and losses, and soil particle size distribution. This model, rigorously calibrated against a large body of observations, is useful both as a descriptive and analytical tool. Comparisons of ecosystem models such as CENTURY with physiologically based models such as GRASS, across a gradient of sites, will provide a better framework to understand the processes controlling the sensitivity of grasslands to environmental change.

11.8 ACKNOWLEDGEMENTS

Data analyses conducted by Mr Brian Newkirk and Ms Song Bo (visiting scholar sponsored by the China Committee of Scholarly Exchange). Graphics prepared by Becky Techau and Michele Nelson. Manuscript prepared by Kay McElwain.

Measurement of monthly biomass dynamics, NPP and SOM at the tropical grassland sites in Kenya, Thailand and Mexico was carried out under United Nations Environment Programme (UNEP) Project FP/6108-88-01(2855) 'Environment Changes and Productivity of Tropical Grasslands' (1989–92), and more recently at the Kenya and Mexico sites under the UK Overseas Development Administration (ODA) Project R4744 'Productivity of Tropical Grasslands' (1991–94). Analysis of data from the sites in the former USSR was carried out under the Russian National Scientific and Technical Programme No. 18 'Changes of Natural Environment and Climate'. Data synthesis and model validation were made possible by the SCOPE (Scientific Committee on Problems of the Environment) Project 'Effects of Climate Change on Production and Decomposition in Coniferous Forests and Grasslands' (1989–92); and the National Institute for Global Environmental Change (NIGEC) Midwestern Region Center, US-DOE project on 'Biological Hysteresis in Climate Change Models for Grasslands: Implications of plant community dynamics on biogeochemical feedbacks'. Model development was primarily funded by the US National Science Foundation (NSF) project BSR 9013888 'Coupling Ecosystem Processes and Vegetation Patterns across Environmental Gradients' and the US NASA Earth Observing System project NACW-2662 'Using Multi-Sensor Data to Model Factors Limiting Carbon Balance in Global Grasslands', and US NSF project BSR 9011659 'Long Term Ecological Research Program: Shortgrass Steppe' and the Ecological Research Division, Office of Health and Environmental Research, US Department of Energy (DOE) project, 'Grassland–Atmosphere Response to Changing Climate: Coupling Local and Regional Scales'.

11.9 REFERENCES

Abrams, M. D., Knapp, A. K. and Hulbert, L. C. (1986) A ten-year record of aboveground biomass in a tallgrass prairie: effects of fire and topographic position. *Am. J. Bot.* **73**, 1509–1515.

Anderson, J. M. (1991) The effects of climate change on decomposition processes in grassland and coniferous forests. *Ecol. Appl.* **1**, 326–347.

Archer, S., Schimel, D. S. and Holland, E. A. (1994) Mechanisms of shrubland expansion: land use, climate or CO_2? *Clim. Change* **29**, 91–99.

Ball, J. T., Woodrow, I. E. and Berry, J. A. (1987) A model predicting stomatal conductance and its contribution to the control of photosynthesis under different environmental conditions. In: Biggins, I. (Ed.) *Progress in Photosynthesis Research*, pp. 221–224. Vol. IV. Martinus Nijhoff, Dordrecht.

Bónis, K., Mészáros, E. and Putsay, M. (1980) On the atmospheric budget of nitrogen compounds over Europe. *Időjárás* **84**, 57–68.

Briggs, J. M., Seastedt, T. R. and Gibson, D. J. (1989) Comparative analysis of temporal and spatial variability in aboveground production in a deciduous forest and prairie. *Holarct. Ecol.* **12**, 130–136.

Burke, I. C., Yonker, C. M., Parton, W. J., Cole, C. V., Flach, K. and Schimel, D. S. (1989) Texture, climate, and cultivation effects on soil organic matter context in U.S. grassland soils. *Soil Sci. Soc. Am. J.* **53**, 800–805.

Chen, D. and Coughenour, M. B. (1994) GEMTM: a general model for energy and mass transfer of land surfaces and its application at the FIFE sites. *Agric. For. Meteorol.* **68**, 145–171.

Chen, D., Coughenour, M. B., Knapp, A. K. and Owensby, C. E. (1993) Mathematical simulation of C4 grass photosynthesis in ambient and elevated CO_2. *Ecol. Model.* **73**, 63–80.

Christie, E. K. and Detling, J. K. (1982) Analysis of interference between C_3 and C_4 grasses in relation to temperature and soil nitrogen supply. *Ecology* **63**, 1277–1284.

Coughenour, M. B. (1984) A mechanistic simulation analysis of water use, leaf angles, and grazing in east African graminoids. *Ecol. Model.* **26**, 203–230.

Eddy, J. A., Malone, T. F., McCarthy, J. J. and Rosswall, T. (Eds) (1991) *Global Change Systems for Analysis, Research and Training (START)*. IGBP Report No. 15, Stockholm.

Esser, G. (1986) The carbon budget of the biosphere – structure and preliminary results of the Osnabruck Biosphere Model. *Veroff. Naturf. Ges. zu Emden von 1814* **7**, 1–160.

Farquhar, G. D., von Caemmerer, S. and Berry, J. A. (1980) A biochemical model of photosynthetic CO_2 assimilation in leaves of C_3 species. *Planta* **149**, 78–90.

Gupta, S. C. and Larson, W. E. (1979) Estimating soil water retention characteristics from particle size distribution, organic matter content and bulk density. *Water Resour. Res.* **15** (6), 1633–1635.

Hall, D. O. and Scurlock, J. M. O. (1991) Climate change and productivity of natural grasslands. *Ann. Bot.* **67** (suppl.), 49–55.

Hao, W. M., Liu, M. and Crutzen, P. J. (1990) Estimates of annual and regional releases of CO_2 and other trace gases to the atmosphere from fires in the tropics. In: Goldammer, J.G. (Ed.) *Fire in the Tropical Biota*, pp. 440–462. Ecological Studies 84. Springer-Verlag, Berlin.

Holland, E. A., Parton, W. J., Detling, J. K. and Coppock, D. L. (1992) Physiological responses of plant populations to herbivory and their consequences for ecosystem nutrient flows. *Am. Natur.* **140**(4), 685–706.

Kinyamario, J. I. and Imbamba, S. K. (1992) Savanna at Nairobi National Park, Nairobi. In: Long, S.P., Jones, M. B. and Roberts, M. J. (Eds) *Primary Productivity of Grass Ecosystems*, Chapter 2. Chapman & Hall, London.

Knapp, A. K., Hamerlynk, E. P. and Owensby, C. E. (1993) Photosynthetic and water relations responses to elevated CO_2 in the C_4 grass *Andropogon gerardii*. *Int. J. Plant Sci.* **154**, 459–466.

Lauenroth, W. K. (1978) Grassland primary production: North American grasslands in perspective. In: French, N. (Ed.) *Perspectives in Grassland Ecology*, pp. 3–21. Springer-Verlag, New York.

Lauenroth, W. K. and Sala, O. E. (1992) Long-term forage production of North American shortgrass steppe. *Ecol. Appl.* **2**, 397–403.

Lieth, H. (1972) Modelling the primary productivity of the world. UNESCO, Paris. *Nature and Resource* **8**, 5–10.

Linacre, E. T. (1977) A simple formula for estimating evaporation rates in various climates, using temperature data alone. *Agric. Meteorol.* **18**, 409–424.

Long, S. P., Garcia Moya, E., Imbamba, S. K., Kamnalrut, A., Piedade, M. T. F., Scurlock, J. M. O., Shen, Y. K. and Hall, D. O. (1989) Primary productivity of natural grass ecosystems of the tropics: a reappraisal. *Plant Soil* **115**, 155–166.

Long, S. P., Jones, M. B. and Roberts, M. J. (1992) *Primary Productivity of Grass Ecosystems of the Tropics and Sub-tropics*. Chapman & Hall, London. 267 pp.

McGuire, A. D., Melillo, J. M., Joyce, L. A., Kicklighter, D. W., Grace, A. L., Moore, B. III and Vrsmarty, C. J. (1992) Interactions between carbon and nitrogen dynamics in estimating net primary productivity for potential vegetation in North America. *Glob. Biogeochem. Cyc.* **6**, 101–124.

Menaut, J. -C. and César, J. (1979) Structure and primary productivity of Lamto savannas, Ivory Coast. *Ecology* **60**, 1197–1210.

Monson, R. K., Littlejohn, R. O. Jr and Williams, G. J. III (1983) Photosynthetic adaptation to temperature in four species from the Colorado shortgrass steppe: a physiological model for coexistence. *Oecologia* **58**, 43–51.

Monson, R. K., Stidham, M. A., Williams, G. J. III and Edwards, G. E. (1982) Temperature dependence of photosynthesis in *Agropyron smithii Rydb. Plant Physiol.* **69**, 921–928.

OIES (Office for Interdisciplinary Earth Studies) (1991) *Arid Ecosystems Interactions: Recommendations for Drylands Research in the Global Change Research Program*. OIES Report 6, Boulder.

Ojima, D. S., Parton, W. J., Schimel, D. S. and Owensby, C. E. (1990) Simulated impacts of annual burning on prairie ecosystems. In: Collins, S. L. and Wallace, L. (Eds) *Fire in the North American Prairies*, pp. 118–132. Univ. of Oklahoma Press, Norman.

Ojima, D. S., Schimel, D. S., Parton, W. J. and Owensby, C. E. (1994) Long- and short-term effects of fire on nitrogen cycling in tallgrass prairie. *Biogeochemistry* **24**, 67–84.

Parton, W. J. (1978) Abiotic section of ELM. In: G. S. Innis (Ed) *Grassland Simulation Model*, pp. 31–53. Ecol. Studies, Vol. 26, Springer-Verlag, Inc., New York.

Parton, W. J. (1984) Predicting soil temperatures in a shortgrass steppe. *Soil Sci.* **138**, 93–101.

Parton, W. J., Cole, C. V., Stewart, J. W. B., Ojima, D. S. and Schimel, D. S. (1989a) Simulating regional patterns of soil C, N, and P dynamics in the U.S. central grasslands region. In: Clarholm, M. and Bergstrom, L. (Eds) *Ecology of Arable Land*, pp. 99–108. Kluwer Academic Publishers, Dordrecht.

Parton, W. J., McKeown, B., Kirchner, V. and Ojima, D. (1992) *CENTURY Users' Manual.* Natural Resource Ecology Laboratory, Colorado State University, Fort Collins.

Parton, W. J., Sanford, R. L., Sanchez, P.A. and Stewart, J.W.B. (1989b) Modelling soil organic matter dynamics in tropical soils. In: Coleman, D. C., Oades, J. M. and Uehara, G. (Eds) *Dynamics of Soil Organic Matter in Tropical Ecosystems*, pp. 153–171. NifTAL Project, Univ. of Hawaii Press, Honolulu.

Parton, W. J., Schimel, D. S., Cole, C. V. and Ojima, D. S. (1987) Analysis of factors controlling soil organic matter levels in great plains grasslands. *Soil Sci. Soc. Am. J.* **51**, 1173–1179.

Parton, W. J., Schimel, D. S., Ojima, D. S. and Cole, C.V. (1994) A general model for soil organic matter dynamics: sensitivity to litter chemistry, texture and management. In: Bryant, R. B. and Arnold, R. W. (Eds) *Quantitative Modelling of Soil Forming Processes.* Soil Sci. Soc. Am. J. Spec. Publ. ASA, CSSA and SSA, Madison, Wis. (in press).

Parton, W. J., Singh, J. S. and Coleman, D. C. (1978) A model of production and turnover of roots in shortgrass prairie. *J. Appl. Ecol.* **47**, 515–542.

Parton, W. J., Stewart, J. W. B. and Cole, C. V. (1988) Dynamics of C, N, P and S in grassland soils: a model. *Biogeochemistry* **5**, 109–131.

Polley, H. W., Norman, J. M., Arkebauer, T. J., Walter-Shea, E. A., Greegor, D. H. Jr and Bramer, B. (1992) Leaf gas exchange of *Andropogon gerardii Vitman, Panicum virgatur*, L., and Sorgastrum nutans (L.) *Nash* in a tallgrass prairie. *J. Geophys.Res.* **97**, 18, 837–18, 444.

Ryan, M. (1991) The effects of climate change on plant respiration. *Ecol. Appl.* **1**, 157–167.

Sala, O. E., Parton, W. J., Joyce, L. A. and Lauenroth, W. K. (1988) Primary production of the Central Grassland Region of the United States. *Ecology* **69**, 40–45.

Sanford, R. L. Jr, Parton, W. J., Ojima, D. S. and Lodge, D. J. (1991) Hurricane effects on soil organic matter dynamics and forest production in the Luquillo Experimental Forest, Puerto Rico: results of simulation modelling. *Biotropica* **23**, 364–372.

Schimel, D. S., Kittel, T. G. F. and Parton, W. J. (1991) Terrestrial biogeochemical cycles: global interactions with the atmosphere and hydrology. *Tellus*, Spec. Issue AB **43**, 188–203.

Sellers, P. J., Berry, J. A., Collatz, G. J., Field, C. B. and Hall, F. G. (1992) Canopy reflectance, photosynthesis, and transpiration. III. A reanalysis using improved leaf models and a new canopy integration scheme. *Remote Sens. Environ.* **42**, 187–216.

Wedin, D. A. and Tilman, D. (1990) Species effects on nitrogen cycling: a test with perennial grasses. *Oecologia* **84**, 433–441.

11.10 APPENDIX

Site-specific parameters relating to plant and soil physiology, and management regimes used for short-term model runs
Site-specific parameters found in the SITENAME.DAT file may be divided into a number of categories:

(a) basic site identification (e.g. SITLAT, SITLNG);
(b) 'minimum input' data required to run the CENTURY model;
   (i) environmental information such as PRECIP, TMN2M, TMX2M;
   (ii) physical features such as SAND, SILT, CLAY;
(c) parameters relating mostly to plant and soil physiology.

Parameters from categories (bii) and (c) are listed below for all the grassland sites reported here, with notes on the differences between sites.

	KENYA.DAT	THAI.DAT	MEXICO.DAT	LAMTO.DAT	CPER.DAT	KONZA.DAT	KHOM.DAT	KURSK.DAT	OTRAD.DAT	SHORTA.DAT	TUVA.DAT
SOM1CI(2,1)	30	30	30	30	20	30	30	30	30	30	30
SOM1CI(2,1)	59	190	190	60	60	190	270	333	150	170	100
SOM2CI(1)	1460	1230	3050	850	1250	2100	3700	5506	3100	2200	1500
SOM3CI(1)	830	570	2100	610	740	2800	2150	5415	2300	3530	2530
(initial C content of SOM pools, obtained from long-term model runs; suitable initial values here greatly reduce the time taken for the model to reach equilibrium. In the case of relatively recently converted grasslands such as Mexico and Thailand, values are based on observed soil C)											
BULKD	1.2	1.2	1.2	1.5	1.5	1.2	1.2	1.2	1.2	0.98	1.25
(soil bulk density; measured value used if available; this controls erosion and is important in calculating actual soil C and N content from% values)											
EPNFA(1)	0.21	0.21	0.21	0.21	0.21	0.21	0.21	0.21	0.21	0.21	0.21
EPNFA(2)	0.0028	0.0028	0.0028	0.0028	0.0028	0.0028	0.0028	0.0028	0.0028	0.0028	0.0028
EPNFS(1)	−1.32	−0.92	−1.32	−1.32	−1.32	−1.32	−0.86	−0.86	−0.86	−0.86	−0.86
EPNFS(2)	0.033	0.01	0.033	0.033	0.033	0.033	0.07	0.07	0.07	0.07	0.07
(intercept and slope for equations describing atmospheric N deposition and soil (symbiotic and nonsymbiotic) N fixation, respectively, as a function of annual precipitation in cm. Mean annual rainfall at Kenya, Thailand, Mexico, Lamto, CPER, Konza, Khomutov, Kursk, Otradnoye, Shortandy and Tuva, is 68, 154, 59, 117, 30, 82, 44, 56, 54, 35 and 21 cm, respectively; hence total N inputs (g m^{-2}) of 1.32, 1.26, 1.00, 3.08, 0.29, 1.83, 2.55, 3.43, 3.28, 1.90 and 0.88, respectively)											
NLAYER	3	5	5	5	5	5	5	5	5	6	6
NLAYPG	2	3	4	4	4	4	3	4	3	4	3
(NLAYER is the number of 'layers' in the soil water/nutrient flow submodel; this simulates actual soil depth over bedrock or hard pan. NLAYPG is the number of these layers which the plants use for active growth; this is equivalent to the depth of the majority of roots)											
PH	7.5	5.5	7.5	6.0	6.8	6.3	7.4	6.3	5.2	7.4	7.3
(soil pH; measured value used if available; mainly of importance to the phosphorus submodel, if used)											

PTYPE	2	2	2	2	2	2	1	1	1	1	1

(plant type; determines temperature response of primary production. '2' denotes plants with higher temperature optimum, i.e. mainly C_4 photosynthetic type. '1' denotes cooler temperature optimum, i.e. mainly C_3 type. N.B. some C_4 plants in temperate regions show type '2' temperature response)

AGPPA	−40	−40	−40	−40	−40	−40	−40	−40	−40	−40	−40
AGPPB	11	11	7.7	7.7	7.7	7.7	7.7	7.7	7.7	7.7	7.7
BGPPA	60	100	100	100	100	100	100	100	100	100	100
BGPPB	4	4	7	7	7	7	7	7	7	7	7

(intercept and slope for aboveground and belowground maximum NPP as a function of annual precipitation (cm); used only to determine root/shoot ratio. Mean annual rainfall at Kenya, Thailand, Mexico, Lamto, CPER, Konza, Khomutov, Kursk, Otradnoye, Shortandy and Tuva, is 68, 154, 59, 117, 30, 82, 44, 56, 54, 35 and 21 cm, respectively; hence mean root/shoot ratios of 0.47, 0.43, 1.24, 1.07, 1.62, 1.14, 1.37, 1.47, 1.27, 1.50 and 2.00, respectively)

FALLRG	0.1	0.15	0.1	0.1	0.2	0.15	0.2	0.2	0.2	0.2	0.2

(monthly fraction of standing dead matter which falls as litter. Note that CENTURY assumes that decomposition occurs only to litter, so this parameter should also take into account direct decomposition of standing dead)

FDFREM(1)	0.85	0.95	0.85	0.90	0.85	0.85	0.85	0.85	0.85	0.5	0.85
FDFREM(2)	0.85	0.95	0.85	0.95	0.60	0.60	0.60	0.60	0.60	0.60	0.60
FLFREM	0.85	0.90	0.85	0.80	0.80	0.80	0.80	0.80	0.80	0.80	0.80

(fraction of standing dead matter, litter and live matter which is consumed by fire. Note that these values act in combination with FALLRG, litter decomposition and FSDETH, respectively; complete disappearance may therefore be produced by values of 0.95 or lower. In particular, standing dead matter may assume negative values if FDFREM(1) is greater than (1 − FALLRG)).

FDGREM	0.06	N/A	N/A	N/A	0.06	0.06	0.80	0.80	0.80	0.06	0.06
FLGREM	0.12	N/A	N/A	N/A	0.12	0.12	0.80	0.80	0.80	0.12	0.12
GREMON(1)-(7)	4-7,10-11	N/A	N/A	N/A	5–9	5–9	9	9	9	4–10	4–10

(fraction of standing dead matter and live matter which is consumed by a grazing event (these values also act in combination with FALLRG, FDFREM, FSDETH, etc.) N.B. the high consumption in a single month at Khomutov, Kursk, and Otradnoye represents mowing of the grass for forage or thatch)

FRET(1)	0.30	0.30	0.30	0.50	0.25	0.30	0.30	0.15	0.30	0.30	0.30
FFCRET	0	0	0	0	0.05	0.05	0.05	0.05	0.05	0.05	0.05

(fraction of burned N which returns to soil as ash, and fraction of C returned to the soil as charcoal. N.B. since the latter does not normally enter SOM pools, it is often ignored and set to zero)

FLIGNG(1,1)	0.2	0.12	0.2	0.106	0.2	0.2	0.2	0.2	0.2	0.2	0.2
FLIGNG(2,1)	0.0012	0	0.0012	0	0.0012	0.0012	0.0012	0.001	0.0012	0.0012	0.0012
FLIGNG(1,2)	0.26	0.16	0.26	0.106	0.26	0.26	0.26	0.26	0.2	0.26	0.26
FLIGNG(2,2)	−0.0015	0	−0.0015	0	−0.0015	−0.0015	−0.0015	−0.0015	−0.0015	−0.0015	−0.0015

(intercept and slope of equations predicting lignin content as a function of annual precipitation (cm); equations fitted to US Great Plains data, so actual lignin determinations are preferable for wet tropical sites)

FSDETH(1)	0.25	0.45	0.25	0.2	0.2	0.2	0.2	0.2	0.2	0.2	0.2
FSDETH(2)	0.95	0.95	0.95	0.95	0.95	0.95	0.95	0.95	0.95	0.95	0.95

(FSDETH(1) is maximum fraction of aboveground live matter which dies monthly; this is a function of estimated monthly water deficit; FSDETH(2) relates to death in the senescence month for a temperate grassland and is not used for tropical systems)

GFCRET	0.30	0.30	0.30	0.30	0.30	0.30	0.30	0	0	0.30	0.30
GRET(1)	0.80	0.80	0.80	0.80	0.80	0.80	0.30	0	0	0.80	0.80

(fraction of C and fraction of N respectively, returned to soil by a grazing event. 0 indicates mowing (total removal of grass) at Khomutov and Otradnoye sites.)

GPFLAG	2	2	2	2	2	2	2	2	2	2	2

(method used to estimate monthly primary production; '2' means that NPP is calculated as a function of monthly water availability and temperature, limited by the most limiting nutrient and any significant shading by standing dead matter – D. Ojima equations, 10/85)

GRASCH(1)	1	1	1	1	3	3	3	3	3	3	3
GRASCH(2)	12	12	12	12	9	9	9	10	8	8	8
GRASCH(3)	13	13	13	13	10	10	11	11	9	9	9
(grass schedule parameters, describing the start month, end month and senescence month of the growing season, respectively; for tropical systems without a clear annual senescence event, the senescence month is set to 13 and is not used)											
GRDR	0.12	0.12	0.12	0.12	0.06	0.12	0.12	0.12	0.12	0.12	0.12
(monthly turnover rate for grass roots; this determines the total level of root matter and the balance between live and dead belowground)											
PRAMNG(1,1)	20	20	20	30	20	20	20	20	20	20	20
PRAMNG(1,2)	0.20	0.15	0.15	0.25	0.15	0.20	0.15	0.15	0.15	0.15	0.15
PRAMXG(1,1)	30	30	30	35	30	30	30	30	30	30	30
PRAMXG(1,2)	0.20	0.15	0.15	0.25	0.15	0.20	0.15	0.15	0.15	0.15	0.15
PRBMNG(1,1)	50	50	50	50	50	50	50	50	50	50	50
PRBMNG(1,2)	0	0	0	0	0	0	0	0	0	0	0
PRBMXG(1,1)	60	60	60	60	60	60	60	60	60	60	60
PRBMXG(1,2)	0	0	0	0	0	0	0	0	0	0	0
(intercept and slope for equations relating plant C/N ratio to biomass for above and belowground matter, respectively; new growth at low biomass has a low C/N ratio, increasing with biomass so that grasslands with high live biomass have a high C/N ratio)											
SAND	0.40	0.58	0.47	0.85	0.70	0.25	0.20	0.32	0.425	0.36	0.74
SILT	0.25	0.40	0.35	0.10	0.15	0.35	0.28	0.31	0.40	0.27	0.17
CLAY	0.35	0.02	0.18	0.05	0.15	0.40	0.52	0.37	0.175	0.37	0.09
(soil texture at the study site; in particular, sand content affects microbial turnover rate and stabilization of 'slow' organic matter (SOM2); clay soils stabilize more passive organic matter (SOM3); sandy soils are more productive at low rainfall, whereas clay soils are more productive at high rainfall)											
N.B. CPER I (sandy loam) is 0.70/0.15/0.15; CPER II (loam) is 0.40/0.35/0.25. Konza (clay loam) is 0.25/0.35/0.40; Konza II [Aldus] ( ) is 0.200/.55/0.25.											
Year-to-year repeated pattern	nnnnf	nnnnf	nnnnnf	f	g	gnb	n	n	n	n	n
Month in which fire occurs	3	2	2	2	—	?	—	—	—	—	—
Management regimes for short-term model runs (repeated pattern of up to 20 years, where g = grazing year, f = fire year, b = both grazing and fire, n = no grazing or fire)											

12 Impact of Climate and Atmospheric Carbon Dioxide Changes on Grasslands of the World

D. S. OJIMA,[1] W. J. PARTON,[1] M. B. COUGHENOUR,[1] J. M. O. SCURLOCK,[2], T. B. KIRCHNER,[1] T. G. F. KITTEL,[1,11] D. O. HALL,[2] D. S. SCHIMEL,[1,5] E. GARCIA MOYA,[8] T. G. GILMANOV,[3] T. R. SEASTEDT,[7] APINAN KAMNALRUT,[9] J. I. KINYAMARIO,[10] S. P. LONG,[12] J-C. MENAUT,[6] O. E. SALA,[13] R. J. SCHOLES[4], and J. A. van VEEN[14]

[1]*Natural Resource Ecology Laboratory, Colorado State University, Fort Collins, USA*
[2]*Division of Life Sciences, King's College London, UK*
[3]*Department of Vertebrate Zoology and Ecology, Moscow State University, Russia*
[4]*Division of Water, Environment and Forest Technology, CSIR, Pretoria, South Africa*
[5]*Climate System Modelling Program, NCAR, Boulder, USA*
[6]*Laboratoire d'Ecologie, Ecole Normal Supérieure, Paris, France*
[7]*Institute of Arctic and Alpine Research, University of Colorado, Boulder, USA*
[8]*Centro de Botanica, Colegio de Postgraduados, Chapingo, Mexico*
[9]*Faculty of Natural Resources, Prince of Songkla University, Hatyai, Thailand*
[10]*Department of Botany, University of Nairobi, Kenya*
[11]*University Corporation for Atmospheric Research, Boulder, USA*
[12]*Department of Biology, University of Essex, Colchester, UK*
[13]*Catedra de Fisiologia y Ecologia Vegetales, Facultad de Agronomia, Buenos Aires, Argentina*
[14]*DLO Research Institute for Plant Protection, Wageningen, The Netherlands*

12.1 INTRODUCTION GRASSLANDS AND THE GLOBAL CARBON CYCLE

Grasslands are one of the major vegetation types, covering nearly one-fifth of the world's land surface or approximately 24×10^6 km^2. Ranging from the savannas of Africa to the North American prairies and the converted grasslands of Latin

Global Change: Effects on Coniferous Forests and Grasslands.
Edited by A. I. Breymeyer, D. O. Hall, J. M. Melillo and G. I. Ågren

UNEP

America and Southeast Asia, these ecosystems are likely to remain roughly constant in area or even see expansion in their areal extent under modified climate conditions. These ecosystems are large reservoirs of carbon (C) globally, containing approximately 30% of global soil C stocks (Anderson 1991; Eswaran *et al.* 1993) and the soil component is the main C reservoir in these ecosystems. Grasslands have developed under climatic regimes characterized by high climate variability that favor herbaceous or open wooded systems.

Potential effects of climate change on grassland biogeochemistry and soil C stocks have received much less attention compared with forests (Hall and Scurlock 1991; Long 1991a). The effects of changes in factors such as temperature, water and nutrients are relatively well understood, and the interactions with the effects of CO_2 fertilization are now beginning to be sufficiently well-developed to evaluate effects on C fluxes in various ecosystems (Schimel *et al.* 1990, 1991, Long 1991a, b; Owensby *et al.* 1993b). However, most of our current understanding has been necessarily derived from short-term experiments of only one or two of the factors modified simultaneously. To fully understand the impact of global change effects on these grassland ecosystems, it is necessary to study the interactive effects of the biogeochemical, hydrological, and physiological processes to changes in climate and CO_2 in a systematic and integrated fashion. One of the more powerful research tools which has been developed during the past 10 years are ecosystem models robust enough to be used to simulate ecosystem dynamics across a broad range of environmental factors. These models are capable of studying the complex interactions and feedbacks in an integrated plant–soil ecosystem framework, but are also able to provide an evaluation of the long-term implications of change to climate and CO_2 levels on key C storage components, such as soil organic levels. Grassland ecosystems store most of their C in soils, where turnover times of the bulk soil are relatively long (thousands of years), and so changes, though they may occur slowly, will be of significant duration (Schimel *et al.* 1994). Changes in grassland C storage can thus have a significant and long-lived effect on global C cycles. Several estimates of soil organic matter response to environmental change have recently been developed. Jenkinson *et al.* (1991) estimated substantial and rapid release of C from soils as a result of global warming (61 Pg over 60 years), while Schlesinger (1990) argued that soils have a low potential for added C storage and can only change slowly.

Most previous studies have had recognized limitations: Schlesinger's (1990) study used inferential and potentially confounded soil chronosequence studies, while the model of Jenkinson *et al.* (1991) was unable to take account of feedback between soil processes and climate change effects on primary production, a key feature of CENTURY. Jenkinson *et al.* (1991, p. 306) noted that in order to determine how climate change affects the flux of CO_2 between soil and atmosphere, their model 'would have to be coupled to another which specifies how global plant production is altered by increasing concentrations of greenhouse gases'.

Under the aegis of a collaborative SCOPE project (Breymeyer and Melillo 1991), the CENTURY model of plant–soil interactions has been further developed in order to apply it to a wide range of grasslands worldwide (Parton *et al.* 1993; Chapter 11 this volume). The SCOPEGRAM group has assembled a unique synthesis of detailed and long-term data from both temperate and tropical grasslands, to evaluate the robustness of the CENTURY model and simulated response to climate change scenarios. A comparative analysis of ecosystem dynamics as simulated by a more detailed ecophysiological model, GRASS (Coughenour 1984) and by CENTURY was also conducted for two of these grassland systems.

The analysis of climate and CO_2 changes on biogeochemical processes needs to incorporate the feedbacks linking plant production system (C sequestering processes) and the decomposition processes (C releasing processes). The net change in these two sets of processes will determine the source–sink response of the ecosystem to climate and CO_2 changes. Climate and CO_2 changes may enhance plant production by increasing water or N availability, but these changes may also increase decomposition rates which would offset the gains in storage of C due to elevated CO_2 levels. In addition, physiological changes due to altered water or N use efficiencies under altered CO_2 levels may affect plant productivity and C turnover rates. These changes may often counterbalance changes in system C storage resulting from climate warming.

Over the last few years, it has been suggested that the majority of the 'missing sink' for C in the global CO_2 balance may be accounted for by terrestrial vegetation rather than by oceanic uptake, and, furthermore, that such a sink may be located in both tropical and temperate regions (Tans *et al.* 1990; Enting and Mansbridge 1991). It is, therefore, imperative to include grasslands with other major terrestrial ecosystems in attempts to model the global C cycle, in order to arrive at a well-founded policy toward climate change. Many approaches to the role of natural ecosystems in climate change involve computer modelling of the C flows between vegetation, soil and atmosphere, and can predict the effects of changes over many years. Here, we demonstrate that a grassland model which has been tested or 'validated' for a wide range of specific study sites can be scaled up to regional levels, and then interfaced with atmospheric general circulation models (GCMs). These GCMs are currently being used to develop national and international policy responses to the threat of global climate change, and it is imperative that these include the latest information on plant/soil response and feedback to climate change.

The primary objective of this chapter is to determine how sensitive grassland ecosystems are to the factors controlling the ecosystem responses (e.g. plant productivity, soil C changes, decomposition, N mineralization) to climate change and increasing atmospheric CO_2 in grasslands found in different ecoregions of the world. This analysis was made using CENTURY and GRASS ecosystem models. A secondary objective is to assess the difficulty of observing change in ecosystem dynamics, especially in arid to semiarid regions where precipitation typically shows great variance. We also analyze the differences between the transient response over 50 years as well as the long-term 'equilibrium' response to CO_2 and climate change for comparison with previous global analyses such as presented by Melillo *et al.* (1993). In our analysis, we did not attempt to model climate-driven redistribution of grassland regions, although it is recognized that changes in land cover due to climate or land use changes may overshadow climate-induced changes in existing grassland regions.

12.2 MODELLING STRATEGY

12.2.1 Grassland ecoregions

Sites were selected to represent a broad range of grassland regions based on climate factors and ecosystem properties (Figure 12.1). These include 16 sites for which detailed monthly data were available on vegetation biomass and dead matter dynamics, in order to test the model (Parton *et al.* 1993; Chapter 11this volume). We assembled at least 25 years of monthly weather data for these 16 sites, plus an additional 15 grassland sites which have been described in the literature (Table 12.1). The regions ranged from temperate ecosystems with highly seasonal temperature patterns to tropical regions where temperature varied little. Precipitation at all of the sites displayed a marked seasonality, and in most cases a high degree of variability between years.

On the basis of temperature, rainfall and seasonality of climate, we grouped 31 sites into 7 bioclimatic regions based on Bailey ecoregions as follows (Table 12.1, Bailey, 1989):

1. Cold desert steppe – dry domain
2. Temperate steppe – dry domain
3. Humid temperate – humid temperate domain
4. Mediterranean – humid temperate domain
5. Dry savanna – dry domain
6. Savanna – humid tropical domain
7. Humid savanna – humid tropical domain

Climatic characteristics of the steppe regions are relatively low annual precipitation (PPT), potential evapotranspiration (PET) rates of around 1000 mm, and a drought index defined by PPT/PET ratios of around 0.3 (Table 12.2). The difference between the cold desert steppes and the temperate steppes are primarily temperature (0 °C vs 10 °C) and the inclusion of some C_4 grasses in the temperate steppes. Steppe ecoregions are characterized by relatively low plant productivity, N mineralization, and abiotic decomposition rates. The cold desert steppes have slightly higher production and N mineralization rates, resulting from higher N deposition and soil N fixation rates. Higher soil C levels in the cold desert steppe are primarily a result of substantially lower decomposition rates.

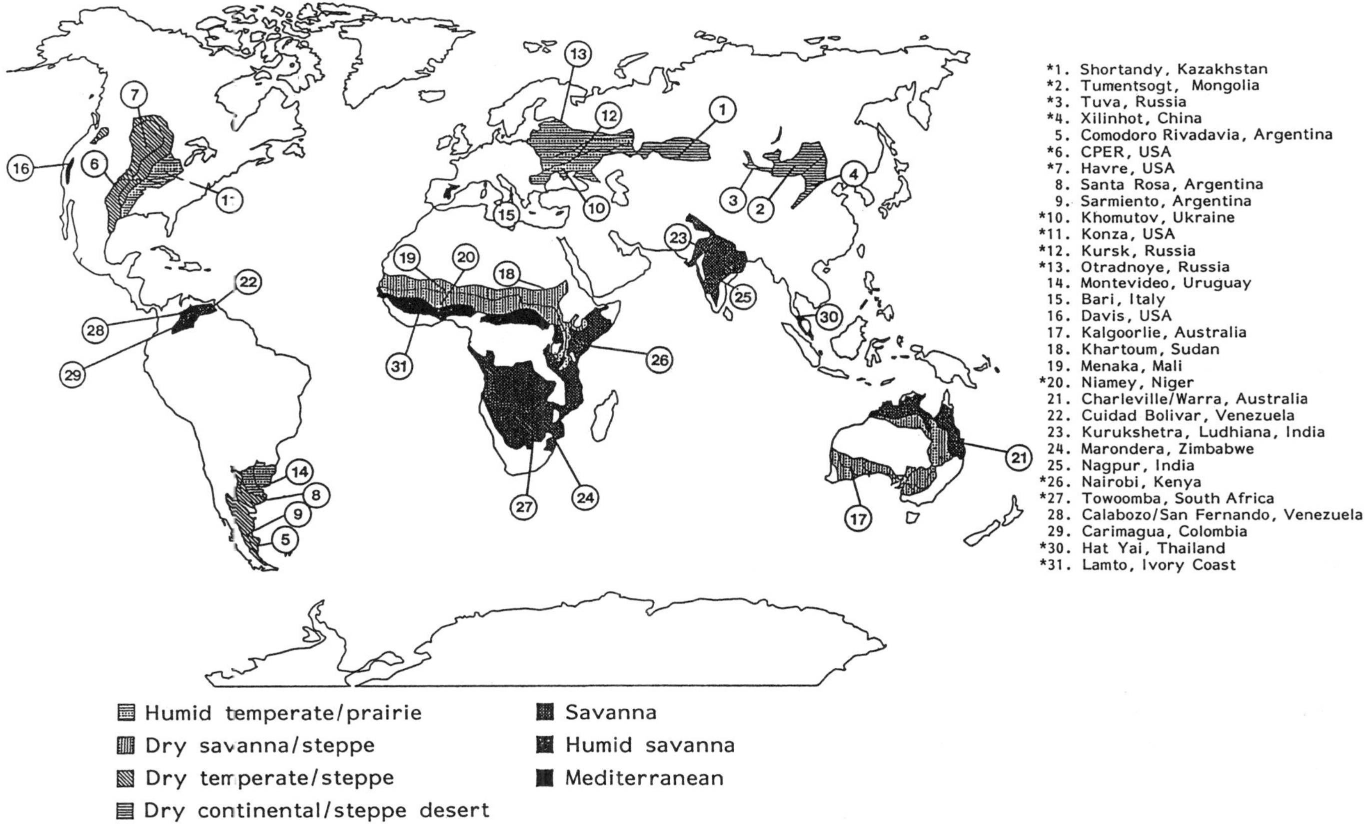

Figure 12.1 Map showing location of grassland sites and boundaries of Bailey ecoregions representing grasslands worldwide. *indicates sites used in the intial parmeterization of different grassland ecosystems in various ecoregions (see Parton *et al.* 1996, Chapter 11 this volume)

Table 12.1 Grassland sites modeled in the present study. Each site is assumed to be representative of its surrounding area on the Bailey (1989) ecoregion map

Site, Country	Ecoregion (Bailey 1989)	Latitude/ Longitude (approx.)	Land area represented (10^6 km^2)	Annual precipitation (mm)	Mean annual temperature (°C)	Soil C (kg m^{-2})	Above-ground C (g m^{-2})	PET	Decomposition factor	N_{min}
Dry domain (300)										
1. Cold desert steppe division (331/333)										
Shortandy, Kazakhstan [1]	(331)	52° N 71° E	0.804	351	1.4	7.37	73.1	96.3	0.003	3.34
Tumentsogt, Mongolia [2]	(333)	46° N 113° E	0.740	269	1.5	4.15	35.3	90.6	0.102	1.92
Tuva, Russia [1]	(333)	52° N 94° E	0.109	214	−3.4	4.02	46.2	107.8	0.074	2.88
Xilinhot, China [3]	(333)	44° N 117° E	0.441	360	−0.1	6.56	84.3	89.3	0.102	4.30
2. Dry steppe division (330) and temperate desert division (340)										
Comodoro Rivadavia, Aargentina[4]	(342)	49° S 68° W	0.454	222	12.7	1.13	19.1	90.5	0.172	1.74
CPER, USA [5]	(311/315)	40° N 105° W	0.581	300	8.7	2.31	45.9	107.8	0.127	2.30
Havre, USA [6]	(331/332)	49° N 110° W	1.450	312	5.5	3.23	33.9	107.0	0.087	1.74
Santa Rosa, Argentina [4]	(311/312)	37° S 64° W	0.335	532	13.6	1.80	66.5	133.4	0.286	5.89
Sarmiento, Argentina [7]	(331/332)	46° S 69° W	0.124	141	10.8	1.63	15.9	81.9	0.120	1.66
Humid temperature domain (200)										
3. Warm continental division (210) and prairie division (250)										
Khomutov, Ukraine [1]	(332)	47° N 38° E	0.860	441	9.5	7.56	112.3	112.1	0.110	3.04
Konza, USA [8]	(251/255)	39° N 97° W	0.781	818	12.9	5.78	184.8	137.8	0.271	5.77
Kursk, Russia [1]	(252)	52° N 37° E	0.607	560	5.5	9.95	175.8	82.5	0.137	5.22
Otradnoye, Russia [1]	(212)	61° N 30° E	1.181	543	3.8	6.09	106.3	72.4	0.104	3.27
Uruguay [9]	(254/255)	35° S 56° W	0.529	936	15.1	6.48	89.5	93.5	0.346	6.76
4. Mediterranean division (260)										
Bari, Italy [4]	(262)	41° N 17° E	0.084	574	15.8	3.17	107.6	118.0	0.233	4.53
Davis, USA [4]	(262)	39° N 122° W	0.076	420	15.6	2.84	50.1	132.6	0.136	2.35

Dry domain (300) – dry savanna									
5. Tropical/subtropical steppe division (310)									
Kalgoorlie, Australia [4]	(312/315)	31° S 122° E	1.813	255	17.2	3.70	40.1	141.8	0.143	2.12
Khartoum, Sudan [4]	(314)	16° N 33° E	1.006	138	25.8	2.15	23.2	220.3	0.101	1.68
Menaka, Mali [10]	(314)	16° N 2° E	0.846	290	29.7	1.23	35.5	225.3	0.156	1.93
Niamey, Niger [4]	(415)	14° N 2°E	1.444	467	23.6	2.03	119.5	203.5	0.290	4.60
Humid tropical domain (400)										
6. Savanna division (410)										
Charleville/Warra, Australia [11]	(411/416)	26° S 146° E	1.138	489	210.5	4.88	92.6	169.6	0.199	3.84
Ciudad Bolivar, Venezuela [4]	(416)	8° N 64° W	0.129	981	27.5	4.51	330.6	143.7	0.506	8.86
Kurukshetra/Ludhiana, India [12]	(412)	31° N 76° E	0.353	715	24.4	3.33	143.4	214.2	0.305	5.10
Marondera, Zimbabwe [13]	(411/415)	18° S 31° E	2.995	819	18.3	4.60	221.8	106.7	0.294	6.54
Nagpur, India [4]	(413/415)	21° N 79° E	1.184	1203	26.9	5.12	228.6	191.2	0.398	7.51
Nairobi, Kenya [14]	(413/416)	1° S 36° E	1.565	680	19.0	6.24	209.5	87.1	0.322	4.14
Towoomba, South Africa [15]	(314)	25° S 29° E	.636	630	26.5	2.47	119.9	131.2	0.339	3.91
7. Humid savanna (414/415) and rainforest division (420)										
Calabozo/San Fernando, Venezuela [16]	(414/415)	9° N 67° W	0.044	1318	28.1	7.52	366.5	138.6	0.505	9.07
Carimagua, Colombia [17]	(414)	4° N 72° W	0.185	2338	26.6	1.58	226.1	126.6	0.737	3.77
Hat Yai, Thailand [14]	(423)	6° N 101°E	0.797	1540	27.6	2.22	461.6	144.7	0.703	7.76
Lamto, Ivory Coast [18]	(414)	6° N 5° W	0.543	1170	27.9	1.78	305.0	151.8	0.628	5.69

1.T. G. Gilmanov *et al.* (in press); Parton *et al.* (1993). 2. Mongolia, Chuluun *et al.* (1996). 3. Wang and Jiang (1982); Chen (1988). 4. Weather data only; 'World Weather Disc' (Weather Disk Associates, Inc., National Climate Data Center, USA). 5. Parton *et al.* (1993). 6. Haas *et al.* (1957). 7. Fernandez *et al.* (1991). 8. Ojima *et al.* (1990), Towne and Owensby (1984), Ojima *et al.* (1994). 9. W. Bethgen (pers. comm., Uruguay). 10. P. Hiernaux (pers. comm.). 11. Tropical Soil Biology and Fertility (TSBF) site. 12. Singh and Yadava (1974). 13. TSBF. 14. Long *et al.* (1989, 1992); Parton *et al.* (1993). 15. R.J. Scholes (pers. comm.). 16. San José, and Medina (1976). 17. CIAT Center. 18. Menaut and César (1979); Parton *et al.* (1993).

Table 12.2 Long-term mean annual precipitation, temperature, potential evapotranspiration rate and ratios of annual precipitation to potential evapotranspiration rate for the seven grassland regions

Biome type	Mean annual precipitation (mm)	Mean annual temperature (°C)	Mean annual potential evapotranspiration (mm)	PPT/PET	Global area (10^6 km^2)	Vegetation type
Cold desert steppe	299	−0.3	260	0.334	2.095	C_3
Temperate steppe	301	10.2	1041	0.294	2.943	C_3/C_4
Humid temperate	700	9.3	997	0.680	3.958	C_4
Mediterranean	497	15.7	1253	0.401	0.161	C_3/C_4
Dry savanna	387	24.1	1977	0.165	5.109	C_4
Savanna	788	23.4	1491	0.554	7.990	C_4
Humid savanna	1555	27.4	1404	1.195	1.708	C_4

The humid temperate ecoregions have similar air temperature and PET to the temperate steppes, but the precipitation is substantially higher. Vegetation type ranges from C_3 in the cooler Russian sites to C_4 in the warmer American sites. Plant productivity is three to four times that of the temperate steppes. Soil C is higher as a result of increased C inputs and only slightly increased abiotic decomposition rate.

The Mediterranean ecoregion is characterized by a warm mild climate with winter rainfall. Annual PET rates are higher than the continental grasslands but the PPT/PET ratio is high during the winter and spring and low during the summer. The vegetation in these regions is dominated by C_3 plants. Plant production is about 50% the level of the humid temperate regions, with reduced N mineralization rates. Decomposition rates are reduced by water stress during the summer period but enhanced during the wet winter period by warmer temperatures. This lower production results in soil C levels about half those of the humid temperate ecoregions.

The savanna ecoregions all have similar temperature (24 °C) but the mean annual precipitation ranges from 400 to 1500 mm, and PET ranges from 2000 mm in the dry savannas to 1400 mm in the savannas and humid savannas. The PPT/PET ratio, therefore, increases markedly from the dry savannas to the humid savannas, mainly as a result of the length of the rainy season. All the savannas are dominated by C_4 grasses. Plant production and N mineralization rates increase rapidly from the dry to the humid savannas, but abiotic decomposition also increases, so the net effect on soil C is only a modest increase from drier to more humid ecoregions. The reduced soil C for the humid tropical ecoregion is a result of the sandy textures for many of the sites considered here.

Overall, in accordance with actual field data, plant production, N mineralization, and the ratio of above- to belowground productivity, all increase with annual precipitation (Long *et al.* 1992). Differences in soil C levels reflect changes in C inputs to the system and the abiotic decomposition rate, with the highest levels in the humid temperate ecoregion and the lowest in the temperate steppes and dry savannas.

12.2.2 Modelling the effect of climate change and increased CO_2

12.2.2.1 CENTURY model

CENTURY is a simulation model of plant–soil interactions in grasslands, forests, crops and savannas (Parton *et al.* 1987, 1988, 1992, 1993; 1995, Chapter 11 this volume). Version 3.0 of CENTURY was used for these model runs (Parton *et al.* 1992, 1993).

The effect of increased atmospheric concentrations of CO_2 on the photosynthetic pathway has been well documented in C_3 plants. In addition to the direct effects of CO_2, it has been observed to increase water and nitrogen use efficiency

(NUE) in both C_3 and C_4 plants (Owensby *et al.* 1993a, b). In CENTURY, we modified the plant production parameters under a 'double-CO_2' climate by changing potential evapotranspiration (PET) and NUE, for both C_3 and C_4 grasslands uniformly. We allowed a 20% decrease in total PET and a 20% decrease in N content with the change in atmospheric CO_2 concentration from 350 to 700 ppm, via a simple linear function based on results from a tallgrass prairie (Owensby *et al.* 1993a, b).

For each grassland site, the model was run to equilibrium for 5000 years, using a repeated 25-year pattern of observed current weather data. Site-specific patterns of grassland management (i.e. burning and grazing) were incorporated in these long-term runs as described in Parton *et al.* (1993).

12.2.2.2 GRASS model

The GRASS model simulates physiological and morphological details of plant growth. Plant growth is simulated with a two-day time step while photosynthesis and transpiration are simulated with two-hour time steps on alternate days (Coughenour *et al.* 1984; Coughenour 1984). Its photosynthesis submodel is derived from the C_3 model of Farquhar *et al.* (1980), with extensions to simulate C_4 photosynthesis (Chen *et al.* 1993), and linkages with the stomatal conductance submodel of Ball *et al.* (1987). The GRASS plant growth submodels are linked with a daily time step implementation of the decomposition and nutrient cycling submodel of CENTURY (daily CSOM).

Simulated photosynthetic responses to atmospheric CO_2 involve biochemical and stomatal processes. Rubisco, the main carboxylating enzyme, responds to CO_2 concentration inside the leaf. The optimum temperature for photosynthesis is shifted upwards by elevated CO_2 in C_3 species due to reduced photorespiration (Long 1991a). In C_4 species, CO_2-concentrating processes prevent Rubisco from being CO_2-limited at low CO_2, thus weakening the response of photosynthesis to CO_2. However, stomata close in response to elevated CO_2 (Ball *et al.* 1987; Collatz *et al.* 1992). Stomatal closure decreases water loss through transpiration while elevated CO_2 increases gaseous CO_2 flux through stomates. The overall result of stomatal closure responses to elevated CO_2 in both C_3 and C_4 species is an increase in water-use efficiency (WUE).

Increased WUE of C_3 and C_4 plants enhances photosynthesis. Since less water is lost through transpiration, more photosynthate is produced from each millimetre of precipitation. Therefore, increased WUE will increase net primary production in water-limited environments. However, increased WUE will have little or no effect when water is not limiting. The increase in WUE leads to increased soil moisture levels, which ameliorates negative effects of water stress on other processes. Increased soil moisture content increases C allocation to shoots, reduces photosynthesis rate, reduces leaf senescence rate, and increases decomposition and N mineralization rates.

12.2.3 Climate change scenarios

The doubled CO_2 climatologies used to drive the CENTURY model in the present study were derived from the Canadian Climate Centre (CCC) and Geophysical Fluid Dynamics Laboratory, High Scenario (GFHI) models of global climate change, which provide monthly projections of temperature and precipitation for 2.5° grid cells worldwide. We took the values projected for a doubling of atmospheric CO_2, and assumed that these would be reached within 50 years from the present, corresponding roughly to IPCC Scenario A (Houghton *et al.* 1990). Projected changes in mean annual temperature and precipitation are shown in Figure 12.2.

The two GCMs simulate similar changes to mean annual air temperature (2–5 °C); changes in precipitation were different between the models. The GFHI model showed increases in precipitation for all of the regions whilst the CCC model showed decreases in precipitation for the temperate steppe and humid savanna regions and increases for the other regions (Figure 12.2).

We assumed that changes in precipitation and temperature at each site were linear over a 50-year period, at which point they stabilized at the modified 'double CO_2' climate, as did atmospheric CO_2 (Figure 12.3). The results from the 25-year period immediately following this represent the transient response to climate change. To determine a 'near-equilibrium' long-term response to change, we continued with 150 years of stable modified weather data. Model results were analyzed for changes in the level of soil organic matter, plant productivity, N mineralization and the model-calculated effect of climate on decomposition (abiotic decomposition factor).

The GRASS model was subjected to $2 \times CO_2$ climate change scenarios generated from the CCC and GFHI models (Table 12.2). The GRASS model was used to simulate C_4 (*Bouteloua gracilis*) and C_3 (*Pascopyrum smithii*) ecosystems in Colorado and a C_4 grass (*Themeda triandra*) ecosystem in Kenya, with the same repeating patterns of base weather data that were used in the CENTURY experiments, with and without CO_2 doubling. Day and night temperatures were increased equally. Relative humidity and cloud cover were not changed. The model was run for an initial 50 years with no climatic changes. Then, the climate was gradually ramped up to the CCC and GFHI $2 \times CO_2$ scenarios over a 50-year period, followed by a 25-year period over which the model was allowed to adjust to the scenarios.

12.3 RESULTS

12.3.1 Regional response to climate and CO_2 Changes

Figures 12.4–12.10 show CENTURY simulations of the impact of the two GCM climate change scenarios and the effect of increasing atmospheric CO_2

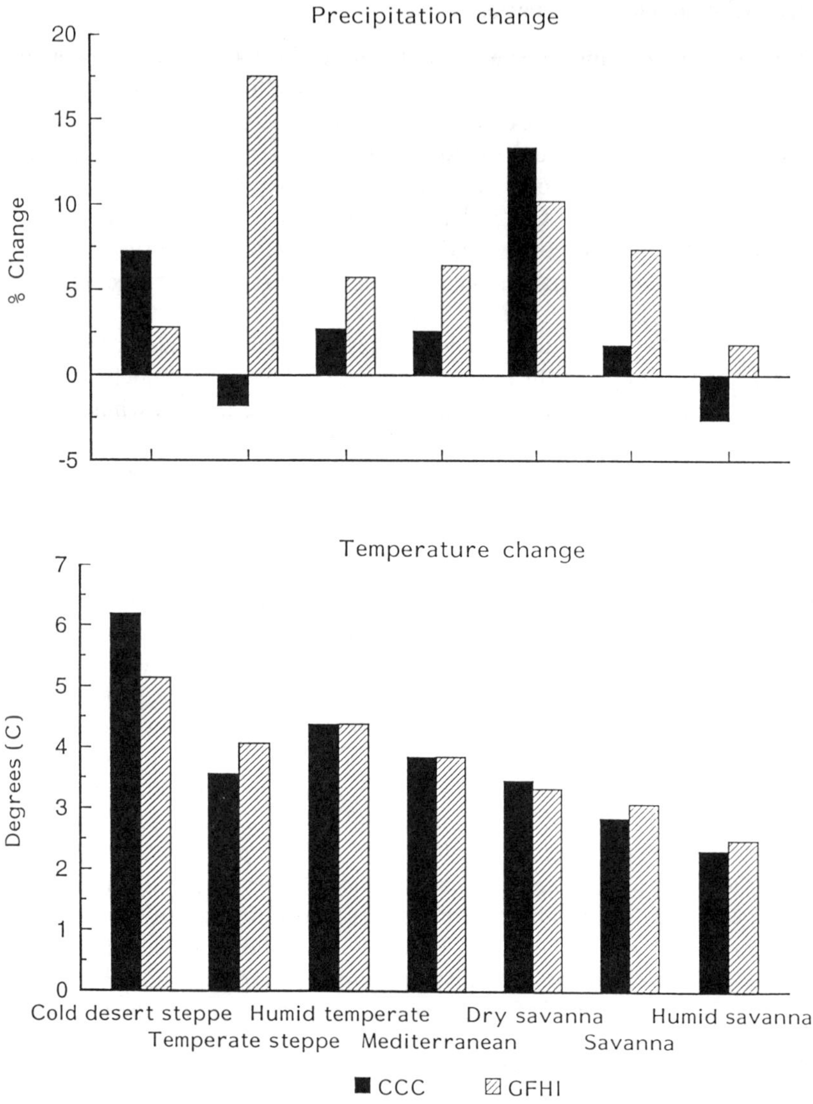

Figure 12.2 Illustration of contrast between changes in temperature and precipitation for each ecoregion under CCC and GFHI scenarios

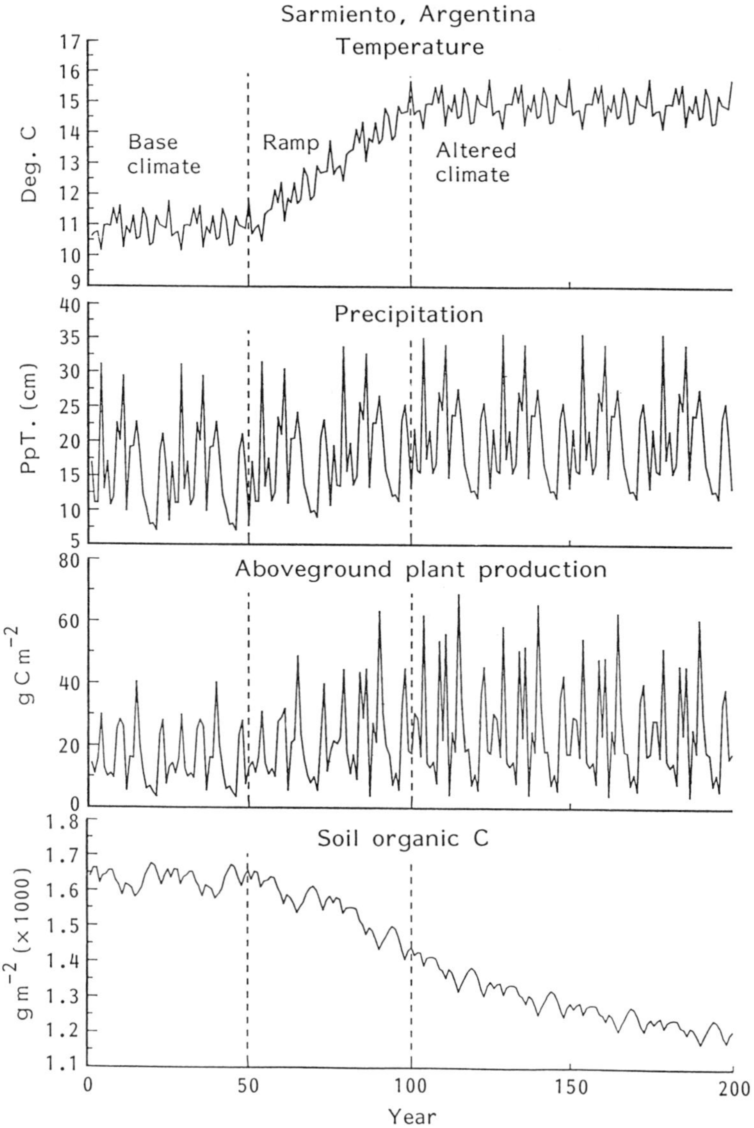

Figure 12.3 Time course of 200-year climate transient and elevated CO_2 effects on above ground plant production and soil organic C predicted by the CENTURY model

independently, for each of the ecoregions considered. These values represent the means for the 25 years of 'double-CO_2' climate immediately following the 50-year climate change period; the results for each site are shown, together with the numerical average for each ecoregion, not taking into account any area weighting.

Statistical analysis of these results shows that changes in total plant production (TPP) are positively correlated to changes in precipitation (PPT) and N mineralization (N_{min}), with N mineralization being the most important term [$TPP = -6.62 + 1.39\ (PPT) + 19.62\ (N_{min})$; $r^2 = 0.64$; $P = 0.0001$]. The response to N mineralization is consistent with the general observations that grasslands respond positively to addition of N fertilizer (Rains *et al.* 1975; Lauenroth and Dodd 1978). The response to increased precipitation is due to reduced drought stress and increase in N deposition.

The changes in N mineralization are positively correlated with changes in decomposition and precipitation. Increased decomposition rates in the model are associated with increased mineralization of soil organic N, and increased precipitation increases N availability over the long term. Changes in C are negatively related to changes in decomposition and positively related to changes in productivity. Changes in abiotic decomposition rates are positively correlated with changes in soil temperature and precipitation.

12.3.1.1 Cold desert steppes

For the majority of sites, both GCMs show an increase in precipitation and temperature, with the CCC scenario showing a greater effect (Figure 12.4). This may be attributed to the common geographical area shared by all sites. The increase in temperature results in higher decomposition rates, with the CCC scenario showing a greater effect on decomposition. Plant production is reduced at three of the four sites, and is only weakly correlated with N mineralization, possibly due to the relatively high N input rates for this region. The mineralization of N shows no clear pattern of change. Soil C tends to decrease as a result of reduced production and increased decomposition. The Tumentsogt site (Mongolia) is an exception, showing little climate-induced change in soil C due to smaller increases in decomposition and a small increase in production.

Increased atmospheric CO_2 independently causes a substantial increase in production (30%). The cold desert steppes are characterized by low annual rainfall levels and high PET values; and changes in soil moisture status due to increased WUE appears to accelerate decomposition rates by 25%. This is the only ecoregion where CO_2 also causes a net increase in N mineralization. Together, these effects produce a modest stimulation of soil C storage (1–3%).

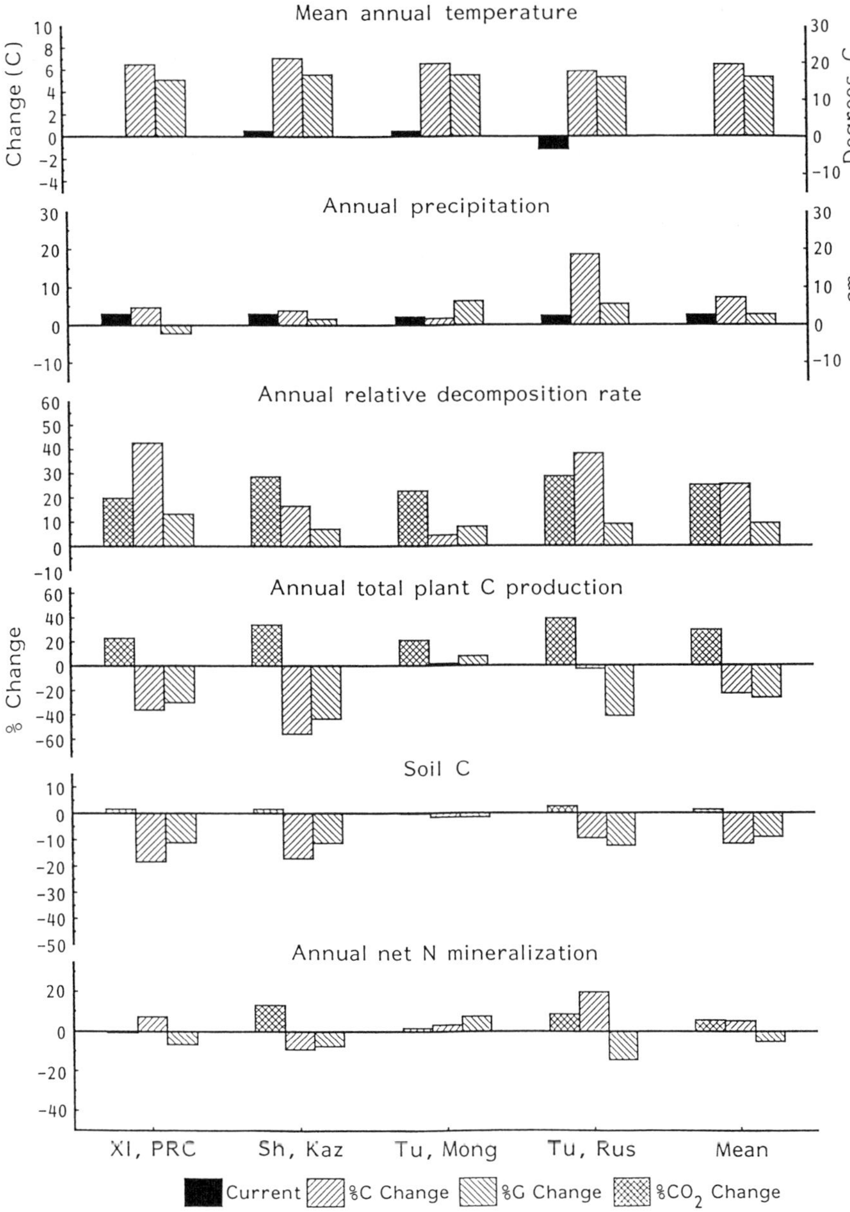

Figure 12.4 Cold desert steppe. Figures 12.4–12.10. are ecoregion summary figures showing % change at each site + ecoregion mean, for temperature precipitation, decomposition, NPP, SOM, and N_{min}. Each histogram cluster of three bars shows CCC, GFDL, CO_2 effect

12.3.1.2 Dry steppes and temperate desert

Annual precipitation increases in the GFHI scenario for these regions, but shows both increases and decreases for the CCC scenario (Figure 12.5). Mean annual temperature increases are slightly lower (3.5–4 °C increases relative to 5 to >6 °C in the cold desert steppe) than those observed for the cold desert steppes, with the GFHI scenario showing slightly higher increases. The combined effect of temperature and precipitation changes is to increase decomposition rates, with larger increases for the GFHI scenario. Increases in plant production are correlated with N mineralization, which itself is correlated with precipitation for most of the sites. Soil C tends to decrease or show little change, correlated with increases in decomposition rate. The Santa Rosa site (Argentina), which has the highest present-day precipitation, shows exceptional climate-induced decreases in N mineralization, plant production and soil C.

The CO_2 effect is a modest increase in production, but also a substantial (30%) increase in decomposition. Carbon dioxide causes a small decrease in N mineralization, and overall, the effect is a slight stimulation of soil C storage (1–2%).

12.3.1.3 Humid temperate

Precipitation is generally increased, with the GFHI scenario showing a greater effect (Figure 12.6). Temperature increases are similar to those for the temperate steppes with no clear difference between the GCMs. Together, these result in about 20% increase in decomposition rates. Changes in plant production are correlated with N mineralization, which appears to be controlled mainly by changes in decomposition rate. Soil C tends to decrease consistently for all sites.

Elevated CO_2 independently causes a 10–15% increase in production, and a 10–15% increase in decomposition. Nitrogen mineralization shows a notable decrease (5–10%) due to elevated CO_2, and overall the effect is a larger stimulation of soil C storage (2–4%) compared with the steppe ecoregions.

12.3.1.4 Mediterranean

Changes in precipitation and temperature were similar to the humid temperate ecoregion, producing a similar increase in decomposition rates (Figure 12.7). Changes in plant production are correlated with N mineralization, which is associated with changes in decomposition rate. Soil C decreases as a result.

As for the humid temperate ecoregion, elevated CO_2 causes a 10–15% increase in production, and about a 10% increase in decomposition. The mineralization of N shows a notable decrease (5–10%) due to elevated CO_2, and overall the effect is a moderate stimulation of soil C storage (3–5%).

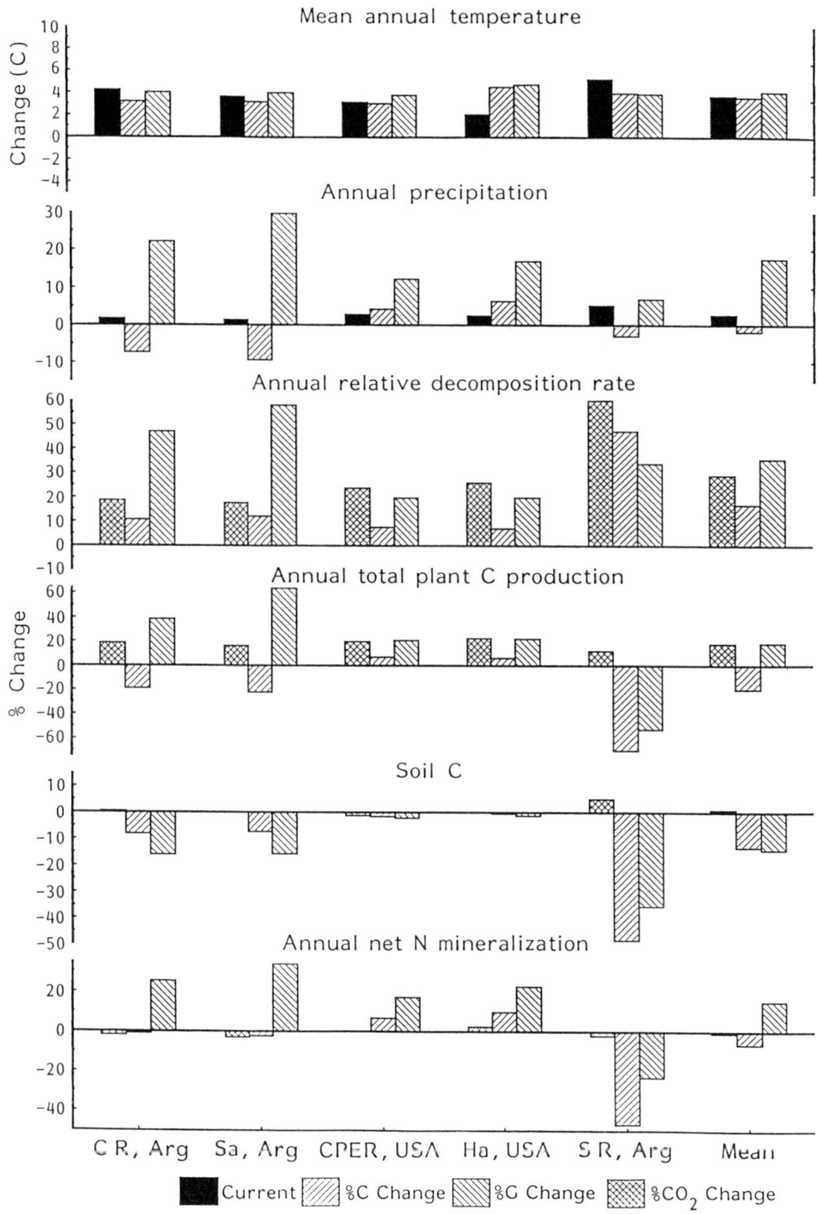

Figure 12.5 Temperate steppe ecoregion

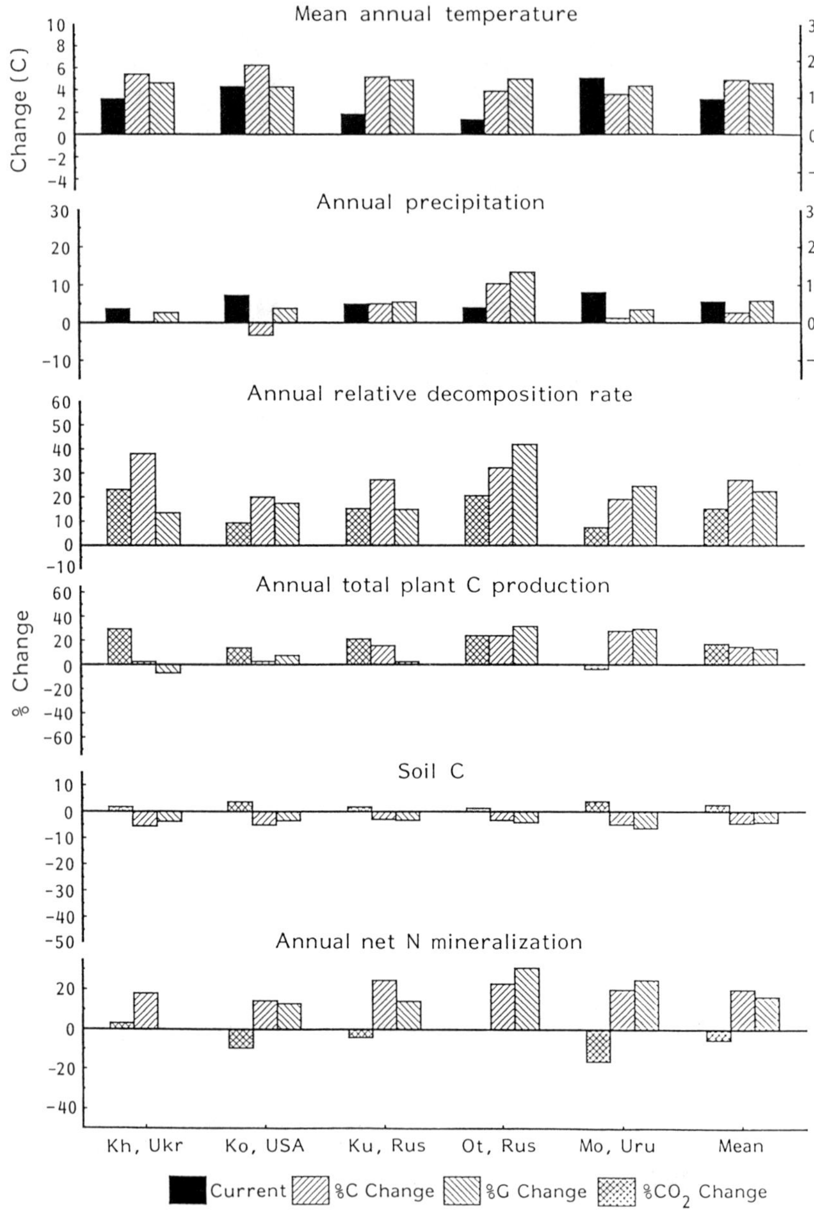

Figure 12.6 Humid temperate ecoregion

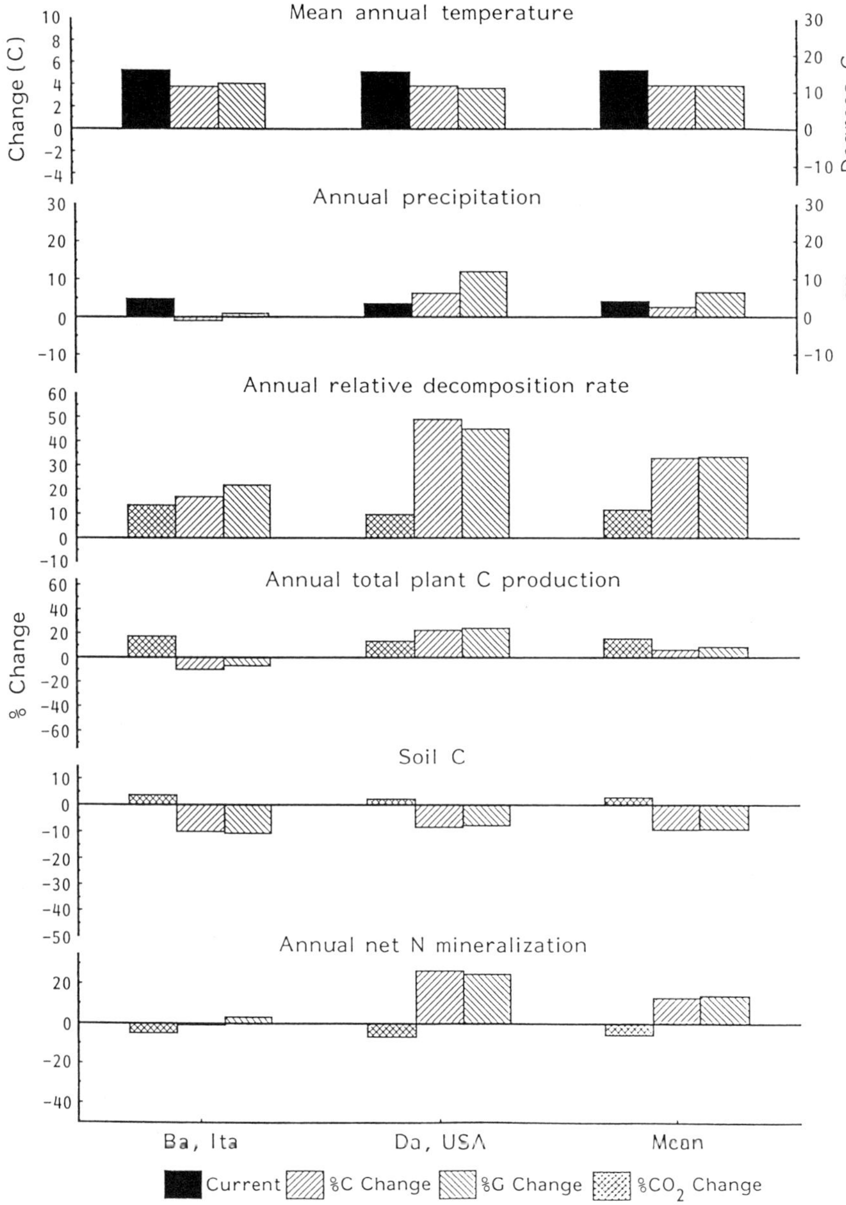

Figure 12.7 Mediterranean ecoregion

12.3.1.5 Dry savannas

Substantial increases in precipitation are predicted by both GCMs, with temperature increases slightly lower than for the temperate ecoregions, but the resulting changes in decomposition are low to modest (−10% to +10%), tending to increase for the CCC and decrease for the GFHI scenario (Figure 12.8). Plant production generally increases and is highly correlated with N mineralization. However, changes in N mineralization were only weakly associated with decomposition rate, and resulting changes in soil C were low (−3% to +5%), showing no clear pattern.

The independent effects of elevated atmospheric CO_2 are similar changes in production, decomposition and N mineralization as for the humid temperate ecoregion. However, the net effect of elevated CO_2 on soil C storage balances out as zero.

12.3.1.6 Savannas

Precipitation generally increased, with temperature increases slightly lower than for the dry savannas (Figure 12.9). Projected changes in decomposition are both positive and negative, with a modest increase on average. Plant production shows an increase on balance, and is well correlated with N mineralization. Changes in N mineralization were associated with decomposition rate, and changes in soil C were negative overall.

Elevated CO_2 independently causes a 10–15% increase in production, and a reduced (10%) increase in decomposition relative to the dry savannas. Nitrogen mineralization shows a substantial decrease (10%) due to elevated CO_2 induced reduction in litter quality, and overall the effect is a moderate stimulation of soil C storage (4–6%).

12.3.1.7 Humid savannas

Precipitation changes showed no clear pattern, and temperate increases were the lowest of any ecoregion (Figure 12.10). Changes in decomposition showed no clear pattern, and plant production was weakly correlated with N mineralization and precipitation changes. Soil C tended to decrease overall (0 to −6%).

Elevated CO_2 produces a modest (10%) increase in production, and a much reduced (5%) increase in decomposition compared with the other savanna ecoregions. The CO_2 effect on N mineralization is the largest decrease for any ecoregion (15%), and the resulting stimulation of soil C storage is the highest for any ecoregion (8–12%).

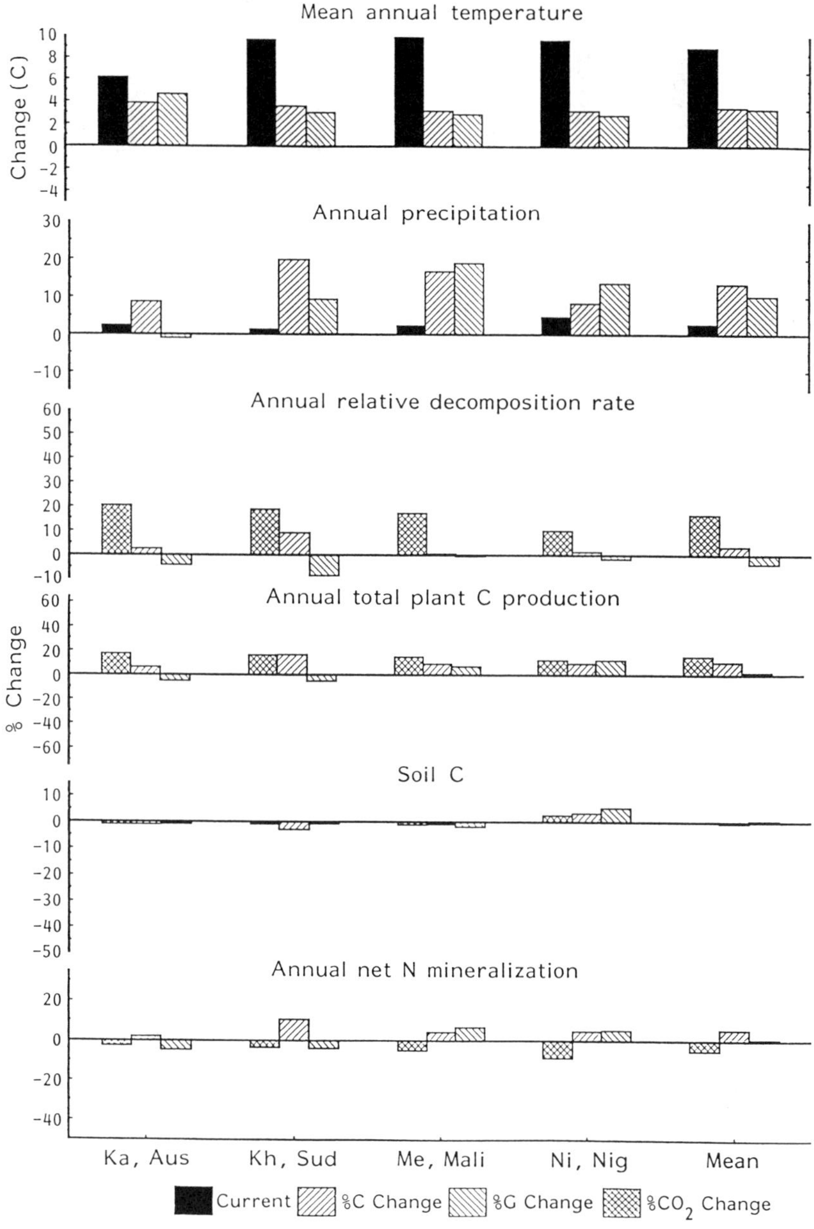

Figure 12.8 Dry savanna ecoregion

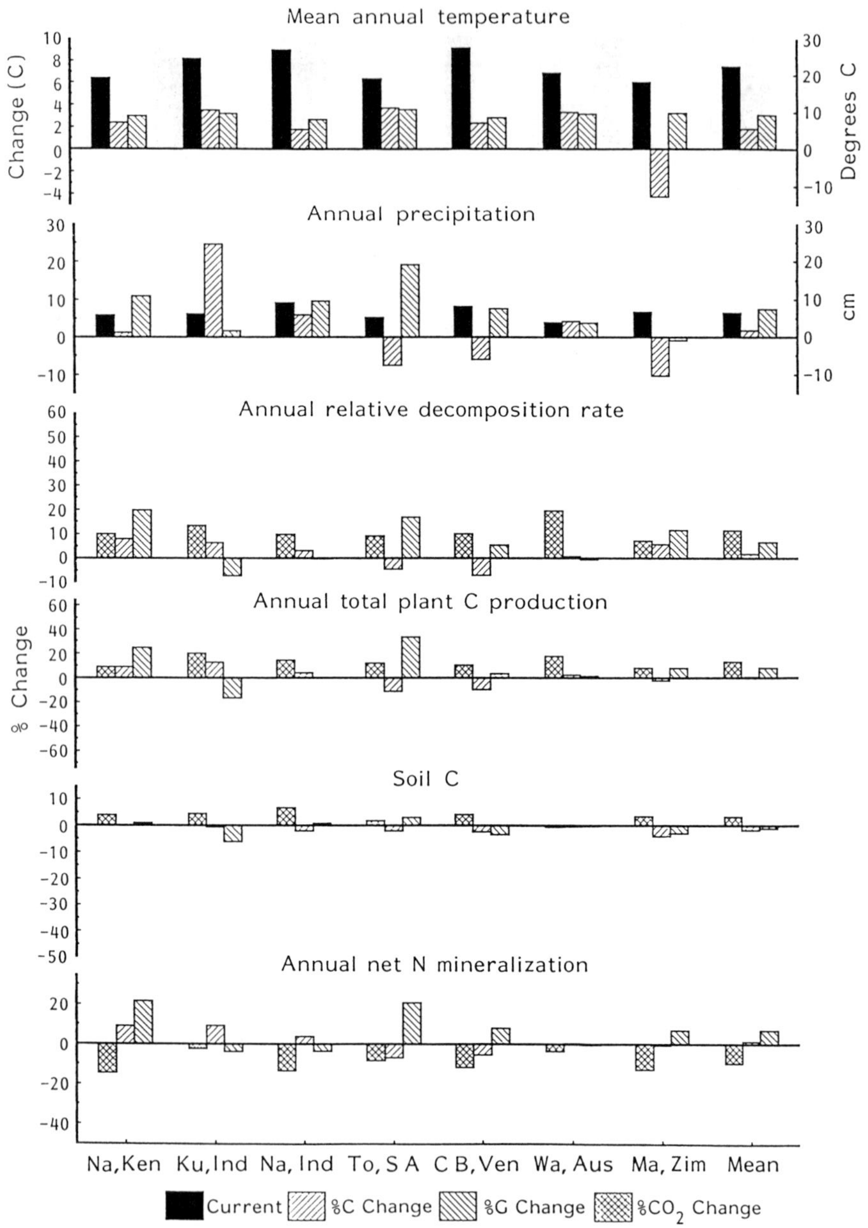

Figure 12.9 Savanna ecoregion

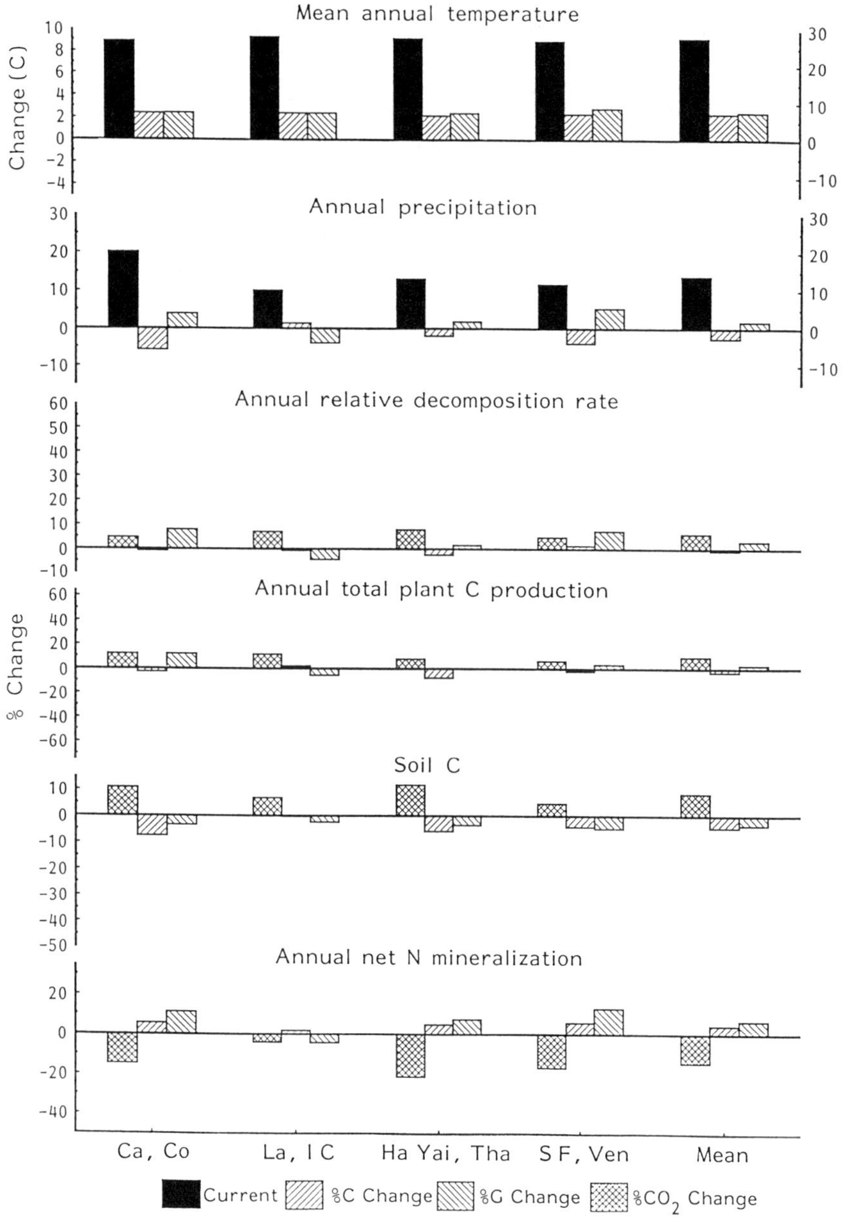

Figure 12.10 Humid savanna ecoregion

12.3.2 Site-level analysis of climate and CO_2 change scenarios

The GRASS model responded quite differently to the CCC and GFHI climate-change-only scenarios at the CPER site (Table 12.3). The GRASS CCC scenario without a CO_2 doubling decreased NPP 52% in the C_3 grass and 48% in the C_4 grass. In contrast, the GFHI scenario increased net primary productivity (NPP) of C_3 plants by 13% and decreased NPP of C_4 plants by 18%. The CENTURY simulations at the CPER site displayed smaller changes to the two GCM projections for NPP and net N mineralization, and no difference in soil organic matter (SOM) values (Table 12.3). At the Kenya site, the differences between scenarios were much less pronounced. The NPP decreased 3% under the CCC scenario and 6% under the GFHI scenario, without CO_2 doubling. At the Kenya site, CENTURY current values of NPP, net N mineralization, and SOM tended to be 20, 25 and 1%, respectively, of the GRASS simulations. The CENTURY ecosystem response for NPP and net N mineralization to climate changes at Kenya tended to be greater than the GRASS simulations, although the difference between GCMs tended to be less than 5% of each other.

The large discrepancies in model responses to the two climate change scenarios in the GRASS CPER simulations can be attributed mainly to precipitation. While the CCC model predicted a 5.2 mm increase in annual precipitation, the GFHI model predicted a 38.0 mm increase. During the March–August growing season, CCC predicted a 9.2 mm decrease while GFHI predicted a 27.5 mm increase. At the Kenya site, CCC predicted a 2.9 mm annual decrease while GFHI predicted a 72.0 mm annual increase. However, during the November–May wet season, both models predicted increased rainfall: CCC added +18.6 mm while GFHI added +36.0 mm.

Doubling CO_2 without climatic change increased NPP 6, 5 and 40% in the CPER C_4 and C_3, and the Kenya C_4 grasslands, respectively for the GRASS simulations. Doubling CO_2 ameliorated negative NPP responses to climate change scenarios in CPER and accentuated positive NPP responses in Kenya for both models. The CCC scenario with CO_2 doubling decreased NPP by 20% in the C_3 grass and 37% in the C_4 grass, relative to the control simulation for the GRASS model. The NPP increased 46% in Kenya under both scenarios of CO_2 doubling and climatic change. The CENTURY simulations at CPER and Kenya under current climate with doubled CO_2 increased NPP by approximately 10%. The impact on SOM at CPER was neutral, whereas at Kenya, doubling CO_2 slightly enhanced SOM levels.

Elevated CO_2 concentrations might favor C_3 over C_4 species (e.g. Carter and Peterson 1983; Bazazz *et al.* 1989). This hypothesis was only supported by the CPER short-term simulations. It was not supported by the long-term simulations. In the long-term simulations, total NPP of C_3 and C_4 grasses responded similarly to CO_2 alone: $2 \times CO_2$ increased NPP 5% in the C_3 grass and 7% in the C_4 grass. The CCC scenario with $2 \times CO_2$ decreased NPP by 25 and 37% in C_3

and C_4 grasses, respectively, which would suggest a competitive advantage for C_4 species under this warm, dry scenario. The GFHI scenario with $2 \times CO_2$ increased NPP 13% in the C_3 and 21% in the C_4 grass, which also indicated that C_4 grasses might be favored. Indeed, the large stimulation of NPP in the tropical C_4 grassland suggested that C_4 species would fare quite well in warm, water-limited ecosystems.

Transpiration rate responses of the GRASS model simulations to climate and CO_2 varied among sites (Figure 12.11). Minor reductions in transpiration rate in response to $2 \times CO_2$ were simulated at the CPER C_4 site. In contrast, transpiration rates were much reduced by elevated CO_2 at the C_3 site, but less so with the increased moisture of the GFHI scenario. Differences in transpiration responses likely contributed to different C_3 and C_4 NPP responses to CO_2. The greater reduction in transpiration rate of the C_3 species was due to greater stomatal and carbon assimilation (A_n) response of C_3 species to elevated CO_2. The magnitude of stomatal opening in response to elevated A_n and closing in response to elevated CO_2 is greater in C_3 species (see Ball *et al.* 1987) than C_4 species (see Collatz *et al.* 1992).

Transpiration rates at the Kenya site were lower than at CPER sites, due to much higher (e.g. $5\times$) leaf area indices. In contrast to the CPER C_4 site, elevated CO_2 reduced transpiration rate significantly in Kenya. The reduction in transpiration rate was much larger than the CPER C_4 response. Stomatal closure due to elevated CO_2 decreased moisture stress on plants and decomposers in Kenya. Responses were not reduced by low temperatures as they were in CPER. This generated a positive feedback as plant growth was accelerated by lower water stress and by increased N availability.

Climate change and CO_2 doubling affected net N mineralization (N_{min}) rates (Table 12.3b). For both CENTURY and GRASS simulations, N_{min} responses contributed to differences in NPP among sites. Doubling CO_2 alone decreased N_{min} for both CPER grass types, but more so for the C_3 type. This response was caused by the negative effect of reduced litter N concentration, which results in greater immobilization of N. Significantly, despite this reduced N availability, NPP was increased by elevated CO_2 due to increased plant N use efficiency (NUE). It is likely that N mineralization would have been stimulated, or that it would not have declined if litter N concentrations were not reduced, since NPP and soil moisture were enhanced by CO_2. Thus, while nutrient limitation did reverse positive responses to CO_2, the reversal could be weakened by increased NUE.

The CCC climate scenario in GRASS simulations alone reduced N mineralization (N_{min}) on both the C_3 and C_4 sites, due to drier soil conditions and smaller litter inputs. The GFHI climate scenario alone in the GRASS simulations increased N_{min} at the C_3 and C_4 sites, due to reduced moisture limitations on decomposition rate. The CENTURY simulations showed greater N_{min} for both GCM projections. Adding $2 \times CO_2$ to the CCC scenario decreased N_{min} very

Table 12.3 Climate change and CO_2 effects on simulated (a) total net primary production, (b) annual net N mineralization, and (c) total soil organic matter for CPER and Kenya using CENTURY and GRASS models. GRASS simulation for CPER are parameterized for C_3 and C_4 grasses, the CENTURY simulations for CPER and Kenya are for C_4 plants. The values are the 25-year means and ranges for 'current' climate, doubled 'CO_2' effects under current climate, climate change effects (CC) for the Canadian Climate Centre GCM (c) and the GFDL GCM (g), and the combined effect of '$CC+CO_2$' for both GCMs (c and g)

				CC		$CC+CO_2$	
CPER		Current	CO_2	c	g	c	g
(a) Total net primary production							
CENTURY	$\bar{x}$	116	139	122	135	150	164
	Range	75–191	88–217	72–213	81–229	92–237	96–263
GRASS-C_3	$\bar{x}$	141	151	75	159	88	171
	Range	49–334	68–286	13–213	92–261	22–218	108–257
GRASS-C_4	$\bar{x}$	136	150	68	161	114	170
	Range	72–288	78–246	13–167	113–217	10–256	108–240
Kenya							
CENTURY	$\bar{x}$	338	368	367	420	398	462
	Range	172–670	216–735	191–694	219–820	223–768	272–912
GRASS	$\bar{x}$	240	316	223	243	305	338
	Range	76–459	97–554	19–397	56–474	21–536	97–643
(b) Annual net N mineralization							
CENTURY	$\bar{x}$	2.3	2.3	2.46	2.70	2.51	2.71
	Range	1.39–3.67	1.52–3.43	1.45–3.70	1.47–4.00	1.55–3.67	1.53–4.00
GRASS-C_3	$\bar{x}$	1.62	1.59	1.08	2.09	1.05	1.98
	Range	0.65–7.85	0.69–2.66	0.37–2.17	1.06–3.15	0.12–2.15	1.19–2.66
GRASS-C_4	$\bar{x}$	1.72	1.47	1.08	2.25	1.25	1.62
	Range	0.57–3.19	0.80–2.11	0–2.43	1.52–2.94	0.05–2.32	1.14–2.34
Kenya							
CENTURY	$\bar{x}$	4.14	3.53	4.52	5.03	3.85	4.38
	Range	2.30–5.85	1.45–5.03	2.68–6.42	2.76–7.18	2.07–5.47	1.95–6.11
GRASS	$\bar{x}$	2.13	2.80	2.07	2.38	2.73	3.14
	Range	−0.11–5.36	0.70–5.72	−0.55–4.62	−0.24–6.08	-0.28–5.22	0.83–6.40

(c) Total soil organic matter							
CENTURY	$\bar{x}$	2.30	2.28	2.27	2.26	2.26	2.25
	Range	2.26–2.36	2.22–2.34	2.23–2.34	2.21–2.29	2.20–2.31	2.20–2.28
GRASS-C_3	$\bar{x}$	2.63	2.63	1.99	2.60	2.06	2.59
	Range	2.56–2.74	2.57–2.73	1.82–2.15	2.53–2.69	1.89–2.23	2.54–2.68
GRASS-C_4	$\bar{x}$	2.56	2.58	1.83	2.49	2.39	2.57
	Range	2.47–2.69	2.53–2.64	1.64–2.05	2.42–2.56	2.25–2.55	2.51–2.60
Kenya							
CENTURY	$\bar{x}$	5.24	5.45	5.24	5.29	5.38	5.41
	Range	5.20–5.29	5.38–5.49	5.19–5.29	5.24–5.34	5.33–5.43	5.34–5.46
GRASS	$\bar{x}$	6.12	6.48	5.75	5.87	6.38	6.37
	Range	5.94–6.28	6.36–6.63	5.65–5.88	5.70–5.98	6.24–6.50	6.21–6.51

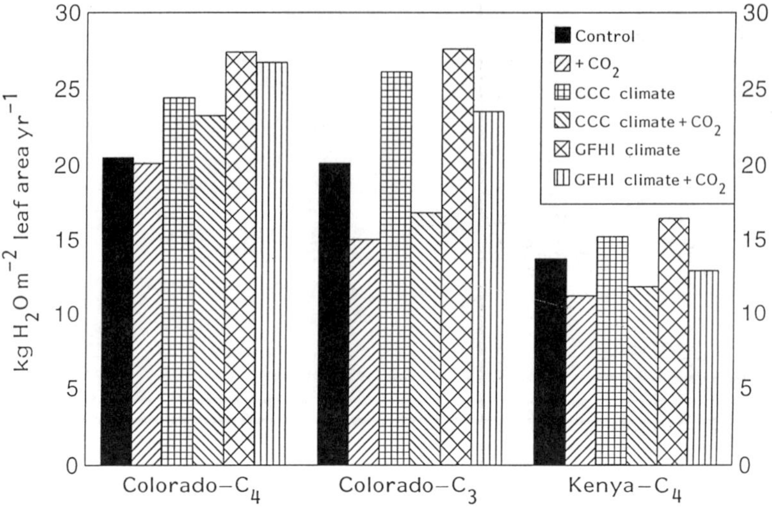

Figure 12.11 GRASS-daily CSOM responses of transpiration rate per unit leaf area index (LAI) in the final 25 years of the climate change scenarios in Figure 12.2. Transpiration rate was calculated by dividing season-long transpiration by peak LAI

slightly at the C_4 site, and increased N_{min} at the C_3 site. This was probably due to the greater NPP response of the C_3 grassland, with consequently higher litter inputs. Adding $2 \times CO_2$ to the GFHI scenario decreased N_{min} slightly at the C_4 site, but greatly decreased Nmin at the C_3 site. The reduction in N_{min} with GFHI suggested increased importance of litter quality when litter inputs are large. The CENTURY simulation at the CPER site with doubled CO_2 plus climatic changes had slight increases in N_{min} due to slightly better soil moisture conditions.

Both climate change scenarios alone elevated N_{min} greatly at the Kenya site, due to increased soil moisture. The CENTURY simulations under elevated CO_2 levels resulted in a 10% decrease in N_{min} relative to climate change N_{min} values alone. Surprisingly, doubling CO_2 in the GRASS simulations increased N_{min} by 24% despite decreases in litter quality. N_{min} at the Kenyan site was primarily limited by soil moisture, in contrast to the CPER site where temperature played a larger role. The increased WUE of the C_4 grass under $2 \times CO_2$ increased soil moisture, which then increased decomposition rate. Similarly, Rice *et al.* (1992) observed elevated microbial biomass and respiration under elevated CO_2 in a dry year, but not in a wet year. Increased N_{min} likely contributed to increased plant growth, which was then recycled through decomposition, thus further elevating N_{min}. This positive feedback loop was undoubtedly enhanced by the high rate of N recycling in this tropical system relative to more temperature-limited systems.

The effect of $2 \times CO_2$ alone at the Kenya site was to increase NPP by 40%, despite only a 24% increase in N_{min} and a 19% increase in plant N uptake. This can be explained by the 15% lower N concentrations in dying roots and shoots, and the associated increase in NUE. The total amount of N in the system was actually slightly lower (5%) in the elevated CO_2 simulation due to greater fire losses. Thus, $2 \times CO_2$ enhanced N recycling rate. Greater total quantities of N uptake (kg N ha^{-1}) have also been observed in a tallgrass prairie, despite reduced %N concentrations in plant biomass (Owensby *et al.* 1993a).

Soil organic matter (SOM) in the GRASS simulation was reduced by 25–27% by the CCC scenario without $2 \times CO_2$ in CPER (Table 12.3c), due to lower C inputs from NPP. With $2 \times CO_2$, SOM was reduced only 5% under C_3 grass, but still declined by 22% under C_4 grass. In Kenya, the CCC scenario without $2 \times CO_2$ decreased SOM 4%, and with $2 \times CO_2$, it increased SOM by 4%. SOM declined by 1–2% under the GFHI scenario in CPER. With $2 \times CO_2$, however, GFHI decreased SOM 1% under C_4 and increased SOM 3% under C_3 grasses. In Kenya, the GFHI scenario decreased SOM by 4% without $2 \times CO_2$, and by 4% with $2 \times CO_2$. Decreased soil C under elevated CO_2 was accompanied by a 2% decrease in total soil N. To summarize, elevated CO_2 alone had little effect on ecosystem C storage in Colorado while it increased C storage 4% in Kenya.

Modeled responses for GRASS and CENTURY to changed climate and CO_2 demonstrated considerable interannual variability (Figure 12.12). For example, under the GFHI climate scenario, Kenya NPP increased by 30–50% in some years, while in other years there were little, no, or even negative responses. Simulations indicated that CO_2 increases may ameliorate the negative effects of certain dry years. For example, periodic dry years in the Kenyan climate data reduced NPP, even under the climate change scenarios (Figure 12.12). However, CO_2 doubling often ameliorated these decreases. Indeed, most of the positive NPP response to CO_2 could be attributed to greater increases in WUE in drier years. Similarly, Owensby *et al.* (1993b) found that NPP of a C_4 grass was stimulated by CO_2 in a dry rainfall year, while it was not stimulated in a wet year.

12.3.2.1 Climate change sensitivity

For all of the temperate ecoregions, the increase in temperature results in an average increase in PET of about 30–40%, and about 15–20% in the tropical ecoregions.

Climate change impacts of the two GCM scenarios on total plant production (above- and belowground), soil C, N mineralization and decomposition are summarized for each ecoregion in Figure 12.13. Results show that the cold desert steppe region had a substantial reduction in plant production (about 25%) while the humid temperate sites had increases greater than 12%. All the savanna ecoregions showed relatively small changes ($<5\%$). Total plant production is positively correlated to changes in precipitation and N mineralization, the latter being more important. The main exception to this is in the cold desert steppe

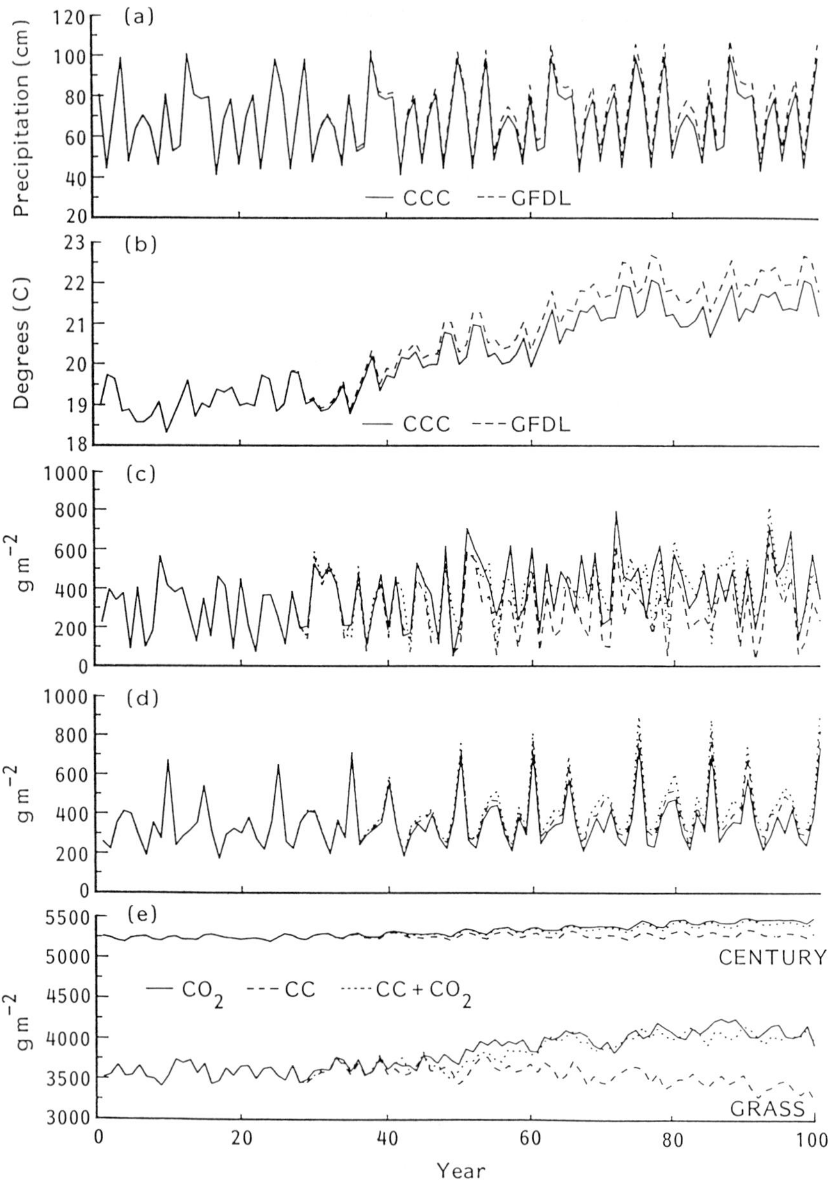

Figure 12.12 Annual dynamics of (a) precipitation, (b) temperature, (c) NPP–GRASS, (d) NPP–CENTURY (e) SOM for the climate change and CO_2 perturbation for Kenya. The climate time series for 100 years are depicted in (a) and (b) for the CCC and GFHI GCMs, the first 25 years is the current observed climate, followed by the 50 year transient ramp, and ending with a 25 year stabilized 2 × CO_2 climate as projected by CCC and GFHI GCMs. Simulated NPP from the (c) CENTURY model and (d) from GRASS are depicted for the 100-year time series. The simulated SOM are depicted in (e)

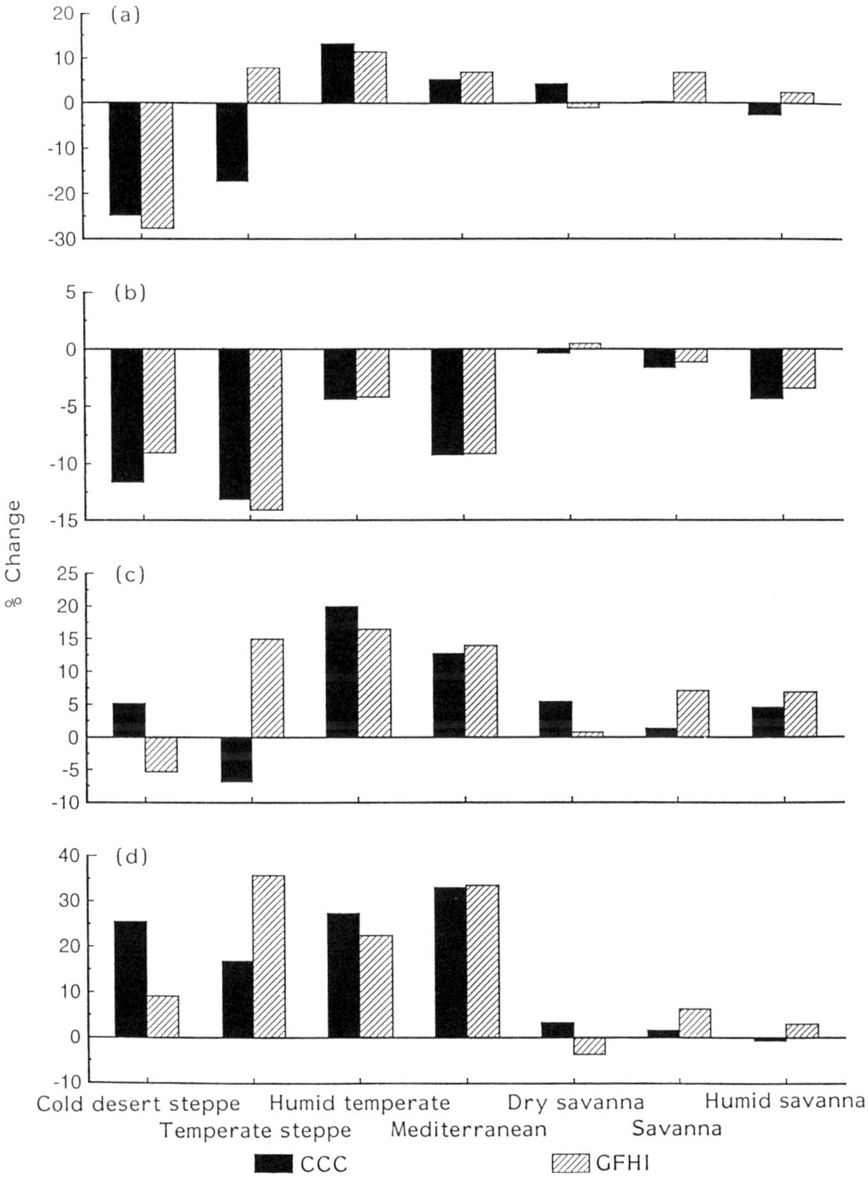

Figure 12.13 Effect of climate change alone on (a) production, (b) SOM, (c) N mineralization, (d) decomposition, for each ecoregion, for the CCC and GFHI scenarios

region. Substantial differences in plant productivity between the GCMs exist for the temperate steppes and dry savannas, and are primarily related to differences in predicted precipitation. These results are in broad agreement with those of Melillo *et al.* (1993), whose terrestrial ecosystem model (TEM) predicted 10–15% increase in grassland NPP for double-CO_2 climate scenarios; however, TEM does not show up differences between contrasting grassland regions.

Soil C results show substantial losses of soil C for all the temperate sites, including the Mediterranean region, with the largest losses occurring in the cold desert steppe and temperate steppe regions. Soil C losses are low in all three savanna regions, with the highest losses being in the humid savannas (7%). The changes in soil C are negatively related to changes in decomposition and positively related to changes in productivity. Soil C changes do not vary much between the two GCMs.

The N mineralization rates tended to increase for all sites, with the largest increases observed for the humid temperate, Mediterranean and temperate steppe ecoregions. The N mineralization changes were positively correlated with changes in abiotic decomposition rates and precipitation, with decomposition rates having the greater impact.

Decomposition rates at the temperate sites are greatly modified under climate change simulations compared with the relatively small changes simulated in the tropical regions of the study. The apparent differences between the temperate and tropical regions in this analysis are the result of the relative changes in the temperature projections; more than 25% increase in mean annual temperature for the temperate regions, compared with approximately 10% increase for the tropics. There is also a trend for abiotic decomposition rates to increase with increasing precipitation.

12.3.2.2 Effect of increased CO_2 and CO_2/climate interaction

We simulated the impact of CO_2 on grassland ecosystems with and without the inclusion of climate change (Figure 12.14). Overall, the effect of increasing CO_2 under current climatic patterns was to increase total plant production and soil C storage. Production increased the most in the cold desert steppe region, and C storage increased most for the humid savannas. There is a trend for soil C changes to be inversely related to changes in abiotic decomposition rate (Figure 12.14). Nitrogen mineralization rates decreased in most regions, resulting from increased input of lower quality litter with higher C/N ratios (Melillo *et al.* 1982), but increased for the cold desert steppe region. The relatively high N inputs for the cold desert steppe regions may have influenced the response to elevated atmospheric CO_2. Changes in relative N mineralization rates appear to be positively related to changes in relative decomposition rates due to the release of N from decomposing SOM. Simulated increases in decomposition under elevated CO_2 resulted from improved soil water relationships at the cold dry sites

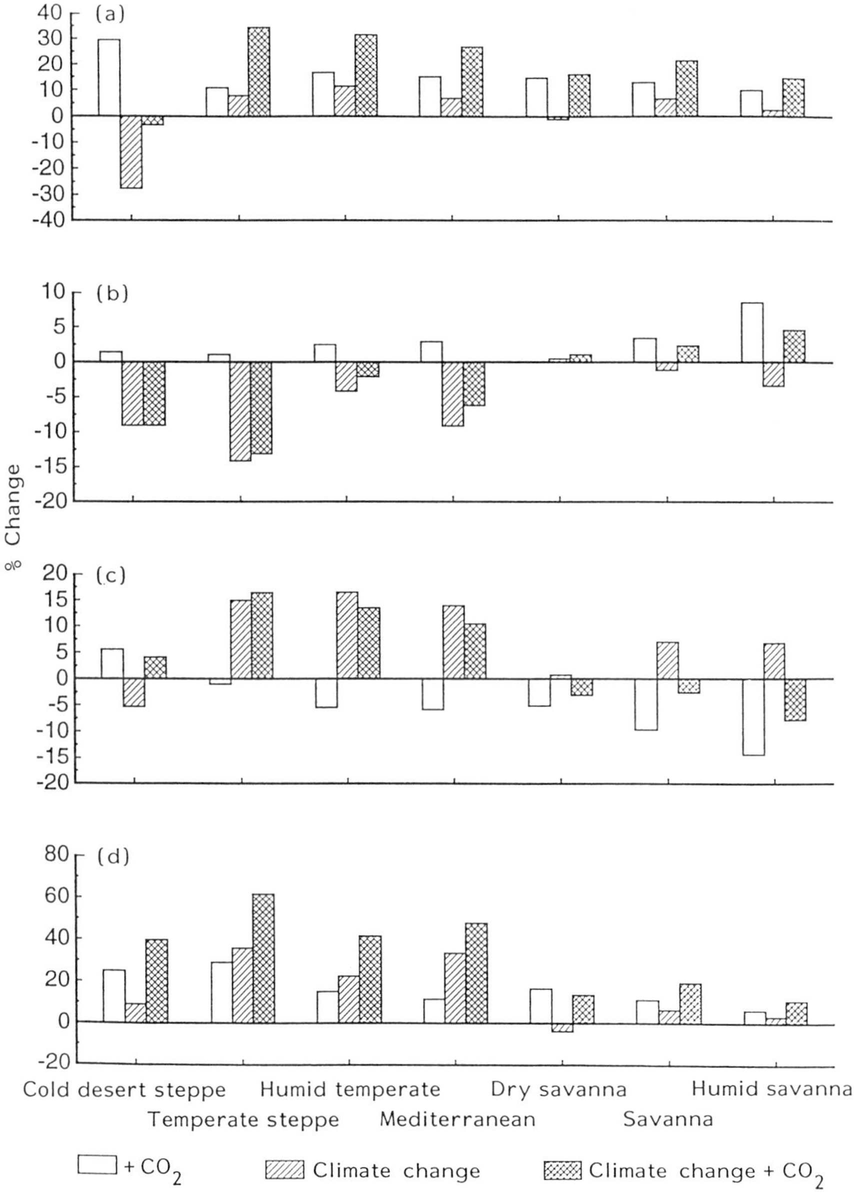

Figure 12.14 Effect of CO_2 alone, climate change alone (GFHI scenario), and combined CO_2 and climate change for each ecoregion, for (a) production, (b) SOM, (c) N mineralization, (d) decomposition

(cold desert steppe, temperate steppe), with smaller increases for the warmer and more humid sites.

The combined effects of CO_2 and climate change were additive for each ecosystem property studied. Relative changes in total plant production, soil C, N mineralization, and decomposition were compared statistically between (i) the combined impact of CO_2 and climate and (ii) the independent impacts of CO_2 and climate change; a linear additive effect was demonstrated ($r_2 = 0.97$, 0.98, 0.97, and 0.92, respectively).

Increased CO_2 enhanced total plant production regardless of the climate change impact, consistent with the results of Melillo *et al.* (1993). This enhancement in plant production reduces soil C losses resulting from climatically driven changes in SOM decomposition. The direct CO_2 effect reduces C losses throughout the grassland systems and, in fact, results in a soil C sink region in the tropical savanna and humid savanna regions.

The impact of these climate and CO_2 changes on current grassland soil C stocks indicates that purely climatic effects resulted in a net loss of over 4 Pg C (Table 12.3), with all regions becoming C sources as a result of increased soil C decomposition. However, the effect of photosynthetic CO_2 increases is to offset the soil C losses due to climatic changes. The CO_2 effect reduced the net global losses from grasslands by approximately 50% in the CCC case and 66% in the GFHI case, suggesting a net C source of 1–2 Pg C over 50 years, or 0.02–0.04 Pg C yr^{-1}. In the tropical regions, savannas and humid savannas actually become soil C sinks with the inclusion of CO_2 and climate changes together, regardless of GCM scenario (Table 12.3; Figure 12.15).

These results are in marked contrast to those of Jenkinson *et al.* (1991), who estimated a net source of 61 Pg C over the next 60 years from SOM worldwide. Grassland soils (temperate steppes and tropical savannas) accounted for about 20% of present soil C stocks in this analysis, suggesting a significant grassland source of about 12 Pg C over 60 years, or 0.2 Pg C yr^{-1}.

12.3.2.3 Transient responses

The discussion to this point has been focused on net changes to system properties after a 'doubled-CO_2' climate impact or atmospheric CO_2 increase has been established. The short-term transient nature of ecosystem responses and the long-term stabilization of ecosystem properties due to the climate and CO_2 effects are also critical to understanding grassland responses to climate and atmospheric perturbations.

Analysis of the dynamics of the NPP and net N mineralization rates during the ramp and the near-term response (i.e. initial 50 years of climate change) indicate a strong correspondence to climatic changes. The NPP is positively correlated to rainfall dynamics. The SOM C and N dynamics appear to be more related to temperature changes; however, in the drier sites, the SOM dynamics appear to also be controlled by moisture conditions. The long-term average increase in NPP is controlled by N availability from SOM, so that changes in litter quality over time has a greater impact on N turnover. This long-term impact of litter quality changes is observed after several 100 years of enriched CO_2 simulations.

Plant production has stabilized 150 years after the start of climate change, SOM has not reached a new steady state, although the rate of change in SOM has decreased. Comparison of transient and long-term (5000-year) equilibrium levels of SOM shows that only 30% of the long-term change in SOM has occurred after 150 years. These comparisons made across the individual ecoregions show that the 50–75-year changes in SOM levels are greatest in the warm humid zones (50, 35, 30, 25, 20, 15, and 10% of the long-term equilibrium SOM changes, for the humid savanna, savanna, humid temperate, dry savanna, Mediterranean temperate steppe, and cold desert steppe ecoregions, respectively). These changes correspond to the regional decomposition rates and agree with analyses indicating that due to faster turnover rates in these sites will approach equilibrium SOM levels more rapidly (Schimel *et al.* 1994).

12.3.2.4 Detecting changes in grasslands

The ability to detect the impact of climate changes on plant production and soil C levels was evaluated using the CENTURY model results. The 25-year average annual plant production and soil C levels for all of the 31 sites prior to climate change were compared with the 25-year average values following the start of climate change for both of the GCM scenarios, and a statistical *t*-test was used to determine if the mean values were significantly different at the 5% level.

Only 8% of the 62 model runs showed significantly different plant production (predominantly those with low productivity), with plant production changes averaging 22% over 25 years (range 17–31%). The nonsignificant comparisons have an average change in plant production of 7% (range 0–16%); 81% of the comparisons showed significant changes in mean soil C, with soil C changes averaging 7% in 25 years (range 1–33%). Nonsignificant comparisons had soil C changes of less than 1%.

These results suggest that soil C and plant production would have to change by at least 1 and 16%, respectively, over 25 years in order to be statistically detectable. Note that it would be impossible to detect a 1% change in soil C in the field because of the large spatial variability in soil C. We repeated the analysis for years 25–50, 50–75, and 75–100 following the start of climate change, but the number of comparisons showing significant differences in mean plant production and soil C did not change with time.

Changes in plant production or soil C induced by climate change and increasing atmospheric CO_2 will, therefore, have to be substantial, overcoming interannual variability and sampling variance, respectively, in order to be detectable. Such changes are likely to occur only at selected grassland locations.

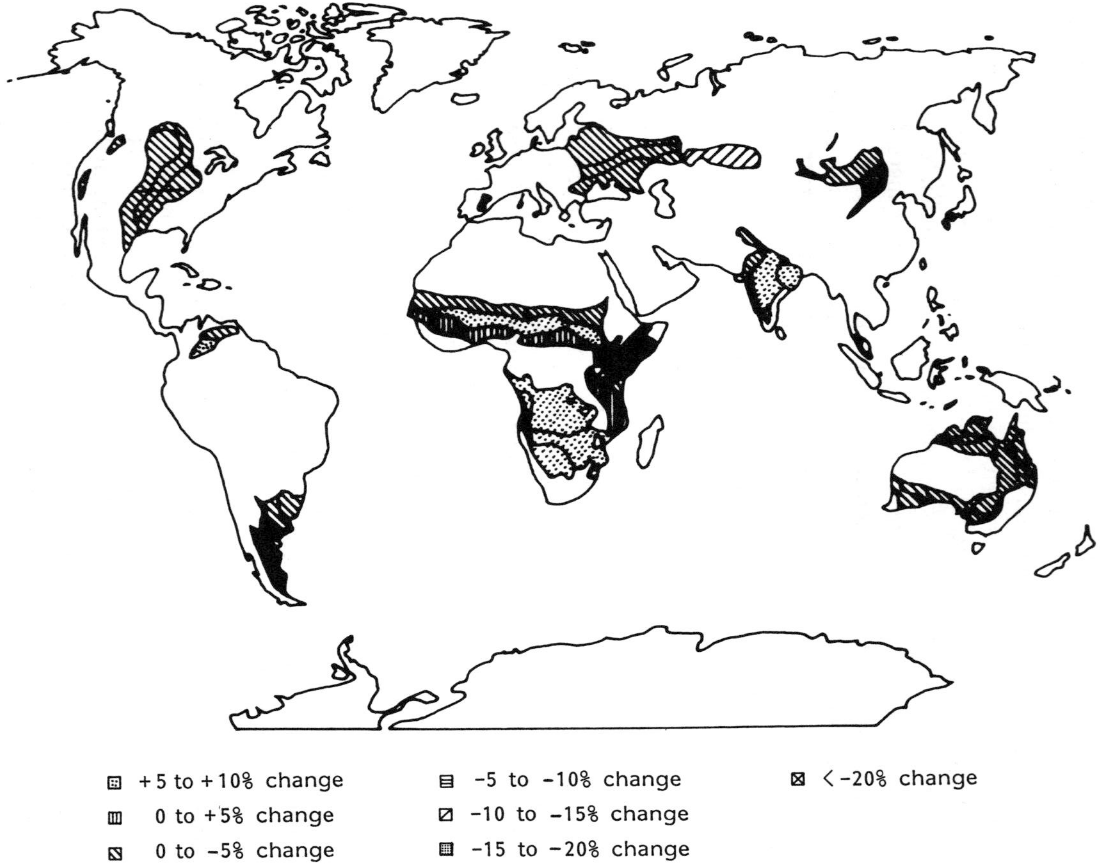

Figure 12.15 Worldwide distribution of changes in soil C for GFHI climate change scenario and CO_2 effect combined. Map showing areal distribution of soil C changes (+5 to +10%, 0 to +5%, 0 to −5%, −5 to −10%, −10 to −15%, −15 to −20%, <−20%)

12.4 CONCLUSIONS

GRASS and the CENTURY modeling experiments indicated that GCM predictions of precipitation have considerable influence on predicted ecosystem responses. The results indicate that climate change alone accounted for an increase in total above- and belowground production for the mesic regions (humid temperate and Mediterranean), mainly attributable to increased N mineralization, and decreased plant production in the cold desert steppe regions. Soil organic matter decreased in all the mesic and colder regions, due to increased decomposition. In line with most GCM predictions, the tropical savanna regions were affected the least. However, there is large uncertainty in GCM precipitation predictions. Since many of the predicted ecosystem responses to changed climate are due to altered precipitation, this uncertainty undermines confidence in ecological model predictions.

Elevated CO_2 may cause relatively small increases in NPP at some sites and large increases at others. Soil organic matter had the greatest proportional increase in tropical savanna regions. Plant responses to CO_2 are modified in complex ways by moisture and nutrient availabilities. In general, the results suggested that CO_2-enrichment has a greater effect with higher soil moisture stress. However, nutrient limitations reduced CO_2 responses. Similarly, response predictions that are based solely on C_3 vs C_4 photosynthetic characteristics are likely to be oversimplifications. Responses of C_4 and C_3 are each modified by moisture stress and temperature.

Elevated CO_2 counteracts negative effects of increased temperatures and decreased precipitation. The reversed negative impact of elevated temperatures on C_3 species is particularly notable and is probably caused by a shift in photosynthetic temperature optima (Long 1991a, b). The net effect of climate change and CO_2 was a significant increase in NPP in mesic regions (attributable to N mineralization) as well as in dry savannas, with little or no net change in cold desert steppe or humid tropical regions. Overall, SOM showed a decrease, especially in temperate steppes and cold desert steppes due to stimulation of decomposition by both climate change and CO_2, but tropical savanna and humid savanna regions were actually soil C sinks, regardless of GCM scenario. The increased importance of CO_2 in dry years is important, in that negative effects of periodic droughts could be ameliorated, thus reducing interannual variability in NPP.

Ecosystem modeling clearly showed that indirect effects and interactions among the many processes involved are important. Responses to single variables were modified by responses to other variables. Positive effects of predicted increases in rainfall are diminished by increased temperature, for example. At the Kenyan site, the GRASS model predicted a 6% decrease in NPP due to the GFHI climate change alone, despite a $+72.0$ mm yr^{-1} increase in rainfall. Interactions among plant and soil processes elicited system-level responses. Thus, while $2 \times CO_2$ increased N_{min} in Kenya it decreased N_{min} in Colorado due to differences in the indirect effects of plants on soil moisture and decomposition rate at the two sites.

Climate change alone predicts a C loss of 3–4 Pg after 50 years of climate change. However, the CO_2 enhancement effect, amounting to 2 Pg over the same time, results in a smaller net loss of 1–2 Pg over 50 years. These numbers are substantially lower than previous estimates of Jenkinson's.

12.5 FURTHER INTER-SITE COLLABORATION

Members and associates of the SCOPE grassland modelling group (SCOPE-GRAM) have contributed data from established grassland study sites in a wide variety of locations, covering most of the major grassland regions worldwide. The SCOPEGRAM group continues to actively seek collaboration with grassland specialists worldwide who have sufficient data to parameterize the CENTURY model. Peak aboveground biomass is required for several years at the same site, or more complete data on monthly biomass and dead matter (above- and belowground). At least 10 years of temperature and precipitation data are needed, together with determinations of plant and soil C and N. Opportunities for exchange of staff and data between collaborating research groups have been encouraged, in keeping with the aims of the International Geosphere-Biosphere Programme (IGBP) strategy of establishing regional research networks. The SCOPEGRAM group is keen to promote cross-model comparisons and the integration of compatible modules from different models, with a view to developing a generic terrestrial community model framework.

12.6 ACKNOWLEDGEMENTS

Data analyses conducted by Mr Brian Newkirk and Ms Song Bo (visiting scholar sponsored by the China Committee of Scholarly Exchange). Graphics prepared by Becky Techau and Michele Nelson. Manuscript prepared by Kay McElwain.

Measurement of monthly biomass dynamics, NPP and SOM at the tropical grassland sites in Kenya, Thailand and Mexico was carried out UNEP Project FP/6108-88-01(2855) 'Environment Changes and Productivity of Tropical Grasslands' (1989–92), and more recently at the Kenya and Mexico sites under the UK Overseas Development Administration (ODA) Project R4744 'Productivity of Tropical Grasslands' (1991–94). Analysis of data from the sites in the former USSR was carried out under the Russian National Scientific and Technical Programme No. 18 'Changes of Natural Environment and Climate'. Data synthesis and model validation were made possible by the SCOPE (Scientific Committee on Problems of the Environment) Project 'Effects of Climate Change on Production and Decomposition in Coniferous Forests and Grasslands' (1989–92); and the National Institute for Global Environmental Change (NIGEC) Midwestern Region Center, US-DOE project on 'Biological Hysteresis in Climate

Change Models for Grasslands: Implications of plant community dynamics on biogeochemical feedbacks'. Model development was primarily funded by the US National Science Foundation (NSF) project BSR 9013888 'Coupling Ecosystem Processes and Vegetation Patterns across Environmental Gradients' and the US NASA Earth Observing System project NACW-2662 'Using Multi-Sensor Data to Model Factors Limiting Carbon Balance in Global Grasslands', and US NSF project BSR 9011659 'Long Term Ecological Research Program: Shortgrass Steppe' and the Ecological Research Division, Office of Health and Environmental Research, US Department of Energy (DOE) project, 'Grassland–Atmosphere Response to Changing Climate: Coupling Local and Regional Scales'.

12.7 REFERENCES

Anderson, J. M. (1991) The effects of climate change on decomposition processes in grassland and coniferous forests. *Ecol. Appl.* **1**, 326–347.

Bailey, R. G. (1989) Explanatory supplement to ecoregions map of the continents. *Environ. Conserv.* **16**, 307–309.

Ball, J. T., Woodrow, I. E. and Berry, J. A. (1987) A model predicting stomatal conductance and its contribution to the control of photosynthesis under different environmental conditions. In: Biggins, I. (Ed.) *Progress in Photosynthesis Research*, pp. 221–224. Vol. IV. Martinus Nijhoff, Dordrecht.

Bazazz, F. A., Garbutt, K., Reekie, E. G. and Williams, W. E. (1989) Using growth analysis to interpret competition between a C_3 and a C_4 annual under ambient and elevated CO_2. *Oecologia* **79**, 223–235.

Breymeyer, A. and Melillo, J. M. (1991) The effects of climate change on production and decomposition in coniferous forests and grasslands. *Ecol. Appl.* **1**, 111.

Carter, D. R. and Peterson, K. M. (1983) Effects of a CO_2-enriched atmosphere on the growth and competitive interaction of a C_3 and C_4 grass. *Oecologia* **58**, 188–193.

Chen, D., Coughenour, M. B., Knapp, A. K. and Owensby, C. E. (1993) Mathematical simulation of C_4 grass photosynthesis in ambient and elevated CO_2. *Ecol. Model.* **73**, 63–80.

Chen, Z. (1988) Topography and climate of the Xilin river basin. In: Inner Mongolia Research Station (Ed.) *Research on Grassland Ecosystem*, Vol. 3, pp. 13–22. Science Press, Beijing (in Chinese with English abstract).

Chuluun, T., Ojima, D. S., Luvsandorjiin, J., Dodd, J., and Williams, S. (1996) Simulation studies of grazing in the Mongolian steppe. In: West, N. E. (Ed.) *Proceedings of the Fifth International Rangeland Congress*, pp. 561–562 (Volume 1). Society of Range Management, Denver, CO, USA.

Collatz, G. J., Ribas-Carbo, M. and Berry, J. A. (1992) Coupled photosynthesis–stomatal conductance model for leaves of C_4 plants. *Austr. J. Plant Physiol.* **19**, 519–538.

Coughenour, M. B. (1984) A mechanistic simulation analysis of water use, leaf angles, and grazing in east African graminoids. *Ecol. Model.* **26**, 203–230.

Coughenour, M. B., McNaughton, S. J. and Wallace, L. L. (1984) Modelling primary production of perennial graminoids: uniting physiological processes and morphometric traits. *Ecol. Model.* **23**, 101–134.

Enting, I. G. and Mansbridge, J. V. (1991) Latitudinal distribution of sources and sinks of CO_2: results of an inversion study. *Tellus* **43B**, 156–170.

Eswaran, H., van den Berg, E. and Reich, P. (1993) Organic carbon in soils of the world. *Soil Sci. Soc. Am. J.* **57**, 192–194.

Farquhar, G. D., von Caemmerer, S. and Berry, J. A. (1980) A biochemical model of photosynthetic CO_2 assimilation in leaves of C_3 species. *Planta* **149**, 78–90.

Fernandez, R. J., Sala, O. E. and Golluscio, R. A. (1991) Woody and herbaceous above-ground production of a Patagonian steppe. *J. Range Manage.* **44**, 434–437.

Gilmanou, T. G., Parton, W. J. and Ojima, D. S. (in press). Testing the CENTURY ecosystem level model on data sets from eight grassland sites in the former USSR representing a wide climatic/soil gradient. *Ecological Modeling* (in press).

Haas, H. J., Evans, C. E. and Miles, E. F. (1957) Nitrogen and carbon changes in Great Plains soils as influenced by cropping and soil treatments. *Tech. Bull. 1164*. United States Department of Agriculture. US Printing Office, Washington, DC, USA.

Hall, D. O. and Scurlock, J. M. O. (1991) Climate change and productivity of natural grasslands. *Ann. Bot.* **67**(suppl.), 49–55.

Houghton, J. T., Jenkins, G. J. and Ephraums, J. J. (Eds) (1990) *Climate Change: The IPCC Scientific Assessment*. Cambridge University Press, Cambridge, UK.

Jenkinson, D. S., Adams, D. E. and Wild, A. (1991) Model estimates of CO_2 emissions from soil in response to global warming. *Nature* **351**, 304–306.

Lauenroth, W. K. and Dodd, J. L. (1978) The effects of water- and nitrogen-induced stresses on plant community structure in a semiarid grassland. *Oecologia* **36**, 211–222.

Long, S. P. (1991a) Modification of the response of photosynthetic productivity to rising temperature by atmospheric CO_2 concentrations: has its importance been underestimated? *Plant Cell Environ.* **14**, 729–739.

Long, S. P. (1991b) Primary production in grasslands and coniferous forests with climate change: an overview. *Ecol. Appl.* **1**, 139–156.

Long, S. P., Garcia Moya, E., Imbamba, S. K., Kamnalrut, A., Piedade, M. T. F., Scurlock, J. M. O., Shen, Y. K. and Hall, D.O. (1989) Primary productivity of natural grass ecosystems of the tropics: a reappraisal. *Plant Soil* **115**, 155–166.

Long, S. P., Jones, M. B. and Roberts, M. J. (1992) *Primary Productivity of Grass Ecosystems of the Tropics and Sub-tropics*. Chapman & Hall, London. 267 pp.

Melillo, J. M., Aber, J. D. and Muratore, J.F. (1982) Nitrogen and lignin control of hardwood leaf litter decomposition dynamics. *Ecology* **63**, 621–626.

Melillo, J. M., McGuire, A. D., Kicklighter, D. W., Moore, B., Vrsmarty, C.J. and Schloss, A.L. (1993) Global climate change and terrestrial net primary production. *Nature* **363**, 234–240.

Menaut, J.-C. and César, J. (1979) Structure and primary productivity of Lamto savannas, Ivory Coast. *Ecology* **60**, 1197–1210.

Ojima, D. S., Parton, W. J., Schimel, D. S. and Owensby, C. E. (1990) Simulated impacts of annual burning on prairie ecosystems. In: Collins, S. L. and Wallace, L. (Eds) *Fire in the North American Prairies*, pp. 118–132. Univ. of Oklahoma Press, Norman.

Ojima, D. S., Schimel, D. S., Parton, W. J. and Owensby, C. E. (1994) Long- and short-term effects of fire on nitrogen cycling in tallgrass prairie. *Biogeochemistry* **24**, 67–84.

Owensby, C. E., Coyne, P. I. and Auen, L. M. (1993a) Nitrogen and phosphorus dynamics of a tallgrass prairie ecosystem exposed to elevated carbon dioxide. *Plant Cell Environ.* **16**, 843–850.

Owensby, C. E., Coyne, P. I., Ham, J. M., Auen, L. M. and Knapp, A. (1993b) Biomass production in a tallgrass prairie ecosystem exposed to ambient and elevated levels of CO_2. *Ecol. Appl.* **3**, 644–653.

Parton, W. J., McKeown, B., Kirchner, V. and Ojima, D. S. (1992) *CENTURY Users' Manual*. NREL Publication. CSU, Fort Collins, Colo.

Parton, W. J., Schimel, D. S., Cole, C. V. and Ojima, D. S. (1987) Analysis of factors controlling soil organic matter levels in Great Plains grasslands. *Soil Sci. Soc. Am. J.* **51**, 1173–1179.

Parton, W. J., Scurlock, J. M. O., Ojima, D. S., Gilmanov, T. G., Scholes, R. J., Schimel, D. S., Kirchner, T., Menaut, J. -C., Seastedt, T., Garcia Moya, E., Kamnalrut, A. and Kinyamario, J. I. (1993) Observations and modelling of biomass and soil organic matter dynamics for the grassland biome worldwide. *Glob. Biogeochem. Cyc.* **7**(4), 785–809.

Parton, W. J., Stewart, J. W. B. and Cole, C. V. (1988) Dynamics of C, N, P and S in grassland soils: a model. *Biogeochemistry* **5**, 109–131.

Rains, J. R., Owensby, C. E. and Kemp, K. (1975) Effects of nitrogen fertilization, burning, and grazing on reserve constituents of big bluestem. *J. Range Manage.* **28**, 358–362.

Rice, C. W., Garcia, F. O., Hampton, C. O. and Owensby, C. E. (1992) Soil microbial biomass and respiration under increased levels of atmospheric CO_2. *Agron. Abstr.* **84**: p. 258.

San José, J. J. and Medina, E. (1976) Organic matter production in the *Trachypogon* savanna at Calbozo, Venezuela. *Trop. Ecol.* **17**, 13–124.

Schimel, D. S., Braswell, B. H., Holland, E. A., McKeown, R., Ojima, D. S., Painter, T. H., Parton, W. J. and Townsend, A.R. (1994) Climatic, edaphic, and biotic controls over storage and turnover of carbon in soils. *Glob. Biogeochem. Cyc.* **8**, 279–293.

Schimel, D. S., Kittel, T. G. F. and Parton, W. J. (1991) Terrestrial biogeochemical cycles: global interactions with the atmosphere and hydrology. *Tellus* **43**, 188–203.

Schimel, D. S., Parton, W. J., Kittel, T. G. F., Ojima, D. S. and Cole, C. V. (1990) Grassland biogeochemistry: links to atmospheric processes. *Clim. Change* **17**, 13–25.

Schlesinger, W. H. (1990) Evidence from chronosequence studies for a low carbon-storage potential of soils. *Nature* **348**, 232–234.

Singh, J. S. and Yadava, P. S. (1974) Seasonal variation in composition, plant biomass and net primary productivity of a tropical grassland at Kurukshetra, India. *Ecol. Monogr.* **44**, 351–376.

Tans, P. P., Fung, I. Y. and Takahashi, T. (1990) Observational constraints on the global atmospheric CO_2 budget. *Science* **247**, 1431–1438.

Towne, G. and Owensby, C. E. (1984) Long-term effects of annual burning at different dates in ungrazed Kansas tallgrass prairie. *J. Range Manage.* **37**, 392–397.

Wang, Y. and Jiang, S. (1982) Effects of arid climate on community structure and aerial biomass of *Stipa grandis* steppe. *Acta Phytoecol. Geobot. Sin.* **13**, 297–308.

13 Comparing Models of Ecosystem Function for Temperate Conifer Forests. I. Model Description and Validation

M. G. RYAN,[1] E. R. HUNT, Jr,[2] R. E. McMURTRIE,[3] G. I. ÅGREN,[4] J. D. ABER,[5] A. D. FRIEND,[6] E. B. RASTETTER,[7] W. M. PULLIAM,[8] R. J. RAISON[9] and S. LINDER[10]

[1]*USDA Forest Service, Rocky Mountain Experiment Station, Fort Collins, USA*
[2]*School of Forestry, University of Montana, Missoula, USA*
[3]*Department of Biological Science, The University of New South Wales, Kensington, Australia*
[4]*Department of Ecology and Environmental Research, Swedish University of Agricultural Sciences, Uppsala, Sweden*
[5]*Complex Systems Research Center, Institute for the Study of Earth, Oceans and Space, University of New Hampshire, Durham, USA*
[6]*Institute of Terrestrial Ecology, Bush Estate, Penicuik, UK*
[7]*The Ecosystems Center, Marine Biological Laboratory, Woods Hole, USA*
[8]*Natural Resource Ecology Laboratory, Colorado State University, Fort Collins, USA*
[9]*CSIRO, Division of Forestry, Canberra, Australia*
[10]*Department of Ecology and Environmental Research, Swedish University of Agricultural Sciences, Uppsala, Sweden*

13.1 INTRODUCTION

Predicting how a given forest will respond to an altered climate, increased CO_2 concentration, pollution stress, or increased nutrient input is a complex problem. If increased temperatures raise decomposition more than productivity, forests may become a source for atmospheric carbon (Woodwell 1987). However, given adequate nutrient availability, forests could store more carbon (C) and slow the rate of increase in atmospheric CO_2 (cf. Rastetter *et al.* 1992). Understanding how climate and vegetation influence the biogeochemistry of ecosystems will allow us to make predictions about how global change will affect forests. This understand-

Global Change: Effects on Coniferous Forests and Grasslands.
Edited by A. I. Breymeyer, D. O. Hall, J. M. Melillo and G. I. Ågren

ICSU UNEP

ing will also allow us to test strategies for managing these ecosystems to mitigate the effects of global change.

Models are an important tool for both understanding ecosystem function and predicting responses to global change. Models can summarize the results of many experiments by incorporating hypotheses and conclusions into a quantitative framework. System-level fluxes are uncertain and very difficult to measure, and models can provide independent estimates. Conducting experiments that alter climate or atmospheric CO_2 of forest ecosystems poses special difficulties, because trees are large and growth and decomposition occur over decades to centuries. Therefore, long-term response to a changed environment may differ from the results of experimentally feasible short-term manipulations. Models can be used to simulate these long-term experiments.

There are a variety of models that simulate element cycling in forest ecosystems (see section 13.2), and consensus about modelling approaches is lacking. Models differ in time scale, linkages between C, nitrogen (N), and water, representation of heterogeneity, and detail of photosynthesis, allocation, and decomposition submodels. These simulation models can be thought of as different syntheses of our understanding of what controls element cycling within ecosystems. By comparing model predictions for the same ecosystems, we hope to make these differences more explicit, and assess the impacts of different processes and the control they exert over the productivity of temperate forests.

In this chapter, we describe the results of a comparison of seven models of ecosystem element cycling. First, we review the different approaches to modelling to provide a context for the models in the comparison. Next, we describe the individual models and compare how the models represent various processes. Then, we compare the ability of the models to predict growth under natural conditions and the response to fertilization and irrigation for young conifer forests in two contrasting climates. Finally, we discuss how the model comparisons help assess the state of our knowledge, and suggest further experimentation to reduce uncertainty in our understanding of ecosystem response to climate change.

13.2 APPROACHES TO MODELLING

An extensive review of the state of the art in modelling ecosystem response to climate changes is given by Ågren *et al.* (1991). The models considered here are separated into two classes by level of resolution. Within each class we discuss how models deal with the various processes likely to be affected by climate changes, and how well those processes are understood.

In addressing the consequences of climate change at the regional or global level, we are faced with the task of scaling up from knowledge of an organ's response to its micro-environment. We recognize four partially overlapping levels where models have been used. These start at the plant physiological level and progress through populations and ecosystems to regional and global scales. At the physiological level, plant processes are described in great depth, down to the biochemistry. The ecosystem scale integrates the physiology to whole-plant or major plant component properties and adds a feedback interaction to the environment, in particular the soil. Population models generally operate at the same scales as ecosystem models but deal with plants in terms of population dynamics. At the coarsest level of resolution, regional and global models are used to explain the distribution of different biomes. Here, we focus only on the physiologically based and ecosystem models, but provide a brief description of population and regional models for context.

13.2.1 Physiologically based models

In recent years significant advances have been made in our capacity to explain and predict plant responses to the environment (e.g. Mooney *et al.* 1987). This progress enables us to predict the response of individual plant processes to climate change. The objectives of physiologically based models of canopy processes is to integrate up from the micro-environmental scale to explain and predict the aggregate element cycling of a forest stand. A model is required to interrelate the various processes and their response to environmental variation and to integrate over space and time. In the model comparison, this class of models is represented by BIOMASS (McMurtrie *et al.* 1990a, 1992), BIOME-BGC (Running and Coughlan 1988; Hunt and Running 1992; Running and Hunt 1993), and HYBRID (Friend *et al.* 1993).

13.2.1.1 Direct effects of CO_2

Models of plant production that include CO_2 enrichment must describe how CO_2 concentration will increase photosynthetic rates at both high and low light levels. Some models do not explicitly include ambient CO_2 concentration but can simulate CO_2 effects, generally by changing parameters corresponding to the photosynthetic rate at saturating irradiance and quantum yield. Others directly incorporate Farquhar and von Caemmerer's (1982) equations for the photosynthesis of C_3 plants, though these models can be computationally demanding if they simulate quantum flux density throughout the canopy (e.g. Wang *et al.* 1992). Nitrogen use efficiency may also affect the response to elevated CO_2 (Jarvis 1989).

Many models calculate transpiration from the Penman–Monteith equation applied to a canopy idealized as a single, so-called big leaf (e.g. Jarvis 1985). The Penman–Monteith equation has a sound theoretical base (but see Finnegan and Raupach 1987) and is general to all vegetation types. However, equations for the dependence of stomatal conductance on environmental variables, including radiation, water status and humidity, are required, and canopy conductance is derived by summation over an idealized canopy. Reduced canopy conductance due to CO_2 enrichment will likely reduce water use for forests with aerodynamically rough canopies and prolong the growing season for forests growing in water-limited environments.

13.2.1.2 Temperature

The biochemical basis of the temperature dependence of photosynthesis has been studied (Kirschbaum and Farquhar 1984) and included in some models (e.g. McMurtrie and Wang 1993). However, temperature is usually included in models of canopy photosynthesis through empirically derived relationships typically representing the variation of photosynthetic parameters about some temperature optimum.

Most models discriminate between growth and maintenance respiration with growth respiration assumed to be in constant proportion to tissue production, and maintenance respiration of various biomass components related to temperature by either the Arrhenius function or an exponential relationship. Approaches adopted to compute stand level maintenance respiration include scaling according to surface area (Woodwell and Botkin 1970), sapwood volume (Ryan and Waring 1992), C mass of tissue (Running and Coughlan 1988) or nutrient content (Ryan 1991b).

Vegetative growth of forests may be temperature-limited in temperate environments (e.g. Cannell *et al.* 1988). Temperature effects on phenology and allocation have often been incorporated in grassland models (usually with degree-day sums) but less frequently in forest models. Temperature dependence also enters models through the effect of water vapour saturation deficit on stomatal conductance which in turn affects both transpiration and evaporation from wet canopies. Increases in the vapour saturation deficit will be diminished if higher temperatures occur in conjunction with increased absolute humidity of air.

13.2.1.3 Water

Several models reliably predict plant available water (e.g. Running *et al.* 1983; McMurtrie *et al.* 1990a; McMurtrie and Landsberg 1992). These models describe rainfall, interception by canopies, soil evaporation, runoff, drainage, tree transpiration and sometimes understorey evapotranspiration and accumulation and melt of snow packs. They require inputs of daily weather data and soil physical characteristics such as soil water retention functions and rooting depth. They should successfully predict the consequences of changes in the frequency and intensity of rainfall on soil moisture.

Various approaches have been adopted to characterize the effects of water stress on photosynthesis, stomatal conductance and tissue production. Recent evidence indicates that it may be possible to relate stomatal conductance directly to plant available water in the rooting zone (Schulze *et al.* 1987), thereby simplifying plant and soil water interactions.

Predicting the consequences of climate change (including effects on nutrient availability) for leaf area production and retention is particularly important because of feedbacks on both water balance and C accumulation. Improved water use efficiency, together with enhanced quantum yield, which benefit C balance of foliage at the base of the canopies, may promote leaf area development. However, in areas with poor summer rainfall, stands with higher leaf area may exhaust soil water reserves earlier in the growing season.

13.2.1.4 Radiation

Both photosynthetic rates and transpiration depend on radiation, which is expected to change in response to altered cloudiness. The relative proportions of the direct and diffuse components of incident radiation will change, indicating the need for caution where canopy assimilation is predicted by models which do not consider the two components separately. A realistic response of simulated transpiration to changes in net radiation is expected from models incorporating the Penman–Monteith equation, with transpiration particularly sensitive to radiation flux for canopies weakly coupled to the atmosphere, such as forest understoreys and grasslands (Jarvis 1989).

13.2.2 Ecosystem/tissue models

Ecosystem/tissue models include those models that consider water, C, and N flows between the plant and soil systems. Within this group of models, the plant compartments are highly aggregated and typically lump all green plant biomass into one compartment. The objectives of the ecosystem/tissue models are quite diverse; however, they generally include the ability to simulate ecosystem responses to changes in the abiotic driving variables (light intensity, soil water, and soil temperature) and the interaction of dry matter production and nutrient cycling. The four models CENTURY (Parton *et al.* 1987; Sanford *et al.* 1991), MBL-GEM (Rastetter *et al.* 1991), PnET-CN (Aber and Federer 1992) and Q (Rolff and Ågren, in preparation) will illustrate this model class.

13.2.2.1 Plant processes

Some ecosystem/tissue models describe the plant processes similarly to the physiologically based models (i.e. growth is calculated from the difference between photosynthesis and respiration and allocated between different plant compartments), but with a lower resolution. In other models, phenomenological concepts such as nutrient productivity (Ågren 1983, 1985, 1988) are used to directly estimate net production. In both cases, nutrient availability is stressed as the important growth regulator. Plants influence nutrient availability through litter quality: poor quality litter decays slowly and immobilizes more nutrients than high quality litter. In these models, soil and plant C and N are explicitly represented and linked by C:N ratios of various plant and soil components.

13.2.2.2 Soil processes

Soil processes respond rapidly to changes in the abiotic driving variables and litter quality. The impact of litter quality on soil processes has been studied for many years (Swift *et al.* 1979) and models have been developed to simulate the impact of litter quality on nutrient cycling and soil organic matter dynamics (Hunt, 1977; McGill *et al.* 1981; Aber *et al.* 1982; Melillo *et al.* 1982; Bosatta and Ågren 1985, 1991a, b; Parton *et al.* 1987; Ågren and Bosatta 1987, 1988; Pastor and Post 1988). The N and lignin content of vegetation are the most commonly used variables for characterizing litter quality in decomposition models. Climate change may alter litter quality through two mechanisms:

(a) the effects of increased atmospheric CO_2 levels on litter quality (Melillo 1983; Peet *et al.* 1985; Strain and Bazzaz 1983);
(b) with changes in plant species composition (litter N and lignin content vary with species).

Models differ in the level of detail used to capture the dynamics of soil organic matter; models with several organic-matter pools may behave differently than single-pool models.

Soil moisture and temperature are the most important abiotic driving variables for the soil system and both would be altered substantially with the proposed climatic scenarios. The effect of soil moisture on decomposition has been represented in different ways in the various models. In more process-oriented models (Hunt 1977; McGill *et al.* 1981) the soil water effect is represented as a function of the soil water potential using non-linear functions. The more empirical models (Parton *et al.* 1987; Pastor and Post 1988) use variables such as relative water content of the soil and the ratio of rainfall to potential evapotranspiration rate to control decomposition.

13.2.3 Other model types

Two other classes of models (population models and regional models) were not included in the model comparison because these model types are more suited to different temporal and spatial scales than those used in this study.

Population models have been developed to simulate tree growth as affected by competition among individuals as well as climate. These models consider birth or recruitment, growth of individuals, stand structure and spacing, and mortality (e.g. Botkin *et al.* 1972; Shugart and West 1977; Shugart 1984). More recent versions of these models also simulate the effects of nutrient limitation on tree growth (Weinstein *et al.* 1982; Aber and Melillo 1982; Aber *et al.* 1982; Pastor and Post 1985, 1986). The HYBRID model (Friend *et al.* 1993) is a physiologically based population model, but the model comparison presented in this chapter concerns stands with a single species.

The regional and global coincidence of vegetation zones and climate patterns has long been noted (Trewartha 1968) and supports the presumption that climate changes will alter the global distribution of vegetation (Emanuel *et al.* 1985a, b). This coincidence suggests that a global vegetation model can be derived and parameterized by overlaying vegetation maps with climate maps. Examples of this type of vegetation map are those that are based on the Holdridge (1947, 1964) life zone classification system. An alternative approach is to use empirically derived regressions between climate and variables such as net primary production (Rosenzweig 1968) or litterfall (Meentemeyer *et al.* 1982). These or similar relationships are assembled in several global models that produce maps of global distribution of net primary production, litterfall, and C storage from global climate data sets (Lieth 1972; Box 1978, 1981; Meentemeyer *et al.* 1982). These regional and global classification and regression approaches may predict the future steady state distributions of vegetation and C pools but lack the ability to model the transient responses shown to be important in this chapter.

To avoid the problems associated with static approaches, simple ecosystem/tissue-type models have been used at regional scales (Raich *et al.* 1991; Burke *et al.* 1991; McGuire *et al.* 1992, 1993; Melillo *et al.* 1993). These models assume uniform conditions over large areas (e.g. 0.5×0.5 grid cells) and simplified physiology but do link plant and soil processes and climate. Another recent approach uses plant physiology, soil, and climate to alter biome distribution (Prentice *et al.* 1992).

13.2.4 Choice of models

A variety of models, each with its specific domain of applicability, will be required to answer the diverse questions about effects of climate change. We want to emphasize the need to select a model that is appropriate with respect to the scale of the perturbation in question, particularly in time but also in space. Viewing the same system from different space and time perspectives makes different phenomena important or unimportant. Models based on plant physiology are ideal to analyse and interpret the detailed short-term reactions of plants to the various components of climate change. On the other hand, for the ecosystem models that emphasize longer time scales, insights from the physiologically based models are required to ensure that no major impact of climate change is neglected and that formulations of growth processes are realistic.

Physiologically based models provide a framework integrating the many direct and indirect effects of elevated CO_2. Effects promoting growth include increased temperature (lengthening growing seasons and increasing photosynthesis in high light environments) and the beneficial direct effects of CO_2 enrichment on photosynthesis and water use efficiency. Effects reducing growth arising from greenhouse warming are increased water vapour saturation deficit and increased respiratory losses. Changes in rainfall patterns will be beneficial in some areas but detrimental in others. Depending on particular circumstances, increased leaf area development may increase or decrease productivity. Where the objective is to assess the net effect of such opposing factors, physiologically based models are appropriate tools (e.g. Booth and McMurtrie 1988).

Ecosystem/tissue models provide a framework for understanding the system-level constraints for the response of productivity to elevated CO_2, and altered temperature and moisture. For example, the response to elevated CO_2 is strongly controlled by nutrient availability (e.g. Oechel *et al.* 1994), but nutrient availability results from complex interactions between climate, litter quality, productivity, and soil. Therefore, information about nutrition is necessary to assess longer-term interactions between the atmosphere and climate, and ecosystem/tissue models are appropriate tools for this task.

13.3 MODEL DESCRIPTIONS

13.3.1 Comparison by process

The models used in this comparison are (Table 13.1, in alphabetical order): BIOMASS (McMurtrie *et al.* 1990b, 1992; McMurtrie and Landsberg 1992); BIOME-BGC (Running and Coughlan 1988; Running and Gower 1991; Hunt and Running 1992); CENTURY (Parton *et al.* 1987, 1988; Sanford *et al.* 1991); HYBRID (Friend *et al.* 1993); MBL-GEM (Rastetter *et al.* 1991); PnET-CN (Aber and Federer 1992), and Q (Rolff and Ågren, in preparation). These models were developed with different objectives; hence, different processes are emphasized. However, because most ecosystem processes are controlled in part by interactions with other processes, each model deals with these interactions, either by simulating the process directly or by using that process as a model input. For example, BIOMASS and HYBRID do not simulate decomposition and N mineralization; both use leaf N concentration as an input circumventing the need to simulate these processes. Also, the model Q uses net primary production as an input with transpiration, photosynthesis and respiration not represented. A brief description of each model, the original modelling objectives and further references are given in sections 13.3.2–13.3.8.

For some processes, the various models have converged on a given algorithm (Table 13.1). For models with a daily time step, transpiration is determined using the Penman–Monteith equation and photosynthesis is simulated using variations of a model originally developed by Farquhar *et al.* (1980). Most of the

Table 13.1 Comparison of model processes

Variable	Models						
	BIOMASS	BIOME-BGC	CENTURY	MBL-GEM	HYBRID	PnET-CN	Q
Time step[a]	D	D/A	M	M/A	D/A	M	A
Transpiration[b]	PM	PM	PM	I	PM	WUE	—
Photosynthesis							
Model[c]	F	F/P	I	P	F	P	I
Scaling method[d]	B	B/N	C	B/N	B/N	N	B/N
Respiration							
Model[e]	G/M	G/M	M	G/M	G/M	G/M	—
Scaling method[f]	N	B	B	N	B	N	—
Allocation[g]	A/V	V	C	V	V	C	V
Turnover[h]	C	C	C	C	C	C	C
Leaf N[i]	I	V	V	V	I	V	V
Litter N[j]	—	P	P	P	—	P	C
Decomposition[k]	—	AET	LCI	LCI	—	R	M
N mineralization[l]	—	P	P	C:N	—	C:N	P

[a]Time step: D = daily, M = monthly, A = annual.
[b]Transpiration: PM = Penman–Monteith, I = soil moisture as input, WUE = water use efficiency, — = not simulated.
[c] Photosynthesis (model): F = Farquhar *et al.* (1980) biochemical model, P = phenonenological model, I = net primary production only.
[d]Photosynthesis (scaling method): B = biomass, N = foliar nitrogen, C = climate.
[e]Respiration (model): G = growth respiration, M = maintenance, respiration, — = not simulated.
[f]Respiration (scaling method): B = biomass, N = tissue nitrogen, — = not simulated.
[g]Allocation of carbon: A = allometric equations, V = variable percentage, C = constant percentage.
[h]Turnover of leaves: C = constant percentage.
[i]Leaf nitrogen: I = input, V = variable.
[j]Litter nitrogen: — = not simulated, P = constant percentage of leaf nitrogen, C = constant.
[k]Decomposition: — = not simulated, AET/L = actual evapotranspiration and lignin concentration, LCI = lignocellulose index, R = residence time for given stage, M = microbial growth rate.
[l]Nitrogen mineralization: — = not simulated, P = proportional to decomposition, C:N = maintains constant carbon to nitrogen ratio.

models distinguish between growth and maintenance respiration (Amthor 1989) but differ over the scaling metric used.

The different approaches to the allocation of C and N, leaf and root turnover, decomposition, and N mineralization reflect both differences in aggregation and a greater level of uncertainty about how to represent the processes. For example, all models represent leaf turnover as a constant percentage of leaf area index, yet Reich *et al.* (1992) show that leaf longevity is affected by many ecosystem processes. Given the importance of these processes on ecosystem function, the analysis of alternative hypotheses is warranted.

13.3.2 BIOMASS

The BIOMASS model is a process-based model of forest growth incorporating submodels for radiation absorption, canopy photosynthesis, partitioning of assimilate between plant organs, litterfall and stand water balance (McMurtrie *et al.* 1990a, 1992; McMurtrie 1993; McMurtrie and Landsberg 1992; McMurtrie and Wang 1993; Figure 13.1). Canopy C gain is calculated from a radiation interception model which uses information about canopy architecture, and a mechanistic model of leaf photosynthesis by C_3 plants (Farquhar and von Caemmerer 1982). Tree crowns are represented by truncated ellipsoids, while the plant community is represented by a two-dimensional random array of trees with a correction introduced to account for spatial nonrandomness. The foliage is divided into three horizontal layers with different photosynthetic parameters.

The radiation interception submodel that is the basis for the calculation of canopy net photosynthesis separately considers the interception of direct and diffuse radiation by foliage. Photosynthesis is evaluated for both sunlit and shaded foliage. Canopy C balance is updated daily after subtracting maintenance respiration calculated using equations from Ryan (1991b). In BIOMASS allocation of C to foliage, branch, stem and root is represented by partitioning coefficients that vary over an annual cycle and are linearly related to foliar N concentration (McMurtrie and Landsberg 1992). Litterfall is subtracted from each biomass component with rates varying over an annual cycle.

Stand water use is calculated from the Penman–Monteith equation, with biological control of the process exerted through a model of stomatal conductance (McMurtrie 1993). Meteorological data required to drive BIOMASS are daily precipitation, short-wave radiation and maximum and minimum air temperatures. The model employs a daily time step, though the diurnal time course of several meteorological variables is incorporated to obtain daily total photosynthesis and transpiration. This is necessary because of the non-linear responses of these processes to meteorological variables.

The BIOMASS model does not incorporate a dynamic model of nutrient cycling and uptake by trees. Instead, N concentrations, which influence simulated photosynthesis, respiration and allocation, are required as model inputs.

13.3.3 BIOME-BGC

The BIOME-BGC (for biogeochemical cycles) is a generalized ecosystem process model (Hunt and Running 1992; Running and Hunt 1993) derived from a model developed for coniferous forests, FOREST-BGC (Running and Coughlan 1988; Running and Gower 1991). A biome is loosely defined as a combination of a lifeform type (conifer, broad-leaved, grass, etc.) in a given climate (Hunt and Running 1992), so BIOME-BGC is strongly controlled by climate, as is also true for FOREST-BGC. The primary objective of BIOME-BGC is to simulate the C,

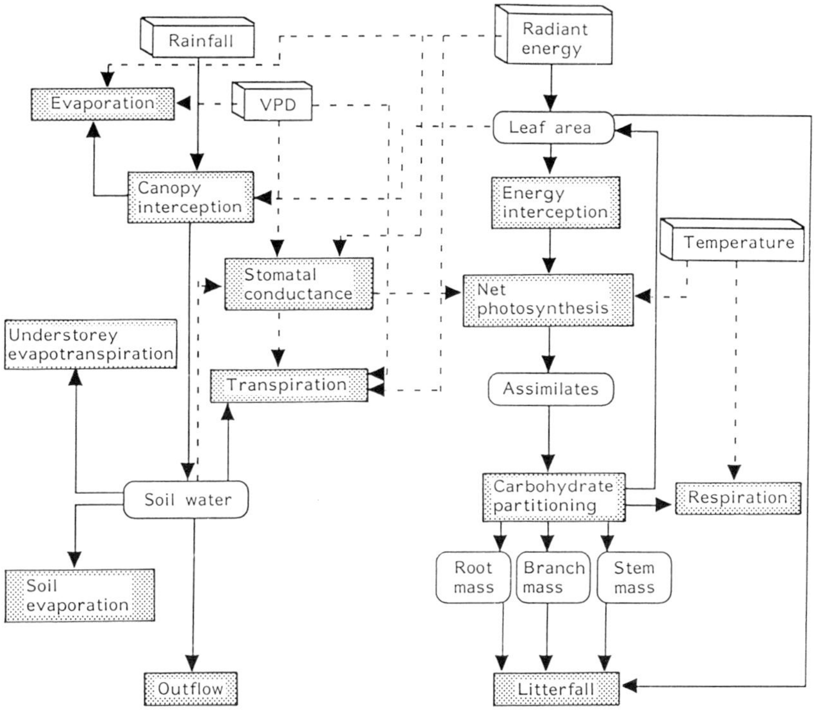

Figure 13.1 Schematic representation of the BIOMASS model. The carbon and water balance models are displayed on the right- and left-hand sides, respectively. Processes are denoted by shaded boxes, meteorological variables by unshaded boxes, state variables by rounded boxes, transfers of material and energy by solid arrows, and influences by broken arrows

N and water budgets for a large homogeneous area of vegetation. The key variable is leaf area index (LAI) which can be estimated by remote sensing. As a consequence, BIOME-BGC treats processes that have a direct relationship to LAI with considerable detail (photosynthesis and transpiration), whereas processes that have an indirect relationship to LAI (e.g. decomposition) are treated with significantly less detail. For coniferous forests, processes represented on a daily basis (photosynthesis and evapotranspiration) for FOREST-BGC, and hence BIOME-BGC, are validated with experimental data on stem growth and soil water content (Nemani and Running 1989; Korol *et al.* 1991; Hunt *et al.* 1991).

The BIOME-BGC has a dual time step, daily for the hydrologic budget, canopy gas exchange and maintenance respiration, and yearly for allocation, decomposition and the N cycle (Figure 13.2). Leaf area index controls the interception and subsequent evaporation of precipitation; precipitation not

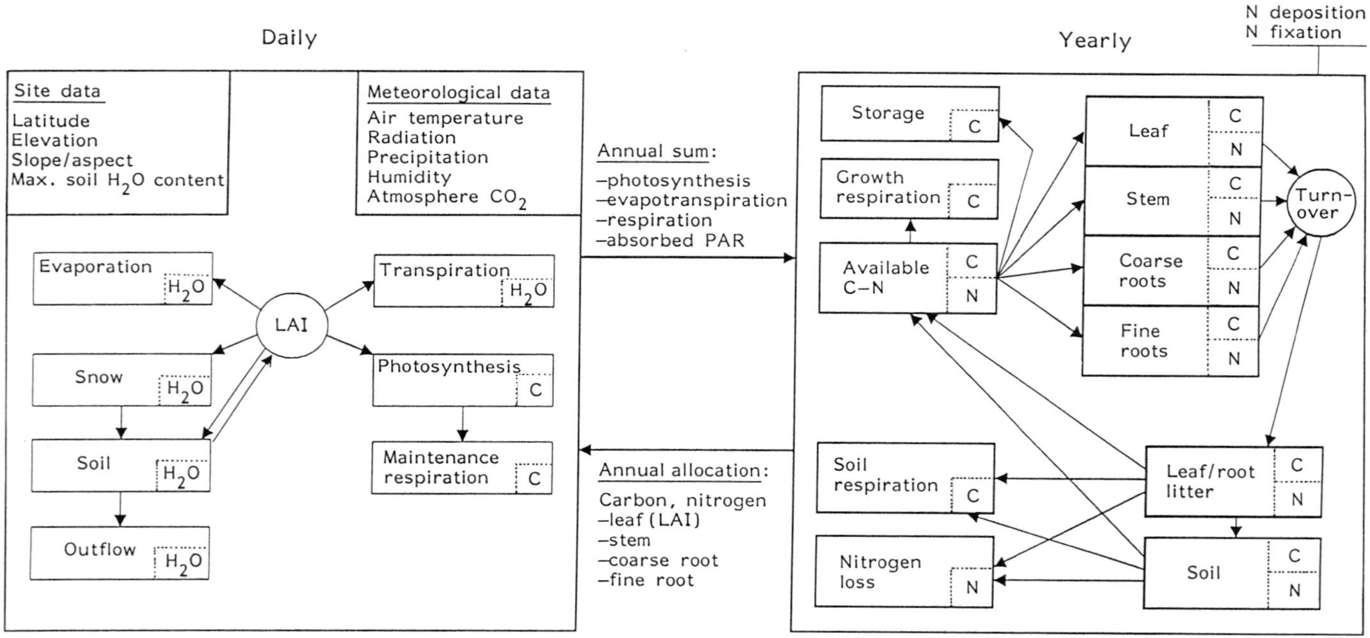

Figure 13.2 Compartment diagram for BIOME-BGC, an ecosystem process model based on data derived by remote sensing

intercepted is routed to a single soil water storage compartment. Based on soil texture, the amount of water in the soil determines predawn leaf water potential, which defines the amount of water stress. Predawn leaf water potential, humidity, temperature and radiation are used to calculate stomatal conductance. Conductance is multiplied by LAI, which is then used in the Penman–Monteith equation to calculate transpiration per ground area. From leaf N concentration and stomatal conductance, a maximum photosynthetic assimilation rate is determined using the model of Farquhar *et al.* (1980) which is coupled to phenomenological equations for light and temperature to calculate canopy photosynthesis (Hunt and Running 1992).

In the annual part of BIOME-BGC, constant fractions of leaf, stem, and root C and N turn over and are added to the litter storage components (Figure 13.2). Litter C decomposition rate is determined by actual evapotranspiration and lignin concentration. Decomposition of soil C and mineralization of soil N are set as proportions of litter C decomposition rate. The amounts of N immobilized or lost from the ecosystem are set as proportions of the N mineralized and N added to the ecosystem through fixation, deposition and fertilization. The net amount of N mineralized and N added to the ecosystem and the amount of N retranslocated from the foliage determine the amount of available N. From the available N and water stress integral (Myers 1988), a leaf C/(leaf C + fine root C) ratio is formed that controls the allocation of C to leaves and fine roots. The difference of C available for growth and C allocated to leaves and fine roots is then allocated to stems and coarse roots.

13.3.4 CENTURY

The grassland version of the CENTURY model is fully described in Chapter 11. The adaptation of this model to forest systems is discussed by Sanford *et al.* (1991). Decomposition and hydrologic components of the forest model are similar to those of the grassland model, but C and nutrient pools representing above- and belowground woody debris have been added. Aboveground woody debris is divided between fine branch and large wood debris. Decomposition rates of woody debris are specified as inputs, and wood decomposition products are partitioned among soil organic matter pools according to initial wood lignin content. Soil C and nutrient components of CENTURY have been extensively validated against field observations of soil organic matter dynamics (Parton *et al.* 1987, 1988; Chapter 11, this volume). Values of parameters that control these routines are generally not altered in calibration of the model to a particular site. Nitrogen mineralization and immobilization are controlled by soil C dynamics and C:N ratios of soil C pools.

The forest version of CENTURY contains new plant production routines that replace those of the grassland version. The forest production model (Figure 13.3) divides trees into leaves, fine branches, large wood, fine roots, and coarse roots.

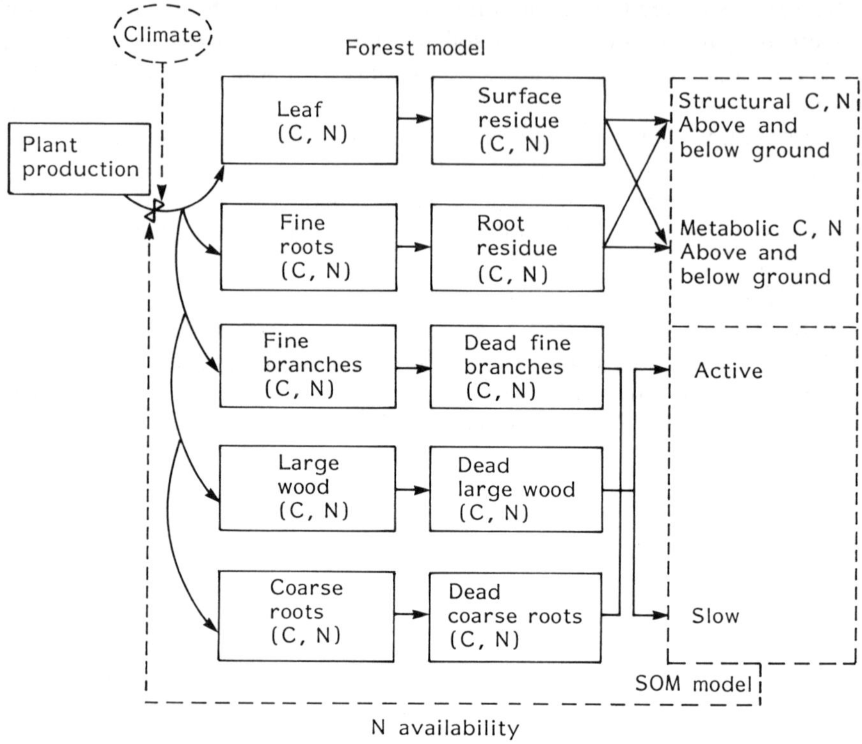

Figure 13.3 Compartment diagram for the Forest CENTURY model. The soil organic matter (SOM) component has several individual pools not shown in this figure

Carbon and nutrients are allocated among these components by a fixed scheme of C allocation fractions and tissue nutrient concentrations. Separate allocation patterns may be used for young and mature forests. Allocation patterns and nutrient concentrations are specified for the system being simulated, and are among the most important calibration parameters in this model. Monthly potential NPP is calculated from a specified maximum value by applying climate and nutrient limitations in the same manner as in the grassland model. Additionally, NPP may be reduced by maintenance respiration demands of sapwood if large wood biomass becomes sufficiently great. Live biomass is transferred to litter and woody debris pools according to component-specific death rates (in deciduous systems, 95% of leaf biomass is transferred to litter at the end of the growing season).

13.3.5 HYBRID

The HYBRID model grew from the merging of the big-leaf model FOREST-BGC (Running and Coughlan 1988) and a version of the FORET class of gap models (Shugart 1984), ZELIG (Urban *et al.* 1991). All of the physiological and allocation routines from FOREST-BGC have been substantially modified and used to replace the less mechanistic growth routine of ZELIG. In HYBRID individual trees are grown in plots, C is taken up and respired and water lost by each individual on a daily time step, and at the end of each year the accumulated C is partitioned to the various structural components (Friend *et al.* 1993; Figure 13.4). Soil water is estimated for a plot with the hydrological and other climate routines taken directly from FOREST-BGC. Individual trees are assigned a set of physiological parameters, depending on their species (Friend *et al.* 1993), and these determine the growth rate of each individual in the plot. Individuals with a low growth rate will become shaded and eventually die.

Photosynthesis and stomatal conductance are calculated using PGEN (Friend 1991), a modified Farquhar *et al.* (1980) model. Leaf water potential is assumed to have a direct effect on photosynthesis (Figure 13.4), and so it is possible to find an optimal stomatal conductance for any set of conditions (Friend 1991). This optimal conductance and the associated rate of net photosynthesis is calculated for each day. Day and night canopy respiration is calculated from foliage N content (Ryan 1991b) and temperature. Woody maintenance respiration is calculated from the total amount of living sapwood and temperature (Ryan 1990). The total amount of living sapwood is calculated from the difference between total woody biomass and dead heartwood biomass, assuming that a species-specific fraction of the sapwood is alive. Each crown is divided into 1 m vertical sections to calculate average daily radiation down through each plot. The daily average radiation in each crown drives daily photosynthesis for that individual. Thus each crown is treated as one layer for the photosynthesis routine. However, the lowest layer of each crown is followed separately because the total annual C balance in this layer, minus the maintenance respiration required by the sapwood to support this layer, is used to calculate annual heartwood growth: if the balance is negative, the heartwood grows by the amount of sapwood that was supporting this layer.

Partitioning logic is based on the pipe-model. A constant ratio of foliage to fine root biomass is assumed. Foliage biomass is linearly related to sapwood area at breast height. Sapwood area is the result of the dynamics of total diameter growth and heartwood growth from year to year. A potential two-thirds of the living sapwood can be used for storage. This storage is used if there is insufficient C available at the end of a year to produce the full complement of foliage, with its associated fine root biomass. A minimum 10% of available C is always partitioned to stem, branch, and coarse root components (this partitioning is calculated allometrically). If the storage is below capacity, this gets preference for

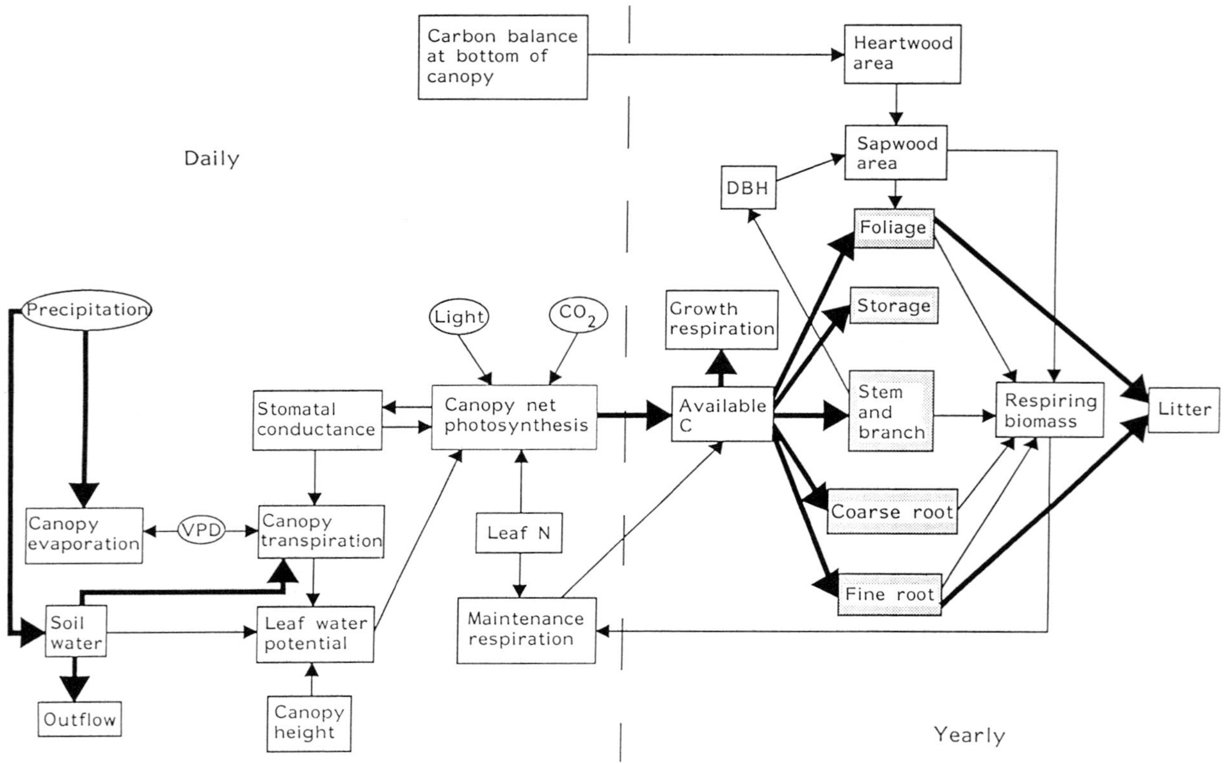

Figure 13.4 Principal physiological and hydrological features of HYBRID at the individual tree level. Temperature affects many of the daily processes, but it is not included for clarity. Narrow lines indicate influences; broad lines indicate fluxes; shaded boxes indicate structural components. DBH is the diameter at 1.4 m

C over any additional stem, branch, and coarse root growth. Foliage litter production is a fixed, species-specific fraction of total foliage biomass, as is fine root litter production.

13.3.6 MBL-GEM

The Marine Biological Laboratory's General Ecosystem Model (MBL-GEM, Figure 13.5) is a process-based, biogeochemical model designed to examine changes in the fluxes and allocation of C and N among plant tissues (foliage, fine roots, stems) and soils in response to changes in atmospheric CO_2 concentration, temperature, soil moisture, irradiance, and inorganic N inputs (Rastetter *et al.* 1991). The model consists of 23 simultaneous ordinary differential equations describing the temporal dynamics of 23 state variables (Figure 13.5). These variables represent the amounts of C and N in several plant tissues and in four soil organic fractions, as well as the amount of soil inorganic N. Within the vegetation, both labile (readily mobilized) and structural (including enzymatic machinery and structural framework) components of foliage, sapwood, and fine roots are represented. The amount of C and N in heartwood (structural only) is also simulated. Younger soil organic matter (including litter) is divided into 'extractives' (extractable in methylene chloride and hot water), 'cellulose' (acid soluble fraction of remaining residue), and 'lignin' (acid insoluble fraction). Older organic matter is converted to humus.

The uptake of C and N and the growth of plant tissues are simulated as enzymatically mediated processes, and the allocation of C and N to the various tissues is controlled by the gradients of labile C and N between tissues (similar to Thornley 1972a, b). This allocation scheme, in conjunction with the co-limitation of tissue growth by labile C and N in the tissues, the stoichiometric shifts in the C:N ratios of leaves and roots, and the dependence of C and N uptake on structural N in leaves and roots, form the mechanisms of N limitation on photosynthesis and C limitation on N uptake.

The turnover of soil organic matter is controlled by the rate of litterfall and the rate and efficiency of the transformations from one soil organic matter fraction to another. Litterfall is partitioned into 'cellulose', 'extractives', and 'lignin' based on the C:N ratio of the litter. Rates of transformation of soil organic matter from one organic matter fraction to another are assumed to be proportional to the amount of C in the donor fraction and are corrected for temperature, soil moisture, and microbial efficiency. Respiratory losses of C during these transformations are calculated from the rate of transformation and the microbial efficiency. Nitrogen is mineralized and immobilized during these transformations to maintain the appropriate C:N ratios of the respective fractions.

The model is implemented through an automated calibration routine. Most of the parameters in the model are set a priori using general, process-based information from the literature (e.g. half-saturation constants, Q_{10}, quantum

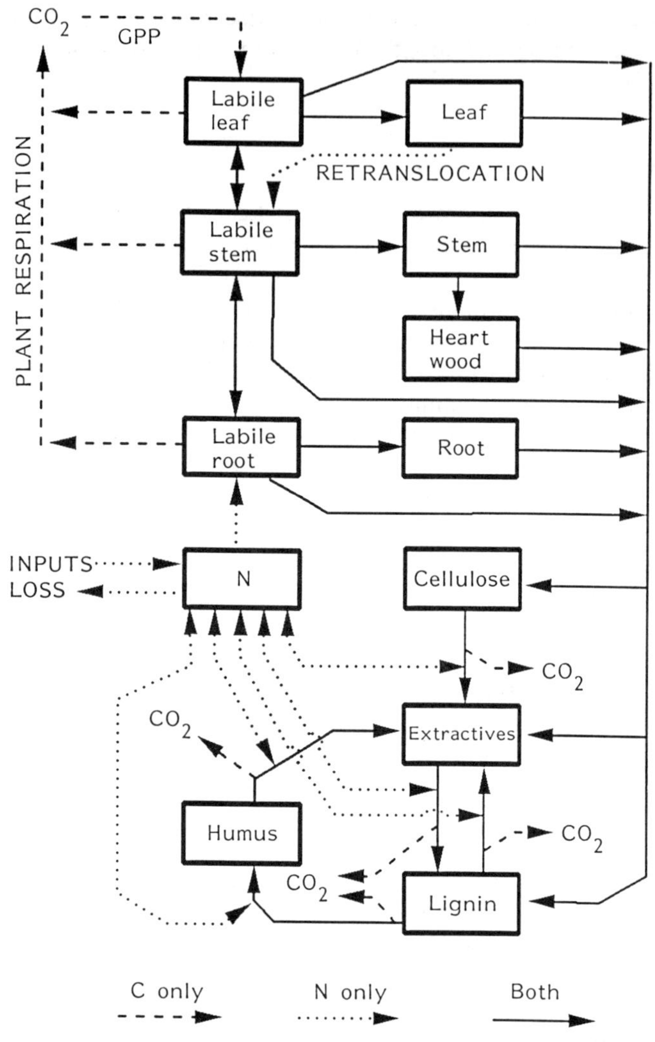

Figure 13.5 Compartment diagram for the MBL-GEM model. Arrows indicate fluxes. GPP is GROSS primary production

yield) or from our own prior experience with the model (e.g. decomposition parameters have been found to be general across a wide range of litter types, Ryan *et al.*, pers. comm.). However, the rate parameter for each of the major fluxes in the model is calibrated by back calculating its value based on the specified rates of several key fluxes. Specifically, the user must specify site-specific rates of gross primary production, litter production for foliage, wood, and fine roots, and the annual growth increment for these tissues. The user must also specify amounts of

C and N in foliage, sapwood, heartwood, and fine roots, and the total amount of C and N in soil organic matter.

13.3.7 PnET-CN

The PnET model (Aber and Federer 1992) is a simple, lumped-parameter, monthly-time-step model of C and water balances of forests built on two principal relationships: (1) maximum photosynthetic rate is a function of foliar N concentration, and (2) stomatal conductance is a function of realized photosynthetic rate. Using these two relationships greatly simplifies the calculation of transpiration and provides a direct link between C gain and water loss. Monthly leaf area display and C and water balances are predicted by combining equations derived from these relationships with standard equations describing light attenuation in canopies and photosynthetic response to diminishing radiation intensity, along with effects of soil water stress and vapour pressure deficit. The PnET model has been successfully validated against field data from 10 well-studied temperate and boreal forest ecosystems, suggesting that the aggregation of climatic data to the monthly scale and biological data such as foliar characteristics to the ecosystem level does not cause a significant loss of information relative to long-term, mean ecosystem responses (Aber and Federer 1992).

The PnET model is similar to the CENTURY model (Parton *et al.* 1988) in that it operates at a monthly time step, uses a single set of parameters to define the physiology of the plant community, produces biomass only by tissue type (e.g. foliage, wood and fine roots), and predicts seasonal changes in leaf area display in response to climatic drivers. It is also similar in structure to the C and water portion of the FOREST-BGC and BIOMASS models (Running and Coughlan 1988; McMurtrie *et al.* 1990a) but differs from these in the methods used to link the photosynthetic and transpiration processes, and in that a monthly, rather than a daily, time step is used.

There are 5 compartments in the model, and 11 fluxes, 3 for C and 8 for water (Figure 13.6, net photosynthesis includes separation of daytime and night-time C fluxes). All fluxes are calculated monthly except C allocation to wood and fine roots, to which excess C accumulated by net photosynthesis is allocated at the end of the year.

The foliage production routine uses radiation, temperature, water stress during the previous month and N content of foliage to derive a potential gross photosynthesis rate and day and night respiration rates for leaves at the top of the canopy. These are combined with a light attenuation coefficient, a photosynthetic light response curve, and foliar longevity to determine the production and shedding of foliar mass, and a potential gross photosynthesis (in the absence of water stress) for the whole canopy.

The C and water balance routine partitions precipitation between rain and

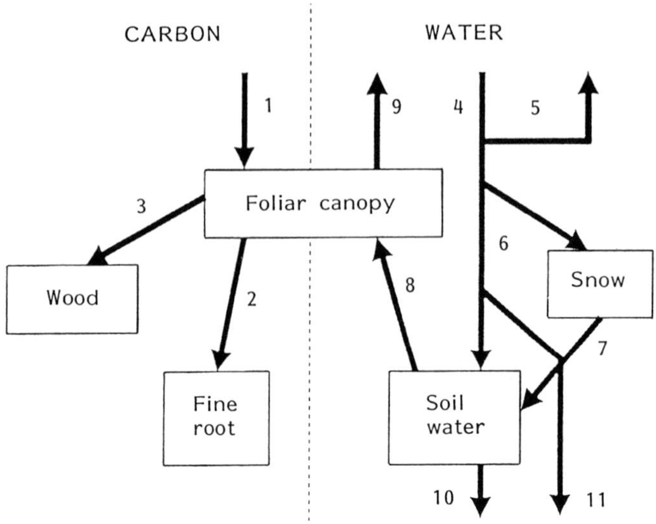

Figure 13.6 Compartment diagram for the PnET-CN model. Numbered arrows represent fluxes: (1) net photosynthesis (a function of light, temperature, water, N, and LAI); (2) fine root allocation (a function of foliage production); (3) wood allocation (by difference); (4) precipitation (input); (5) interception (constant fraction); (6) snow–rain division (a function of temperature); (7) snow melt (a function of temperature); (8) plant uptake (equals evapotranspiration); (9) evapotranspiration (a function of net photosynthesis and vapour-pressure deficit); (10) leaching (a function of soil water content and soil water-holding capacity); (11) 'fast flow' (a function of precipitation and snow melt)

snow, calculates snowmelt, total water input to the soil, and a fast, non-Darcian, drainage volume to determine water availability over the month. It then performs a numerical integration (daily time step) of water inputs to soil and transpirational demand (a function of gross photosynthesis and water use efficiency) over the month to produce changes in water storage, the degree of actual water stress on vegetation, and realized transpiration and net photosynthesis. Water remaining beyond water holding capacity at the end of this integration is drained away. Finally, the C allocation routine accumulates net photosynthesis over the year, subtracts foliage production and allocates the remainder to wood and root tissues.

13.3.8 Q

The forest growth model Q (after one of its central variables, substrate quality (q)) was developed to analyze the N cycle and its relation to forest stand development. The model is based on the interaction between a soil subsystem and a plant subsystem, each being described in terms of their content of C and N, respectively. The first development and application of the model were done in a context of

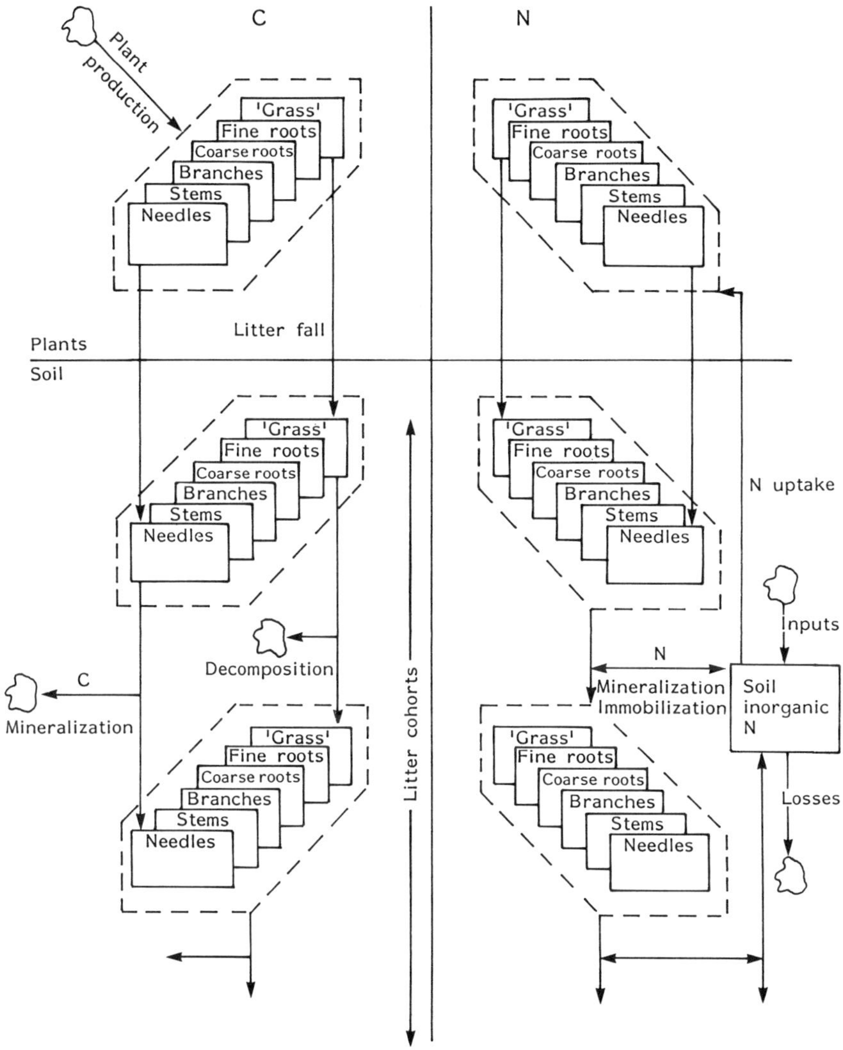

Figure 13.7 Compartment diagram for the Q model

nutritional consequences of whole-tree harvesting (Rolff and Ågren, in preparation). The model has later been extended and adopted to answer other questions as well, including those presented by the current project.

Plants are represented in the model by trees (separated into components) and a 'grass' component that includes all vegetation not in the dominant tree layer. Individuals are not distinguished. Each model component has one C and one N state variable (Figure 13.7). Tree biomass is separated into foliage (by age-class), branches, stems, coarse roots (including stumps), and fine roots. The primary role of the single grass component is to conserve mineralized N after clear-felling. Soil organic matter is divided into litter cohorts, thus keeping track of both the source of the material as well as the time of its appearance in the soil. The time step is one year.

Growth of all tree components (except foliage greater than one year old) is described by the N productivity concept (Ågren 1983, 1985, 1988), where potential maximum growth is reduced as a function of needle biomass and N content. Needles greater than one year old are created by transfer from the next younger year-class. Growth of grass is calculated from N uptake by grass and a constant N concentration, with an assumed maximum biomass. A constant fraction of leaf and root biomass is transferred to the soil in each time step. In addition, plant components are lost at harvests.

The behavior of a litter cohort is described by two variables, amount and quality (Ågren and Bosatta 1987; Bosatta and Ågren 1985, 1991a). Quality measures the degradability of the litter cohort, and decreases with time. Modelling of C mineralization is based on the assumption that decomposers are C limited. Both litter amount and quality decrease over time as a function of microbial utilization. Total soil C is the sum of all litter cohorts.

Uptake of N by trees depends on N availability and the capacity of trees to take it up. Nitrogen availability is the sum of the current year's N mineralization, deposition, and fertilization. Uptake capacity is a negative exponential function of fine root biomass; small trees could assimilate only a small fraction of the available N, while a fully developed stand could assimilate it all. Uptake is restricted by a maximum N concentration in needle biomass. Nitrogen required for growth of all tree compartments but needles is calculated assuming fixed N concentrations; the remainder is transferred to needle biomass. Any N deficit for a given year is drawn from the needles. Nitrogen loss from plants occurs with loss of biomass at constant N concentrations. Nitrogen concentration of senescing branches is lower than that of live ones, creating an internal recycling that is added to the N taken up.

Nitrogen mineralization is derived from changes in the total amount of organically bound N in the soil. The difference between that in the current year and in the previous year is net N mineralization. The N amount of a litter cohort is calculated from the N concentration in microbial biomass and the initial N:C ratio of the litter cohort. Microbial N concentration is an increasing function of available soil inorganic N, which provides a way of immobilizing external inputs of N.

13.4 METHODS

13.4.1 Design

We compared estimates from seven forest models (Table 13.1) against data from two intensively studied sites with contrasting climates (Biology of Forest Growth (BFG) site, 10–14-year-old *Pinus radiata* near Canberra, Australia; and Swedish Coniferous Biome (SWECON) site, 14–30-year-old *P. sylvestris* near Jädraås, Sweden). We selected these sites because:

(a) detailed information about ecosystem processes was available;
(b) both sites had similar treatments to manipulate water and nutrient availability (control, fertilization, irrigation, irrigation plus fertilization);
(c) they represented physiologically similar species growing in very different climates (Table 13.2). Detailed site descriptions are given below (section 13.4.2).

Treatments were started in 1974 for SWECON and in 1983 for BFG. Each model estimated stem biomass, foliar N content, annual foliage and stem wood production for 1984–87 (BFG) and for 1979–90 (SWECON). Additionally, each model estimated as much of a complete C, N, and water budget as possible for a single year for each site. For the time-series comparisons, cumulative errors could affect results. For the complete budgets, all models were initialized with standing crops in 1979 (SWECON) or 1986 (BFG).

Modellers were provided with a summary of the data for the BFG and SWECON sites, assembled from the sources given in Tables 13.3 and 13.4. Climate data (daily values of maximum, minimum, and average temperature,

Table 13.2 Long-term average temperature and precipitation of the BFG (1929–82) and SWECON (1931–60) sites.

	Temperature (°C)		Precipitation (mm)	
	BFG	SWECON	BFG	SWECON
January	21.0	−7.0	67	45
February	20.5	−6.9	62	29
March	18.0	−3.4	68	27
April	14.0	2.9	61	36
May	9.5	8.6	58	36
June	7.0	13.0	57	57
July	6.0	15.8	63	73
August	7.5	14.1	73	77
September	10.0	9.2	67	68
October	13.0	3.9	86	50
November	16.0	−0.5	69	61
December	19.0	−3.8	60	48
Total (precipitation) or average (temperature)	13.5	3.8	791	607

relative humidity, precipitation, and shortwave radiation) for the duration of the studies were also provided. Additional data or parameters required for model runs were the responsibility of the individual modellers. Special adjustments to the models necessary for the BFG and SWECON sites are summarized below.

13.4.1.1 BIOMASS

The early development of the BIOMASS model occurred in conjunction with the BFG field experiment (Benson *et al.* 1992a). Because the model has been parameterized for the BFG site (McMurtrie *et al.* 1990b, 1992; McMurtrie 1993; McMurtrie and Landsberg 1992), comparison of simulations and BFG biomass data did not constitute an independent test of its performance. Adapting the model to describe *Pinus sylvestris* growing at SWECON in the northern hemisphere required modifications to assumed annual patterns of biomass production and litterfall. Aboveground production was restricted to a three-month period, June to August (S. Linder, pers. comm.), with assimilates produced outside these months allocated to either storage or roots. Needle longevity was set at three years with needle fall occurring between August and October.

13.4.1.2 BIOME-BGC

Changes were made to the published version of the model (Running and Gower 1991), mainly to accommodate the heavy fertilization treatments. The maximum ratio of foliage to foliage + fine roots was allowed to increase to 0.67 from the published value of 0.5. Nitrogen loss for these runs increases exponentially with N availability (capped at 75% of available N), compared with a linear loss rate in the published version. Maximum photosynthesis was adjusted to data taken at the sites, by altering the parameter controlling the fraction of leaf N in RubP carboxylase. A Farquhar-type photosynthesis model was used for these simulations (Hunt and Running 1992).

13.4.1.3 CENTURY

The forest version of CENTURY was calibrated to the BFG and SWECON sites in a similar fashion as for other forest sites. Site-specific parameters included C:N ratios of biomass components, leaf mortality, and initial C:N ratios of soil organic matter pools. Parameters controlling production and C allocation were calibrated to match the control and irrigation + fertilization treatments. Biomass C:N ratios and leaf longevity are not normally altered in response to model treatments. However, for these runs, the control parameters for biomass C:N and leaf longevity were manually adjusted to reflect the data for the fertilizer treatments. At the SWECON site, it was necessary to double maximum allowable

C:N ratios for intermediate- and slow-turnover soil C pools in order for CENTURY to realistically simulate observed soil C:N ratios.

13.4.1.4 HYBRID

Species-specific parameters for allometry and the ratio of sapwood area to leaf area were derived from the literature. Optimal temperature for electron transport was set at 31 °C, living sapwood fraction of total sapwood was set at 5.5%, fine root turnover was set at twice biomass. Foliar respiration was calculated using Ryan (1991a), and leaf litter production was set at a constant 33% of foliage biomass. The HYBRID model was run only for the BFG site.

13.4.1.5 MBL-GEM

Parameters for the MBL-GEM model were derived by calibrating (as described in section 13.3.6) to year 1 for the control stands at each site. Annual gross production was estimated using equations in Ryan (1991a). The MBL-GEM model was designed as an equilibrium model where results applied to mature stands. To simulate the stand development of young, rapidly aggrading stands, parameters controlling growth were calibrated to the first year's growth of the control stands, and turnover of 'humus' was set to that of a mature stand.

13.4.1.6 PnET-CN

A simple one-box soil model was added to the published version of PnET (Aber and Federer 1992) for these model runs. The soil model uses the C:N ratio of the active pool to estimate gross and net N mineralization. Plant N availability is estimated as net N mineralization.

13.4.1.7 Q

Model runs used a constant, average climate. Differences in water availability for the irrigated and nonirrigated treatments at BFG were simulated by varying the reduction in N productivity per unit biomass between treatments. Water availability was assumed to have no effect at SWECON.

13.4.2 Site descriptions

13.4.2.1 Biology of Forest Growth experiment

The Biology of Forest Growth (BFG) experiment is located at 630 m elevation in the Pierce's Creek Forest about 20 km west of Canberra in the Australian Capital Territory (35° 21′ S, 148° 56′ E). The site, climate and the experiment have been described in detail by Benson *et al.* (1992a), and key aspects are outlined below. A

Table 13.3 Site and stand conditions and data sources used for simulations of the BFG site

Item	Value	Source
Soil water storage 0–2 m	250 mm	Benson *et al.* (1992a)
Soil water conductivity	125–225 mm d^{-1}	Benson *et al.* (1992a)
Soil bulk density	1.6 g cm^{-3}	Raison *et al.* (1987)
Soil C, initial	3657 g m^{-2}	Raison *et al.* (1987)
Soil N, initial	155 g m^{-2}	Raison *et al.* (1987)
Soil P, initial	48.3 g m^{-2}	Benson *et al.* (1992a)
Soil temperature	6–20 °C	Raison *et al.* (1992a)
Annual rainfall	1982–83: 637 mm 1983–84: 1264 mm 1984–85: 790 mm 1985–86: 680 mm 1986–87: 765 mm 1987–88: 731 mm	Benson *et al.* (1992a)
Specific leaf area	9.4–13.0 m^2 (all sides) kg^{-1}; varies with treatment	Raison *et al.* (1992c)
Maximum photosynthesis (projected leaf area)	14.8-18.5 μmol m^{-2} s^{-1},varies with treatment	Thompson and Wheeler (1992)
Soil and plant water	Predawn water potential was −0.6 MPa to 30% plant available water	Myers and Talsma (1992)
Stem wood N concentration	0.1%	Beets and Pollock (1987)
Bark N concentration	0.5%	Beets and Pollock (1987)
Foliar N concentration	0.7–1.8%,varies with treatment and year	Crane and Banks (1992)
Peak projected LAI	3.5–6.8, varies with treatment and year	McMurtrie *et al.* (1990b)
Roots > 5 mm	18% of total aboveground biomass	Cannell (1982)
Roots < 2 mm	23 g C m^{-2}	Santantonio and Santantonio (1987)
Biomass C, initial	Foliage: 249–260 g m^{-2} Wood: 1220–1399 g m^{-2} Both vary with treatment	Snowdon and Benson (1992)
Biomass N, initial	Foliage: 5.0–6.3 g m^{-2}, varies with treatment	Snowdon and Benson (1992)
Litter layer C, initial	1202 g m^{-2}, control stand in 1985	Raison *et al.* (1990)
Litter layer N, initial	15 g m^{-2}, control stand in 1985	Raison *et al.* (1990)
Soil inorganic N concentration, initial	0.5 g m^{-2}	Raison *et al.* (1992a)
Foliage production, control stand, 1983–84	204 g C m^{-2}	Raison *et al.* (1992b)

Table 13.3 *(continued)*

Item	Value	Source
Litterfall, control stand, 1983–84	95 g C m^{-2}	Raison *et al.* (1992b)
Litterfall N, control stand, 1983–84	1.3 g m^{-2}	Raison *et al.* (1992b)
Net N mineralization, control stand	0.7–3.8 g m^{-2}, varies by year	Raison *et al.* (1987, 1992a, 1992c)
N uptake by vegetation, from soil, control stand	0.6–3.7 g m^{-2}, varies by year	Raison *et al.* (1992a)
N uptake by vegetation, from decomposing litter, control stand	0.3–1.1 g m^{-2}	Crane and Banks (1992
N leaching losses	0 g m^{-2}	Khanna *et al.* (1992)

partial list of values used for modelling and sources for data are given in Table 13.3.

The site was cleared of the original eucalypt woodland in 1934, broadcast burned and planted to *Pinus radiata* in 1935. The pine crop was clear-felled in 1972, residues burned, and the second pine stand planted in 1973. The forest was thus 10 years old when the BFG experiment commenced in winter 1983; at that time tree density was ~ 700 ha^{-1}, height ~ 9.5 m, mean tree diameter (1.4 m) ~ 15 cm and stand basal area ~ 12.5 m^2 ha^{-1} (full details given by Snowdon and Benson 1992).

The soil supporting the forest is a yellow podzolic (Stace *et al.* 1968) or typic albaqualf (Soil Survey Staff 1975) derived from granite. It has a duplex profile comprising a sandy A horizon extending to about 40 cm depth, and a very dense (bulk density 1.7 g cm^{-3}) B horizon of clay and gravel overlying a C horizon of weathering granite (> 100 cm). The soil has a very low N-supplying capacity (Raison *et al.* 1992a) but provides adequate P for tree growth (Raison *et al.* 1990). Soil water storage capacity is only ~ 75 mm in the A horizon where most fine roots are concentrated, and ~ 250 mm to a depth of 2 m (Myers and Talsma 1992). The combination of low soil water storage and high probability of summer drought results in trees suffering severe water stress during the summer–early autumn period (Myers and Talsma 1992).

The site experiences mild winters (mean daily minimum and maximum temperatures in July are 1 and 11 °C) and warm summers (14 and 28 °C in January). Summer maxima may reach 35–40 °C when relative humidity is very low (< 10%). Annual rainfall averages 791 mm (Table 13.2), but this is highly variable with summer droughts of 2–3 months' duration common; during the life of the stand, < 00 mm of rainfall in the combined three summer months occurred in five growing seasons (1978–79, 1981–82, 1982–83, 1984–85, and 1985–86).

During 1983 and 1984, combinations of N fertilization and irrigation were applied to stands (0.25-ha plots) with the objective of markedly modifying growth rates to facilitate study of the mechanisms by which N and water interact to control productivity. The treatments were: control (C), irrigated (I), fertilized (F), irrigated–fertilized (IF) and irrigated–liquid fertilized (IL). The irrigation treatment aimed to maintain soil moisture near field capacity, and approximately doubled annual rainfall. Water was added as often as every 2–3 days, based on a simple model of water use and fortnightly adjustment based on measurement of soil water content. Irrigation was not applied during the summer of 1983–84 because the growing season was too wet. Solid fertilizer (F and IF treatments) was added in spring 1983 and contained 400 kg N, 200 kg P, 100 kg K and 10 kg B ha^{-1}. The IL treatment commenced in spring 1984 and provided a balanced nutrient input weekly in the irrigation water and was varied seasonally to match tree demand; annual inputs of N totalled ~ 300 kg ha^{-1}. The IL treatment aimed to remove both water and nutrients as limitations to growth.

A synthesis of the major research findings from the BFG experiment has recently been published (Raison and Myers 1992a). Irrigation alone increased NPP by about 40–50% (Benson *et al.* 1992b; cf. Figure 13.8), and fertilization increased it by about 25–30% except when the growing season was very dry. A marked interaction between irrigation and fertilization resulted in about 200% increase in NPP. The treatments affected the rate of canopy development and stand foliage carrying capacity (Raison *et al.* 1992c), the period of active photosynthesis (Thompson and Wheeler 1992) and the efficiency of light utilization for biomass production (Raison and Myers 1992b). There is also evidence that a larger proportion of photosynthate may be allocated in belowground processes in those BFG stands having low N status (S. Pongracic and R.J. Raison, pers. comm., 1992).

13.4.2.2 Swedish Coniferous Forest Project

The Swedish Coniferous Forest Project (SWECON) was started in 1972 to study the structure and function of conifer forests in Sweden. As part of this effort, extensive studies on element cycling within young Scots pine (*Pinus sylvestris*) stands were established at Ivantjärnsheden near Jädraås, Sweden (60° 49′ N, 16° 30′ E). Climate, soils, and stand characteristics of the Ivantjärnsheden site have been described by Axelsson and Bråkenhielm (1980) and many results from the study have been described in T. Persson (1980). A partial list of values used for modelling and sources for data are given in Table 13.4.

Ivantjärnsheden is located in the southern end of the 'northern coniferous' region. The site is a sandy plain and soils are sandy and podzolic with a medium sand texture (Axelsson and Bråkenhielm 1980). The A_0 horizon varies from 1 to 7 cm, A_1 from 0.2 to 1.6 cm, and A_2 from 2 to 7 cm; the litter layer is ~ 1 cm. For the Ih II stands at Ivantjärnsheden, the water table is 10 m below the surface and the soil water capacity in the 0–30 cm root zone is 72 mm. A

Table 13.4 Site and stand conditions and data sources used for simulations of the Swedish Coniferous Biome site

Item	Value	Source
Soil water storage 0–60 cm	81 mm	Axelsson and Bråkenhielm (1980)
Soil bulk density	0–10 cm: 0.98 g cm^{-3} 10–20 cm: 1.2 g cm^{-3}	Axelsson and Bråkenhielm (1980)
Soil C, initial	2572 g m^{-2}, from older stand (Ih V)	Bringmark (1980)
Soil N, initial	83 g m^{-2}, from older stand (Ih V)	Bringmark (1980)
Soil temperature	0–20 °C	Axelsson and Bråkenhielm (1980)
Annual Rainfall	607 mm	Axelsson and Bråkenhielm (1980)
Specific Leaf Area	8.6–10.4 m^2 (all sides) kg^{-1}; varies with treatment	Troeng and Linder (1982)
Maximum photosynthesis (projected leaf area)	15.8–17.7 μmol m^{-2} s^{-1}, varies with treatment	Linder and Troeng (1980)
Stem wood N concentration	0.1%	Göran Ågren, unpublished data
Bark N concentration	0.5%	Göran Ågren, unpublished data
Foliar N concentration	1.1–2.0%, varies with treatment and year	Göran Ågren, unpublished data
Understorey biomass	202 g C m^{-2}	Persson, H. (1980)
Roots < 2 mm	67 g C m^{-2}, trees and shrubs	Flower-Ellis and Persson (1980)
Biomass C, initial	Foliage: 4–29 g m^{-2} Wood: 8 g m^{-2} Foliage varies with treatment	Göran Ågren, unpublished data
Biomass N, initial	Foliage: 0.2–1.2 g m^{-2} Wood: 0.2 g m^{-2}, Foliage varies with treatment	Göran Ågren, unpublished data

low, dense understorey of lichen and dwarf shrubs exists in these stands.

The site experiences cool summers with long days, and long, cold winters. Mean monthly air temperature is 15.8 °C in July and −7 °C in January (Table 13.2; Axelsson and Bråkenhielm 1980). Mean annual precipitation is about 600 mm, varying between 413 and 647 mm (1931–60). Because the sandy soils have low water storage capacity, summer droughts occur in this ecosystem.

Data for the model comparison was collected at stands designated Ih II. The Ih II stand was harvested in 1957 and seed trees were felled in 1962; the naturally regenerated stand was about 14 years old at the start of treatments in 1974. In 1973, the stand had 1095 stems ha^{-1}, a basal area of 2.15 m^2 ha^{-1}, and an average height of 2.1 m (Flower-Ellis and Persson 1980). Replicate plots of 20 × 20 m were established for control, irrigated (100–300 mm yr^{-1}), fertilized (8 g N m^{-2} yr^{-1}), and irrigated+fertilized (100–300 mm yr^{-1}; 7–20 g N m^{-2} yr^{-1}) treatments (Aronsson and Elowson 1980). Fertilization resulted in dramatic increases in productivity (Linder and Axelsson 1982; cf. Figures 13.8 and 13.10) and treatment response persisted long after fertilization was discontinued.

13.5 COMPARISONS WITH DATA AT BFG AND SWECON CONIFER SITES

13.5.1 Stem biomass, foliage nitrogen, aboveground productivity

Most models closely estimated stem biomass at both the BFG and SWECON control stands (Figure 13.8). Models that used data values for foliar N (BIOMASS, HYBRID) or a calibration to the fertilized treatment (CENTURY) were also able to closely simulate the response of stem biomass to irrigation and fertilization. Where N fluxes were explicitly modelled (MBL-GEM, PnET-CN, BIOME-BGC, Q), model predictions of treatment response were less successful. Overestimates by Q and PnET-CN for some site and treatment combinations were partially caused by differences in initial conditions (Q grew stands from seedlings, PnET from the initial year of the study). Models differed little in their relative predictions for the two sites.

Foliar N mass is a critical variable for most models because it can affect photosynthesis, respiration, and litter decomposition and nutrient mineralization. The models that simulate N allocation (BIOME-BGC, CENTURY, MBL-GEM, PnET-CN, Q) were generally successful at estimating foliar N mass for the control and irrigated treatments (Figure 13.9), but less so for the fertilizer treatments. Modelling the fate of large fertilizer applications in a mechanistic manner appears to remain problematic.

Foliage production is critical because foliage mass partially controls transpiration and C fixation, and foliage turnover affects decomposition and nutrient mineralization. Stem production is thought to have low priority for fixed C (Waring and Pitman 1985); hence stem production should be a sensitive indicator of shifts in the annual C budget. Models were generally successful at estimating foliage and stemwood production for the control and irrigated treatments (Figures 13.10 and 13.11), but less so for the fertilizer treatments. Most models successfully predicted the twofold difference in the rate of stem production between the two sites. Where foliar N was obtained from data (BIOMASS, HYBRID), response of foliage and stem production closely matched data.

13.5.2 Single-year carbon budgets

Model estimates of an annual C budget for BFG and SWECON are given in Figure 13.12. Differences were apparent between data (where they exist) and

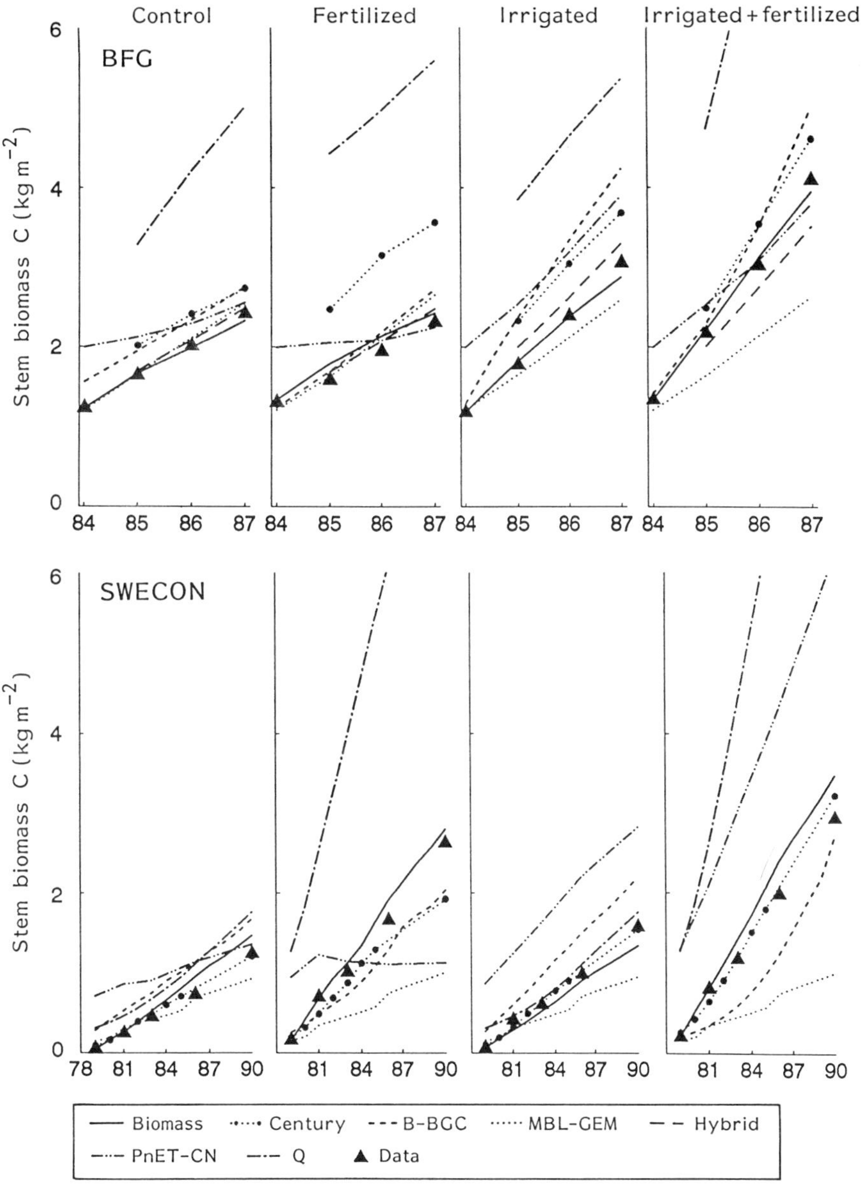

Figure 13.8 Stem biomass C and model predictions for BFG and SWECON sites. For BFG, Q was initialized in 1973 from seedlings and PnET-CN in 1983; the other models started in 1984. For SWECON, Q was initialized in 1960 from seedlings and PnET-CN and GEM in 1974 (the beginning of the experiment); the other models started in 1979. Values for Q for the fertilized + irrigated (BFG) and fertilized (SWECON) treatments are off scale for the latter years of the simulations

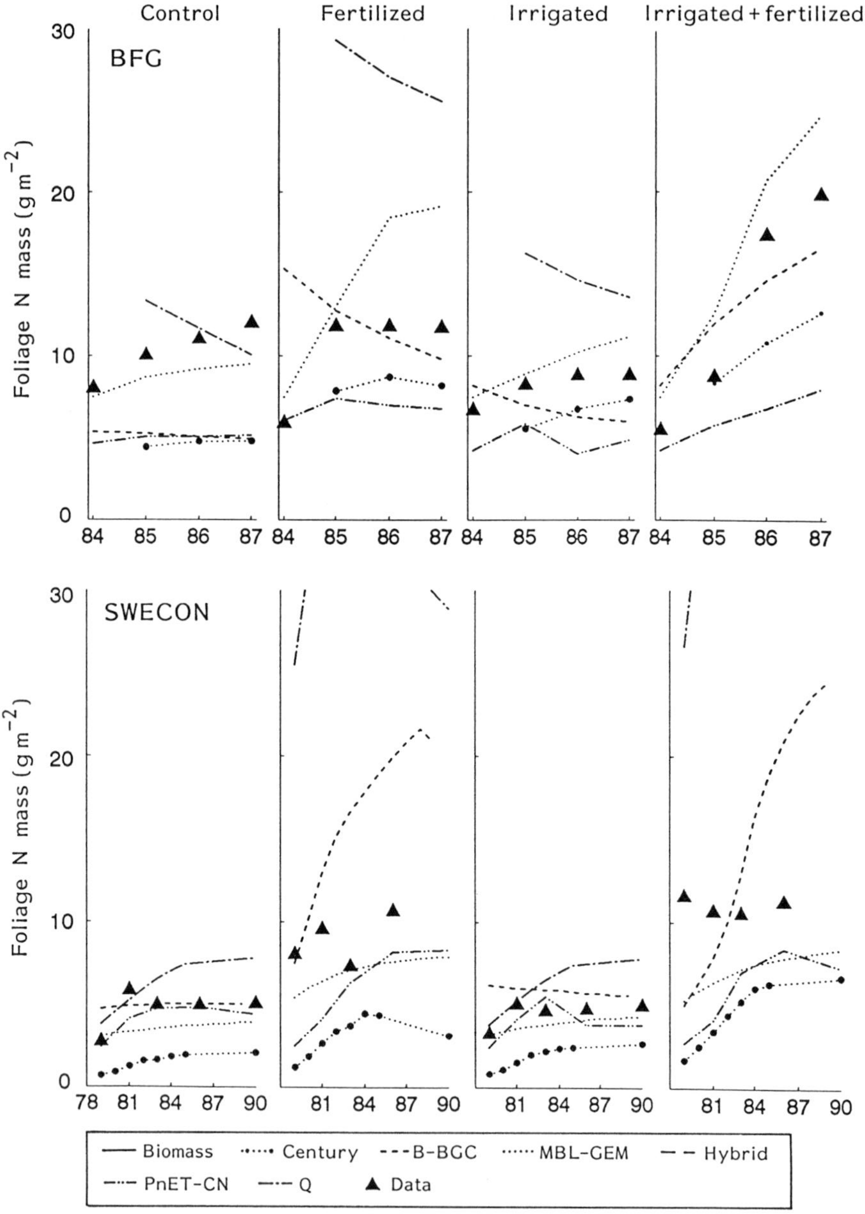

Figure 13.9 Foliage N mass and model predictions for BFG and SWECON sites. For BFG, Q was initialized in 1973 from seedlings and PnET-CN in 1983; the other models started in 1984. For SWECON, Q was initialized in 1960 from seedlings and PnET-CN and GEM in 1974 (the beginning of the experiment); the other models started in 1979. For CENTURY, foliage N mass is N content the foliage would have if it were fresh litter (after retranslocation). Actual foliar N for CENTURY would be roughly double the given values. Foliage N mass for Q for BFG irrigated + fertilized treatment is off scale at about 55 g m^{-2}

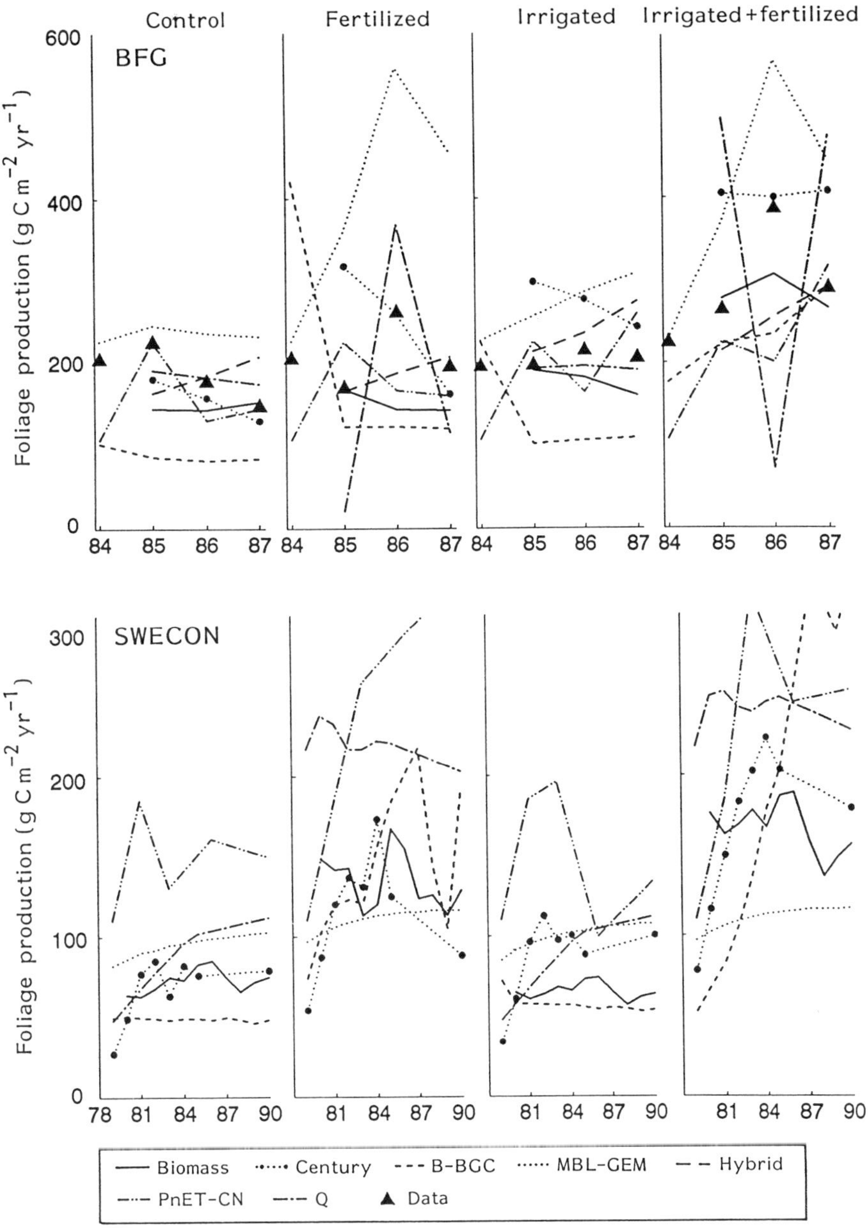

Figure 13.10 Foliage production (C) and model predictions for BFG and SWECON sites. For BFG, Q was initialized in 1973 from seedlings and PnET-CN in 1983; the other models started in 1984. For SWECON, Q was initialized in 1960 from seedlings and PnET-CN and GEM in 1974 (the beginning of the experiment); the other models started in 1979. Note difference in scales for BFG and SWECON sites

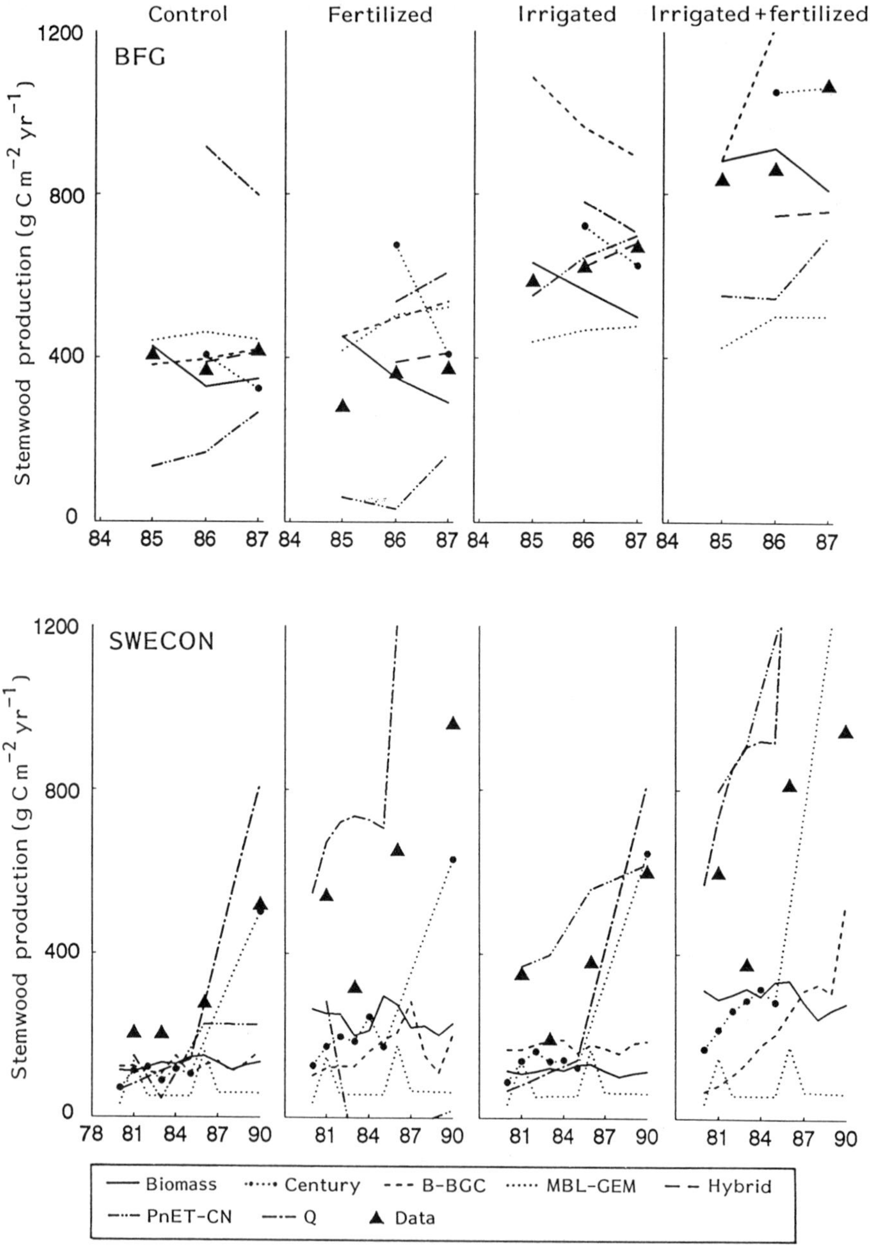

Figure 13.11 Stem biomass production (as C) and model predictions for BFG and SWECON sites. For BFG, Q was initialized in 1973 from seedlings and PnET-CN in 1983; the other models started in 1984. For SWECON, Q was initialized in 1960 from seedlings and PnET-CN and GEM in 1974 (the beginning of the experiment); the other models started in 1979. Stemwood production for Q for the irrigated + fertilized treatment at SWECON is off scale

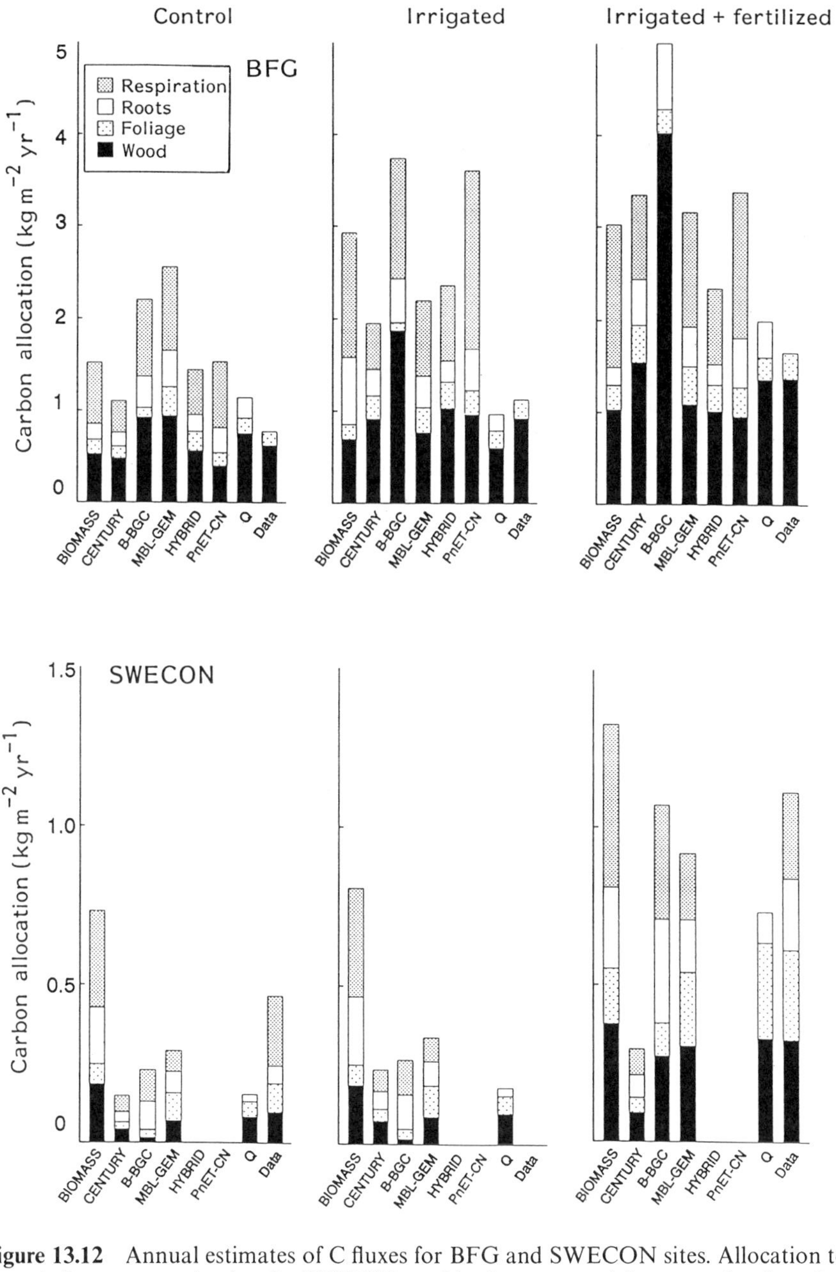

Figure 13.12 Annual estimates of C fluxes for BFG and SWECON sites. Allocation to roots is fine root production for HYBRID, and both coarse and fine roots for the other models. Respiration is growth respiration for all tissues plus maintenance respiration for above- and belowground woody tissues and fine roots. Growth respiration was estimated for CENTURY as 25% of growth. Q does not estimate respiration; BFG data do not include fine root and respiration estimates. Note difference in scale for BFG and SWECON

model predictions, and also among the predictions of different models. Differences among model estimates are not the result of differences in initialization or cumulative errors: these C budgets were estimated starting from the same initial conditions (standing crops in 1979 for SWECON and 1986 for BFG) and climate. Therefore, the differences shown result from different representations of processes, their interactions, and parameterization.

The total height of the bars represents model estimates of annual net canopy fixation (gross C fixation minus any foliar maintenance respiration). Model estimates were generally within 50% of the mean for all treatments at both sites. Differences among models may result from differences in the photosynthesis model, extinction of light within the canopy, aggregation of fine-scale processes to the canopy, aggregation over time, or sensitivity to climate.

The irrigation + fertilization treatments are useful for comparing models, because both ecosystem/tissue and physiologically based models should agree when resources are not limiting. All models except BIOME-BGC matched data for this treatment at the BFG site (Figure 13.12), and all participating models except CENTURY matched data for this treatment at the SWECON site.

Most models assume that wood production has low priority for fixed C (see Waring and Pitman 1985). Therefore, allocation to wood should respond most to treatment and climate. Model predictions of C allocation to wood did differ within a treatment and among treatments. Allocation of net canopy fixation to wood production by the models averaged 24% at the SWECON site but 37% at the BFG site. At SWECON, wood production:net canopy fixation ranged from 6% (BIOME-BGC) to 28% (CENTURY) for the control stand and 25% (BIOME-BGC) to 33% (MBL-GEM) for the irrigated and fertilized stand. At BFG, wood production:net canopy fixation ranged from 26% (PnET-CN) to 44% (CENTURY) for the control stand and 28% (PnET-CN) to 51% (BIOME-BGC) for the irrigated and fertilized stand.

Assumptions about how climate and nutrition affect allocation to root production differ among models. The ratio of foliage production to the total of foliage plus fine root production (F:F + R) estimates shifts in root allocation. At SWECON, F:F + R varied from 22% (BIOME-BGC) to 68% (Q) for the control stand and 25% (BIOME-BGC) to 76% (Q) for the irrigated and fertilized stand. At BFG, F:F + R varied from 23% (BIOME-BGC) to 53% (Q) for the control stand and 20% (BIOME-BGC) to 58% (BIOMASS) for the irrigated and fertilized stand. The F:F + R ratio was most sensitive to treatment for BIOMASS and Q, less sensitive to treatment for MBL-GEM, BIOME-BGC, and HYBRID, and insensitive to treatment for CENTURY and PnET-CN.

13.5.3 Single-year nitrogen budgets

For those models that simulate N budgets, model estimates of net N mineralization and plant uptake were variable, and larger than field estimates (Figure

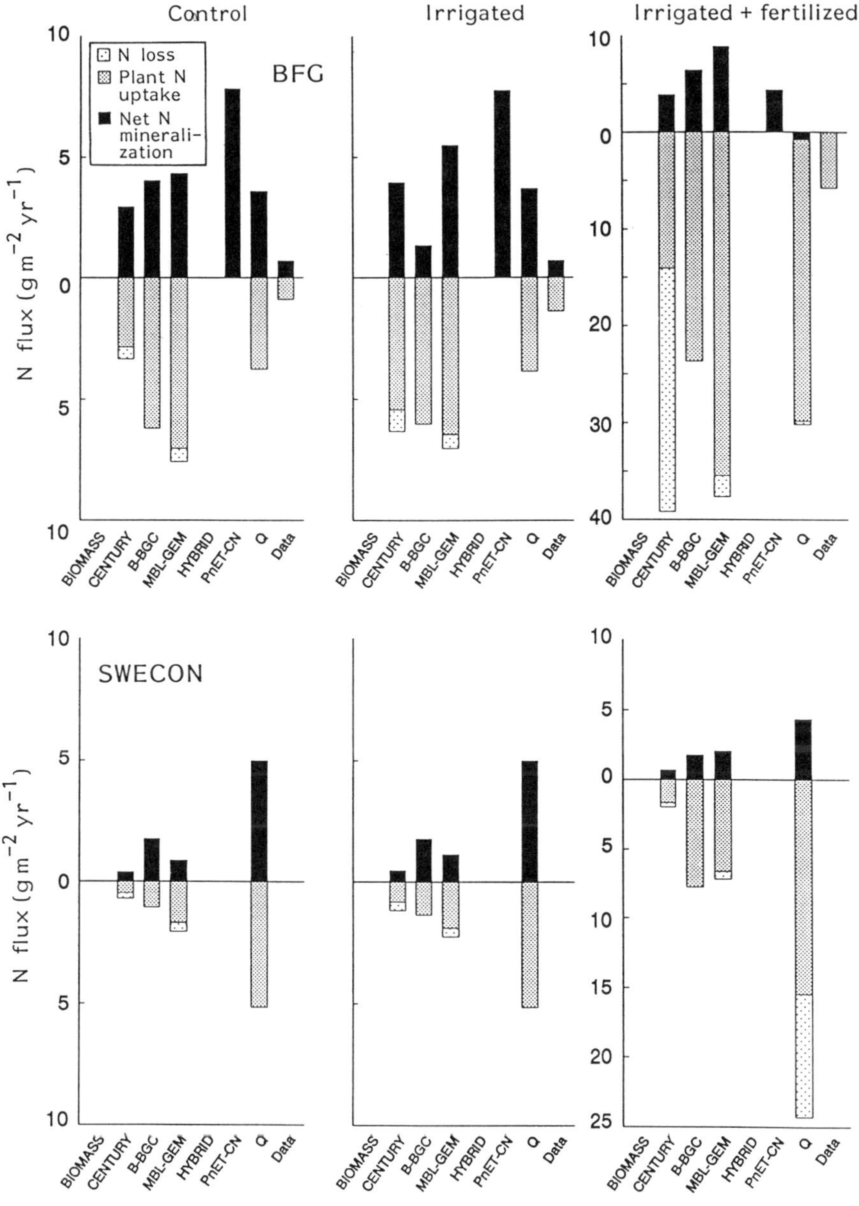

Figure 13.13 Annual estimates of N fluxes for BFG and SWECON sites. N loss includes leaching and gaseous losses. Models shown with data estimate all three components, except PnET-CN, which only provided net N mineralization. Note differences in scale for the irrigated + fertilized treatments

13.13). All models except Q estimated a much greater N mineralization at the BFG site. Fertilization did not greatly affect N mineralization for any model except Q at the BFG site. Differences in plant uptake in turn influenced the C budgets discussed in section 13.5.2, because tissue N concentrations affect photosynthesis and respiration in most models.

13.5.4 Single-year water budgets

Total annual water flux agreed closely with data for all models that simulate water balance, because total flux is nearly equal to total annual precipitation (Figure 13.14). Model predictions for the partitioning of water to transpiration, evaporation, and groundwater loss differed, particularly for the irrigated treatments.

13.6 DISCUSSION

Our comparison and validation of models had two purposes. First, we wished to assess our understanding of temperate conifer forests as captured in state-of-the-art models. Is our understanding adequate to model, from first principles, element budgets of forest ecosystems and their response to dramatic changes in resource availability? Because global change will affect resource availability, models must successfully simulate these changes to reliably predict growth in future climates. Second, we wished to determine if the different hypotheses about controls over processes and their interactions affected model response to changes in resources. Are models that explicitly link production and decomposition more robust simulators than models that consider only climate and production?

The experimental sites represented a severe challenge to the models because dramatic shifts in resources occurred over a short period of time. In general, models had difficulty (1) estimating the fate of applied fertilizer, (2) allocating C and especially N within plants, and (3) estimating actual C and N fluxes. Models that directly simulated the N cycle (all except BIOMASS and HYBRID) had serious problems determining how much N was available to plants, how much was taken up, and where in the tree N accumulated. Because most models estimate photosynthesis and respiration as some function of tissue N content, errors in plant uptake and allocation of N caused errors in other fluxes. However, the agreement between measured and modelled fluxes for the irrigation + fertilization treatments is encouraging.

With the exception of the handling of the N cycle, no consistent differences in model performance or response to treatment were apparent for models that differed greatly in detail, functional equations, and spatial and temporal aggregation. However, because models differed greatly, it was difficult to isolate

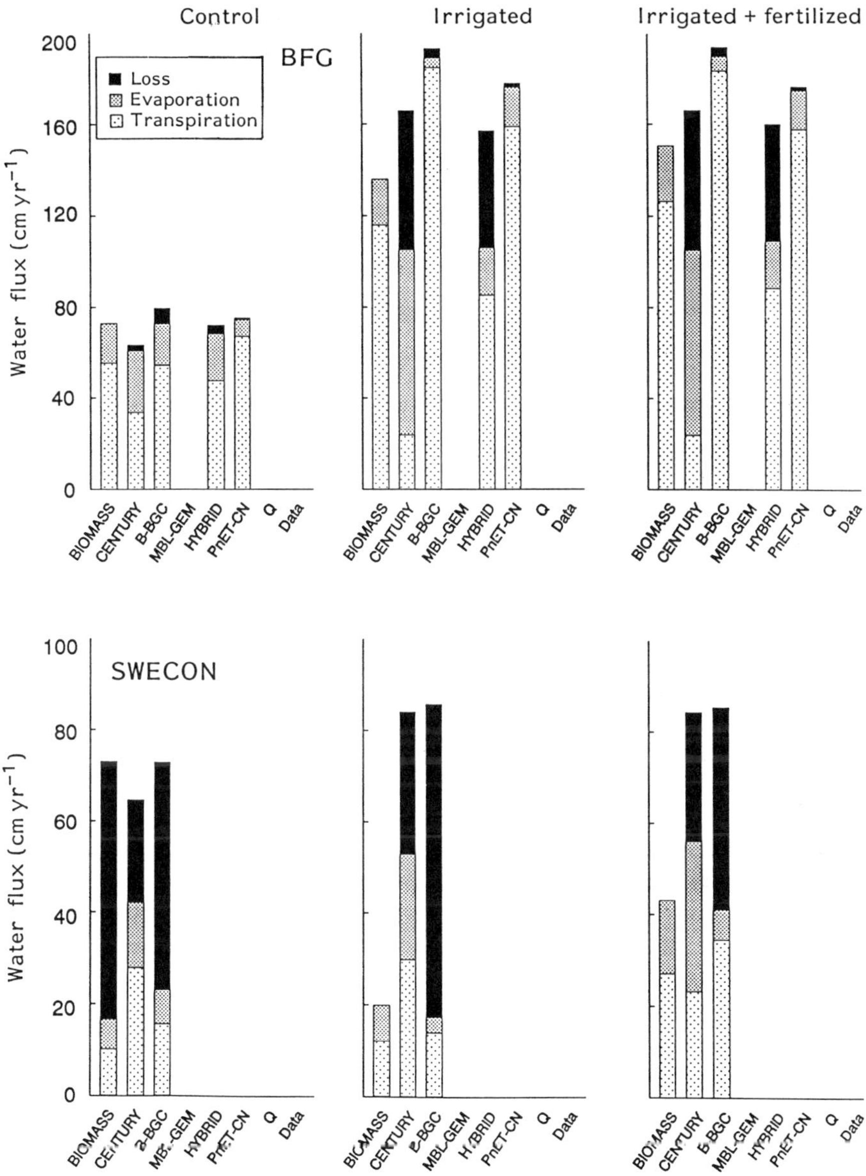

Figure 13.14 Annual estimates of water fluxes for BFG and SWECON sites. Loss is runoff or water transported to groundwater. BIOMASS model does not estimate loss for irrigated treatments

reasons for differences in model estimates. Below, we discuss individual processes in more detail.

13.6.1 Photosynthesis and canopy dynamics

Despite starting from the same initial conditions, and using the same climate, model estimates for net canopy photosynthesis differed substantially when water and nutrients were limiting. Differences in model estimates likely resulted from differences in aggregation, plant N uptake over the year, response to climate, and whether changes in canopy (seasonality) were explicitly modelled. Seasonal changes in the canopy were quite dramatic at the BFG site (Raison *et al.* 1992c), and models that assumed a constant canopy may have difficulty estimating annual photosynthesis. Until we can reliably measure canopy photosynthesis, or reliably sum respiration and dry matter production, our models of photosynthesis will be uncertain. However, model agreement for simulations with non-limiting resources is encouraging, and suggests that we understand the limits to productivity but need more work to understand resource interactions.

13.6.2 Respiration

Because models generally view maintenance respiration as a fixed cost (having the first priority for fixed C), respiration can dramatically affect dry matter production. Estimates for respiration differed greatly, particularly in response to treatment. All models partitioned respiration into the components of growth and maintenance and used a similar exponential response to temperature. Therefore, differences in model estimates were caused by differences in biomass (BIOME-BGC, HYBRID) or tissue N content (BIOMASS, CENTURY, MBL-GEM, PnET-CN) that developed over the course of the year. Differences in model estimates were also caused by differences in respiration rates and the parameters controlling temperature response. No model simulated acclimation of respiration rates to climate (Amthor 1989), a process that could affect the response of productivity to global change.

13.6.3 Carbon and nitrogen allocation

Allocation of C and N to tissues (wood, roots, and foliage) from first principles remains a challenge. For C, correctly estimating allocation depends initially on correctly estimating photosynthesis and respiration. However, relative allocation is also very important because allocation can alter the amount of tissue in foliage, wood, and roots and thus affect future rates of photosynthesis, respiration, and nutrient uptake.

Models disagree about the partitioning of C to foliage and roots for the control stands, and for the response to treatment. Not surprisingly, models with flexible allocation (BIOMASS, BIOME-BGC, HYBRID, MBL-GEM, Q) decreased the relative allocation to roots for the fertilized stands. However, models with fixed allocation (PnET-CN and CENTURY) did not do noticeably worse (compared with data) than the models that altered allocation dynamically.

Allocation of the N taken up by plants to various types of plant tissue has received little experimental attention. However, this understanding is essential for successful modelling, because photosynthesis, respiration, (and perhaps C allocation to fine roots and foliage) are strongly controlled by tissue N contents.

13.6.4 Nitrogen cycle and decomposition

The lack of agreement among model estimates of N mineralization and plant uptake reflects both differences in model structure and our lack of understanding of controls over these processes. Because the N cycle is so important for the linked models of production and decomposition, differences in N mineralization or plant uptake yield differences in photosynthesis, respiration, and litter quality. Also, soil organic matter content and turnover control much of the N mineralization, but change slowly. Therefore, simulated rates of turnover depend more on assumed initial conditions than on the rates of recycling litterfall, and it is difficult to evaluate changes in these processes for short-term experiments.

Non-biological immobilization of N into soil organic matter can strongly affect N availability after fertilization (e.g. Nömmik and Vahtras 1982; Axelsson and Berg 1988). Perhaps because none of these models included an explicit representation of this process, the models had difficulty simulating both N availability in the soil and the distribution of N to plants and soil organic matter after heavy fertilization.

Models that simulate the N cycle agree that litter quality varies as some function of the content of lignin and N, and that N immobilization and mineralization from litter vary with litter quality. However, models have not reached a consensus about what controls C and N mineralization from 'humus' (older, slowly decaying soil organic matter). For example, it is uncertain whether litter quality and N in the soil solution affect decomposition rates and C:N of humus. If litter quality does affect humus quality and turnover, the changes in litter quality expected from increased atmospheric CO_2 and changes in species composition could markedly affect long-term response of ecosystems to global change.

13.6.5 Hydrology

The hydrological cycle is better understood (Landsberg *et al.* 1991) than the C and nutrient cycles. Therefore, modelling hydrology has generally been more successful. Also, ecosystem estimates for precipitation and runoff, while uncertain, are readily available and constrain model predictions.

Models that simulated the hydrologic cycle agreed well for total water flux in this exercise because the total water flux was constrained by precipitation. However, differences in model estimates for the components of transpiration, evaporation, and loss to groundwater are problematic and may affect model predictions of response to a changing climate. Predictions for models that used a daily time step (BIOMASS, BIOME-BGC, HYBRID) agreed reasonably well with each other and with one of the monthly models (PnET-CN). The PnET-CN model's estimates of transpiration derived from monthly estimates of C assimilation agreed well with estimates of transpiration derived with the Penman–Monteith approach.

The CENTURY model showed considerable differences from the other models in the apportioning of water to transpiration versus evaporation. It does not model transpiration directly, and its C and N dynamics are not directly sensitive to it. The water budget, production, and decomposition in this model are controlled by precipitation and estimated evapotranspiration. Estimated evapotranspiration is divided between evaporation and transpiration by using linear relations between evaporation, canopy biomass, and precipitation. Evaporation is given first priority for water, with the remainder allocated to transpiration. These linear relationships produce high estimates of evaporation and low estimates of transpiration (relative to other models) when precipitation (or irrigation) and biomass are large.

13.7 RECOMMENDATIONS FOR EXPERIMENTATION AND MODELLING

13.7.1 Experimentation

Reliable estimates of fluxes at the ecosystem level are problematic for C, water, and nutrients (Landsberg *et al.* 1991), although much recent progress has been in estimating C and water fluxes with eddy-covariance technology. This lack of reliable flux estimates hinders evaluation of models. Additionally, ecosystem-level flux measurements are generally of net fluxes, while component fluxes are most useful for model validation. Because processes change dramatically with stand development and since most of our knowledge of these processes is derived from young stands, we also need more information from mature forests.

To evaluate models of ecosystem processes, a small number of studies where resources are manipulated and all of the components are measured can be as valuable as estimates of a few components (e.g. net primary production) at many sites. Manipulation experiments directly test hypotheses about the controls on processes such as allocation and nutrient retention and uptake. Therefore, the benefits of studies like BFG and SWECON to modelling efforts are enormous. Another imperative is the compilation and publication of integrated, whole-system data sets (such as those used for this comparison) for developing and testing models. The lack of such data, particularly for natural ecosystems, is crippling our efforts to understand the impacts of human activities on the biosphere.

On the process level, we have a critical need to understand the controls over the allocation of C and N within trees. We also need to understand the long-term controls over N mineralization from soil organic matter. Allocation of C has received attention, but we are still uncertain about the responses of belowground allocation and respiration to changes in resource availability. However, because of the importance of nutrition in defining photosynthesis (Field and Mooney 1986), respiration (Amthor 1989; Ryan 1991b) and allocation (Thornley 1972a; Field 1983; Bazzaz *et al.* 1987; Ingestad and Ågren 1991), we need to focus on the controls over the uptake and allocation of N as well. Finally, we need to search for and define the limits for allocation ratios, tissue N concentrations and mineralization rates.

13.7.2 Modelling

This modelling exercise helped identify the strengths and limitations of the participating models. Without the rigorous validation and exposure to new sites in this study, model limitations could not have been identified. Limitations found have generally fostered new work and searches for alternative approaches.

Because this model comparison was useful but difficult to execute, we need to develop better methods for comparing models. We need to strive harder to identify whether differences are trivial (caused by differences in rate constants), or important (caused by differences in model formulation or interactions within a model). Development of a common model framework might be one mechanism to accomplish this task. Such a framework might specify the processes to be modelled, the driving variables, and inputs and outputs to the various submodels. Formulations for the submodels, the interactions between submodels, and level of aggregation over space and time could remain flexible. Under such a system, it would be simple to determine whether another model of photosynthesis or respiration caused real differences in model estimates, or whether different hypotheses about interactions between production and decomposition affected long-term productivity.

However, if such a common model framework were implemented, the model framework should be flexible enough to allow real differences to exist. Some of the most important differences in models may not be in the simulation of individual processes, but how they are linked at a higher level. The overall strategy for assessing the effects of global change should stress a diversity of approaches. It is this diversity of approaches that will either give us confidence in our predictions (if the models agree) or indicate unknowns (if models do not agree). This diversity of approach could be incorporated by making the common structure modular and hierarchical. Thus, submodels might represent simply different formulations for a process (e.g. photosynthesis) or might represent a totally new approach to simulating ecosystems.

Despite some of the differences among model predictions, models were generally successful at estimating the large differences in fluxes resulting from the different climates in Sweden and Australia and in estimating fluxes where water and nutrients are not limiting.

13.8 REFERENCES

Aber, J. D. and Federer, C. A. (1992) A generalized, lumped-parameter model of photosynthesis, evapotranspiration and net primary production in temperate and boreal forest ecosystems. *Oecologia* **92**, 463–474.

Aber, J. D. and Melillo, J. M.(1982) FORTNITE: a computer model of organic matter and nitrogen dynamics in forest ecosystems. *Res. Bull. R3130*, University of Wisconsin.

Aber, J. D., Melillo, J. M. and Federer, C. A. (1982) Predicting the effects of rotation length, harvest intensity, and fertilization on fiber yield from northern hardwood forests in New England. *For. Sci.* **28**, 31–45.

Ågren, G. I. (1983) Nitrogen productivity of some conifers. *Can. J. For. Res.* **13**, 494–500.

Ågren, G. I. (1985) Theory for growth of plants derived from the nitrogen productivity concept. *Physiol. Plant.* **64**, 17–28.

Ågren, G. I. (1988) Ideal nutrient productivities and nutrient proportions in plant growth. *Plant, Cell Environ.* **11**, 613–620.

Ågren, G. I. and Bosatta, E. (1987) Theoretical analysis of the long term dynamics of carbon and nitrogen in soils. *Ecology* **68**, 1181–1189.

Ågren, G. I. and Bosatta, E. (1988) Nitrogen saturation of terrestrial ecosystems. *Environ. Pollut.* **54**, 185–197.

Ågren, G. I., McMurtrie, R. E., Parton, W. J., Pastor, J. and Shugart, H. H. (1991) State-of-the-art of models of production-decomposition linkages in conifer and grassland ecosystems. *Ecol. Appl.* **1**, 118–138.

Amthor, J. S. (1989) *Respiration and Crop Productivity*. Springer-Verlag, New York. 215 pp.

Aronsson, A. and Elowson, S. (1980) Effects of irrigation and fertilization on mineral nutrients in Scots pine needles. In: Persson, T. (Ed.) *Structure and Function of Northern Coniferous Forests – an Ecosystem Study. Ecol. Bull.* **32**, 219–228. Stockholm.

Axelsson, B. and Berg, B. (1988) Fixation of ammonia (^{15}N) to *Pinus sylvestris* needle litter in different stages of decomposition. *Scand. J. For. Res.* **3**, 273–279.

Axelsson, B. and Bråkenhielm, S. (1980) Investigation sites of the Swedish Coniferous Forest Project – biological and physiographical features. In: Persson, T. (Ed.) *Structure and Function of Northern Coniferous Forests – an Ecosystem Study, Ecol. Bull.* **32**, 25–64, Stockholm.

Bazzaz, F. A., Chiariello, N. R., Coley, P. D. and Pitelka, L. F. (1987) Allocating resources to reproduction and defense. *BioScience* **37**, 58–67.

Beets, P. N. and Pollock, D. S. (1987) Uptake and accumulation of nitrogen in *Pinus radiata* stands as related to age and thinning. *New Zeal. J. For. Sci.* **17**, 353–371.

Benson, M. L., Landsberg, J. J. and Borough, C. J. (1992a) The biology of forest growth experiment: an introduction. *For. Ecol. Mgmt* **52**, 1–16.

Benson, M. L., Myers, B. J. and Raison, R. J. (1992b) Dynamics of stem growth of Pinus radiata as affected by water and nitrogen supply. *For. Ecol. Mgmt.* **52**, 117–137.

Booth, T. H. and McMurtrie, R. E. (1988) Climatic change and *Pinus radiata* plantations in Australia. In: Pearman, G.I. (Ed.) *Greenhouse: Planning for Climate Change*, pp. 534–545. E.J. Brill, Leiden.

Bosatta, E. and Ågren, G. I. (1985) Theoretical analysis of decomposition of heterogeneous substrates. *Soil Biol. Biochem.* **17**, 601–610.

Bosatta, E. and Ågren, G. I. (1991a) Dynamics of carbon and nitrogen in the organic matter of the soil: a generic theory. *Am. Nat.* **138**, 227–245.

Bosatta, E. and Ågren, G.I. (1991b) Theoretical analysis of carbon and nutrient interactions in soils under energy-limited conditions. *Soil Sci. Soc. Am. J.* **55**, 728–733.

Botkin, D. B., Janak, J. F. and Wallis, J. R. (1972) Some ecological consequences of a computer model of forest growth. *J. Ecol.* **60**, 849–872.

Box, E. (1978) Geographical dimensions of terrestrial net and gross primary production. *Rad. Environ. Biophysics* **15**, 305–322.

Box, E. (1981) *Macroclimate and Plant Forms: An Introduction to Predictive Modeling in Phytogeography*. Dr W. Junk Publishers, The Hague, Netherlands.

Bringmark, L. (1980) Ion leaching through a podsol in a Scots pine stand. In: Persson, T. (Ed.) *Structure and Function of Northern Coniferous Forests – an Ecosystem Study. Ecol. Bull.* **32**, 341–361, Stockholm.

Burke, I. C., Kittel, T. G. F., Lauenroth, W. K., Snook, P., Yonker, C. M. and Parton, W. J. (1991) Regional analysis of the central Great Plains. *BioScience* **41**, 685–692.

Cannell, M. G. R. (1982) *World Forest Biomass and Primary Production Data.* Academic Press, London.

Cannell, M. G. R., Grace, J. and Booth, A. (1988) *Possible Impacts of Climatic Warming on Trees and Forests in UK: A Review.* Report to the Department of Environment and Natural Environment Research Council, Penicuik, United Kingdom.

Crane, W. J. B. and Banks, J. C. G. (1992) Accumulation and retranslocation of foliar nitrogen in fertilised and irrigated *Pinus radiata. For. Ecol. Mgmt* **52**, 201–223.

Emanuel, W. R., Shugart, H. H. and Stevenson, M. P. (1985a) Climatic change and the broad-scale distribution of terrestrial ecosystem complexes. *Clim. Change* **7**, 29–43.

Emanuel, W. R., Shugart, H. H. and Stevenson, M. P. (1985b) Response to comment: climatic change and the broad-scale distribution of terrestrial ecosystem complexes. *Clim. Change* **7**, 457–460.

Farquhar, G. D. and von Caemmerer, S. (1982) Modelling of photosynthetic response to environment. In: Lange, O. L. Nobel, P. S., Osmond, C. B. and Ziegler, H. (Eds) *Encyclopaedia of Plant Physiology*, New Series Vol. 12B, pp. 547–587. Springer-Verlag, Berlin.

Farquhar, G. D., von Caemmerer, S. and Berry, J. A. (1980) A biochemical model of photosynthetic CO_2 assimulation in leaves of C_3 species. *Planta* **149**, 78–90.

Field, C. (1983) Allocating leaf nitrogen for the maximization of carbon gain: leaf age as a control on the allocation program. *Oecologia* **56**, 341–347.

Field, C. and Mooney, H. A. (1986) The photosynthesis–nitrogen relationship in wild plants. In: Givnish, T.J. (Ed.) *On the Economy of Plant Form and Function*, pp. 25–55. Cambridge University Press, Cambridge.

Finnegan, J. J. and Raupach, M. R. (1987) Transfer processes in plant canopies in relation to stomatal characteristics. In: Zeiger, E., Farquhar, G.D. and Cowan, I.R. (Eds) *Stomatal Function*, pp. 385–429. Stanford University Press, Stanford.

Flower-Ellis, J. G. K. and Persson, H. (1980) Investigation of structural properties and dynamics of Scots pine stands. In: Persson, T. (Ed.) *Structure and Function of Northern Coniferous Forests – an Ecosystem Study. Ecol. Bull.* **32**, 125–138. Stockholm.

Friend, A. D. (1991) Use of a model of photosynthesis and leaf microenvironment to predict optimal stomatal conductance and leaf nitrogen partitioning. *Plant, Cell*

Environ. **14**, 895–905.

Friend, A. D., Shugart, H. H. and Running, S. W.(1993) A physiology-based gap model of forest dynamics. *Ecology* **74**, 792–797.

Holdridge, L. R. (1947) Determination of world plant formations from simple climate data. *Science* **105**, 367–368.

Holdridge, L. R. (1964) *Life Zone Ecology*. Tropical Science Center, San José, Costa Rica.

Hunt, E. R. Jr, Martin, F. C. and Running, S. W. (1991) Simulating the effects of climatic variation on stem carbon accumulation of a ponderosa pine stand: comparison with annual growth increment data. *Tree Physiol.* **9**, 161–171.

Hunt, E. R. Jr and Running, S. W. (1992) Simulated dry matter yields for aspen and spruce stands in the North American boreal forest. *Can. J. Remote Sens.* **18**, 126–133.

Hunt, H. W. (1977) A simulation model for decomposition in grasslands. *Ecology* **58**, 469–484.

Ingestad, T. and Ågren, G. I. (1991) The influence of plant nutrition on biomass allocation. *Ecol. Appl.* **1**, 168–174.

Jarvis, P. G. (1985) Transpiration and assimilation of tree and agricultural crops: the 'omega factor'. In: Cannell, M. G. R. and Jackson, J. E. (Eds) *Attributes of Trees as Crop Plants*, pp. 460–480. Institute of Terrestrial Ecology, Huntington.

Jarvis, P. G. (1989) Atmospheric carbon dioxide and forests. *Phil.Trans.R.Soc.Lond.B* **324**, 369–392.

Khanna, P. K., Raison, R. J., Falkiner, R. A., Willett, I. R. and Connell, M. J. (1992) Effects of N P K fertilisation on the chemistry of a yellow podzolic soil under *Pinus radiata*. *For. Ecol. Mgmt* **52**, 65–85.

Kirschbaum, M. U. F. and Farquhar, G. D. (1984) Temperature dependence of whole-leaf photosynthesis in *Eucalyptus pauciflora* Sieb. ex Spreng. *Aust. J. Plant Physiol.* **11**, 519–538.

Korol, R. L., Running, S. W., Milner, K. S. and Hunt, E. R. (1991) Testing a mechanistic carbon balance model against observed tree growth. *Can. J. For. Res.* **21**, 1098–1105.

Landsberg, J. J., Kaufmann, M. R., Binkley, D., Isebrands, J. and Jarvis, P. G. (1991) Evaluating progress toward closed forest models based on fluxes of carbon, water, and nutrients. *Tree Physiol.* **9**, 1–15.

Lieth, H. (1972) Modeling the primary productivity of the world. *Nat. Resour.* **8**, 5–10.

Linder, S. and Axelsson, B. (1982) Changes in carbon uptake and allocation patterns as a result of irrigation and fertilization in a young *Pinus sylvestris* stand. In: Waring, R.H. (Ed.) *Carbon Uptake and Allocation in Subalpine Ecosystems as a Key to Management: Proceedings of an IUFRO Workshop*, pp. 38–44. Forest Research Laboratory, Oregon State University, Corvallis.

Linder, S. and Troeng, E. (1980) Photosynthesis and transpiration of 20-year-old Scots pine. In: Persson, T. (Ed.) *Structure and Function of Northern Coniferous Forests – an Ecosystem Study*. *Ecol. Bull.* **32**, 165–181, Stockholm.

McGill, W. B., Hunt, H. W., Woodmansee, R. G. and Reuss, J. O. (1981) PHOENIX, a model of the dynamics of carbon and nitrogen in grassland soils. *Ecol. Bull.* **33**, 49–115, Stockholm.

McGuire, A.D., Joyce, L.A., Kicklighter, D.W., Melillo, J.M., Esser, G. and Vörösmarty, C.J. (1993) Productivity response of climax temperate forests to elevated temperature and carbon dioxide: a North American comparison between two global models. *Clim. Change* **24**, 287–310.

McGuire, A. D., Melillo, J. M., Joyce, L. A., Kicklighter, D. W., Grace, A. L., Moore, B. III and Vörösmarty, C. J.(1992) Interactions between carbon and nitrogen dynamics in estimating net primary productivity for potential vegetation in North America. *Global Biogeochem. Cyc.* **6**, 101–124.

McMurtrie, R.E. (1993) Modelling canopy carbon and water balance. In: Hall, D.O., Scurlock, J.M.O., Bolar, H., Leegood, R.C. and Long, S.P. (Eds) *Photosynthesis and Production in a Changing Environment: A Field and Laboratory Manual*, pp. 220–231. Chapman & Hall, London.

McMurtrie, R.E., Benson, M.L., Linder, S., Running, S.W., Talsma, T., Crane, W.J.B. and Myers, B.J. (1990a) Water/nutrient interactions affecting productivity of stands of *Pinus radiata*. *For. Ecol. Mgmt* **30**, 15–423.

McMurtrie, R.E. and Landsberg, J.J. (1992) Using a simulation model to evaluate the effects of water and nutrients on the growth and carbon partitioning of *Pinus radiata*. *For. Ecol. Mgmt* **52**, 243–260.

McMurtrie, R.E., Leuning, R., Thompson, W.A. and Wheeler, A.M. (1992) A model of canopy photosynthesis and water use incorporating a mechanistic formulation of leaf CO_2 exchange. *For. Ecol. Mgmt* **52**, 261–278.

McMurtrie, R.E., Rook, D.A. and Kelliher, F.M. (1990b) Modelling the yield of Pinus radiata on a site limited by water and nitrogen. *For. Ecol. Mgmt* **30**, 381–413.

McMurtrie, R.E. and Wang, Y.-P. (1993) Mathematical models of photosynthetic response of plant stands to rising CO_2 concentrations and temperatures. *Plant, Cell Environ.* **16**, 1–16.

Meentemeyer, V., Box, E.O. and Thompson, R. (1982) World patterns and amounts of terrestrial plant litter production. *BioScience* **32**, 125–128.

Melillo, J.M. (1983) Will increases in atmospheric CO_2 concentrations affect decay rates? In: *The Ecosystems Center 1983 Annual Report*, pp. 10–11, Marine Biological Laboratory, Woods Hole, Mass.

Melillo, J.M., Aber, J.D. and Muratore, J.F.(1982) Nitrogen and lignin control of hardwood leaf litter decomposition. *Ecology* **63**, 621–626.

Melillo, J.M., McGuire, A.D., Kicklighter, D.W., Moore, B.III, Vörösmarty, C.J. and Schloss, A.L. (1993) Global climate change and terrestrial net primary production. *Nature*, **363**, 234–240.

Mooney, H.A., Pearcy, R.W. and Ehrlinger, J. (1987) Plant physiological ecology today. *BioScience* **37**, 18–20.

Myers, B.J. (1988) Water stress integral – a link between short-term stress and long-term growth. *Tree Physiol.* **4**, 315–323.

Myers, B.J. and Talsma, T. (1992) Site water balance and tree water status in irrigated and fertilized stands of *Pinus radiata*. *For. Ecol. Mgmt* **52**, 17–42.

Nemani, R.R. and Running, S.W. (1989) Testing a theoretical climate–soil–leaf area hydrologic equilibrium of forests using satellite data and ecosystem simulation. *Agric. For. Met.* **44**, 245–60.

Nömmik, H. and Vahtras, K. (1982) Retention and fixation of ammonium and ammonia in soils. In: Stevenson, F.I.(Ed.) *Nitrogen in agricultural soils*, Agronomy Series 22, pp. 123–171. University of Wisconsin, Madison.

Oechel, W.C., Cowles, S., Grulke, N., Hastings, S.J., Lawrence, B., Prudhomme, T., Reichers, G., Strain, B., Tissue, D. and Vourlitis, G. (1994) Transient nature of CO_2 fertilization in Arctic tundra. *Nature* **371**, 500–503.

Parton, W.J., Schimel, D.S., Cole, C.V. and Ojima, D.S. (1987) Analysis of factors controlling soil organic matter levels in Great Plains grasslands. *Soil Sci. Soc. Amer. J.* **51**, 1173–1179.

Parton, W.J., Stewart, J.W.B. and Cole, C.V. (1988) Dynamics of C, N, P, and S in grassland soils: a model. *Biogeochemistry* **5**, 109–131.

Pastor, J. and Post, W.M. (1985) Development of a linked forest productivity – soil process model. ORNL/TM-9519, Oak Ridge National Laboratory, Oak Ridge, Tenn.

Pastor, J. and Post, W.M. (1986) Influence of climate, soil moisture, and succession on forest carbon and nitrogen cycles. *Biogeochemistry* **2**, 3–27.

Pastor, J. and Post, W.M. (1988) Response of northern forests to CO_2-induced climate change. *Nature*, **334**, 55–58.

Peet, M.M., Huber, S.C. and Patterson, D.T. (1985) Acclimation to high CO_2 in monoecious cucumber. II. Carbon exchange rates, enzyme activities, and starch and nutrient concentrations. *Plant Physiol.* **80**, 63–67.

Persson, H. (1980) Structural properties of the field and bottom layers at Ivantjärnsheden. In: Persson, T. (Ed.) *Structure and Function of Northern Coniferous Forests – an Ecosystem Study. Ecol. Bull.* **32**, 153–163. Stockholm.

Persson, T. (Ed.) (1980) *Structure and Function of Northern Coniferous Forests – an Ecosystem Study, Ecol. Bull.* **32**, 609 pp. Stockholm.

Prentice, I.C., Cramer, W., Harrison, S.P., Leemans, R., Monserud, R.A. and Solomon, A.M. (1992) A global biome model based on plant physiology and dominance, soil properties and climate. *J. Biogeog.* **19**, 117–134.

Raich, J.W., Rastetter, E.B., Melillo, J.M., Kicklighter, D.W., Steudler, P.A., Peterson, B.J., Grace, A.L., Moore, B. III and Vörösmarty, C.J.(1991) Potential net primary productivity in South America: application of a global model. *Ecol. Appl.* **1**, 399–429.

Raison, R.J., Connell, M.J. and Khanna, P.K. (1987) Methodology for studying fluxes of soil mineral-N *in situ. Soil Biol. Biochem.* **19**, 521–530.

Raison, R.J., Connell, M.J., Khanna, P.K. and Falkiner, R.A. (1992a) Effects of irrigation and nitrogen fertilization on fluxes of soil mineral nitrogen in a stand of *Pinus radiata. For. Ecol. Mgmt* **52**, 43–64.

Raison, R.J., Khanna, P.K., Benson, M.L., Myers, B.J., McMurtrie, R.E. and Lang, A.R.G. (1992b) Dynamics of *Pinus radiata* foliage in relation to water and nitrogen stress. II. Needle-loss and temporal changes in total foliage mass. *For. Ecol. Mgmt* **52**, 159–178.

Raison, R.J., Khanna, P.K., Connell, M.J. and Falkiner, R.A. (1990) Effects of water availability and fertilization on N cycling in a stand of *Pinus radiata. For. Ecol. Mgmt* **30**, 31–43.

Raison, R.J. and Myers, B.J. (Eds) (1992a) The Biology of Forest Growth Experiment. *For. Ecol. Mgmt* **52**, 1–311.

Raison, R.J. and Myers, B.J. (1992b) The Biology of Forest Growth Experiment: linking water and nitrogen availability to the growth of *Pinus radiata. For. Ecol. Mgmt* **52**, 279–308.

Raison, R.J., Myers, B.J. and Benson, M.L. (1992c) Dynamics of *Pinus radiata* foliage in relation to water and nitrogen stress. I. Needle production and properties. *For. Ecol. Mgmt* **52**, 139–158.

Rastetter, E.B., McKane, R.B., Shaver, G.R. and Melillo, J.M. (1992) Changes in C storage by terrestrial ecosystems: how C–N interactions restrict responses to CO_2 and temperature. *Water, Air, Soil Pollut.* **64**, 327–344.

Rastetter, E.B., Ryan, M.G., Shaver, G.R., Melillo, J.M., Nadelhoffer, K.J., Hobbie, J.E. and Aber, J.D. (1991) A general biogeochemical model describing the responses of the C and N cycles in terrestrial ecosystems to changes in CO_2, climate, and N deposition. *Tree Physiol.* **9**, 101–126.

Reich, P.B., Walters, M.B. and Ellsworth, D.S. (1992) Leaf life-span in relation to leaf, plant, and stand characteristics among diverse ecosystems. *Ecol. Monog.* **62**, 365–92.

Rosenzweig, M.L. (1968) Net primary production of terrestrial communities: prediction from climatological data. *Am. Nat.* **102**, 67–74.

Running, S.W. and Coughlan, J.C. (1988) A general model of forest ecosystem processes for regional applications. I. Hydrologic balance, canopy gas exchange, and primary production processes. *Ecol. Mod.* **42**, 125–154.

Running, S.W. and Gower, S.T. (1991) FOREST-BGC, a general model of forest ecosystem processes for regional applications. II. Dynamic carbon allocation and nitrogen budgets. *Tree Physiol.* **9**, 147–160.

Running, S.W. and Hunt, E.R. (1993) Generalization of a forest ecosystem process model for other biomes, BIOME-BGC, and an application for global-scale models. In: Ehleringer, J.R. and Field, C.(Eds) *Scaling Physiological Processes from Leaf to Globe*, pp. 141–158. Academic Press, Orlando.

Running, S.W., Knight, D.H. and Fahey, T.J. (1983) Description and application of H2OTRANS, a stand level hydrologic model for western coniferous forests. In: Lauenroth, W.K., Skogerboe, G.V. and Flug, M. (Eds) *Analysis of Ecological Systems: State-of-the- Art in Ecological Modelling.* Developments in Environmental Modelling 5, pp. 489–495. Elsevier Scientific Publishing, New York.

Ryan, M.G. (1990) Growth and maintenance respiration in stems of *Pinus contorta* and *Picea engelmannii. Can. J. For. Res.* **20**, 48–57.

Ryan, M.G. (1991a) A simple method for estimating gross carbon budgets for vegetation in forest ecosystems. *Tree Physiol.* **9**, 255–266.

Ryan, M.G. (1991b) Effects of climate change on plant respiration. *Ecol. Appl.* **1**, 157–167.

Ryan, M.G. and Waring, R.H.(1992) Maintenance respiration and stand development in a subalpine lodgepole pine forest. *Ecology* **73**, 2100–2108.

Sanford, R.L. Jr, Parton, W.J., Ojima, D.S. and Lodge, D.J. (1991) Hurricane effects on soil organic matter dynamics and forest production in the Luquillo Experimental Forest, Puerto Rico: results of simulation modelling. *Biotropica* **23**, 364–372.

Santantonio, D. and Santantonio, E. (1987) Seasonal changes in live and dead fine roots during two successive years in a thinned plantation of *Pinus radiata* in New Zealand. *New Zeal. J. For. Sci.* **17**, 315–328.

Schulze, E.-D., Turner, N.C., Gollan, T. and Shackel, K.A. (1987) Stomatal responses to air humidity and to soil drought. In: Zeiger, E., Farquhar, G.D. and Cowan, I.R.(Eds) *Stomatal Function*, pp. 311–321. Stanford University Press, Stanford.

Shugart, H.H. (1984) *A Theory of Forest Dynamics.* Springer-Verlag, New York. 278 pp.

Shugart, H.H. and West, D.C. (1977) Development of an Appalachian deciduous forest succession model and its application to assessment of the impact of the chestnut blight. *J. Environ. Mgmt.* **5**, 161–179.

Snowdon, P. and Benson, M.L.(1992) Effects of combinations of irrigation and fertilization on the growth and above-ground biomass production of *Pinus radiata. For. Ecol. Mgmt* **52**, 87–116.

Soil Survey Staff (1975) *Soil Taxonomy.* Agric. Handb. No. 436, US Dept. Agric., Soil Conserv. Serv., Washington DC, USA.

Stace, H.C.T., Hubble, G.D., Brewer, R., Northcote, K.H., Sleeman, J.R., Mulcahy, M.J. and Hallsworth, E.G.(1968) *A Handbook of Australian Soils.* Rellim Technical Publications, Glenside, South Australia. 429 pp.

Strain, B.R. and Bazzaz, F.A. (1983) Terrestrial plant communities. In: Lemon, E.R. (Ed.) *CO_2 and Plants. The Response of Plants to Rising Levels of Atmospheric Carbon Dioxide.* American Assoc. Advancement of Science Selected Symposium 84, pp. 177–222. Westview Press, Inc., Boulder, Colo.

Swift, M.J., Heal, O.W. and Anderson, J.M.(1979) *Decomposition in Terrestrial Ecosystems.* Blackwell Scientific Publications, Oxford. 372 pp.

Thompson, W.A. and Wheeler, A.M. (1992) Photosynthesis by mature needles of field grown *Pinus radiata. For. Ecol. Mgmt* **52**, 225–242.

Thornley, J.H.M. (1972a) A model to describe the partitioning of photosynthate during vegetative growth. *Ann. Bot.* **36**, 419–430.

Thornley, J.H.M. (1972b) A balanced quantitative model for root:shoot ratios in vegetative plants. *Ann. Bot.* **36**, 431–441.

Trewartha, G.T. (1968) *An Introduction to Climate*. McGraw-Hill, New York.

Troeng, E. and Linder, S. (1982) Gas exchange in a 20-year-old stand of Scots pine. *Physiol. Plant.* **54**, 7–14.

Urban, D.L., Bonan, G.B., Smith, T.M. and Shugart, H.H. (1991) Spatial applications of gap models. *For. Ecol. Mgmt* **42**, 95–100.

Wang, Y.-P., McMurtrie, R.E. and Landsberg, J.J. (1992) Modelling canopy photosynthetic productivity. In: Baker, N.R. and Thomas, H. (Eds) *Crop Photosynthesis: Spatial and Temporal Determinants*, Topics in Photosynthesis, Vol. 12, pp. 43–67. Elsevier, Amsterdam.

Waring, R.H. and Pitman, G.B. (1985) Modifying lodgepole pine stands to change susceptibility to mountain pine beetle attack. *Ecology* **66**, 889–897.

Weinstein, D.A., Shugart, H.H. and West, D.C. (1982) *The Long-term Nutrient Retention Properties of Forest Ecosystems: a Simulation Investigation*. ORNL/TM-8472, Oak Ridge National Laboratory, Oak Ridge, Tenn.

Woodwell, G.M. (1987) Forests and climate: surprises in store. *Oceanus* **29**, 71–75.

Woodwell, G.M. and Botkin, D.B. (1970) Metabolism of terrestrial ecosystems by gas exchange techniques: the Brookhaven approach. In: Reichle, D.E.(Ed.) *Analysis of Temperate Forest Ecosystems*, pp. 73–85. Springer-Verlag, New York.

14 Comparing Models of Ecosystem Function for Temperate Conifer Forests. II. Simulations of the Effect of Climate Change

M. G. RYAN,[1] R. E. McMURTRIE,[2] G. I. ÅGREN,[3] E. R. HUNT, JR,[4] J. D. ABER,[5] A. D. FRIEND,[6] E. B. RASTETTER[7] and W. M. PULLIAM[8]

[1] *USDA Forest Service, Rocky Mountain Experiment Station, Fort Collins, USA*
[2] *Department of Biological Science, The University of New South Wales, Kensington, Australia*
[3] *Department of Ecology and Environmental Research, Swedish University of Agricultural Sciences, Uppsala, Sweden*
[4] *School of Forestry, University of Montana, Missoula, USA*
[5] *Complex Systems Research Center, Institute for the Study of Earth, Oceans and Space, University of New Hampshire, Durham, USA*
[6] *Institute of Terrestrial Ecology, Bush Estate, Penicuik, UK*
[7] *Ecosystems Center, Marine Biological Laboratory, Woods Hole, USA*
[8] *Natural Resource Ecology Laboratory, Colorado State University, Fort Collins, USA*

14.1 INTRODUCTION

This text is the second of two companion chapters. In the first (Chapter 13), we described seven models of forest biogeochemistry and compared model predictions against field data. Here, we compare model predictions for simple scenarios of altered climate at two sites with similar species, but contrasting natural climates. The simulations for altered climates considered both one-year scenarios and scenarios run for the normal expected life-spans of the species (60 or 120 years). The one-year simulations highlight the effect of climate on short-term physiological changes, while the long-term simulations focus on interactions between production and decomposition, and physiological changes that occur with forest development. We assumed that no species replacement would occur during the simulations.

Global Change: Effects on Coniferous Forests and Grasslands.
Edited by A. I. Breymeyer, D. O. Hall, J. M. Melillo and G. I. Ågren

14.2 METHODS

We compared predictions for scenarios of altered climate for seven models of forest element cycling at two sites with contrasting climates. The models were: BIOMASS (McMurtrie *et al.* 1990, 1992b; McMurtrie 1993; McMurtrie and Landsberg 1992; McMurtrie and Wang 1993); BIOME-BGC (Running and Coughlan 1988; Running and Gower 1991; Hunt and Running 1992); CENTURY (Parton *et al.* 1987, 1988; Sanford *et al.* 1991); HYBRID (Friend *et al.* 1993); MBL-GEM (Rastetter *et al.* 1991, 1992); PnET-CN (Aber and Federer 1992), and Q (Rolff and Ågren, in preparation). See section 13.3. Models differed greatly in the level of detail included, functional equations, temporal and spatial aggregation, and in interactions among different processes. The BIOME-BGC, CENTURY, MBL-GEM, PnET-CN, and Q models explicitly link production and decomposition (carbon (C) and nitrogen (N) cycling), while BIOMASS and HYBRID do not. The HYBRID simulations were run only for the BFG site.

14.2.1 Sites

Models were run using stand conditions and climate at two intensively studied sites: the Biology of Forest Growth (BFG) site near Canberra, Australia, and the Swedish Coniferous Forest Project (SWECON) site near Jädraås, Sweden. We selected these sites because they (1) differed in climate and resource availability, (2) had ecosystem-level data available for model initialization, parameterization, and validation, and (3) had species with similar physiology.

The BFG site is warm, somewhat limited by N availability, and can experience drought in the summer months (section 13.4.2.1; Raison and Myers 1992). In contrast, the SWECON site is cool, strongly limited by N availability, and rarely experiences drought in the growing season (section 13.4.2.2; Persson, 1980). The BFG site receives an average of 791 mm precipitation distributed evenly throughout the year (Table 13.2); the coldest month is July (mean temperature is 6 °C) and the warmest is January (mean temperature of 21 °C). The SWECON site receives an average of 607 mm precipitation; mean monthly temperature is −7 °C in January and 15 °C in July. At each site, similar irrigation and fertilization treatments manipulated water and nutrient availability. Many results from the studies are presented in Raison and Myers (1992) for BFG and in Persson (1980) for SWECON.

At BFG, *Pinus radiata* (Monterey pine) was planted in 1973 after harvest. At SWECON, *P. sylvestris* (Scots pine) was grown from stands naturally regenerated after harvest in 1957. Closed-canopy stands for the two species differ in leaf area; however, rates of photosynthesis and respiration per unit leaf area are similar, as are N concentrations for foliage and woody tissues (Chapter 13).

14.2.2 Modelling protocols

Modellers were provided with a summary of field data from the BFG and SWECON sites, assembled from sources given in Chapter 13 (Tables 13.3 and 13.4). Data included (1) the stand characteristics, (2) the storage of C and N in vegetation, litter, and soil, (3) net production, and (4) physiological parameters. Climate data (daily values of maximum, minimum, and average temperature, relative humidity, precipitation, and shortwave radiation) for the duration of the studies were also provided. Additional data or parameters required for model runs were the responsibility of the individual modellers. Special adjustments to the models necessary for the BFG and SWECON sites are given in sections 13.4.1.1–13.4.1.7, and modifications required to simulate altered climates are discussed in section 14.2.3.

Because our objective was to compare model predictions for different modelling approaches, we selected 3 simple scenarios of altered climate:

(a) increased temperature (+4 °C above ambient);
(b) increased atmospheric CO_2 concentration (700 ppm CO_2);
(c) increased temperature combined with high CO2 (+4 °C, 700 ppm CO_2).

Climate for +4 °C scenarios was constructed by simply adding 4 °C to maximum and minimum daily temperatures. Relative humidity was calculated assuming that daily minimum temperature was dew point–effectively raising absolute humidity slightly for the +4 °C scenarios. Precipitation and shortwave radiation were not altered. The scenarios are useful tools for exercising the models, but do not represent our expectations of changes in climate for the sites.

For the one-year scenarios, models were run using the actual meteorological data for the site until the target year, when the modified climate was substituted. For BFG, models were initialized in 1983 (at the start of the experimental treatments) and run until July 1986 under actual climate. For SWECON, models were initialized in 1979 and run for January 1979 to December 1979 with the modified climate. At both sites, simulations were run for conditions in the control, irrigation, and irrigation + fertilization treatments, using actual water and nutrient additions to the stands as model inputs. Models simulated the entire C, water, and N budgets for the year. We report results as changes in model estimates relative to those for actual climate.

For the long-term scenarios, models were run using a single year of either the actual or altered climate (1979 for SWECON, July 1986–June 1987 for BFG) repeated throughout the simulation. We chose the length of the simulation to represent the approximate age of mature (but not decadent) stands typical for the species at the study sites. For BFG, stands were initialized using actual values in 1983, and simulations were run for 60 years. For SWECON, models were initialized using actual values in 1979 and simulations were run for 120 years. The long-term simulations were run only for the control stands, and we assumed no changes in the pine species at either site. We report simulated stem biomass, aboveground net primary production, and soil C.

14.2.3 How models implemented changes in climate

Most of the models presented in Chapter 13 were designed to respond to climate. However, because the original objectives of the models dictated different model structures and temporal scales, some models were changed from the published versions to run the simulations presented here. The BIOME-BGC, HYBRID, and PnET-CN models made no changes to simulate the altered climates.

The BIOMASS model assumed that an atmospheric CO_2 concentration of 700 ppm would reduce stomatal conductance by 30% and that the effect would occur throughout the simulation. To estimate respiration costs for the long-term simulations, BIOMASS calculated stem maintenance respiration from sapwood volume, and assumed that sapwood volume decreased linearly from 100% of total wood volume at age 10 to 25% at stand maturity.

Temperature effects were implemented in CENTURY by increasing each monthly temperature value by +4 °C. The CENTURY model does not explicitly simulate direct effects of atmospheric CO_2 levels on tree physiology and production. Therefore, CO_2 effects on nutrient and water use efficiencies were simulated by manually increasing foliar C to N ratios and maximum net primary productivity by 30%. These changes allowed higher rates of production under a given nutrient and/or water status.

The MBL-GEM explicitly models the effects of CO_2 on C gain, growth, allocation, litter quality and N mineralization. Soil moisture affects C fixation, growth and decomposition, but is a model input. The MBL-GEM assumed that the +4 °C scenario decreased soil moisture by 15% at BFG and SWECON. The 700 ppm CO_2 scenario was assumed not to affect soil moisture at either site.

The Q model assumed that a doubling of CO_2 increased N productivity (biomass production per unit of plant-available N (Ågren 1983, 1985)) by 20%, and temperature increased all process rates exponentially assuming a Q_{10} of 2.

14.3 SIMULATION RESULTS: ONE-YEAR RESPONSE TO ALTERED CLIMATES

One-year simulations examine the effect of changes in atmospheric CO_2 and temperature on the processes of photosynthesis, respiration, N mineralization, transpiration, and evaporation integrated over a year. In one year, interactions between plants, litter, and soil are generally less important than changes in process rates. However, interactions within the plant system and between plant and soil can occur within a year because of rapid changes in leaf mass. Shedding of foliage under water or nutrient stress (or increased retention of foliage) can

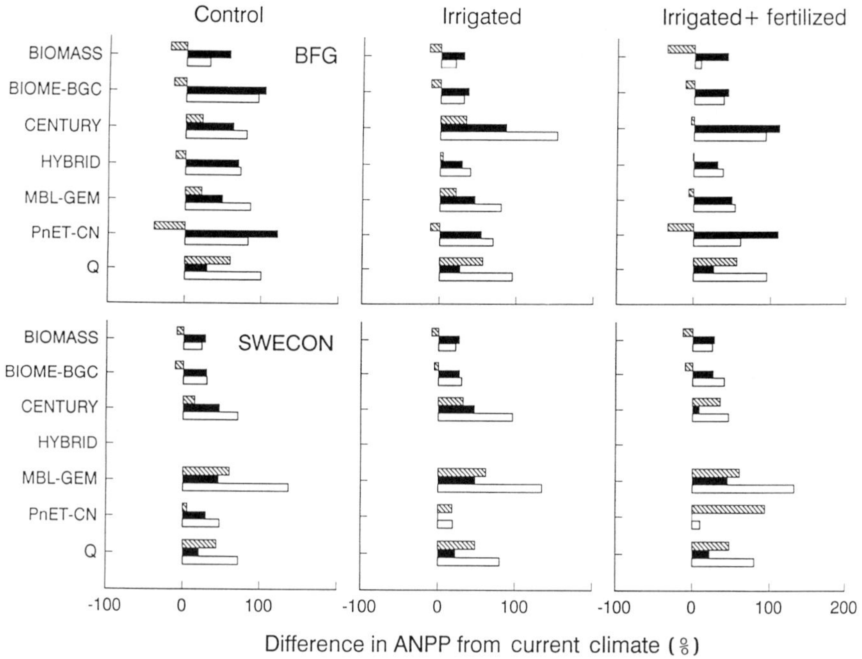

Figure 14.1 The effect of one-year changes in climate and/or CO_2 level on annual aboveground net primary production (hatched bars = +4 °C; black bars = 700 ppm CO_2; and open bars = +4 °C plus 700 ppm CO_2). Bars are percent difference from current climate for control, irrigated, and irrigated + fertilized stands of *Pinus radiata* in Australia (BFG) and *P. sylvestris* in Sweden (SWECON)

affect net C gain by the canopy, C allocation within trees, and N mineralization or immobilization from the litter.

14.3.1 Carbon cycle

Aboveground net primary production (ANPP) is the annual sum of all dry matter production for aboveground plant parts. Model predictions for ANPP differed substantially (in direction) for the +4 °C scenario. However, model predictions for the two scenarios with increased CO_2 were remarkably consistent in both direction and magnitude (Figure 14.1).

The response of net canopy photosynthesis (photosynthesis minus 24-hour foliage maintenance respiration) closely matched that of ANPP (data not shown). Additionally, all models showed that photosynthesis was more responsive to temperature and CO_2 than was respiration. Consequently, at the BFG site, when aboveground production for BIOMASS, BIOME-BGC, HYBRID, and PnET-

CN decreased, the decrease was caused by lower canopy photosynthesis, not by greater respiration. Where models estimated an increase in productivity for the +4 °C scenarios, the increase occurred because photosynthesis increased more than respiration increased. Where models simulated increased photosynthesis, the photosynthetic response was caused by increased N content of foliage – derived, in turn, from the greater N available because of higher decomposition rates.

14.3.1.1 Response to increased temperature

Model predictions differed in response to increases in temperature (Figure 14.1). The CENTURY, MBL-GEM, and Q models predicted that ANPP would increase 21–59% for the BFG control stands, while the other models predicted a 12–40% decrease in ANPP. At the cooler SWECON site, only BIOME-BGC and BIOMASS estimated decreases (about 10%) in ANPP, while the other models predicted increases of 5–60%. Simulations for irrigated and irrigated + fertilized plots were similar to those for control plots, indicating that the availability of water and nutrients had little effect on the increased-temperature simulations.

Differences in model predictions for the +4 °C climate were caused by differences in model assumptions. Models with a strongly linked plant and soil model (CENTURY, MBL-GEM, Q) predicted increased ANPP in a warmer climate, where models with detailed canopy physiology (BIOMASS, BIOME-BGC, HYBRID) predicted decreased ANPP. The physiologically based models predicted lower photosynthesis in the warmer climate, because of reduced leaf area and increased stomatal closure caused by low soil and plant water potentials. Most of the linked plant–soil models predicted greater production in the warmer climate because of higher decomposition rates and greater N availability. The PnET-CN is also a linked plant–soil model, but it estimates water use from a C assimilation using a relationship between stomatal conductance and assimilation (Aber and Federer 1992). Interestingly, PnET-CN predicted lower ANPP for the BFG site and higher ANPP for the SWECON site for the +4 °C climate. The other models simulated differences in the same direction for both sites.

14.3.1.2 Response to increased CO_2

All models simulated that a doubled atmospheric CO_2 concentration (alone or together with a temperature increase) would increase annual ANPP (Figure 14.1). Estimated increases ranged from 23% for BIOMASS to 137% for MBL-GEM. Models estimated that doubled atmospheric CO_2 would increase annual ANPP less at the cooler SWECON site (average increase = 34%) than for the warmer BFG site (average increase 70%). Compared with simulations for the control plots, models predicted less response in annual ANPP to increased CO_2 for the irrigation (average response = 44%) and irrigation + fertilization (average response = 60%) treatments at BFG. At SWECON, model predictions differed little between the control and treated stands.

Aboveground net primary production increased in response to doubled atmospheric CO_2 because of higher photosynthesis rates (BIOMASS, BIOME-BGC, HYBRID, MBL-GEM, PnET-CN), greater water-use efficiency (BIOMASS, BIOME-BGC, HYBRID, PnET-CN), dilution of the available nutrients by increasing C:N ratios in tissues (MBL-GEM), increased N productivity (Q), and increased net production (CENTURY). Acclimation of photosynthesis to high CO_2 (cf. Eamus and Jarvis 1989) was not explicitly simulated in any of the models. However, MBL-GEM does acclimate to increased CO_2 by estimating photosynthesis from foliar N stocks and diluting those stocks at high CO_2.

No model simulated the direct effects of CO_2 on respiration rates. Several studies have that dark respiration declines at elevated CO_2 (Amthor 1991), perhaps because of higher energy-use efficiency in the short term, coupled with lower tissue N concentrations in the long term. If included, a respiration–CO_2 response is likely to increase ANPP.

14.3.1.3 Response to increased CO_2 and temperature

The average change in annual ANPP was much greater for the simulations with increased CO_2 (with or without a temperature increase) than for the +4 °C simulations (Figure 14.1). Therefore, simulations for increased CO_2 were similar to simulations for increased CO_2 and temperature. There were essentially no temperature/CO_2 interactions: annual ANPP simulated for the combined +4 °C, 700 ppm CO_2 scenarios was simply the sum of annual ANPP simulated for the individual +4 °C and 700 ppm CO_2 scenarios.

Interactions between temperature and CO_2 would likely be expressed though changes in nutrient cycling (e.g. N availability, litter quality) and associated changes in foliar biomass and photosynthetic capacity. However, because these changes take longer than one year to occur, the lack of a CO_2/temperature interaction for these one-year simulations is understandable.

14.3.2 Nitrogen cycle

The response of N mineralization to altered temperature and atmospheric CO_2 concentration differed considerably from that of ANPP (Figure 14.2). Models generally simulated increased N mineralization in a +4 °C climate, but predicted that doubled CO_2 would have little impact. However, CENTURY estimated elevated CO_2 would decrease N mineralization for the BFG plots. Increased temperature enhanced N mineralization more in the irrigated plots, but otherwise, climate and water availability had little effect on N mineralization. In the

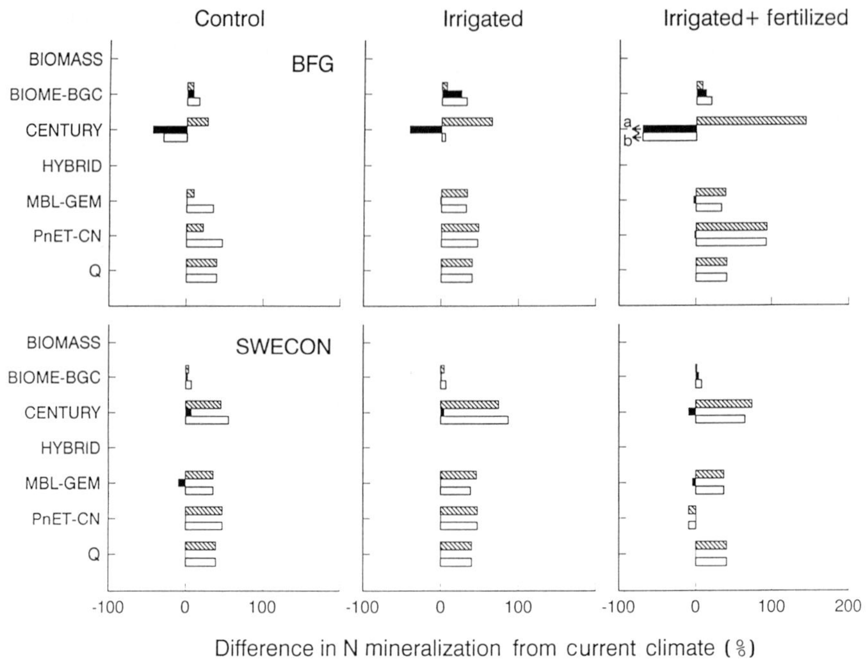

Figure 14.2 The effect of one-year changes in climate and/or CO_2 on annual net N mineralization (hatched bars = +4 °C; black bars = 700 ppm CO_2; and open bars = +4 °C plus 700 ppm CO_2). Bars are percent difference from current climate for control, irrigated, and irrigated + fertilized stands of *Pinus radiata* in Australia (BFG) and *P. sylvestris* in Sweden (SWECON). Arrows and letters indicate values off scale ($a = -186$, $b = -109$)

irrigation + fertilization treatments, BIOME-BGC and MBL-GEM predicted that N mineralization would differ little from the control plots, while Q, CENTURY, and PnET-CN predicted a very different response to altered temperature and CO_2.

The response of the decomposer community to altered soil temperature and moisture will largely determine how N mineralization will respond to changes in climate over a one-year period. However, changes in litter amount or quality could also have a short-term impact. In general, models predicted that increased soil temperature had a greater influence on decomposition than did any reduction in soil moisture. These predictions are consistent with those found using CENTURY to estimate climate-change responses for grassland soils (Burke *et al.* 1991).

Predictions of N mineralization by CENTURY were quite different from other models. In CENTURY, litter decomposition and N mineralization are very sensitive to the C:N ratio of litter. In the elevated CO_2 simulations, higher litter

C:N causes an immediate increase in microbial N immobilization at the BFG site, resulting in a sharp drop in predicted net N mineralization rates. This effect is enhanced further in the fertilization trials because of the elevated level of available N. At the SWECON site, litter C:N is already high and microbial N immobilization occurs at near maximum values. Thus, changes in litter C:N or N availability have little effect on net N mineralization simulated by CENTURY at SWECON.

Aboveground productivity and N mineralization have been strongly linked (cf. Nadelhoffer *et al.* 1985). The non-parallel response of ANPP and N mineralization found for the one-year simulations suggests that the large changes in climate temporarily decoupled the plant and soil systems. Production clearly increases more in a high-CO_2 atmosphere than does N mineralization. Such an increase is possible if the available nutrients are diluted (lowering C:N ratios in tissues), and nutrient-use efficiency increases. However, the increases may not be sustainable over the long term if element ratios within tissues have an upper limit, or if lower quality litter increases N immobilization (Rastetter *et al.* 1992; Oechel *et al.* 1994; see Long and Drake 1991, and Mooney *et al.* 1991, for a different argument). These considerations may cause differences between short- and long-term predictions.

14.3.3 Water cycle

All models (except CENTURY) that simulated transpiration predicted that transpiration would decrease in an atmosphere of 700 ppm in any treatment, location, or air temperature simulated (Figure 14.3). The CENTURY model predicted that transpiration would increase for irrigated stands at BFG and for all stands in the +4 °C–700 ppm CO_2 scenario at SWECON. All models predicted that transpiration would increase for the +4 °C scenario, but the magnitude of the response differed for CENTURY and the other models at BFG and for BIOME-BGC and the other models at SWECON. Model predictions did not differ substantially by resource availability (among treatments) or climate (between sites), except for BIOME-BGC and CENTURY.

All models that estimated evaporation (BIOMASS, BIOME-BGC, CENTURY, HYBRID) predicted little change over current climate for any treatment at the BFG site (data not shown; estimates for BIOMASS ranged from −16 to 23%; estimates for the other models ranged from −6 to 6%). At SWECON, BIOME-BGC and CENTURY predicted a 24–64% increase in evaporation for the +4 °C scenarios; BIOMASS predicted less than 7% change in evaporation for all altered climates.

A decrease in stomatal conductance with increased CO_2 is well documented at the leaf level (cf. Eamus and Jarvis 1989). Because most models include this response, stomatal conductance and transpiration decreased in a higher CO_2 atmosphere. However, no model considered acclimation of conductance, which

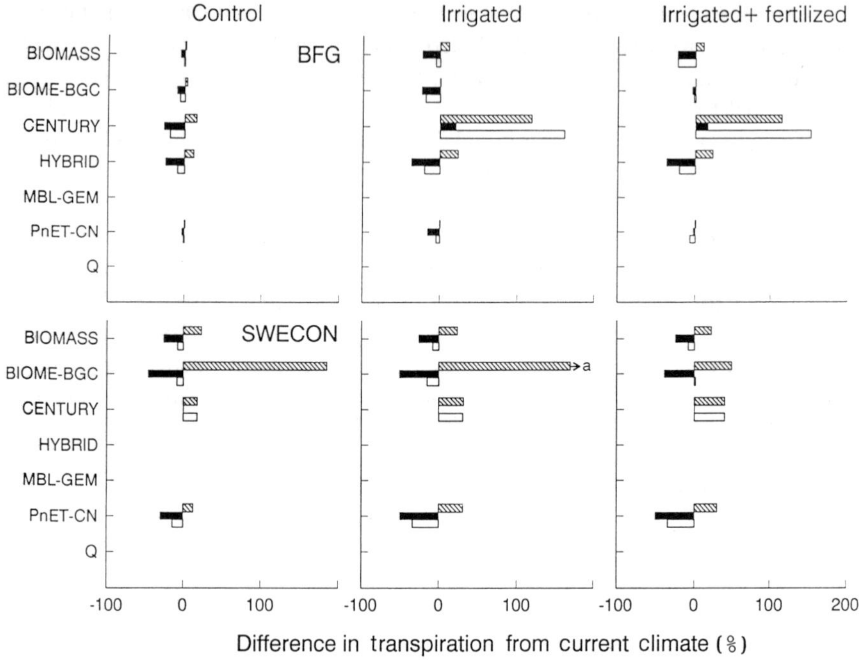

Figure 14.3 The effect of one-year changes in climate and/or CO_2 level on annual water flux for transpiration (hatched bars = +4 °C; black bars = 700 ppm CO_2; and open bars = +4 °C plus 700 ppm CO_2). Bars are percent difference from current climate for control, irrigated, and irrigated + fertilized stands of *Pinus radiata* in Australia (BFG) and *P. sylvestris* in Sweden (SWECON). The arrow and letter indicates a value off scale ($a = -222$)

can occur in conifers (Eamus and Jarvis 1989) and would mitigate any CO_2-induced decreases in stomatal conductance. Also, CO_2 may have a minimal effect on site water balance, if energy (McNaughton and Jarvis 1991), not stomatal conductance, controls water fluxes from ecosystems without water stress.

14.4 SIMULATION RESULTS: LONG-TERM RESPONSE TO ALTERED CLIMATES

Longer simulations integrate the response of processes that operate at the fine temporal scale (e.g. photosynthesis, respiration, stomatal conductance, and decomposition), with the interactions that occur between the plant and soil systems. Interactions between plants and soil considered by these models include:

(a) the effects of atmospheric CO_2 and plant N availability on litter quality;
(b) the effects of litter quality on decomposition, plant N availability, and soil organic matter turnover and storage;
(c) the effect of N availability on decomposition rates.

Not all interactions are represented in all models (see section 13.3), and BIOMASS and HYBRID do not simulate nutrient interactions.

Ecosystem response to persistent changes in climate can be influenced by the normal changes that occur with stand development as biomass and leaf area increase, litter and woody debris accumulate, and the conversion efficiency of energy to biomass decreases (Ryan *et al.* 1996). We recognize that such interactions exist and will complicate the interpretation of the long-term simulations. However, several of the models (BIOMASS, BIOME-BGC, HYBRID, PnET-CN, and Q) do not approach equilibrium; simulations of stand behaviour beyond mature stands would require assumptions or processes not included in the models. Because we could not do a comparison for equilibrium conditions, and because much of the temperate forest landscape consists of stands that are accumulating biomass, we decided to run the simulations for these 'aggrading' stands. Because the changes in ecosystem processes that occur with forest development can exceed those expected from an altered climate, we discuss aggrading stands under current climate before we present the results for altered climate.

14.4.1 Biomass accumulation and stand development under current climate

14.4.1.1 Stem biomass

Model predictions for both the endpoints and trajectories of stem biomass C differed substantially for both the BFG and SWECON sites (Figure 14.4). After a 60-year simulation at BFG, estimates of stem biomass C ranged from 6 kg m^{-2} (HYBRID) to 27 kg m^{-2} (Q) for the control stands; after a 120-year simulation at SWECON, estimates of stem biomass C ranged from 5 kg m^{-2} (MBL-GEM) to 15 kg m^{-2} (BIOMASS). Model predictions for the rate of stem biomass accumulation declined rapidly with age for CENTURY, MBL-GEM, BIOME-BGC, and HYBRID, but not for BIOMASS and Q. For the models where the rate of stem biomass accumulation declined with age, the decline was more pronounced at BFG than at SWECON.

14.4.1.2 Primary production and ecosystem production

Despite the large differences in predictions for stem biomass, model estimates of ANPP were remarkably consistent for the control stands (Figure 14.5). In all cases (except CENTURY at SWECON), ANPP declined with stand age, and the decline with age was more pronounced at BFG.

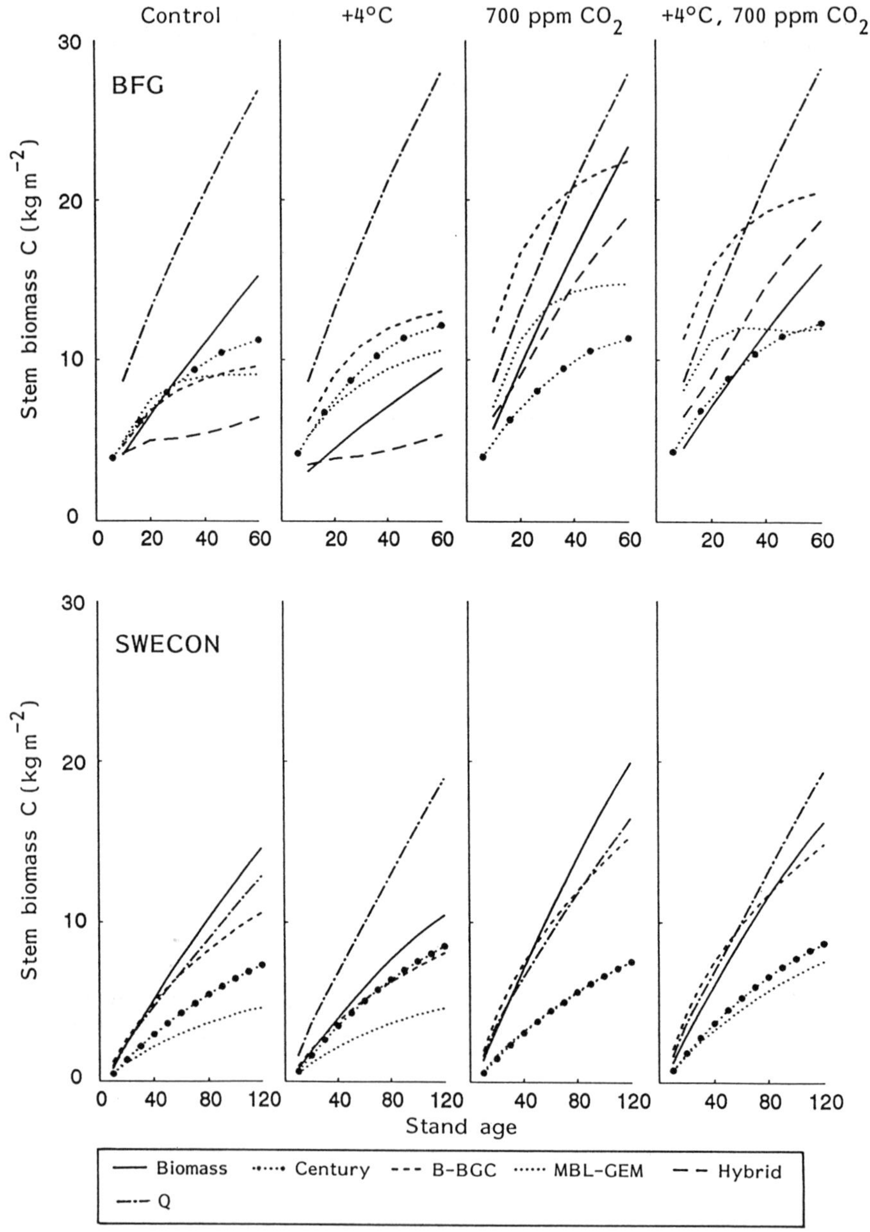

Figure 14.4 Stem biomass for control stands of *Pinus radiata* in Australia (BFG) and *P. sylvestris* in Sweden (SWECON) grown under altered climates (+4 °C, 700ppm CO_2, +4 °C and 700 ppm CO_2) for 60 (BFG) or 120 (SWECON) years. Model runs were started using data for stands in 1984 (BFG) or 1979 (SWECON)

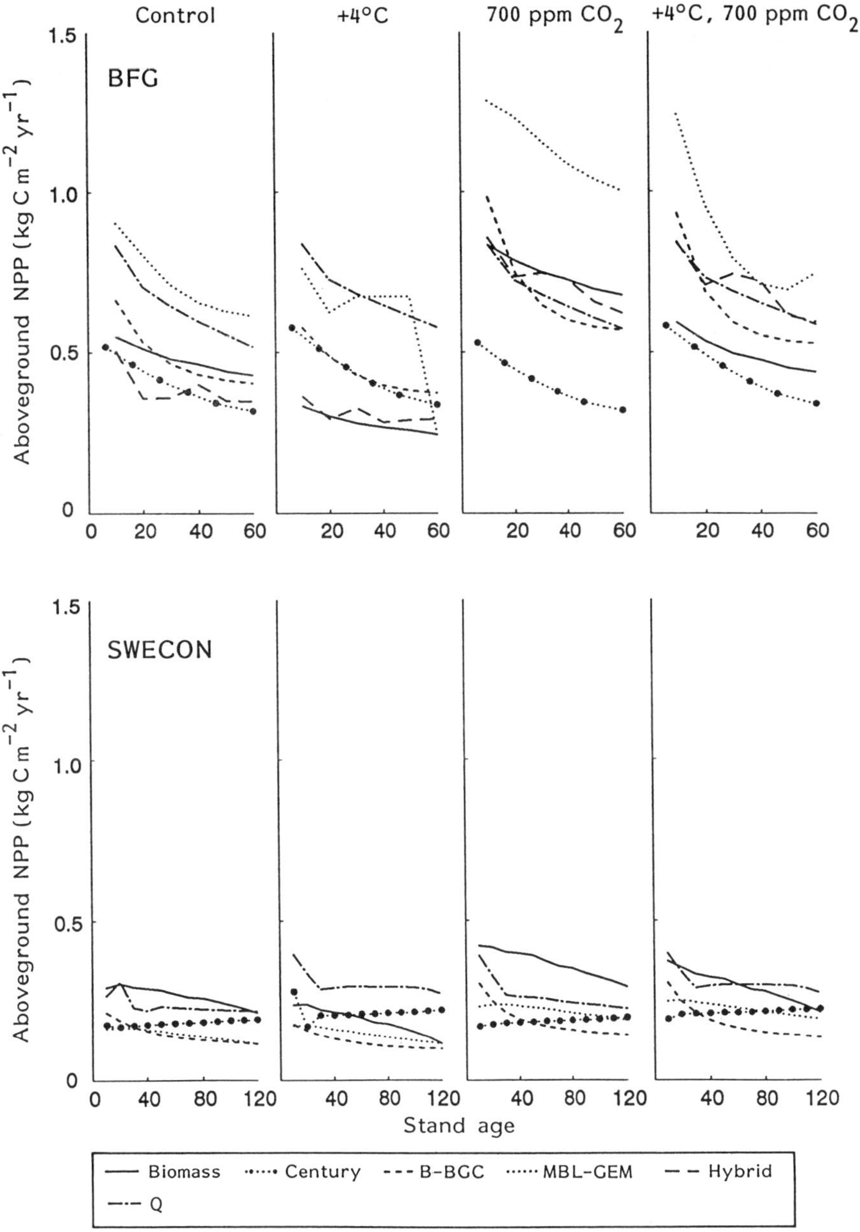

Figure 14.5 Aboveground net primary production for control stands of *Pinus radiata* in Australia (BFG) and *P. sylvestris* in Sweden (SWECON) grown under altered climates (+4 °C, 700 ppm CO_2, +4 °C and 700 ppm CO_2) for 60 (BFG) or 120 (SWECON) years. Model runs were started using data for stands in 1984 (BFG) or 1979 (SWECON)

Net ecosystem productivity (gross photosynthesis minus autotrophic and heterotrophic respiration) can be estimated from the simulation results by summing the change in C storage in vegetation and soil over time. Model predictions for net ecosystem production (NEP) (and, hence C storage) for mature stands differed substantially. At the BFG site after a 60 year simulation, NEP (in $g\,C\,m^{-2}\,yr^{-1}$) was 310 for BIOME-BGC, 299 for Q, 77 for CENTURY, and 19 for MBL-GEM. At the SWECON site after a 120 year simulation, NEP (in $g\,C\,m^{-2}\,yr^{-1}$) was 96 for Q, 76 for BIOME-BGC, 49 for CENTURY, and 12 for MBL-GEM.

14.4.1.3 Soil organic carbon

Model estimates of soil organic C varied, and model predictions differed by site (Figure 14.6). For those models that estimated soil organic C at BFG, only BIOME-BGC showed soil C increasing with stand age; the other models predicted a slight decrease. At SWECON, however, BIOME-BGC and CENTURY predicted soil C would increase with stand age, and Q and MBL-GEM predicted a decrease. Initial conditions differ for the models because BIOME-BGC and MBL-GEM include litter as part of soil organic C, while CENTURY and Q do not.

14.4.1.4 Explanation for differences in model behavior

Differences in model predictions for stand development can be partly explained by differences in the estimates of photosynthesis, net production, or allocation integrated over a long simulation (see section 13.5). Smaller differences in annual production (Figure 14.5) accumulate to larger differences in stem biomass (Figure 14.4) or soil C (Figure 14.6). Additionally, interactions within models change in different ways over time. For example, the cause of the decline in ANPP with stand development (shown in Figure 14.5) differed among models. For CENTURY, MBL-GEM, and Q, ANPP declined with stand development because nutrient availability declined as nutrients were sequestered in biomass. For BIOMASS, BIOME-BGC, and HYBRID, ANPP declined with stand development because of higher respiration costs (because of greater amounts of wood and foliage). Because both respiration and nutrient uptake are enhanced in a warmer climate, the decline in ANPP with stand development was similar for all models, and more pronounced at the warmer BFG site.

Causes of a decline in productivity with stand age are not well understood. Several sources attribute the decline to increased respiration costs associated with the amount of woody biomass (Kramer and Kozlowski 1979; Waring and Schlesinger 1985) or surface area (Whittaker and Woodwell 1967). However, woody-tissue respiration did not explain productivity decline in a lodgepole pine forest (Ryan and Waring 1992). Other possible explanations are a decrease in

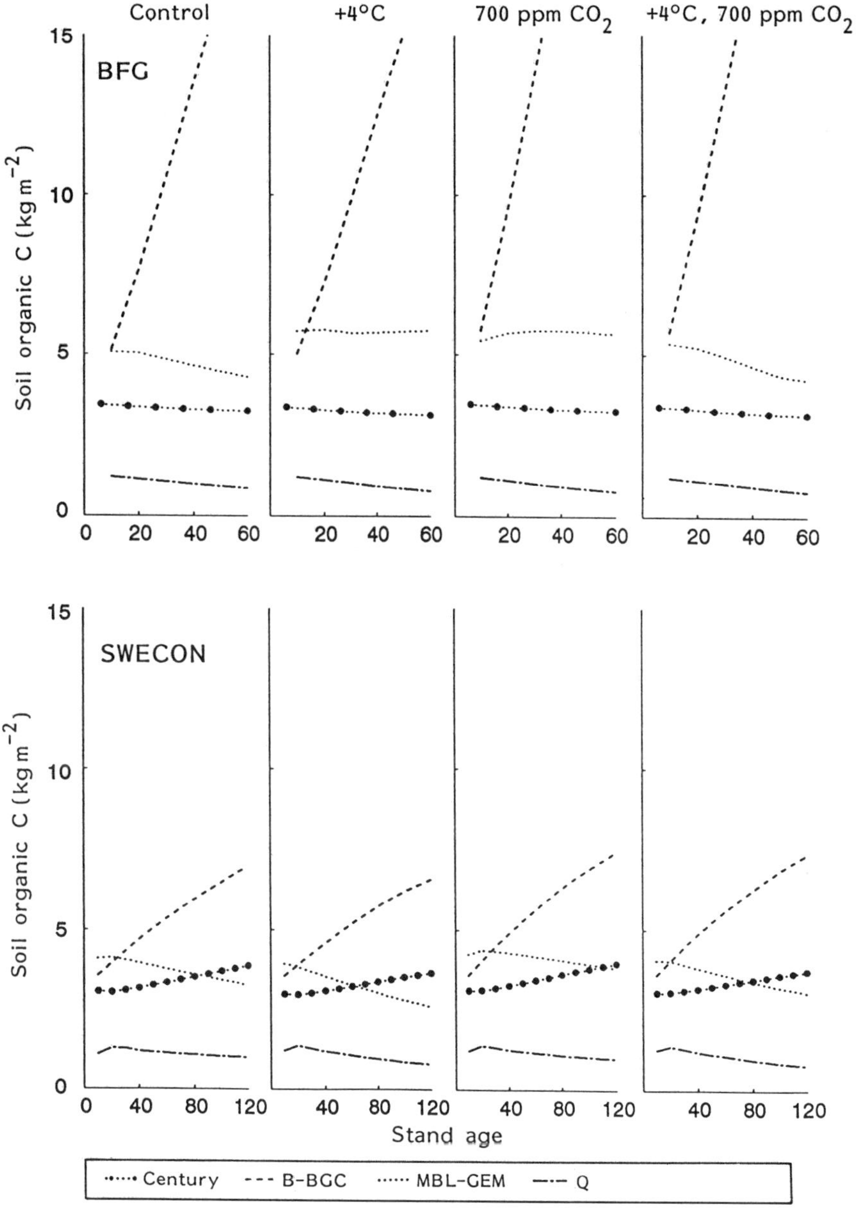

Figure 14.6 Soil C for control stands of *Pinus radiata* in Australia (BFG) and *P. sylvestris* in Sweden (SWECON) grown under altered climates (+4 °C, 700 ppm CO_2, +4 °C and 700 ppm CO_2) for 60 (BFG) or 120 (SWECON) years. Model runs were started using data for stands in 1984 (BFG) or 1979 (SWECON)

plant-available nutrients (Grier *et al.* 1981), increased mortality, or decreased photosynthesis (Yoder *et al.* 1994). Understanding the reasons for the changes that occur with stand development will be critical to building realistic models.

Differences in model predictions for soil organic C are caused by differences in decomposition rates, and in the amount of woody and fine litter estimated over time. Litter production from foliage (as a fraction of standing crop) is unlikely to change greatly with stand development. But, woody litter tends to increase as a stand reaches maturity and branches and trees die. Soil organic C declined with stand age for Q and MBL-GEM because litter production from wood was set at a constant, low fraction.

At the BFG site, CENTURY predicts a slow decline in soil organic C in response to the gradual decline in production and litter inputs. At SWECON, the reverse effect takes place, as the model simulates that atmospheric N inputs accumulate over time and allow for higher production. Higher production in turn results in higher soil C stocks. The difference between the two sites may be attributed to the initial conditions of the two sites. SWECON begins at a very N deficient state, while BFG starts with N only moderately limiting to production.

BIOME-BGC estimates that soil organic C will increase fourfold for the BFG site and twofold for the SWECON site. In BIOME-BGC, woody litter, because of its high lignin content, is transferred directly into the soil organic C compartment, and C turnover rates are unrealistically low because N immobilization was not modelled. As a result, BIOME-BGC is being revised.

14.4.2 Biomass accumulation and stand development under altered climates

For the longer simulations, we considered only the C balance. All models predicted that stem biomass C, ANPP, soil organic C, and net ecosystem production would differ after growth under altered temperature and atmospheric CO_2 concentration (Figures 14.4–14.8). For increases in CO_2, all models except CENTURY and Q predicted that stem biomass and ANPP (at the end of the simulation) would increase at least 25% over the control climate (Figure 14.7). Similar predictions were found in net ecosystem production for those models that estimated NEP. For the +4 °C scenario, models simulations differed in the direction and magnitude of the response of ANPP, stem biomass, and NEP (Figures 14.7 and 14.8). In the scenario with increased temperature and atmospheric CO_2, all models predicted a positive response in stem biomass, ANPP (except BIOMASS for ANPP at SWECON) and NEP. No model predicted that the altered climates would induce a change in soil organic C greater than 36% (Figure 14.7).

The responses of ANPP, stem biomass, and NEP to altered climate were similar, though the magnitude of the response was generally lower for ANPP and NEP. The direction of the predicted responses to climate did not change with site for any of the models, but most models predicted less response to altered climates

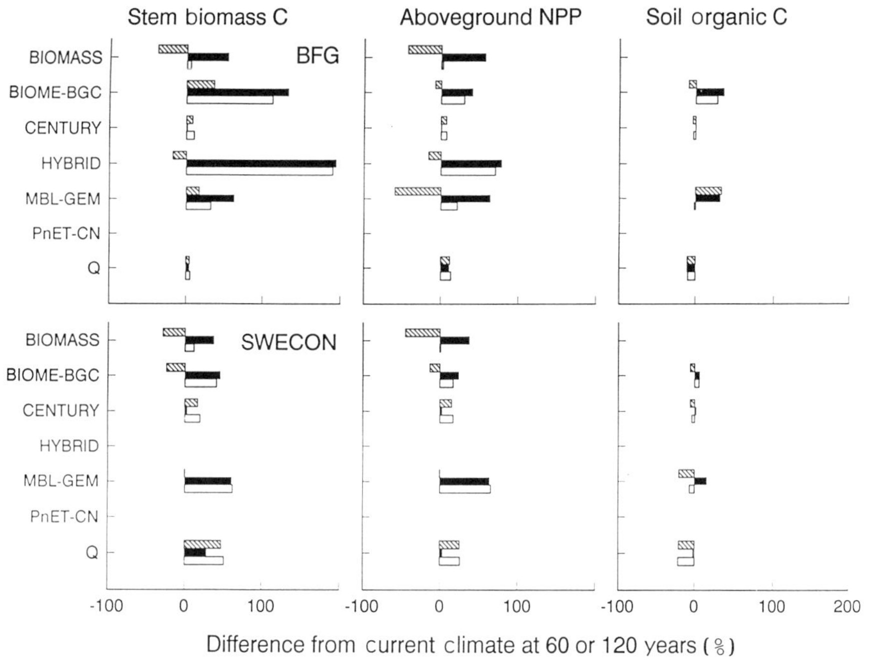

Figure 14.7 Percentage difference from current climate at 60 or 120 years for stem biomass, ANPP, and soil C. Simulations were for control stands of *Pinus radiata* in Australia (BFG) and *P. sylvestris* in Sweden (SWECON) grown under altered climates (hatched bars = +4 °C, black bars = 700 ppm CO_2, open bars = +4 °C, 700 ppm CO_2) for 60 (BFG) or 120 (SWECON) years. Model runs were started using data for stands in 1984 (BFG) or 1979 (SWECON)

(as a percentage difference from current climate) at the cooler SWECON site. Altered climate had little effect on model predictions for the pattern of stem biomass C, ANPP, and soil organic C over time (Figures 14.4–14.6).

14.4.2.1 Response to increased CO_2

Models generally agreed that greater atmospheric CO_2 (alone, or in combination with temperature) would increase aboveground net production and biomass accumulation and have a minor effect on soil C. Models disagreed about the effects of the 700 ppm CO_2 and the +4 °C scenarios on NEP. The difference in model predictions between the increased-temperature scenario and the combined temperature and CO_2 increase suggests that climate will strongly influence the response of ANPP to increased CO_2.

The patterns observed in the elevated-CO_2 simulations with aggrading stands were caused by factors similar to those discussed for the one-year simulations

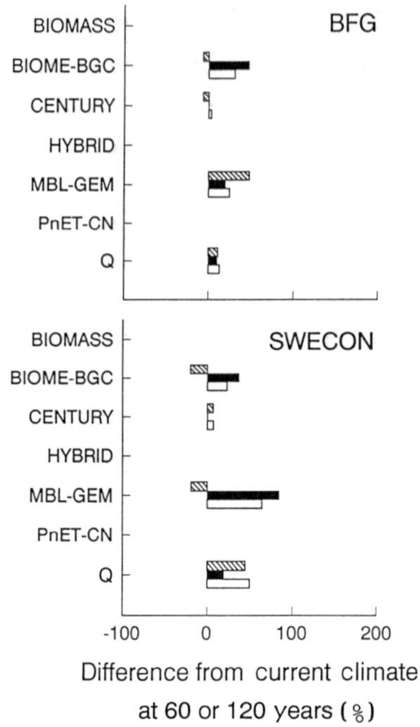

Figure 14.8 Percentage difference from current climate at 60 or 120 years for net ecosystem production. Simulations were for control stands of *Pinus radiata* in Australia (BFG) and *P. sylvestris* in Sweden (SWECON) grown under altered climates (hatched bars = +4 °C, black bars = 700 ppm CO_2, open bars = +4 °C, 700 ppm CO_2) for 60 (BFG) or 120 (SWECON) years. Model runs were started using data for stands in 1984 (BFG) or 1979 (SWECON)

(section 14.3.1). In an elevated-CO_2 atmosphere, photosynthesis and water use efficiency generally increase (Eamus and Jarvis 1989), and incremental changes accumulate over time. Models that strongly link plant and soil processes (CENTURY, MBL-GEM, Q) tend to limit the productivity gains induced by CO_2 with the constraint that nutrients must be available (Rastetter *et al.* 1991). However, these linked models can sequester more C and sustain a productivity response if C to element ratios within tissues can increase at elevated CO_2. For this reason, MBL-GEM (with flexible tissue element ratios) showed an increase in ANPP and stem biomass for the long-term simulation while CENTURY and Q (with fixed tissue element ratios) did not. The CENTURY and Q models predicted little or no long-term effect of CO_2 on NEP, perhaps because the response to CO_2 is constrained by nutrients (or because CO_2 effects are not modelled mechanistically).

14.4.2.2 Response to increased temperature

As with the one-year runs, the increased-temperature scenario provoked the greatest differences among model predictions of ANPP, stem biomass, and NEP. At the BFG site, BIOMASS, BIOME-BGC, HYBRID, and MBL-GEM estimated that the increased temperature would lower plant-available water, forcing stomatal closure, and hence reduced photosynthesis and reduced leaf area. The CENTURY, Q, BIOME-BGC, and MBL-GEM predicted that temperature would increase N availability; for CENTURY and Q, this increase in N availability more than offset the effect of a water limitation. The same factors influenced model predictions at the SWECON site, but for MBL-GEM, increased N availability offset the effect of a water limitation.

14.4.2.3 Soil organic carbon

No model predicted that the altered climates would greatly impact on soil organic C (Figure 14.7). BIOME-BGC and MBL-GEM predicted increased CO_2 would increase soil C 32–36% at the BFG site, but only 6–16% at the SWECON site. In contrast, CENTURY and Q predicted little response to high CO_2 (−11% to 2%) at either site. Because C stocks in the soils and litter for both sites are low relative to C stocks in aboveground biomass, the small relative changes shown in Figure 14.7 represent small absolute changes.

Model predictions for soil C are similar to those for aboveground production, which suggests that the increases in soil C predicted by BIOME-BGC and MBL-GEM are caused by increased litter input. Any difference in litter quality induced by the high CO_2 apparently has little effect on soil C in 60- or 120-year simulations. What these simulations do highlight is the much greater effect of forest growth on soil C stocks, compared with the effect of elevated CO_2 or increased temperature. However, the small differences in soil organic C observed for the elevated CO_2 simulations might be important in forests that contained greater soil C.

14.4.3 Comparison with one-year responses

To compare the one-year and long-term (60- or 120-year) simulations, we will compare ANPP for the control stand in Figure 14.1 with ANPP for the long term shown in Figure 14.7. Model predictions for altered climates generally shift in the same direction, but the magnitude is reduced for the longer simulations. However, there were a few important exceptions. For BIOMASS, the decrease in ANPP in response to the +4 °C scenario was greater for the longer simulation. At the BFG site, MBL-GEM predicted a positive response to a +4 °C scenario for one year, but a strong negative response for the 60-year simulation. The Q model (also at the BFG site) predicted a negative response to increased CO_2 for the one year, but a positive response for the 60-year simulation. In general, climate had a larger effect on the one-year simulations than on the long-term simulations.

The attenuation of the response to altered climate for the long-term simulations compared with the one-year simulations was similar for models both with and without explicit plant and soil interactions. This similarity occurs not because the mechanisms are similar for the models, but because all these models have some constraint that limits productivity, particularly at the end of the life cycle of a stand. For most models, canopy density limits production and any increase in resource availability is more important for a young stand with low leaf area, than for a mature closed-canopy stand. Also, changes in leaf area caused by changes in water use efficiency or water stress promoted by an altered climate are similar to changes in leaf area (and hence productivity) caused by reduced nutrient availability.

14.5 IMPLICATIONS

14.5.1 Carbon budgets for temperate conifer forests

The models in this study represent the state of the art for understanding the influence of climate on element cycling, C storage, and the interactions between plant and soil processes in temperate conifer forests (Ågren *et al.* 1991). Because model predictions for altered climates differed, large uncertainties still remain about the response of temperate conifer ecosystems to changes in air temperature and atmospheric CO_2. Modelling stand growth and C storage for the expected life of a stand using a mechanistic approach remains problematic. Therefore, our current knowledge, as incorporated into these mechanistic models, is not sufficient to provide an unambiguous answer to the question of how C storage will respond to altered climate.

Model predictions differed for:

(a) the amount of C stored over a stand's lifetime;
(b) the absolute response of productivity and C to increased temperature and CO_2;
(c) the magnitude of the response relative to the same model's predictions under current climate.

However, there were some bright spots. As expected, models predicted that growing forests are strong sinks for C. Additionally, most models did predict that productivity would increase for growing forests under elevated CO_2 and all models estimated C storage would increase. Model predictions also showed that changes in fluxes and C storage that occur over a forest's lifetime are far greater than those expected under elevated CO_2 or increased temperature.

14.5.2 Research needs identified

This study identified four critical areas for research emphasis. First, the model predictions differed for the increased-temperature scenarios, because the models made different assumptions about how temperature affects productivity. Second, because the models made different assumptions about the long-term controls over productivity, model predictions for stand development disagreed. Third, the biogeochemical feedbacks constraining plant response to elevated CO_2 are poorly understood. Finally, there is a critical need for flux measurements made at the ecosystem level, and an understanding of acclimation processes. We discuss each of these areas in greater detail below.

14.5.2.1 Plant and soil response to increased temperature

Some of the greatest differences in model predictions were for the +4 °C scenarios and for the interaction between increased temperature and increased CO_2. The predictions differed because the physiologically based models assumed that increased temperature would decrease stomatal conductance and lower productivity. In contrast, the nutrient-oriented models assumed that increased temperature would not only affect the water budget, but would also increase plant-available N and hence productivity. Results from fertilizer trials suggests that fertilization can increase productivity (even in dry forests), but there is a strong interaction with water availability (e.g. Bensen *et al.* 1992). Experiments should be designed to resolve this question. Also, linking a detailed canopy model (such as BIOMASS) with a detailed decomposition model (such as CENTURY or MBL-GEM) can help explore the trade-offs (e.g. McMurtrie *et al.* 1992a; Comins and McMurtrie 1993).

14.5.2.2 Changes in ecosystem function with forest growth

Model predictions showed important differences for total C stored in biomass and for the decline of productivity with age. Because simulated C storage and productivity differed more with forest development than with increased CO_2 or temperature, we must first understand the changes that occur with forest growth before we can understand response to a different climate. Additionally, simulated responses to climate change would likely differ for equilibrium stands.

Causes for changes in productivity that occur with stand development are poorly understood (Ryan and Waring 1992; Yoder *et al.* 1994) and mechanistically modelling stand development is problematic. However, aggrading stands are a major feature of temperate conifer landscapes because management and disturbance frequently initiate new forests. Therefore, understanding the changes in physiology, element cycling, and mortality that occur with succession or stand development (and any interactions with climate) will be critical to answering the C storage question (Ryan *et al.* 1996).

14.5.2.3 Biogeochemical feedbacks that constrain plant response to CO_2

Model simulations showed that C and nutrient allocation and plant and soil interactions can affect productivity and C storage over the long term. Reliable models exist that describe the response of gross primary production to temperature and CO_2. However, knowledge gaps exist for:

(a) the allocation of the C fixed in photosynthesis to different plant parts;
(b) the effects of temperature on the phenology of tissue production;
(c) the response of gross primary production to elevated CO_2 when water and nutrients are limiting.

These factors may cause differences between initial responses to climate (primarily physiological) and longer-term responses (e.g. Oechel *et al.* 1994). However, for some systems, physiological responses dominate both the short- and long-term response of the ecosystem (e.g. Long and Drake 1991).

Comins and McMurtrie (1993) suggest how plant and soil interactions with elevated CO_2 might be treated for nutrient-limited sites, but interactions for water-limited sites are poorly understood (but see Pastor and Post 1988). For water-limited sites, understanding the interactions among water use efficiency, photosynthesis, soil moisture, N availability, and leaf area is necessary to understand productivity and C storage.

14.5.2.4 Ecosystem flux measurements and acclimation

The two sites considered in this study provide the most complete data sets available for mechanistically analysing how climate and resource availability affect forest productivity, C allocation, and element cycling. However, even these studies lack several key measurements for certain treatments. To further our mechanistic understanding, there is a critical need for data taken at the ecosystem scale coupled with data on the components. Measurements of C, water, and energy fluxes with eddy correlation (Desjardins 1992; Baldocchi 1993) will provide system-level flux estimates, but they must be coupled with measurements at a smaller scale.

Unfortunately, whole-ecosystem experiments to increase CO_2 or alter temperature for a forest have not been done, and may not be done because of the cost. Results from this comparison suggest that for aggrading stands, long-term responses are similar to those for shorter periods. But climate did influence the relative responses to increased atmospheric CO_2 and temperature. Therefore, we could learn much from a small number of CO_2 fertilization experiments performed in different climates.

No model included an explicit description of how photosynthesis and autotrophic respiration might acclimate to increased CO_2 and temperature.

However, acclimation does occur for these processes, and might affect long-term response.

14.6 ACKNOWLEDGEMENTS

We thank Dennis Ojima and David Schimel for their constructive reviews.

14.7 REFERENCES

Aber, J.D. and Federer, C.A. (1992) A generalized, lumped-parameter model of photosynthesis, evapotranspiration and net primary production in temperate and boreal forest ecosystems. *Oecologia* **92**, 463–474.

Ågren, G.I. (1983) N productivity of some conifers. *Can. J. For. Res.* **13**, 494–500.

Ågren, G.I. (1985) Theory for growth of plants derived from the N productivity concept. *Physiol. Plant* **64**, 17–28.

Ågren, G.I., McMurtrie, R.E., Parton, W.J., Pastor, J. and Shugart, H.H. (1991) State-of-the-art of models of production–decomposition linkages in conifer and grassland ecosystems. *Ecol. Appl.* **1**, 118–138.

Amthor, J.S. (1991) Respiration in a future, higher-CO_2 world. *Plant Cell. Environ.* **14**, 13–20.

Baldocchi, D.D. (1993) Scaling water vapor and C dioxide exchange from leaves to canopy: rules and tools. In: Ehleringer, J.R. and Field, C.B. (Eds) *Scaling Physiological Processes: Leaf to Globe*, pp. 77–114. Academic Press, San Diego.

Benson, M.L., Myers, B.J. and Raison, R.J. (1992) Dynamics of stem growth of *Pinus radiata* as affected by water and N supply. *For. Ecol. Mgmt.* **52**, 117–137.

Burke, I.C., Kittel, T.G.F., Lauenroth, W.K., Snook, P., Yonker, C.M. and Parton, W.J. (1991) Regional analysis of the central Great Plains. *BioScience* **41**, 685–692.

Comins, H.N. and McMurtrie, R.E. (1993) Long-term response of nutrient-limited forests to CO_2 enrichment; equilibrium behavior of plant–soil models. *Ecol. Appl.* **3**, 666–681.

Desjardins, R.L. (1992) Review of techniques to measure CO_2 flux densities from surface and airborne sensors. In: Stanhill, G.(Ed.) *Advances in Bioclimatology*, pp. 1–23. Springer-Verlag, Berlin.

Eamus, D. and Jarvis, P.G. (1989) The direct effects of increase in the global atmospheric CO_2 concentration on natural and commercial temperate trees and forests. *Advan. Ecol. Res.* **19**, 1–55.

Friend, A.D., Shugart, H.H. and Running, S.W. (1993) A physiology-based gap model of forest dynamics. *Ecology* **74**, 792–797.

Grier, C.C., Vogt, K.A., Keyes, M.R. and Edmonds, R.L. (1981) Biomass distribution and above- and below-ground production in young and mature *Abies amabilis* zone ecosystems of the Washington Cascades. *Can. J. For. Res.* **11**, 155–167.

Hunt, E.R. and Running, S.W. (1992) Simulated dry matter yields for aspen and spruce stands in the North American boreal forest. *Can. J. Remote Sens.* **18**, 126–133.

Kramer, P.J. and Kozlowski, T.T. (1979) *Physiology of Woody Plants*. Academic Press, New York. 811 pp.

Long, S.P. and Drake, B.G. (1991) Effect of the long-term elevation of CO_2 concentration in the field on the quantum yield of photosynthesis of the C_3 sedge, *Scirpus olneyi*. *Plant. Physiol.* **96**, 221–226.

McMurtrie, R.E. (1993) Modelling canopy C and water balance. In: Hall, D.O., Scurlock, J.M.O., Bolar, H., Leegood, R.C. and Long, S.P. (Eds) *Photosynthesis and Production in a Changing Environment: A Field and Laboratory Manual*, pp. 220–231. Chapman & Hall, London.

McMurtrie, R.E., Comins, H.N., Kirschbaum, M.U.F. and Wang, Y.-P. (1992a) Modifying existing forest growth models to take account of effects of elevated CO_2. *Aust. J. Bot.* **40**, 657–677.

McMurtrie, R.E. and Landsberg, J.J. (1992) Using a simulation model to evaluate the effects of water and nutrients on the growth and C partitioning of *Pinus radiata*. *For. Ecol. Mgmt.* **52**, 243–260.

McMurtrie, R.E., Leuning, R., Thompson, W.A. and Wheeler, A.M. (1992b) A model of canopy photosynthesis and water use incorporating a mechanistic formulation of leaf CO_2 exchange. *For. Ecol. Mgmt* **52**, 261–278.

McMurtrie, R.E., Rook, D.A. and Kelliher, F.M. (1990) Modelling the yield of *Pinus radiata* on a site limited by water and N. *For. Ecol. Mgmt* **30**, 381–413.

McMurtrie, R.E. and Wang, Y.-P. (1993) Mathematical models of photosynthetic response of plant stands to rising CO_2 concentrations and temperatures. *Plant Cell Environ.* **16**, 1–16.

McNaughton, K.G. and Jarvis, P.G. (1991) Effects of spatial scale on stomatal control of transpiration. *Agric. For. Meteorol.* **54**, 279–301.

Mooney, H.A., Drake, B.G., Luxmore, R.J., Oechel, W.C. and Pitelka, L.F. (1991) Predicting ecosystem responses to elevated CO_2 concentrations. *BioScience* **41**, 96–104.

Nadelhoffer, K.J., Aber, J.D. and Melillo, J.M. (1985) Fine roots, net primary production, and soil N availability: a new hypothesis. *Ecology* **66**, 1377–1390.

Oechel, W.C., Cowles, S., Grulke, N., Hastings, S.J., Lawrence, B., Prudhomme, T., Reichers, G., Strain, B., Tissue, D. and Vourlitis, G. (1994) Transient nature of CO_2 fertilization in Arctic tundra. *Nature* **371**, 500–503.

Parton, W.J., Schimel, D.S., Cole, C.V. and Ojima, D.S. (1987) Analysis of factors controlling soil organic matter levels in Great Plains grasslands. *Soil Sci. Soc. Amer. J.* **51**, 1173–1179.

Parton, W.J., Stewart, J.W.B. and Cole, C.V. (1988) Dynamics of C, N, P, and S in grassland soils: a model. *Biogeochemistry* **5**, 109–131.

Pastor, J. and Post, W.M. (1988) Response of northern forests to CO_2-induced climate change. *Nature* **334**, 55–58.

Persson, T. (Ed.) (1980) *Structure and Function of Northern Coniferous Forests – an Ecosystem Study*, *Ecol. Bull.* (Sweden) **32**, 609 pp.

Raison, R.J. and Myers, B.J. (Eds) (1992) The Biology of Forest Growth Experiment. *For. Ecol. Mgmt* **52**, 1–311.

Rastetter, E.B., McKane, R.B., Shaver, G.R. and Melillo, J.M. (1992) Changes in C storage by terrestrial ecosystems: how C–N interactions restrict responses to CO_2 and temperature. *Water, Air, Soil Pollut.* **64**, 327–344.

Rastetter, E.B., Ryan, M.G., Shaver, G.R., Melillo, J.M., Nadelhoffer, K.J., Hobbie, J.E. and Aber, J.D. (1991) A general biogeochemical model describing the responses of the C and N cycles in terrestrial ecosystems to changes in CO_2, climate, and N deposition. *Tree Physiol.* **9**, 101–126.

Running, S.W. and Coughlan, J.C. (1988) A general model of forest ecosystem processes for regional applications. I. Hydrologic balance, canopy gas exchange, and primary production processes. *Ecol. Model.* **42**, 125–54.

Running, S.W. and Gower, S.T. (1991) FOREST-BGC, a general model of forest ecosystem processes for regional applications. II. Dynamic C allocation and N budgets.

Tree Physiol. **9**, 147–160.

Ryan, M. G., Binkley, D. and Fownes, J. H. (1996). Age-related decline in forest productivity: pattern and process. *Adv. Ecol. Res.*, (in press).

Ryan, M.G. and Waring, R.H. (1992) Maintenance respiration and stand development in a subalpine lodgepole pine forest. *Ecology* **73**, 2100–2108.

Sanford, R.L., Jr, Parton, W.J., Ojima, D.S. and Lodge, D.J. (1991) Hurricane effects on soil organic matter dynamics and forest production in the Luquillo Experimental Forest, Puerto Rico: results of simulation modelling. *Biotropica* **23**, 364–372.

Waring, R.H. and Schlesinger, W.H. (1985) *Forest Ecosystems: Concepts and Management*. Academic Press, Orlando, Fla, 340 pp.

Whittaker, R.H. and Woodwell, G.M. (1967) Surface area relations of woody plants and forest communities. *Am. J. Bot.* **54**, 31–939.

Yoder, B., Ryan, M.G., Waring, R.H., Schoettle, A.W. and Kaufmann, M.R. (1994) Evidence of reduced photosynthetic rates in old trees. *For. Sci.* **40**, 513–527.

15 Global Climate Change and Carbon Cycling in Grasslands and Conifer Forests

A. D. McGUIRE, D. W. KICKLIGHTER and J. M. MELILLO
The Ecosystems Center, Marine Biological Laboratory, Woods Hole, USA

15.1 INTRODUCTION

Most climate models predict that the buildup of greenhouse gases such as carbon dioxide (CO_2) and methane are likely to lead to surface air temperature rises of 1.5–4.5 °C and changes in precipitation and cloud patterns over the next century (Houghton *et al.* 1992). The simultaneous changes in the chemistry of the atmosphere and the climate are expected to affect both the function and the structure of terrestrial ecosystems (Melillo *et al.* 1990). Functional changes may include changes in processes such as photosynthesis, plant respiration and decomposition. Structural changes may be of various types, including changes in the distribution of carbon and nitrogen between the plant and soil pools, changes in the species composition within an ecosystem, and changes in the distribution of major vegetation groups or biomes.

Process-based models have been used in regional studies to evaluate the functional responses of terrestrial ecosystems to changes in climate (Schimel *et al.* 1990; Burke *et al.* 1991; Running and Nemani 1991; McGuire *et al.* 1992, 1993). One of these models, the terrestrial ecosystem model (TEM), has been used at the global scale (Melillo *et al.* 1993) to explore how carbon and nitrogen cycling in terrestrial ecosystems might change according to predictions of climate change made by several general circulation models (GCMs). In this chapter, we use version 3 of TEM (McGuire *et al.* 1993; Melillo *et al.* 1993) to investigate how carbon cycling of grasslands and conifer forests might respond to GCM-predicted climate change. We focus our analysis on the equilibrium response of carbon pools in vegetation and soils and on the response of net primary production (NPP) for the potential vegetation distribution that is appropriate to contemporary climate. Our analysis does not consider how the response of vegetation distribution to climate change (Emanuel *et al.* 1985; Prentice 1990;

Global Change: Effects on Coniferous Forests and Grasslands.
Edited by A. I. Breymeyer, D. O. Hall, J. M. Melillo and G. I. Ågren

Woodward and McKee 1991; and Prentice *et al.* 1992) affects carbon pools or NPP. We also investigate how assumptions about soil carbon at the sites used to calibrate TEM influence the response of soil carbon to changes in CO_2 and climate. Finally, we compare the responses predicted by TEM to those of other models and evaluate the relevance of spatial scale for the comparison of model responses.

A dominant feature of GCM-predicted climate for a doubled CO_2 atmosphere is an increase in mean surface temperature of the globe (Mitchell *et al.* 1990); precipitation and cloudiness are expected to increase in some areas and decrease in others, and there is disagreement among the output of GCMs about the spatial distribution of these changes (Mitchell *et al.* 1990). Elevated temperature may affect carbon cycling in ecosystems in a variety of ways. It may enhance decomposition of soil organic matter to increase the loss of carbon from the soil. Enhanced decomposition may also increase nitrogen availability through higher rates of nitrogen mineralization. Uptake of this nitrogen by vegetation may enhance NPP. Decreases in NPP may result from elevated temperature by reducing soil moisture or by increasing respiration. Because the ability of vegetation to incorporate elevated CO_2 into production depends on the nutrient and water status of the vegetation, climate change may influence carbon cycling in ecosystems by altering nutrient availability and soil moisture. The TEM has been developed to evaluate how climate change influences simultaneous interactions among the carbon, nitrogen, and water cycles in ecosystems.

15.2 THE TERRESTRIAL ECOSYSTEM MODEL

The terrestrial ecosystem model (TEM; Figure 15.1) is a process-based ecosystem simulation model that uses spatially referenced information on climate, elevation, soils, vegetation, and water availability to make monthly estimates of important carbon and nitrogen fluxes and pool sizes (Raich *et al.* 1991; McGuire *et al.* 1992, 1993; Melillo *et al.* 1993). Because we use TEM to make equilibrium predictions, its estimates of carbon and nitrogen dynamics apply only to mature, undisturbed vegetation; they do not include the effects of land use.

For each monthly time step in a model run, NPP is calculated as the difference between gross primary production (GPP) and plant respiration (R_A). The calculation of GPP considers simultaneous interactions among temperature, light, CO_2, water, and nitrogen availiability (McGuire *et al.* 1992, 1993; Melillo *et al.* 1993). Therefore, the response of GPP to elevated CO_2 is potentially constrained by the availability of light, water, and nitrogen. The calculation of R_A considers both maintenance respiration (McGuire *et al.* 1992, 1993) and construction respiration (Raich *et al.* 1991).

The data sets used to drive TEM are gridded at a resolution of 0.5° latitude by 0.5° longitude. The sources for the climate data (air temperature, precipitation,

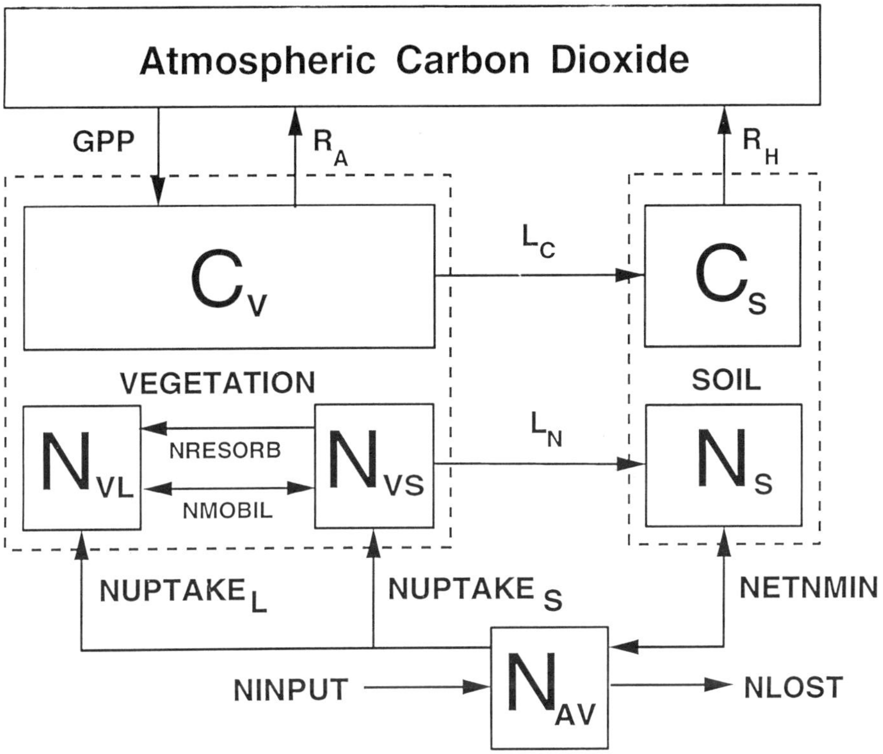

Figure 15.1 The Terrestrial Ecosystem Model (TEM). The state variables are: carbon in the vegetation (C_V); structural nitrogen in the vegetation (N_{VS}); labile nitrogen in the vegetation (N_{VL}); organic carbon in soils and detritus (C_S); organic nitrogen in soils and detritus (N_S); and available soil inorganic nitrogen (N_{AV}). Arrows show carbon and nitrogen fluxes; GPP, gross primary production; R_A, autotrophic respiration; R_H, heterotrophic respiration; L_C, litterfall carbon; L_N, litterfall nitrogen; $NUPTAKE_S$, nitrogen uptake into the strutural nitrogen pool of the vegetation; $NUPTAKE_L$, nitrogen uptake into the labile nitrogen pool of the vegetation; NRESORB, nitrogen resorption from dying tissue into the labile nitrogen pool of the vegetation; NMOBIL, nitrogen mobilized between the structural and labile nitrogen pools of the vegetation; NETNMIN, net nitrogen mineralization of soil organic nitrogen; NINPUT, nitrogen input from outside the ecosystem; and NLOST, nitrogen losses from the ecosystem

and cloudiness), elevation, vegetation, and soil texture are described elsewhere (Raich *et al.* 1991; Melillo *et al.* 1993); the climate data represent long-term averages. Hydrological inputs for TEM are determined with a water balance model (Vörösmarty *et al.* 1989) that uses the climate, elevation, soils, and vegetation data.

The application of TEM to a grid cell requires the use of monthly climatic and hydrological data and the soil- and vegetation-specific parameters appropriate to

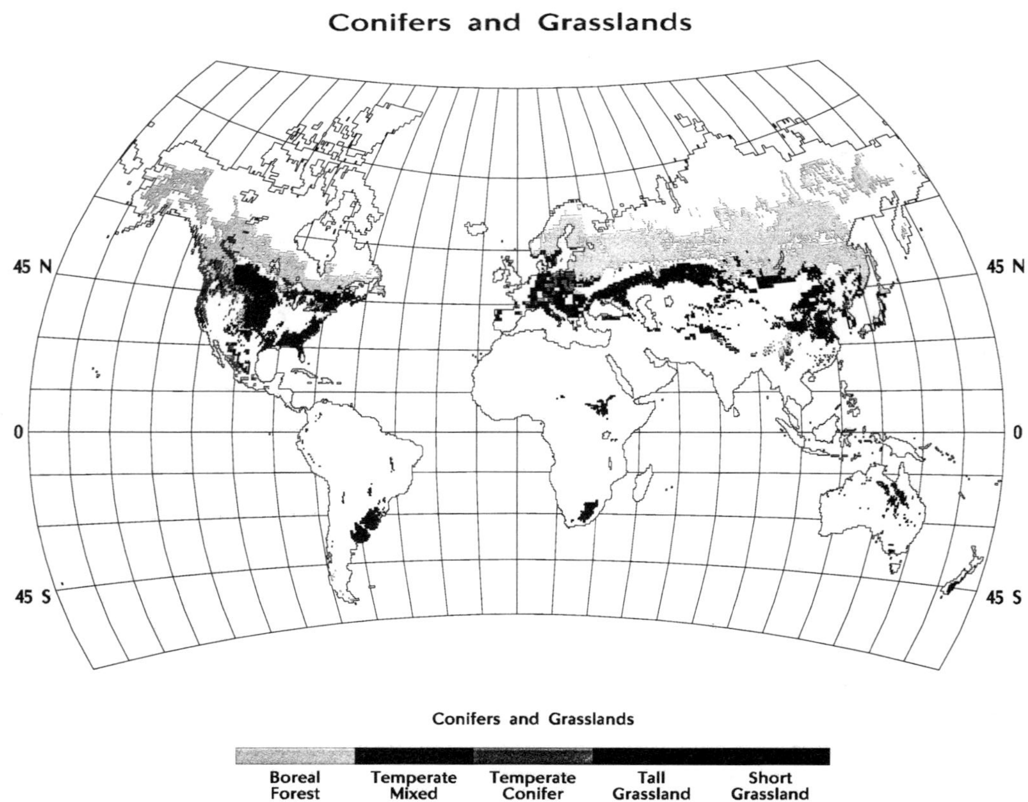

Figure 15.2 The global distribution of grasslands (tall and short) and conifer forests (boreal, temperate, and temperate mixed)

Table 15.1 Areal extent of grasslands and coniferous forests

	Area (10^6 km^2)	Cells
Grasslands		
Tall	3.6	1 557
Short	4.7	2 050
Total	8.3	3 607
Conifer forests		
Boreal Forest	12.2	7 406
Temperate conifer	2.4	1 081
Temperate mixed	5.1	2 250
Total	19.7	10 737
All ecosystems	127.3	56 090

the grid cell. Although many of the vegetation-specific parameters in the model are defined from published information (Raich *et al.* 1991; McGuire *et al.* 1992), some are determined by calibrating the model to the steady-state fluxes and pool sizes of an intensively studied field site. Most of the data used to calibrate the model for the vegetation types considered in this study are documented elsewhere (McGuire *et al.* 1992). In most of the analyses we present, TEM has been calibrated to the soil organic carbon and nitrogen found to approximately 1 m depth at the calibration site, the '1 m' calibration. We also calibrated the model to 20 cm soil carbon and nitrogen (the '20 cm' calibration) to examine how different assumptions about the actively decomposing soil organic carbon affect responses of NPP and carbon pools to changes in climate and CO_2.

15.3 POTENTIAL DISTRIBUTION OF GRASSLANDS AND CONIFER FORESTS

The global distribution of grasslands and conifer forests (Figure 15.2), which represents 22% of the area occupied by the terrestrial biosphere (Table 15.1), is determined from a global georeferenced data base (0.5° spatial resolution) of potential vegetation developed from extant maps (Melillo *et al.* 1993). To develop this data base we modified the potential vegetation data of Matthews (1983), which has 1° spatial resolution, by incorporating information from several regional sources (White 1981; Institute of Geography of the Siberian Department of USSR Academy of Science *et al.* 1990; Hou *et al.* 1979; Australian Surveying and Land Information Group 1990; Rowe 1972; Joint Federal–State Land Use Planning Commission for Alaska 1973; Küchler 1964; UNESCO 1981).

The global grassland communities are aggregated into two major vegetation types, tall and short grasslands, based on the relative height of the dominant

Table 15.2 Estimates by the TEM of annual NPP for grasslands and conifer forests in the terrestrial biosphere at an atmospheric CO_2 concentration of 355 ppmv (parameterized for 1 m soil carbon)

	Total NPP[a]	Mean NPP[b]	Max. NPP[b]	Min. NPP[b]
Grasslands				
Tall	1.2	335	756	136
Short	1.0	214	438	72
Total	2.2	267	756	72
Conifer forests				
Boreal forest	2.9	238	434	124
Temperate conifer	1.1	465	704	208
Temperate mixed	3.4	669	1066	231
Total	7.4	378	1066	124

[a]Units are Pg C (10^{15} g C) yr^{-1}.
[b]Units are g C m^{-2} yr^{-1}.

vegetation. Tall grasslands contain grasses with heights greater than 1 m and occur in more mesic sites than short grasslands. Although the northern meadow-steppe of Euro-Asia contains relatively 'short' species, the dynamics of this community are thought to be similar to the tallgrass prairie of North America (Walter 1979). Therefore, northern meadow-steppes have been classified as tall grassland. Both grassland types are found throughout the temperate and tropical zones (Figure 15.2). We do not consider grasslands found in savannas of the temperate and tropical regions.

Conifer forests are aggregated into three major vegetation types based on physiognomy and climate: boreal forests, temperate mixed forests, and temperate conifer forests. Boreal forests (located mainly in Canada and Alaska, northern Europe, and the Commonwealth of Independent States) contain both conifer and deciduous dominant species and represents 62% of all potential conifer forests (Table 15.1). The next abundant forest type, temperate mixed forests (26%), also contains both conifer and deciduous dominant species and is found mainly in the United States, Europe, and China. Temperate conifer forests have a global distribution similar to temperate mixed forests, but are also found in many mountainous regions.

15.4 EXTRAPOLATION OF TEM FOR CONTEMPORARY CLIMATE

To estimate fluxes and pools of grasslands and conifer forests for 'contemporary' conditions we applied TEM at 355 ppmv CO_2 using the long-term climate data with the 1 m calibration. For grasslands, TEM estimates an annual NPP of 2.2 Pg C (10^{15} g C; Table 15.2). The vegetation and soil carbon estimates for grassland

Table 15.3 Estimates by the TEM of vegetation carbon for grasslands and conifer forests in the terrestrial biosphere at an atmospheric CO_2 concentration of 355 ppmv (parameterized for 1 m soil carbon)

	Total carbon[a]	Mean carbon[b]	Max. carbon[b]	Min. carbon[b]
Grasslands				
Tall	1.8	512	1 156	208
Short	1.6	337	690	113
Total	3.4	413	1 156	113
Conifer forests				
Boreal forest	118.5	9 739	17 739	5 033
Temperate conifer	90.7	37 803	57 196	16 939
Temperate mixed	77.5	15 224	24 269	5 261
Total	286.7	14 586	57 196	5 033

[a]Units are Pg C (10^{15} g C) yr^{-1}.
[b]Units are g C m^{-2} yr^{-1}.

Table 15.4 Estimates by the TEM of soil carbon for grasslands and conifer forests in the terrestrial biosphere at an atmospheric CO_2 concentration of 355 ppmv (parameterized for 1 m soil carbon)

	Total carbon[a]	Mean carbon[b]	Max. carbon[b]	Min. carbon[b]
Grasslands				
Tall	58.4	16 211	23 714	8 039
Short	17.3	3 701	5 129	1 689
Total	75.7	9 156	23 714	1 689
Conifer forests				
Boreal forest	132.3	10 878	12 189	4 578
Temperate conifer	45.1	18 804	28 219	5 097
Temperate mixed	54.0	10 612	15 606	6 285
Total	231.4	11 777	28 219	4 578

[a]Units are Pg C (10^{15} g C) yr^{-1}.
[b]Units are g C m^{-2} yr^{-1}.

are 3.4 Pg C (Table 15.3) and 75.7 Pg C (Table 15.4). Although conifer forests in our analysis occupy 2.4 times the area of grasslands, the estimated annual NPP is 3.4 times that of grasslands (7.4 Pg C; Table 15.2), vegetation carbon is 84.3 times greater (286.7 Pg C; Table 15.3), and soil carbon is 3.1 times greater (231.4 Pg C; Table 15.4). The higher vegetation carbon of conifer forests reflects the ability of forests to store carbon in woody tissue. Per unit area, TEM estimates that total carbon storage in global conifer forests is 2.8 times that in global grasslands.

Because of the way we aggregated grasslands and forests, our estimates of total NPP and carbon pools for grasslands are not always directly comparable to other published estimates. However, the area-based estimates by TEM may be appropriately compared. The grassland NPP estimate of 267 g C m^{-2} yr^{-1} by TEM (Table 15.2) is similar to the estimate of 225 g C m^{-2} yr^{-1} by Whittaker and Likens (1973) for temperate grasslands. However, the vegetation carbon estimate of 413 g C m^{-2} (Table 15.3) is substantially lower than their estimate of 700 g C m^{-2}. The soil carbon estimate of 9156 g C m^{-2} by TEM for grasslands (Table 15.4) is similar to 10 692 g C m^{-2} (417 Pg/3900 10^6 ha) reported by Ojima *et al.* (1993a) for soil carbon to 1 m depth in potential world grasslands.

The TEM estimate of 238 g C m^{-2} yr^{-1} for boreal forest NPP (Table 15.2) is substantially lower than the Whittaker and Likens (1973) estimate of 360 g C m^{-2} yr^{-1}. However, the estimate for boreal forest vegetation carbon (9739 g C m^{-2}; Table 15.3) is similar to their estimate of 9000 g C m^{-2}. The TEM estimate of 10 878 g C m^{-2} for boreal forest soil carbon (Table 15.4) is lower than the 14 900 g C m^{-2} estimate of Schlesinger (1977); the TEM estimate is lower because the model does not represent anaerobic processes that cause higher carbon storage to occur in northern peatlands (see Schlesinger 1977).

The Whittaker and Likens (1973) NPP estimate of 585 g C $m^{-2}yr^{-1}$ for temperate evergreen forest is intermediate between the TEM estimates for temperate conifer and temperate mixed forest (Table 15.2). Their estimate of 16 000 g C m^{-2} for vegetation carbon of temperate evergreen forest is similar to the TEM estimate of 15 224 g C m^{-2} for temperate mixed forest (Table 15.3); the TEM estimate for temperate conifer forest (37 803 g C m^{-2}; Table 15.3) is much higher because the calibration site is an old-growth forest in the Pacific Northwest of the United States (McGuire *et al.* 1992). The Schlesinger (1977) estimate of 11 800 g C m^{-2} for temperate forest soil carbon is similar to the TEM estimate for temperate mixed forest (10 612 g C m^{-2}; Table 15.4). Again the TEM estimate for temperate conifer forest (18 804 g C m^{-2}; Table 15.4) may be higher because the calibration site is an old-growth forest.

15.5 DEVELOPMENT OF FUTURE CLIMATE SCENARIOS FOR USE WITH TEM

We obtained the output of two general circulation models (GCMs) from the National Center for Atmospheric Research (Jenne 1992). The simulations estimate equilibrium climates that correspond to a doubling of the atmospheric CO_2 concentration and include one from the Geophysical Fluid Dynamics Laboratory (GFDL; Manabe and Wetherald 1987) and one from Oregon State University (OSU; Schlesinger and Zhao 1989). The GFDL GCM represents a high temperature impact scenario and predicts increases of 4.9 °C for the grassland biome and 6.2 °C for the conifer forest biome. The OSU GCM represents a low impact scenario and predicts increases of 3.2 and 3.4 °C for the grassland and conifer forest biomes. Precipitation for grasslands is predicted to increase by both GCMs with larger increases predicted by OSU (82.0 mm) than by GFDL (46.5 mm). Precipitation is also predicted to increase for conifer forests, with similar increases predicted by GFDL (67.7 mm) and OSU (75.9 mm). Cloudiness is predicted by GFDL and OSU to decrease 1.4 and 2.1% for grasslands, respectively. For conifer forests, cloudiness is predicted by GFDL to increase 0.6%, but by OSU to decrease 2.5%.

Because we were interested in the implications of climate change for NPP and carbon pools, we generated 'GCM climates' for TEM by using the output variables of surface air temperature, precipitation, and total cloud cover for the current and $2 \times CO_2$ simulations of each GCM to modify the contemporary climate data for TEM. We organized each of the output variables of each GCM simulation at 0.5° spatial resolution with a spherical interpolation procedure (Willmott *et al.* 1985). Next, similar to the method used in a study of the potential effects of climate change on US agriculture (Adams *et al.* 1990), we calculated for each grid cell the ratio of the monthly output of the $2 \times CO_2$ simulation to that of the $1 \times CO_2$ simulation for each of the three output variables; temperature was converted to Kelvin before calculating monthly temperature ratios. We then multiplied each ratio by the corresponding variable in our data for contemporary climate to determine the input data for TEM that represent the $2 \times CO_2$ climate for each GCM.

To help separate the effects of changes in CO_2 concentration from those of the GCM climates on estimates of NPP and carbon pools we performed a factorial experiment with the 1 m calibration of TEM involving two levels of CO_2 (312.5 and 625.0 ppmv) and three climate scenarios (contemporary and the two GCM climates). We chose the CO_2 level of 312.5 ppmv because it was the average baseline concentration of four GCMs that we used in a previous study (Melillo *et al.* 1993).

15.6 RESPONSES TO DOUBLED CARBON DIOXIDE

For doubled CO_2 and no climate change, TEM predicts that NPP, vegetation carbon, and soil carbon for individual grid cells of grasslands and conifer forests either do not change or increase (Figures 15.3–15.5). For global grasslands, TEM predicts that NPP increases 0.2 Pg C (9.1%; Table 15.5), vegetation carbon increases 0.3 Pg C (9.1%; Table 15.6), and soil carbon increases 3.4 Pg C (4.3%; Table 15.7). For global conifer forests, TEM predicts NPP to increase 0.4 Pg C (5.5%; Table 15.5), vegetation carbon to increase 17.5 Pg C (6.2%; Table 15.6), and soil carbon to increase 9.0 Pg C (3.9%, Table 15.7). Thus, for doubled CO_2 and no climate change conifer forests are potentially a more responsive carbon sink than grasslands.

The response of conifer forest NPP and soil carbon to elevated CO_2 predicted by TEM depends on the degree to which NPP is limited by nitrogen availability

(McGuire *et al.* 1993). In moist regions of temperate conifer forest, where NPP is predicted by TEM to be limited by nitrogen availability more than by soil moisture, there is little response to elevated CO_2 (McGuire *et al.* 1993). In dry regions elevated CO_2 promotes enhanced water-use efficiency in TEM which translates into increased NPP (McGuire *et al.* 1993). Because most of the conifer forest region considered in this study is moist conifer forest, i.e. in the boreal and temperate–mixed regions, the enhancement of NPP and soil carbon in response to elevated CO_2 is small.

15.7 RESPONSES TO CHANGES IN CLIMATE

With no change in CO_2 concentration, TEM predicts for both the GFDL and OSU climates that responses of NPP, vegetation carbon, and soil carbon for individual grid cells of grasslands and conifer forests can be either positive or negative (Figures 15.3–15.5). For global grasslands, TEM predicts annual NPP to increase 0.4 Pg C (18.2%) for the GFDL climate and 0.2 Pg C (9.1%) for the OSU climate (Table 15.5). Vegetation carbon is predicted to increase 0.7 Pg C (21.2%) for the GFDL climate and 0.6 Pg C (18.2%) for the OSU climate (Table 15.6). For the GFDL climate, soil carbon decreases 3.8 Pg C (5.1%; Table 15.7). In contrast, soil carbon is predicted to increase slightly for the OSU climate (0.7 Pg C, 0.9%; Table 15.7).

For global conifer forests, TEM predicts annual NPP to increase 1.0 Pg C (13.7%) for the GFDL climate and 0.9 Pg C (12.3%) for the OSU climate (Table 15.5). Increases in vegetation carbon are slightly higher for the GFDL climate (38.1 Pg C, 13.5%; Table 15.6) than for the OSU climate (34.4 Pg C, 12.2%; Table 15.6). The decreases in soil carbon predicted for the GFDL climate (31.3 Pg C, 13.7%; Table 15.7) are more than three times those predicted for the OSU climate (9.2 Pg C, 4.0%; Table 15.7). Thus, for the climates considered with no change in CO_2, grasslands are potentially either carbon sources or sinks and conifer forests are potentially sinks. The sink strength is greater for conifer forests than grasslands because of the ability of forests to store carbon in woody tissue.

In grasslands, and in boreal and cool–moist temperate regions of conifer forest, elevated temperature generally increases the NPP predicted by TEM through enhanced nitrogen availability (McGuire *et al.* 1992, 1993; Melillo *et al.* 1993). Schimel *et al.* (1990), based on 50-year simulations of climate change with the CENTURY model for sites in the Great Plains, attributed increased NPP to elevated nitrogen availability because of enhanced decomposition, but indicated that nitrogen losses related to higher decomposition could decrease NPP in the long term. Burke *et al.* (1991) applied the doubled-CO_2 climate predicted by the Goddard Institute for Space Studies (GISS) GCM to the central Great Plains and reported that aboveground NPP for the region increased less than 10% after 50 years of simulation with CENTURY. This result is similar to the equilibrium

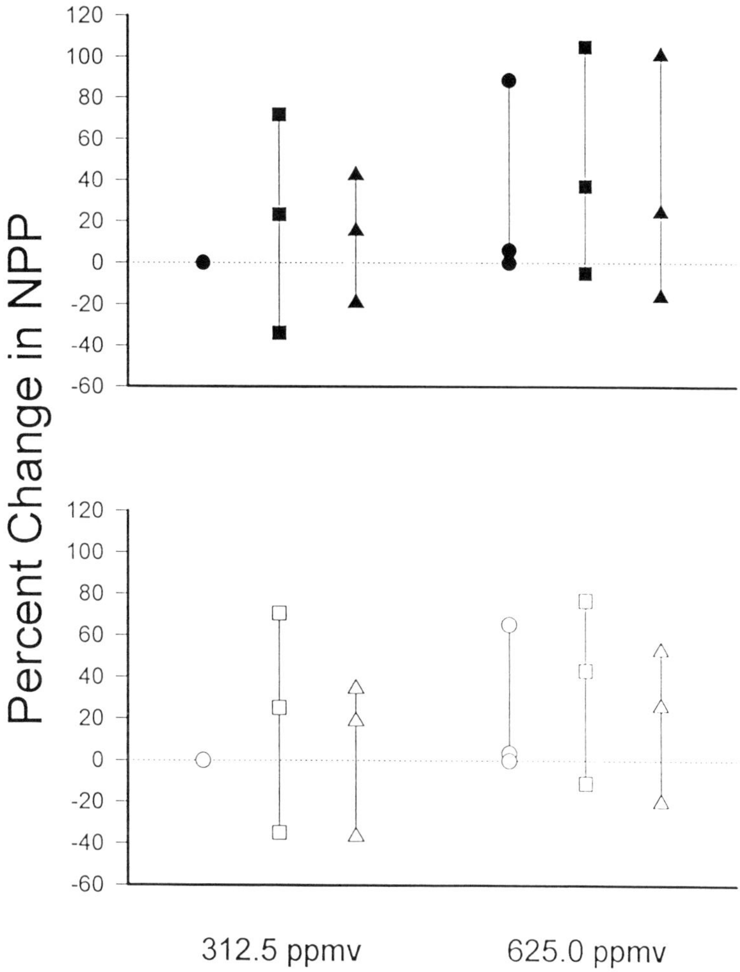

Figure 15.3 The range and mean of responses in annual NPP for grasslands (filled symbols) and conifer forests (open symbols) at two levels of atmospheric carbon dioxide (312.5 and 625.0 ppmv) and three levels of climate (contemporary, circles; GFDL, squares; OSU, triangles). The response is relative to NPP for contemporary climate and atmospheric carbon dioxide of 312.5 ppmv

response predicted by TEM for the OSU climate applied to global grasslands. Linked models of forest productivity and soil processes have also predicted that elevated temperature enhances conifer growth through increased nitrogen availability for simulations at specific sites (Pastor and Post 1988; Bonan and Van Cleve 1992).

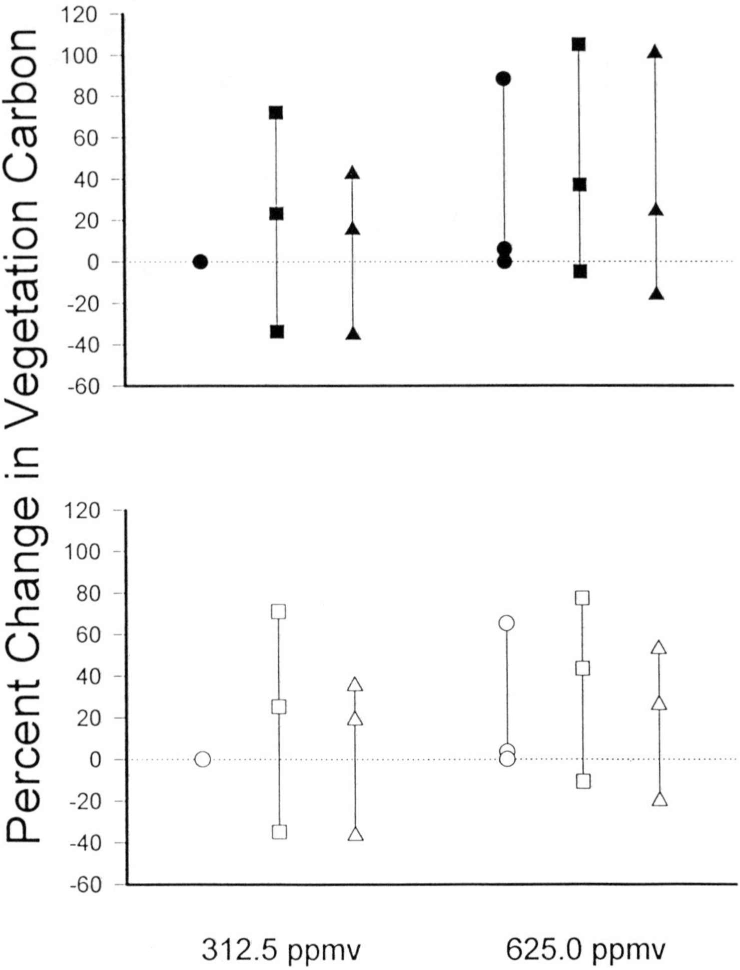

Figure 15.4 The range and mean of responses in vegetation carbon for grasslands (filled symbols) and conifer forests (open symbols) at two levels of atmospheric carbon dioxide (312.5 and 625.0 ppmv) and three levels of climate (contemporary, circles; GFDL, squares; OSU, triangles). The response is relative to vegetation carbon for contemporary climate and atmospheric carbon dioxide of 312.5 ppmv

Decreases in NPP may result from elevated temperature by reducing soil moisture. This mechanism primarily affects the NPP response of TEM in dry regions of conifer forest (McGuire *et al.* 1993). Other models have predicted that elevated temperature may increase evapotranspiration to decrease forest growth in both dry and wet regions of present-day conifer forest (Pastor and Post 1988;

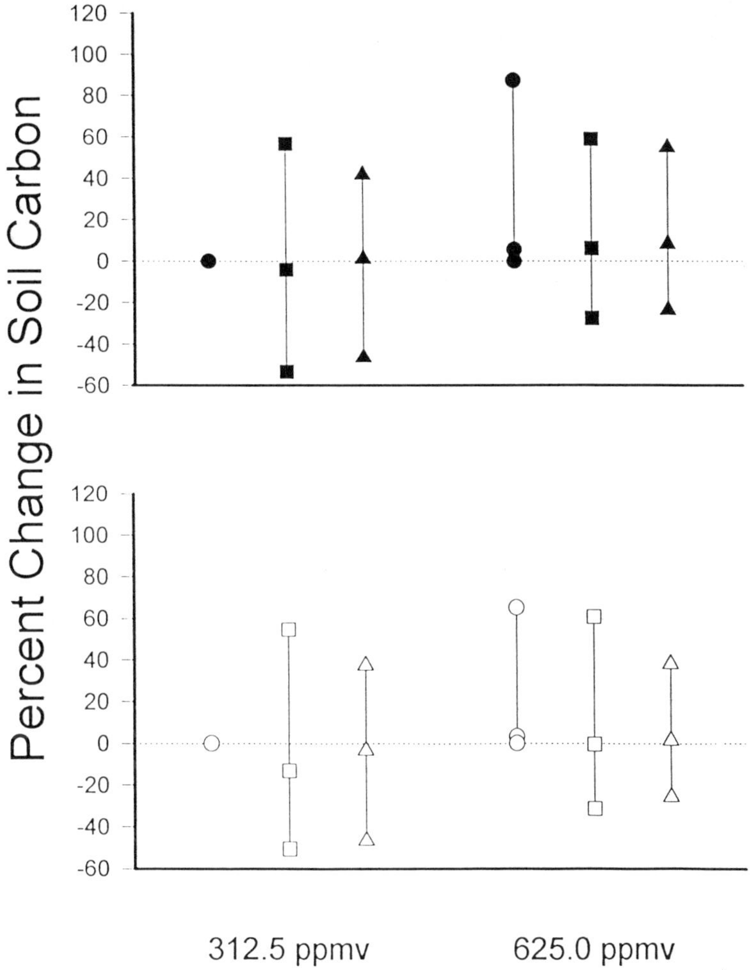

Figure 15.5 The range and mean of responses in soil carbon for grasslands (filled symbols) and conifer forests (open symbols) at two levels of atmospheric carbon dioxide (312.5 and 625.0 ppmv) and three levels of climate (contemporary, circles; GFDL, squares; OSU, triangles). The response is relative to soil carbon for contemporary climate and atmospheric carbon dioxide of 312.5 ppmv

Bonan *et al.* 1990; Running and Nemani 1991). Elevated temperature may also decrease NPP by enhancing respiration costs relative to carbon uptake. This mechanism has been observed to primarily affect the NPP response of TEM in warm moist regions of temperate conifer forest (McGuire *et al.* 1993); it has also been observed to be an important factor influencing the NPP response predicted

Table 15.5 Response of NPP (10^{15} g C yr^{-1}) by region for experiment involving two levels of atmospheric CO_2 and three levels of climate (parameterized for 1 m soil carbon)

Climate:	Contemporary		GFDL1		OSU	
CO_2 concentration (ppm):	312	625	312	625	312	625
Grasslands						
Tall	1.2	1.3	1.4	1.5	1.3	1.4
Short	1.0	1.1	1.2	1.4	1.1	1.2
Total	2.2	2.4	2.6	2.9	2.4	2.6
Conifer forests						
Boreal forest	2.9	2.9	3.8	4.4	3.5	3.7
Temperate conifer	1.1	1.2	1.1	1.3	1.1	1.3
Temperate mixed	3.3	3.6	3.4	4.0	3.6	4.0
Total	7.3	7.7	8.3	9.7	8.2	9.0

Table 15.6 Response of vegetation carbon (10^{15} g C) by region for experiment involving two levels of atmospheric CO_2 and three levels of climate (parameterized for 1 m soil carbon)

Climate:	Contemporary		GFDL1		OSU	
CO_2 concentration (ppm):	312	625	312	625	312	625
Grasslands						
Tall	1.8	1.9	2.1	2.3	2.1	2.2
Short	1.5	1.7	1.9	2.2	1.8	2.0
Total	3.3	3.6	4.0	4.5	3.9	4.2
Conifer forests						
Boreal forest	117.9	120.5	154.9	178.2	144.7	151.5
Temperate conifer	88.3	98.0	88.6	108.9	89.7	104.7
Temperate mixed	75.7	80.9	76.5	91.9	81.9	91.7
Total	281.9	299.4	320.0	379.0	316.3	347.9

by the Forest-BGC model to elevated temperature at a site in warm–moist temperate forest (Running and Nemani 1991). Because elevated temperature influences processes that can both enhance or decrease the NPP predicted by TEM, the NPP response to the high-temperature GFDL climate did not substantially differ from the NPP response to the low-temperature OSU climate.

The NPP response of TEM to elevated temperature influences the response of vegetation carbon; increased NPP translates into increased vegetation carbon for both the GFDL and OSU climate scenarios. Linked models of forest productivity and soil processes have also predicted increased aboveground vegetation carbon in response to climate warming for conifers of both the temperate and boreal region (Pastor and Post 1988; Bonan and Van Cleve 1992), although the

Table 15.7 Response of soil carbon (10^{15} g C) by region for experiment involving two levels of atmospheric CO_2 and three levels of climate (parameterized for 1 m soil carbon)

Climate:	Contemporary		GFDL1		OSU	
CO_2 concentration (ppm):	312	625	312	625	312	625
Grasslands						
Tall	58.0	60.0	54.9	59.6	58.7	61.7
Short	17.0	18.1	16.3	18.4	17.0	18.4
Total	75.0	78.1	71.2	78.0	75.7	80.1
Conifer forests						
Boreal forest	131.9	134.2	114.1	129.5	128.6	133.9
Temperate conifer	44.3	48.0	37.3	45.0	40.8	46.4
Temperate mixed	53.1	56.1	46.7	54.7	50.7	55.6
Total	229.3	238.3	198.1	229.2	220.1	235.9

response depends on whether or not soil moisture is affected (Pastor and Post 1988; Bonan *et al.* 1990).

The response of soil carbon to elevated temperature will depend on the NPP response, which influences inputs into the soil, and on the decomposition response per unit soil carbon, which influences CO_2 losses from the soil organic pool. In TEM, if elevated temperature does not substantially decrease available soil moisture, then it will increase decomposition per unit soil carbon. Elevated temperature is predicted by TEM to decrease soil carbon for the high-temperature GFDL climate, but not for the low-temperature OSU climate. Schimel *et al.* (1990) indicated that soil carbon levels decreased in response to elevated temperature at sites in both the northern and southern Great Plains of the United States. Similarly, Burke *et al.* (1991) reported that soil carbon levels of the central Great Plains decreased approximately 3% after running CENTURY for 50 years with the GISS climate. This decrease is intermediate to the responses predicted by TEM for the GFDL and OSU climates; the temperature increase predicted by the GISS climate is intermediate between the GFDL and OSU climates. Elevated temperature is predicted by TEM to decrease the soil organic pool of conifer forests for both the GFDL and OSU climates, with greater decreases predicted for the high-temperature GFDL scenario. A linked model of boreal forest productivity and soil processes also predicts that soil organic carbon of boreal conifers decreases in response to climatic warming (Bonan and Van Cleve 1992).

15.8 RESPONSES TO CHANGES IN CLIMATE AND CARBON DIOXIDE

With changes in both climate and CO_2 concentration, TEM predicts that responses of NPP, vegetation carbon, and soil carbon for individual grid cells of grasslands and conifer forests may be positive or negative (Figures 15.3–15.5). For global grasslands, TEM estimates annual NPP to increase 0.6 Pg C (27.3%) for the GFDL climate (Table 15.5). The predicted increases for the OSU climate are slightly less (0.4 Pg C, 18.2%; Table 15.5). Vegetation carbon increases 1.2 Pg C (36.4%) for the GFDL climate and 0.9 Pg C (27.3%) for the OSU climate (Table 15.6). The predicted increases in soil carbon are less for the GFDL climate (3.0 Pg C, 4.0%; Table 15.7) than for the OSU climate (5.1 Pg C, 6.8%; Table 15.7).

For global conifer forests, TEM predicts annual NPP to increase 2.4 Pg C (32.9%) for the GFDL climate, which is about 40% more than the 1.7 Pg C (23.3%) increase for the OSU climate (Table 15.5). Increases in vegetation carbon predicted for the two climates show a similar pattern; for the GFDL climate, the 97.1 Pg C increase (34.4%) is more than 40% higher than the 66.0 Pg C increase (23.4%) predicted for the OSU climate (Table 15.6). Soil carbon decreased slightly for the GFDL climate (0.1 Pg C; $< 0.1\%$; Table 15.7), but increased for the OSU climate (6.6 Pg C; 2.9%; Table 15.7). Thus, for the climates considered with elevated CO_2 concentration, conifer forests are potentially much stronger carbon sinks than grasslands because of the ability to store carbon in woody biomass.

The response of NPP to elevated CO_2 and temperature predicted by TEM is influenced by moisture availability. In moist regions of temperate forest, elevated temperature enhances decomposition to increase nitrogen availability. The increased nitrogen availability allows the vegetation to generally incorporate elevated CO_2 into production, but the overall effect on NPP is sensitive to the plant respiration response (McGuire *et al.* 1993). In contrast, the FOREST-BGC model predicts a slight decrease in NPP for a site in warm–moist conifer forest because the enhanced respiration costs of elevated temperature more than offset the photosynthetic gains from elevated CO_2 (Running and Nemani 1991). In dry regions of temperate forest, NPP is predicted by TEM to increase because enhanced carbon uptake in response to elevated CO_2 generally more than compensates for decreased soil moisture or increased plant respiration caused by elevated temperature (McGuire *et al.* 1993). Similarly, for a site in a dry conifer forest the Forest-BGC model predicts increased NPP in response to elevated temperature and CO_2 because photosynthetic gains more than offset respiration costs (Running and Nemani 1991). The increases in NPP predicted by TEM for conifer forests in response to elevated temperature and CO_2 translate into increased vegetation carbon for both the GFDL and OSU climates. However, the enhanced NPP is able to compensate for the increased decomposition per unit

Table 15.8 Response of soil carbon (10^{15} g C) in tall grasslands and temperate conifer forest for the 1 m and 20 cm calibrations in the experiment involving two levels of atmospheric CO_2 and three levels of climate

Climate:	Contemporary		GFDL1		OSU	
CO_2 concentration (ppm):	312	625	312	625	312	625
Tall grasslands						
1 m calibration	58.0	60.0	54.9	59.6	58.7	61.7
20 cm calibration	17.4	18.0	16.6	18.0	17.6	18.6
Temperate conifer forests						
1 m calibration	44.3	48.0	37.3	45.0	40.8	46.4
20 cm calibration	12.7	13.8	10.7	12.9	11.7	13.3

soil carbon for the low-temperature OSU climate, but not for the GFDL climate. Thus, soil carbon increases are predicted for the OSU climate and decreases for the GFDL climate.

15.9 SENSITIVITY OF SOIL CARBON RESPONSE TO CALIBRATION DEPTH

The responses of NPP and vegetation carbon to changed climate or CO_2 do not demonstrate any sensitivity to the calibration depth of soil carbon for either tall grasslands or temperate conifer forests. However, the absolute responses of soil carbon for the 1 m calibrations of tall grasslands and temperate conifer forest are approximately three to four times larger than for the 20 cm calibrations (Table 15.8). Although the absolute response of soil carbon is always greater for the 1 m calibration, the proportional responses are essentially identical for both the 1 m and 20 cm calibrations. For models that make equilibrium estimates, these results indicate the importance of identifying at the calibration site the soil carbon that is likely to be actively decomposing over the time frame of interest. For climate change studies involving a doubled CO_2 atmosphere, the appropriate time scale is decades to centuries. The inclusion of soil carbon that turns over on the time scale of millennia ('old carbon') will overestimate the response of soil carbon. The CENTURY model (Schimel *et al.* 1990; Burke *et al.* 1991; Ojima *et al.* 1993b) estimates soil carbon to a depth of 20 cm. This depth may be approximately appropriate for identifying the relevant soil carbon in grasslands where most inputs are near the surface, but in forests the rooting zone may be much deeper than 20 cm. Because both recent and old carbon may occur at all depths in a forest soil, depth may not be the best metric to identify the actively decomposing soil carbon that is appropriate to doubled CO_2 climate studies.

Table 15.9 Comparison of the responses of grasslands and conifer forest to climate change and elevated CO_2 among various models

Models	Scale	Grasslands	Conifer forests	Climate response			CO_2 response			Climate and CO_2 response		
				NPP	Vegetation C	Soil C	NPP	Vegetation C	Soil C	NPP	Vegetation C	Soil C
CENTURY (Schimel *et al.* 1990)	Site	X		+/−	NR[a]	−	NR	NR	NR	NR	NR	NR
CENTURY (Burke *et al.* 1991)	Regional	X		+	NR	−	NR	NR	NR	NR	NR	NR
CENTURY (Ojima *et al.* 1993b)	Site	X		+/−	NR	+/−	NR	NR	NR	+/−	NR	+/−
TEM (this study)	Site	X		+/−	+/−	+/−	+/0	+/0	+/0	+/−	+/−	+/−
TEM (this study)	Regional	X		+	+	+/−	+	+	+	+	+	+
LINKAGES (Pastor and Post 1988)	Site		X	+/−	+/−	NR	NR	NR	NR	NR	NR	NR
FOREST-BGC (Running and Nemani 1991)	Site		X	−	NR	NR	+	NR	NR	+/−	NR	NR
Boreal forest model (Bonan and Van Cleve 1992)	Site		X	+	+	−	NR	NR	NR	NR	NR	NR
TEM (this study)	Site		X	+/−	+/−	+/−	+/0	+/0	+/0	+/−	+/−	+/−
TEM (this study)	Regional		X	+/0	+	−	+/0	+	+	+	+	+/−

[a]NR indicates no response.

15.10 RELEVANCE OF SPATIAL SCALE FOR THE COMPARISON OF MODEL RESPONSES

During the next century, substantial simultaneous changes are predicted to occur in several climatic variables including CO_2 temperature, precipitation, and cloudiness (Mitchell *et al.* 1990; Watson *et al.* 1992). To assess the influence of these changes on regional carbon cycling, it is desirable to represent how the interactive effects of climate change and elevated CO_2 influence ecosystem processes in a spatially continuous manner. Although several models have been used to study the potential effects of climate changes on carbon cycling in grasslands and conifer forests, few have been used to study the interactive effects of climate with elevated CO_2. Also, most investigations have focused on potential responses at specific sites rather than the responses at larger spatial scales; the results of site-specific investigations can appear contradictory so that responses at the regional scale are difficult to assess. In this discussion we compare the directions of response of our study to those of other modeling studies with respect to the changes in climate and CO_2 and with respect to the spatial scale of response (Table 15.9). The spatial scale is considered at two resolutions. The fine resolution is 'site', which may include a grid cell or polygon for which the climate variables were treated in the study as having no spatial variability. The coarse spatial scale is 'regional', which we define as an aggregation of grid cells or polygons.

The CENTURY model has been used to study potential responses of NPP and soil carbon in the grassland biome at both the site (Schimel *et al.* 1990; Ojima *et al.* 1993b) and regional (Burke *et al.* 1991) scales. The region considered by Burke *et al.* (1991) is the central Great Plains of the United States, which is contained within the total grassland region considered by TEM. The climatic responses predicted by both models are consistent at both the site and the regional spatial scales, although the regions considered by CENTURY and TEM are not identical. Of the CENTURY studies, only Ojima *et al.* (1993b) consider responses to changes in both climate and CO_2; these are consistent with the TEM results at the site scale.

The application of models to study the potential responses of conifer forests has primarily focused on the effects of climate change at the site scale (Pastor and Post 1988; Running and Nemani 1991; Bonan and Van Cleve 1992). The results of these studies are consistent with responses predicted by TEM at the site scale, but only one of the studies has examined the response of soil carbon (Bonan and Van Cleve 1992). Also, only one of these studies has examined the potential response of conifer forest NPP to changes in both CO_2 and climate (Running and Nemani 1991); the responses are consistent with those of TEM at the site scale.

Although there is much consistency among models for the results reported in Table 15.9, there are limitations in comparing the results. Most of the studies have focused on the responses of carbon cycling to climate change without considering elevated CO_2. Responses at both site and regional scales are sensitive to the interaction of elevated CO_2 with climate change, and few studies are available for this comparison. Also, the responses at the site scale depend on the sites chosen. Studies that use few sites are more likely to miss some variability in potential responses across a biome. By examining site responses in a spatially continuous manner, e.g. grid cells or polygons, models can document the range of responses to a particular climate scenario for the region of interest; the aggregation of responses can be used to examine the overall response for the region. The aggregation error associated with regional estimates developed from a spatially continuous model application is generally less than for regional estimates developed from sparse site-specific applications (see Kicklighter *et al.* 1994).

15.11 CONCLUSIONS

The potential distribution of grasslands and conifer forests in our analysis represents 22% of the area occupied by the terrestrial biosphere. Globally, TEM estimates that conifer forests are 1.4 times more productive per unit area than grasslands. Carbon storage per unit area in global conifer forests is estimated by TEM to be approximately 2.8 times that of global grasslands, primarily because of carbon storage in woody tissue. In response to climate change and elevated CO_2, conifer forests are potentially much stronger carbon sinks than grasslands because of the ability to store carbon in woody tissue. The responses of NPP and vegetation carbon in both grasslands and conifer forests appear insensitive to assumptions about the soil carbon at the calibration site. However, the response of soil carbon to changes in climate or CO_2 is three to four times greater for carbon considered to a 1 m depth than for carbon considered to a 20 cm depth. For models that make equilibrium predictions, these results indicate the importance of identifying at the calibration site the soil carbon that is likely to be actively decomposing over the time frame of interest.

The application of process-based models in a spatially continuous manner represents an important tool for making assessments of ecosystem response to climate change because it provides scientists and policy makers with the capability to investigate the potential effects in a quantitative and geographically specific way. However, it is important to recognize that there is uncertainity surrounding the responses predicted by any model. There is a need to compare different models applied in a spatially continuous fashion to help quantify some of the uncertainity. Progress in understanding the potential responses of grasslands and conifer forests to changes in climate and CO_2 requires this activity. Also, there is a need to link models of vegetation distribution with models of ecosystem function to more fully understand the potential effects of climate change on carbon cycling in grasslands and conifer forests.

15.12 REFERENCES

Adams, R. M., Rosenzweig, C., Peart, R. M., Ritchie, J. T., McCarl, B. A., Glyer, J. D., Curry, R. B., Jones, J. W., Boote, K. J. and Allen, L. H. Jr (1990) Global climate change and US agriculture. *Nature* **345**, 219–224.

Australian Surveying and Land Information Group (1990) *Vegetation*. Atlas of Australian Resources, Third Series, Vol. 6, Commonwealth of Australia, Canberra.

Bonan, G. B., Shugart, H. H. and Urban, D. L. (1990) The sensitivity of some high-latitude boreal forests to climatic parameters. *Clim. Change* **16**, 9–29.

Bonan, G. B. and Van Cleve, K. (1992) Soil temperature, nitrogen mineralization, and carbon source–sink relationships in boreal forests. *Can. J. For. Res.* **22**, 629–639.

Burke, I. C., Kittel, T. G. F., Lauenroth, W. K., Snook, P., Yonker, C. M. and Parton, W. J. (1991) Regional analysis of the Central Great Plains. *BioScience* **41**, 685–692.

Emanuel, W. R., Shugart, H. H. and Stevenson, M. P. (1985) Climatic change and the broad-scale distribution of terrestrial ecosystem complexes. *Clim. Change*, **7**, 29–43.

Hou, H. Y., Sun, S. C., Chang, J. W., Ho, M. K., Wang, Y. F., Kong, D. C. and Wang, S. C. (1979) *Vegetation Map of China*. The Laboratory of Plant Ecology and Geobotany, Institute of Botany, Chinese Academy of Sciences. Map Publisher of the People's Republic of China, Beijing, China.

Houghton, J. T., Callander, B. A. and Varney, S. K. (Eds) (1992) *Climate Change 1992 – The Supplementary Report to the IPCC Scientific Assessment*. Cambridge University Press, Cambridge.

Institute of Geography of the Siberian Department of USSR Academy of Sciences, Botanical Institute of USSR Academy of Sciences, and Moscow State University Geography Department (1990) *Vegetation of the USSR*. GVGK, Minsk, USSR.

Jenne, R. L. (1992) Climate model description and impact on terrestrial climate. In: Majumdar *et al.* (Eds) *Global Climate Change: Implications, Challenges and Mitigation Measures*, pp. 145–164. Pennsylvania Academy of Science, Easton, Pa.

Joint Federal-State Land Use Planning Commission for Alaska (1973) *Major Ecosystems of Alaska*. United States Geological Survey, Fairbanks, Ala.

Kicklighter, D. W., Melillo, J. M., Peterjohn, W. T., Rastetter, E. B., McGuire, A. D., Steudler, P. A. and Aber, J. D. (1994) Aspects of spatial and temporal aggregation in estimating regional carbon dioxide fluxes from temperate forest soils. *J. Geophys. Res.* **99**, D1, 1303–1315.

Küchler, A. W. (1964) *Potential Natural Vegetation of the Conterminous United States* (map). American Geographical Society, New York.

McGuire, A. D., Joyce, L. A., Kicklighter, D. W., Melillo, J. M., Esser, G. and Vrsmarty, C. J. (1993) Productivity response of climax temperate forests to elevated temperature and carbon dioxide: a North American comparison between two global models. *Clim. Change* **24**, 287–310.

McGuire, A. D., Melillo, J. M., Joyce, L. A., Kicklighter, D. W., Grace, A. L., Moore, B. III and Vörösmarty, C. J. (1992) Interactions between carbon and nitrogen dynamics in estimating net primary productivity for potential vegetation in North America. *Glob. Biogeochem. Cyc.* **6**, 101–124.

Manabe, S. and Wetherald, R. T. (1987) Large scale changes in soil wetness induced by an increase in carbon dioxide. *J. Atmos. Sci.* **44**, 1211–1235.

Matthews, E. (1983) Global vegetation and land use: new high resolution data bases for climate studies. *J. Clim. Appl. Meteorol.* **22**, 474–487.

Melillo, J. M., Callaghan, T. V., Woodward, F. I., Salati, E. and Sinha, S. K. (1990) Effects on ecosystems. In: Houghton, J. T. *et al.* (Eds) *Climate Change: The IPCC Scientific Assessment*, pp. 283–310. Cambridge University Press, Cambridge.

Melillo, J. M., McGuire, A. D., Kicklighter, D. W., Moore, B. III, Vörösmarty, C. J. and Schloss, A. L. (1993) Global climate change and terrestrial net primary production. *Nature* **363**, 234–240.

Mitchell, J. F. B., Manabe, S., Meleshko, V. and Tokioka, T. (1990) Equilibrium climate change – and its implications for the future. In: Houghton, J. T. *et al.* (Eds) *Climate Change: The IPCC Scientific Assessment*, pp. 131–172. Cambridge University Press, Cambridge.

Ojima, D. S., Dirks, B. O. M., Glenn, E. P., Owensby, C. E. and Scurlock, J. O. (1993a) Assessment of C budget for grasslands and drylands of the world. *Water Air Soil Pollut.* **70**, 95–109.

Ojima, D. S., Parton, W. J., Schimel, D. S., Scurlock, J. M. O. and Kittel, T. G. F. (1993b) Modeling the effects of climatic and CO_2 changes on grassland storage of soil C. *Water Air Soil Pollut.* **70**, 643–657.

Pastor, J. and Post, W. M. (1988) Response of northern forests to CO_2-induced climate change. *Nature* **334**, 55–58.

Prentice, I. C., Cramer, W., Harrison, S. P., Leemans, R., Monserud, R. A. and Solomon, A. M. (1992) A global biome model based on plant physiology and dominance, soil properties and climate. *J. Biogeogr.* **19**, 117–134.

Prentice, K. C. (1990) Bioclimatic distribution of vegetation for general circulation model studies. *J. Geophys. Res.* **95**, D8, 11 811–11 830.

Raich, J. W., Rastetter, E. B, Melillo, J. M., Kicklighter, D. W., Steudler, P. A., Peterson, B. J., Grace, A. L., Moore, B. III and Vörösmarty, C. J. (1991) Potential net primary productivity in South America: application of a global model. *Ecol. Appl.* **1**, 399–429.

Rowe, J. S. (1972) *Forest Regions of Canada.* Department of Fisheries and the Environment, Canadian Forestry Service, Publication No. 1300, Ottawa.

Running, S. W. and Nemani, R. R. (1991) Regional hydrologic and carbon balance responses of forests resulting from potential climate change. *Clim. Change* **19**, 349–368.

Schimel, D. S., Parton, W. J., Kittel, T. G. F., Ojima, D. S. and Cole, C. V. (1990) Grassland biogeochemistry: links to atmospheric processes. *Clim. Change* **17**, 13–25.

Schlesinger, M. E. and Zhao, Z.-C. (1989) Seasonal climatic changes induced by doubled CO_2 as simulated by the OSU atmospheric GCM/mixed-layer ocean model. *J. Clim.* **2**, 459–495.

Schlesinger, W. H. (1977) Carbon balance in terrestrial detritus. *Ann. Rev. Ecol. Syst.* **8**, 51–81.

UNESCO (1981) *Vegetation Map of South America.* UNESCO, Paris.

Vörösmarty, C. J., Moore, B. III, Grace, A. L., Gildea, M. P., Melillo, J. M., Peterson, B. J., Rastetter, E. B. and Steudler, P. A. (1989) Continental scale models of water balance and fluvial transport: an application to South America. *Glob. Biogeochem. Cyc.* **3**, 241–265.

Walter, H. (1979) *Vegetation of the Earth and Ecological Systems of the Geo-biosphere.* Springer-Verlag, New York.

Watson, R.H., Filho, L.G.M., Sanhueza, E. and Janetos, A. (1992) Greenhouse gases: sources and sinks. In: Houghton, J.T. *et al.* (Eds) *Climate Change 1992: The Supplementary Report to the IPCC Scientific Assessment*, pp. 25–46, Cambridge University Press, New York.

White, F. (1981) *Vegetation of Africa.* UNESCO, Paris, France.

Whittaker, R. H. and Likens, G.E. (1973) Carbon in the biota. In: Woodwell, G. W. and Pecan, E. V. (Eds) *Carbon and the Biosphere*, pp. 281–302. United States Atomic Energy Commission, Washington, DC.

Willmott, C. J., Rowe, C. M. and Philpot, W. D. (1985) Small-scale climate maps: a

sensitivity analysis to some common assumptions associated with a grid-point interpolation and contouring. *J. Am. Cartog.* **12**, 5–16.
Woodward, F. I. and McKee, I. F. (1991) Vegetation and climate. *Environ. Int.* **17**, 535–546.

16 Prediction of Global Biome Distribution Using Bioclimatic Equilibrium Models

R. LEEMANS,[1] W. CRAMER,[2] and J. G. VAN MINNEN[1]

[1]*Global Change Department, Dutch National Institute of Public Health and the Environment, Bilthoven, The Netherlands*

[2]*Potsdam Institute of Climate Impact Research, Potsdam, Germany*

16.1 INTRODUCTION

In this more general chapter, we will present different methodologies to delimit the distributions of ecosystems under current and future climate. This is important, because large shifts in vegetation patterns can occur under a changing climate (Cramer and Leemans 1993). Current grasslands and coniferous forests can be replaced regionally by other ecosystems. If these shifts are not taken into account, impact assessment for specific ecosystems under changed climate will be misleading. The objective of this chapter is to present a global framework for impact assessment, which could define the boundary conditions of specific ecosystem assessments, such as those presented in earlier chapters.

Terrestrial ecosystems play a major role in the global C cycle. The total C content of both vegetation and soil is about three times as high as that of the atmosphere. The exchange of C between the terrestrial biosphere and the atmosphere is about 20 times larger than the anthropogenic emissions resulting from fossil fuel use. This exchange of C is influenced by a multitude of feedback processes, such as CO_2-fertilization (e.g. Bazzaz 1990), climatic change on both plant growth (e.g. Fitter and Hay 1981; Larcher 1980) and soil respiration (e.g. Parton *et al.* 1987) and vegetation distribution (Leemans 1992; Cramer and Leemans 1993). The C dynamics of ecosystems are mainly determined by net primary productivity (NPP) in plants, followed by C partitioning over different compartments and losses through respiration and decomposition. Every ecosystem has its characteristic C budget and dynamics, and this is often used to parameterize C cycle models (e.g. Emanuel *et al.* 1981; Goudriaan and Ketner 1984; Esser 1991; McGuire *et al.* 1992; Smith *et al.* 1992a; Melillo *et al.* 1993; Klein-Goldewijk *et al.* 1994).

An adequate description of vegetation and its global patterns is important for

Global Change: Effects on Coniferous Forests and Grasslands.
Edited by A. I. Breymeyer, D. O. Hall, J. M. Melillo and G. I. Ågren

the initialization of C cycle models. The earlier C cycle models (e.g. Goudriaan and Ketner 1984) used a simple ecosystem-specific C density, which combined with ecosystem extent, allowed for a straightforward characterization of its C budget. The extents of different ecosystems types were typically taken from statistical, highly aggregated sources. Shifts in vegetation patterns, driven by climate or changing land use, were simply prescribed or simulated using transition probabilities. Recently, several modelling groups have taken a more realistic approach by using geographically explicit data bases to drive their C models (e.g. Esser 1991; McGuire *et al.* 1992; Klein-Goldewijk *et al.* 1994). Feedback processes are implemented in these models in such a way that they account for local environmental differences, such as heterogeneity in topography, climate and soils. The geographic explicit ecosystem patterns are often based on global vegetation data bases, such as Matthews (1983), Olson *et al.* (1985), Melillo *et al.* (1993) and Küchler (1949, in Espenshade and Morrison 1991) or different climate classifications (Leemans 1992; Prentice *et al.* 1992; Cramer and Leemans 1993). One of the currently most widely used classifications in C cycle modelling is the life zone classification by Holdridge (1967; e.g. Prentice and Fung 1990; Smith *et al.* 1992a; Smith and Shugart 1993).

Our approach is to review the different global vegetation, ecosystem or land cover data bases that have been used for C cycle modelling. We will distinguish between data bases derived from 'observational records' and different types of models. All data bases will be compared with each other and the major differences will be explained in terms of their underlying assumptions, limitations and origins. We will attempt to rank the global cover data bases according to their applicability to global C models. We will further review a series of applications of these data bases and models to analyse different aspects of global change issues, such as the missing C sink, feedbacks in the C cycle and vegetation response to climatic change.

16.2 REVIEW OF GLOBAL LAND COVER DESCRIPTIONS

16.2.1 Global land cover classifications and data bases

Land cover classifications are often regarded as synonymous with vegetation classifications and several institutes are devoted (e.g. The International Institute of Vegetation Mapping in Toulouse, France) to create such classifications and prepare maps. Many different aspects of vegetation have been used, either individually or in combination (Mueller-Dombois and Ellenberg 1974). The criteria used in most schemes can be characterized by:

- *Structure and physiognomy*. Vegetation attributes such as height, growth form, cover, deciduousness and leaf form are used as a basis of the classification. Examples of this approach are Küchler (1967), Küchler and Zonneveld (1988), and also the hierarchical UNESCO classification (1973). The life form classification by Raunkiær (1934) also fits in this category.

- *Floristics.* The taxonomic affinities of plants can also be used as the basis of a classification. The lower levels of the UNESCO (1973) classification are based on floristics. Other examples include Braun-Blanquet (1964) and Tahktajan (1973). Most of these classification systems are based on key plant taxa which occur in that vegetation type. Grabherr and Kojima (1993) recently stated that the floristic approach allows for the solid basis for a globally applicable and implementable land cover classification system. However, although ecological function is implicit in most taxonomic units, the lack of detailed information on the species of the world and on their function, makes it impossible to apply the floristic approach directly to global land cover mapping.
- *Bioclimatic characterization.* Bioclimatic schemes are not based on actual vegetation, but on the climatic regime that prevails in that region. Many of these classification schemes are based on the observation that both vegetation physiognomy and species composition are a function of climate. The close correlation was recognized in the early nineteenth century (von Humboldt 1807) and many vegetation and climate maps were published on the basis of this relationship. Even in recently published vegetation maps (e.g. Bartholomew *et al.* 1988) one can tell that the major vegetation patterns are derived from climate, rather than vegetation observations. The relation, however, is only obvious for the broad scale, global vegetation patterns. At the regional and local scales, other environmental (e.g. soil and topography) and historical (e.g. disturbance and succession) factors also strongly influence the actual vegetation at any place (Leemans 1992).

  The approach is frequently used because comprehensive data on climate are more often available than good comparable vegetation data. Examples of this approach are developed by Box (1981), Budyko (1986), Köppen (1936), Holdridge (1967), Prentice *et al.* (1992), Thornthwaite (1948), Walter (1985) and Whittaker (1975). The different approaches are reviewed by Tuhkanen (1980). This type of classification is most frequently used for climate change impact assessments and global change studies.
- *Hybrid classification schemes.* Several hybrid classifications schemes that include elements of structural, physiognomic, floristic and bioclimatic attributes have been developed. This is apparent in many of the commonly used names for vegetation categories: 'secondary tropical moist montane rain forests' or temperate short grass steppe'. Unfortunately, almost no standardization and harmonization in defining these categories is followed among different classifications schemes, so that large differences can occur between different sources (UNEP/GEMS 1993).

  The most well-known hybrid classification is the UNESCO (1973) classification. The UNESCO classification was developed for the description of the potential vegetation at a climax stage. This has probably been one of the major reasons that the classification has never resulted in a global assessment of land cover. The different categories often do not refer to actual vegetation cover and this has led to major confusion. This confusion becomes very obvious in the global, highly aggregated implementation of the classification by Matthews (1983). Despite its problems, several UNESCO vegetation maps have been produced for different regions (e.g. White 1983). Another regional example of a successful hybrid classification is presented by Whitmore (1984) for tropical forests.

Many more land cover classification schemes have been developed and implemented for specific ecosystems (e.g. Whitmore 1984), large regions (e.g. Bailey 1980; White 1983; Anderson *et al.* 1976) or political entities (e.g. Whitmore 1984; CORINE 1989). A recent review of different approaches to vegetation classification and mapping is presented in Küchler and Zonneveld (1988). Currently there are also several global land cover data bases available that describe the current patterns. However, the origin of the data included in the compilations is not always clear. Furthermore, the data bases are often developed for a specific purpose, such as C cycling or land surface parameterizations of climate models. Here we will describe the most frequently used data bases. The legends of these data bases are given in Appendix 16.1.

16.2.1.1 The physiognomic vegetation data base (Küchler 1949)

This data base is derived from structural characteristics of potential vegetation (trees; shrubs; grasses; deserts) and major physiognomic features, such as deciduousness, and needle versus broad-leaved. Additional attributes used in the classification involve the percentage canopy cover (no, sparse, open and closed). A global map as presented in Espenshade and Morrison (1991) has been digitized at a resolution of 1° longitude and latitude. The data base consists of 34 classes. Although the data base gives a good representation of the potential vegetation cover, it has not (yet) been used for global change studies. This is mainly due to its coarse resolution and its incompatibilities of legends with other global tabular data sets, such as UN-ECE/FAO (1992), which are mainly based on mixed classifications. For example, the southern pine forests in Florida are not distinguished from the boreal forests. To make these necessary distinctions for linkage with other data sets, additional information on climate, soil and topography is needed.

16.2.1.2 The major ecosystems of the world (Olson et al. 1985)

This is a global data base with a resolution of 0.5° longitude and latitude, that was developed primarily to describe the C content of the major ecosystems of the world. Its documentation (Olson *et al.* 1985) is comprehensive with short descriptions of each class with a list of dominating species, ecosystem structure, data sources and C densities. The data base consists of 48 classes. This data base is the only land cover data base that includes explicitly natural vegetation categories, such as taiga and tropical montane rain forests, non-vegetated land categories, such as ice and stony desert, and land use categories, such as arable land, irrigated drylands and paddylands. An updated version of this data base was recently developed by Olson (included in Kineman 1992), but it lacks the detailed documentation which made the earlier version so valuable. The improvements were made only for a few regions of the world.

The C values given in this data base have frequently been used to parameterize C budget models using bioclimatic schemes to allow for shifting vegetation zones under a changed climate (e.g. Prentice and Fung 1990; Smith *et al.* 1992a). The IMAGE 2 model uses the data base to initialize the current land cover patterns for its dynamic simulations of changing future land use and C cycling (Alcamo *et al.* 1994; Leemans and van den Born 1994).

16.2.1.3 The global vegetation data base (Matthews 1983)

The global vegetation data base uses the UNESCO (1973) classification. Only the highest hierarchical levels of this classification are used here, so that the data base uses mainly the functional attributes. The data base has a relatively coarse resolution of 1° longitude and latitude and consists of 32 cover classes. The vegetation data base can be linked to a set of compatible data bases on cultivation intensity and albedo (Matthews 1983). These have been used in many different global change assessments of the NASA-GISS group (e.g. Matthews and Fung 1989; Kaufman *et al.* 1990; Bouwman *et al.* 1993). One of the problems of this data base is that it is not clear how the many different data sources (*ca.* 70 atlases), from which it actually was developed, are used and translated into the UNESCO system. Furthermore, the legend is a difficult to interpret combination of actual, potential and man-induced vegetation (cf. Appendix 16.1).

16.2.1.4 The global potential vegetation data base (Melillo et al. 1993)

This data base is an extension and improvement of Matthews' (1983) global vegetation data base. First, the resolution was increased to 0.5° longitude and latitude. Next, the data base was overlaid with regional implementations of the UNESCO vegetation classification, such as the vegetation map of Africa (White 1983). Ambiguous classes have been removed, which resulted in a more comprehensive data base. Despite the improvements achieved by this approach, there are large regional differences within the data base, because regional data bases that are compatible with the UNESCO classification do not exist for all parts of the world. The data base has been used to initialize an equilibrium global C cycle model (Melillo *et al.* 1993).

From this short review of global vegetation data bases, it becomes apparent that no satisfactory and comprehensive data base on global land cover has yet been developed. However, several international research programmes aim to produce improved data sets with modern technologies, such as remote sensing (e.g. Townshend 1992) and several of these, probably more reliable, data bases will become available to the global change research community within this decade. These approaches have already led to an improved assessment of deforestation patterns in Brazil (Skole and Tucker 1993).

16.2.2 Using environmental characteristics to predict vegetation distributions

Due to the low reliability of the current global land cover data bases, other approaches have been applied to define the global land cover patterns (including coniferous forests and grasslands). The most important one to assess changes in land use and cover has been the use of tabular statistical data to determine the extent of each vegetation category within political boundaries. Straightforward transition probability matrices were applied to determine the impact of changing patterns (e.g. Houghton *et al.* 1983; Goudriaan and Ketner 1984). The main limitation of this approach is that georeferenced determinants of vegetation patterns are not used to their full extent, which has resulted in large discrepancies between different analyses.

Observation-based land cover data bases suffer from scarcity of observation and from problems associated with the classifications. The first of these problems can be overcome by predicting the dominant ecosystem types from descriptors of basic physical habitat characteristics, such as climate, soils or hydrology. This, however, cannot reflect the overwhelming influence of human land use on land cover.

These habitat predictors allow for the development of land cover change scenarios driven by climate change in a geographically comprehensive way. Previously, such changes have been subscribed as one-to-one changes (e.g. Holten 1990). Many of such studies use only climatic parameters that can be computed from readily available weather station data (such as Müller 1982) and global climate data bases (such as Leemans and Cramer 1991). Some studies use simple climatic parameters directly to delimit specific vegetation types (e.g. Hulme *et al.* 1992) or agricultural crops (e.g. Parry 1992), but most studies use existing bioclimatic classifications or have developed more adequate, new classification schemes.

Among the first researchers who used this approach globally was Walter (1970). The patterns of monthly temperature and precipitation in his climate diagrams are used to define vegetation. These climate diagrams are popular, because they give a straightforward visual impression of seasonality and the moisture balance at a given site. They are problematic, however, because the intersections of the temperature and precipitation 'curve' are only remotely related to evapotranspiration. The beginning and end of the dry season is therefore not precisely defined. Further, it requires expert judgement to link such a climate diagram to the proper vegetation type. Therefore, this approach has not been used for global change studies.

The more frequently used approaches are comprehensive climate–vegetation classifications. The classifications are usually defined by the climatic categories only. Therefore, they can be implemented on a computer using climate data bases and geographic information systems (GIS). When the spatial pattern of current climate has been captured by the GIS, anomalies in climate change scenarios can be overlaid and new boundaries for the bioclimatic classes can be derived. This approach has mainly been used for impact studies on natural vegetation (e.g. Leemans 1992) and terrestrial C models (e.g. Prentice and Fung 1990; Smith *et al.* 1992a; Melillo *et al.* 1993).

Here we will present and discuss some of the most frequently used bioclimatic classifications (Appendix 16.2). We have implemented all these classifications on a global grid of 0.5° longitude and latitude using a global data base with climatic normals for the period 1931–60 (Leemans and Cramer 1991). The classifications are further used to analyse the impacts of a changed climate. We have used the simulated climate anomalies for an equilibrium climate for an atmosphere with doubled CO_2 atmospheric conditions. The approach taken to define a future climate is to overlay current climate with the anomalies. The precise methodology is given in Leemans (1992) and conforms to the standardized IPCC approach (Carter *et al.* 1992).

16.2.2.1 Life zone classification (Holdridge 1967)

One of the most frequently used climate–vegetation classifications is the Holdridge life zone classification (Holdridge 1967, Figure 16.1). This classification is based on two climatic indices: annual precipitation and biotemperature. The latter is the average annual positive temperature. Although an axis marked PET is added, no additional information on this climatic parameter is needed, because it is simply a linear function of biotemperature, and only the balance between the two main indices determines moisture conditions. Different logarithmic combinations of the indices are used to delimit different life zones or biomes (Figure 16.1). A considerable limitation of this model is that it is based on only two annual indices. The moisture balance is therefore not properly described and seasonal aspects are lacking completely. Biomes characterized by strong seasonality, such as monsoonal forests, cannot be captured satisfactorily.

Nevertheless, the life zone classification was the first scheme to be used for the analysis of the impacts of climatic change on global vegetation patterns (Emanuel *et al.* 1985). Large shifts in vegetation zones can be observed when the

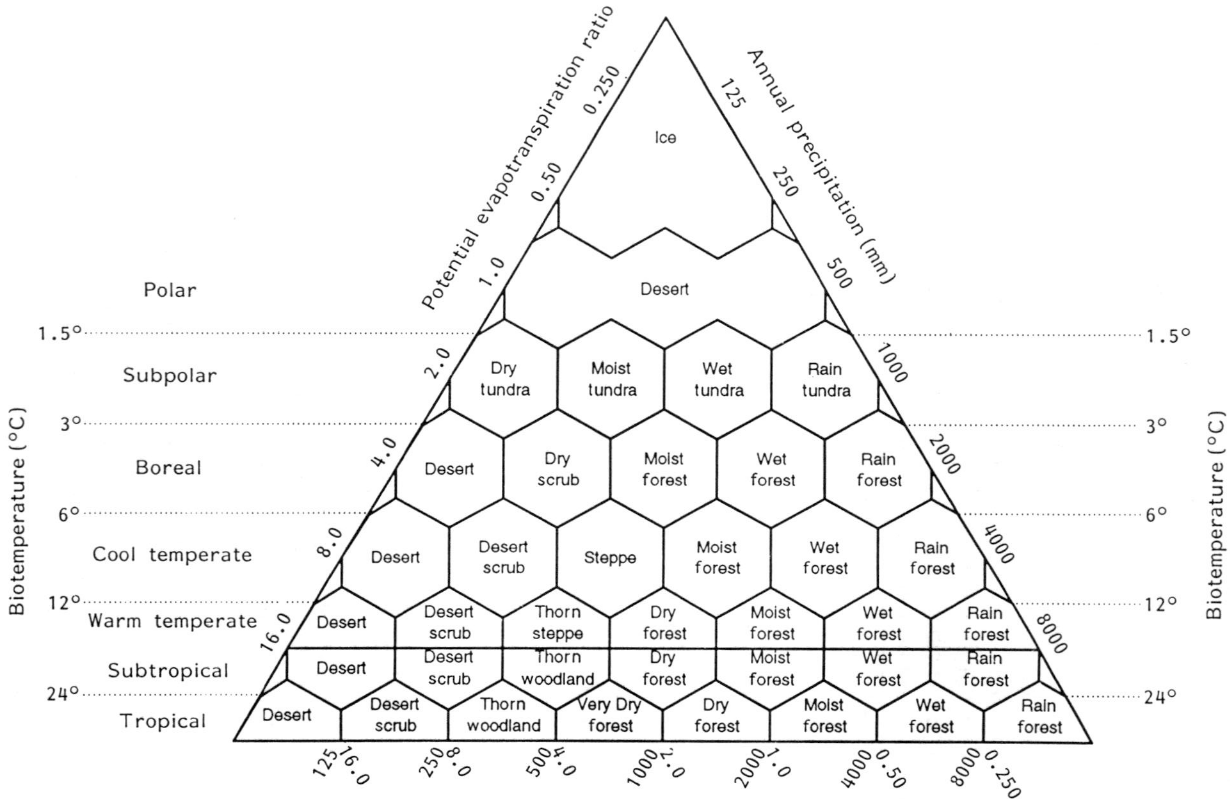

Figure 16.1 The life zone classification (Holdridge 1967) is determined from biotemperature and annual mean precipitation. Potential evapotranspiration is a linear function of biotemperature. The hexagons delimit the different life zones

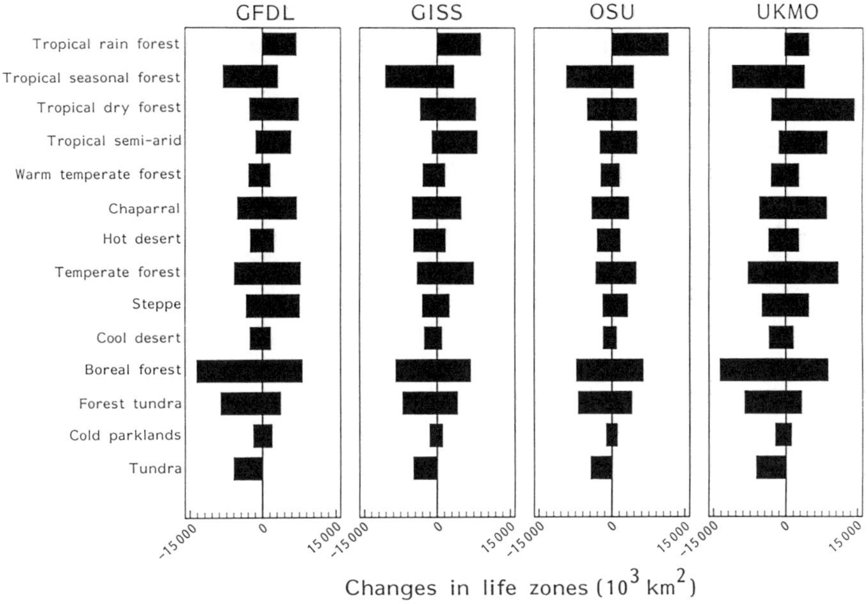

Figure 16.2 Shifts in life zones under several doubled CO_2-derived climates as determined by the life zone classification (adapted from Leemans 1992). The left part of the histogram presents a decrease in extent, while the right part presents an increase, both in respect to the current extent

classification is combined with different climate scenarios (Figure 16.2). The largest changes occur in high latitude areas. This is not only due to the higher temperature increase in these regions by the scenarios (for a discussion see Mitchell *et al.* 1990), but also to the specific sensitivities of the life zone classification.

16.2.2.2 Global climate classification (Köppen 1936)

Köppen (1936) tried to capture the annual cycles of temperature and precipitation in a climate classification. He designed his classification so that the categories approximately resemble global vegetation patterns. It is the first empirical and objective climate classification that is established with a limited set of climatic parameters, defining five major climate categories (Table 16.1; Strahler and Strahler 1987). The strength of the Köppen (1936) climate classification is that it includes both the different latitudinal zones (based on extreme temperatures) and seasonality in both precipitation and temperature. It therefore should theoretically perform better than the life zone classification by Holdridge (1967). However, the extent of arid climates is largely underestimated due to an inadequate description of the hydrological cycle by only comparing extreme temperatures with precipitations. The classification system neglects evapotranspiration as an important limiting factor for plant growth (Thornthwaite 1943). Its advantage, however, is that it does not use land cover terms to label climatic zones, but a hierarchy of symbols (Table 16.1). The possibility of misinterpreting the nature of these climatic zones by less experienced investigators is therefore reduced.

The Köppen (1936) climate classification has been used by the US Forest Service to delineate ecosystem regions (Bailey 1983, 1989). Bailey (1983) defines ecoregions as large ecosystems of regional extent that contain a number of smaller ecosystems. They define major geographical zones that represent associations of similarly functioning vegetation or potential land covers. Bailey's (1983) purpose was to develop a land cover classification that divided the landscape into variously sized ecosystem units that have significance both for resource development or environmental conservation. The major problem with such an ecoregion approach is that only climatological parameters are emphasized and that the resulting cover classification will strongly focus on potential cover class and neglect human use. A slightly modified Kppen (1936) climate classification (Trewartha 1968) that addressed some of the shortcomings of the aridity definitions has been used by Guetter and Kutzbach (1990) to analyse the impacts of changing climate on land cover patterns during the last 18 000 years. They clearly illustrate the large changes in land cover, especially in mid and high latitudes, that have occurred since the last glaciation. Figure 16.3 illustrated the potential shifts of the climate classes under several doubled CO_2 climates.

16.2.2.3 Biogeographical zones (Budyko 1986)

The biogeographical zones (Budyko 1986) are based on an ordination of moisture- and temperature-related indices. As such it has strong similarities to the classifications of Thornthwaite (1948) and Whittaker (1975). However, the latter two are based on climatic indices that are less functional in their definition, and therefore less likely to reflect the major correlations between climate and vegetation. Budyko (1986) uses an approach that delimits vegetation classes through the computed energy (or radiation) and moisture balance (Figure 16.4). The moisture balance is characterized by a dryness index, which is based on an elaborate evapotranspiration scheme, that accounts for latitude, humidity and energy provided by the local radiation balance. If climate data are available, this approach is more reliable than the empirical evapotranspiration indicators developed by Holdridge (1959) or Thornthwaite and Mather (1957). It clearly separates tundra, forest, steppes, semi-deserts and deserts of the main zones. Within each zone there are large differences in the energy balance for the forested zone, but these become much smaller with increasing aridity (Figure 16.4).

The Budyko scheme (1986) has frequently been used by investigators from the

Table 16.1 The different categories in the global climate classification (Köppen 1936)

Major climatic zones	Climatic types	Climatic specifiers
A Tropical rainy climates		
	f	No drought periods: evergreen tropical rain forest
	w	Drought period: tropical deciduous forests and savannas
	m	Pronounced drought period: monsoonal deciduous forests
B Arid climates		
	S	Semi-arid climates such as grasslands, dry savannas and low shrubs
	W	Desert climates with sparse vegetation cover
		h Dry and hot
		k dry and cold
C Temperate rainy climates		
	f	No drought periods: deciduous and evergreen forests
	s	Summer droughts: oak and eucalyptus forests
	w	Winter droughts
		a Hot summer
		b Warm summer
		c Cool, short summer
D Boreal climates		
	w	Winter drought: deciduous coniferous forests
	f	No drought periods: deciduous and evergreen coniferous forests
		a Hot summer
		b Warm summer
		c Cool, short summer
		d Very cold winters
E Snow climates		
	T	Climates with a short growing season: treeless tundra
	F	Climates with no growing season: ice

former Soviet Union in their global change impacts studies (e.g. Izrael *et al.* 1990) and has recently been coupled to climate change scenarios derived from climate models, by Tchebakova *et al.* (1993a, b; Figure 16.5). They distinguished 16 land cover classes world-wide and illustrated the potential shifts of vegetation patterns under a changed climate. Although the results compare well with other studies and the performance of the ordination is relatively good (see below), the approach still has some major disadvantages for global change studies. First, the radiation

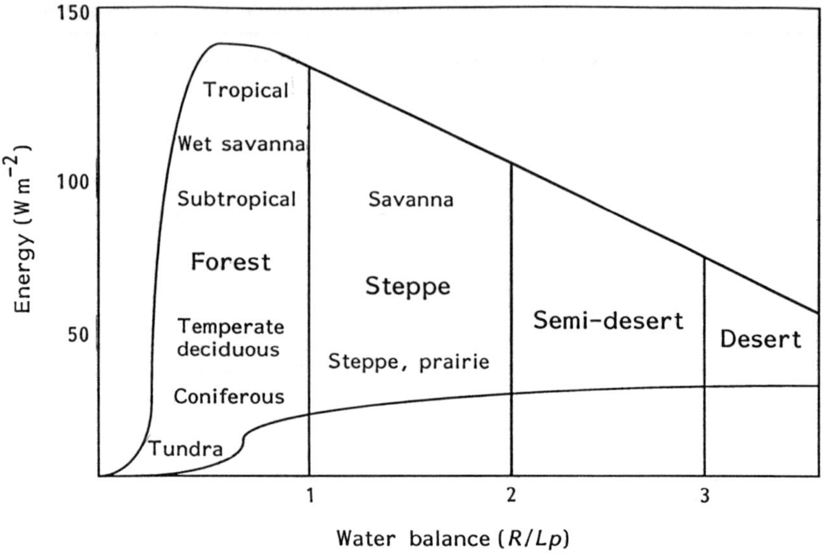

Figure 16.3 The ordination for biogeographical zones as defined by the energy balance (R, W m^{-2}) and the relative dryness index (R/Lp, L = the latent heat of evaporation, p = annual precipitation) and the delineations of the major biogeographical zones (Budyko 1986)

balance is used to compute some vegetation characteristics, such as albedo. Such characteristics should not be used for vegetation prediction, because of the circularity in the approach. Second, seasonality is not adequately covered, so that no linkages can be made to the more physiognomic and structural vegetation types like those of Küchler (1949). Third, the boundaries are not very clearly defined. The table containing limits for the different zones (Budyko 1986: p. 94) has to be modified subjectively to be implemented unambiguously on global climate data bases. Finally, the classification depends on a series of not readily available climatic and physical parameters (such as humidity and surface albedo). The lower quality of data bases with these parameters limits the suitability of the biogeographical zones for global change studies.

16.2.2.4 The BIOME plant functional type model

A different approach to model global vegetation patterns has recently been developed by Prentice *et al.* (1992) with the BIOME model. Their aim was to develop a conceptually simple model which primarily defined the climatic limits of the most important plant types, rather than biomes. Hence, the model captures aspects of ecological function in limiting the distribution of plant types. The bioclimatic limits are defined such that they can be based on physiological

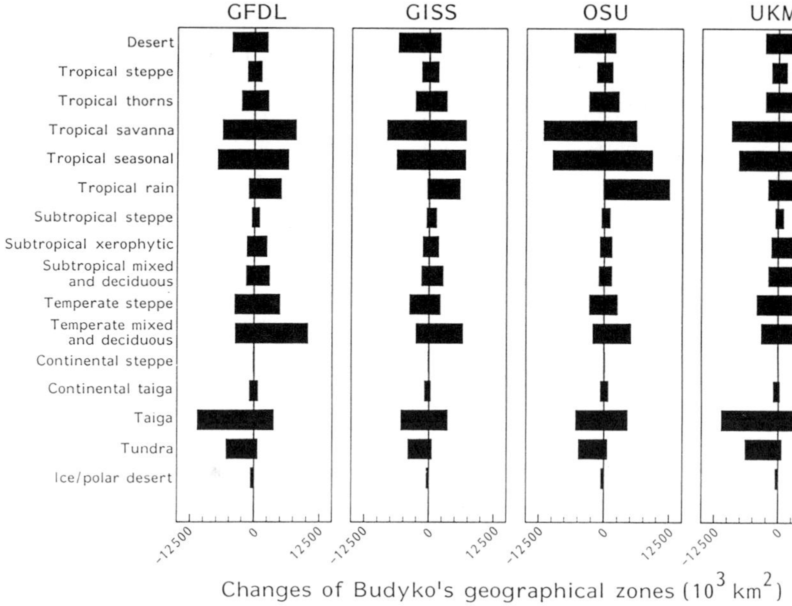

Figure 16.4 Shifts in biogeographic zones (Budyko 1986) under several doubled CO_2-derived climates. The left part of the histogram presents a decrease in extent, while the right part presents an increase, both in respect to the current extent

processes and their physical limits. The approach was based on the model of plant-forms (Box 1981). The BIOME approach also has similarities with the rule-based model for the North American vegetation types (Neilson *et al.* 1992; Neilson 1993).

Plant functional types (PFTs) are defined for distinct temperature zones (tropical, temperate and boreal or warm, cool and cold). These zones coincide with the major latitudinal zones, but the labels for these zones are just for convenience and do not imply a rigid boundary. Secondly, for each zone the major physiognomic adaptations relating to limiting climatic factors such as evergreen vs deciduous, broad-leaved vs needle-leaved, and woody vs herbaceous are described. The PFTs show similarities to Küchler's (1949) physiognomic classification. Although the list (cf. Table 16.2) represents an oversimplification of the variety of plant types that exist, it can capture the major features of major biomes and their transient zones rather well.

The BIOME model relates the distributions of the PFTs to climate indices such as growing degree days, mean temperature of the coldest and warmest month, and the *a*-moisture index (Figure 16.6). The temperature-based indices defined the cold tolerances, chilling and heat requirements for each PFT. The *a*-moisture index is defined as the ratio between actual and potential evaporation

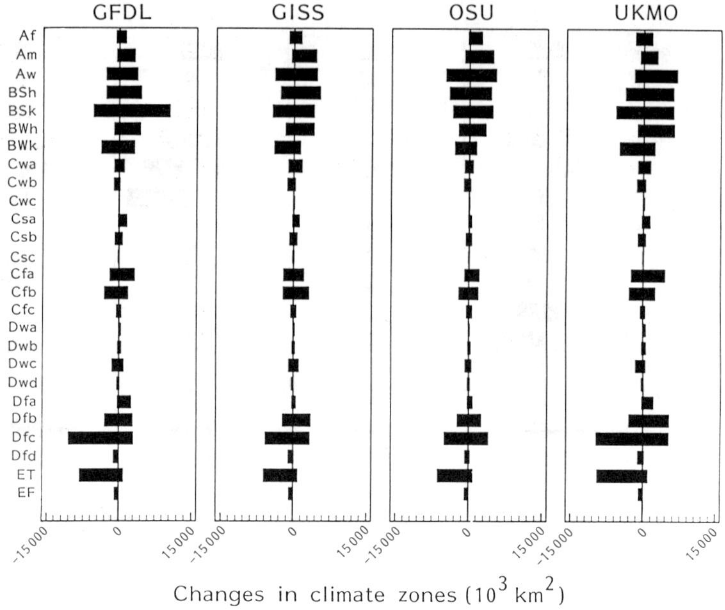

Figure 16.5 Shifts in climatic zones under several doubled CO_2-derived climates as determined by the modified global climate classification (Köppen 1936; Trewartha 1968). The left part of the histogram presents a decrease in extent, while the right part presents an increase, both in respect to the current extent

and determined from a plain bucket-type soil water balance model. This model computes actual evapotranspiration by accounting for precipitation, potential evapotranspiration and a soil-specific water supply to plants. It explicitly includes soil characteristics and is able to carry forward moisture into dry seasons (Prentice *et al.* 1992). Prentice *et al.* (1993) give a complete description of the algorithm. Only a small number of the critical values for the climatic variables are given (based on ecophysiological principles. Table 16.2). The criterion for use of certain limits was that it proved to be necessary to match the vegetation patterns given by the major ecosystems of the world data base (Olson *et al.* 1985). For example, tropical evergreen trees are assumed to tolerate no frost and to have a high moisture requirement. Based on a world-wide regression of annual minimum temperature against mean coldest-month temperatures, 'no frost' implies a coldest month temperature of $> 15.5\,°C$. Based on map comparisons, a 'high moisture requirement' means an *a*-coefficient of at least 0.8. Limits for all other plant types are determined in a similar way. Finally, a 'dominance hierarchy' in which PFTs dominate over others (e.g. trees dominate over grasses) was defined. This hierarchy was strictly applied so that only PFTs from the highest level present were retained.

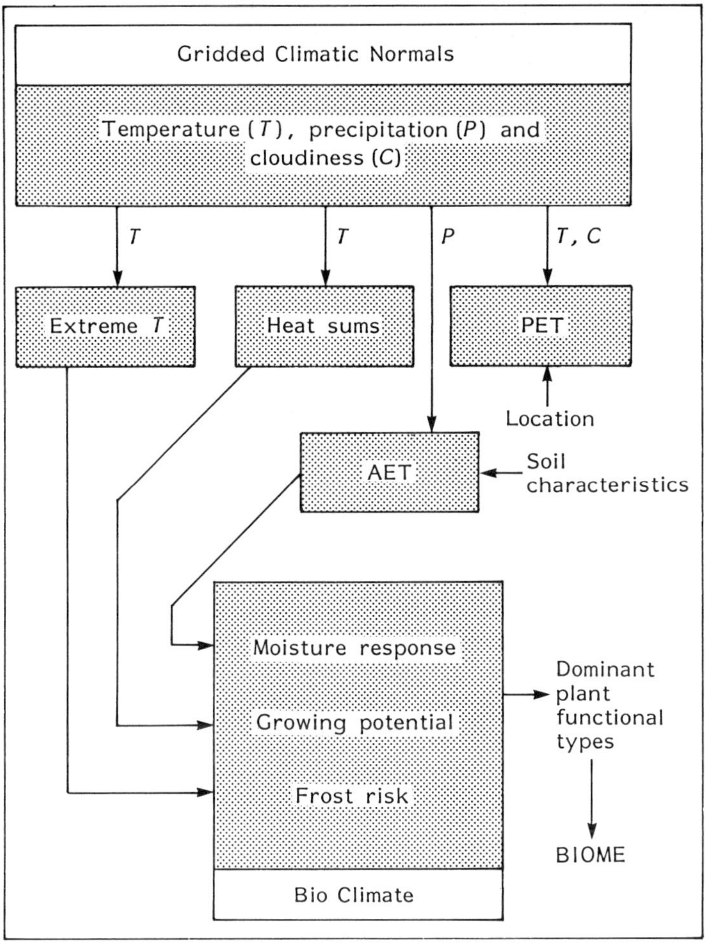

Figure 16.6 Structural diagram of the BIOME model (Prentice *et al.* 1992)

At each location, an array of PFTs is determined to occur. Unique combinations of PFTs define implicitly different biomes, which therefore emerge from the analysis, rather than being defined a priori. The model should be quite capable of producing novel combinations under a changed climate. The model generated 17 different combinations for current climate (Table 16.3). The BIOME model has been used to assess C dynamics in past, current and future climates (Leemans 1992; Prentice *et al.* 1994). Because it is driven only by climatic and soil characteristics, the BIOME model cannot simulate any effect of changing land use. In an attempt to assess changes in global C storage caused by changes in

Table 16.2 The parameters for each plant type of the BIOME model: 1. Growing degree days, base 0 °C; 2. Growing degree days, base 5 °C; 3. Mean temperature of the coldest month; 4. Mean temperature of the warmest month; 5. α moisture index. The last column (6) gives the dominance hierarchy for each plant functional type

	1	2	3	4	5	6
Trees:						
1. Tropical evergreen trees	None	None	> 15.5	None	> 0.80	1
2. Tropical raingreen trees	None	None	> 15.5	None	0.45 to 0.95	1
3. Warm temperate evergreen trees	None	None	> 5.0	None	> 0.65	2
4. Temperate summergreen grees	None	> 1200	−15.0 to 15.5	None	> 0.65	3
5. Cool temperate conifers	None	> 900	−19.0 to 5.0	None	> 0.65	3
6. Boreal evergreen conifers	None	> 350	−35.0 to −2.0	None	> 0.75	3
7. Boreal summergreen trees	None	> 350	< 5.0	None	> 0.65	3
Non-trees:						
8. Sclerophyll shrubs/succulents	None	None	5.0 to 15.5	None	> 0.33	4
9. Warm grasses and shrub	None	None	None	> 21.0	> 0.28	5
10. Cool grasses and shrub	None	> 500	None	None	> 0.33	6
11. Cold grasses and shrub	> 120	None	None	None	> 0.33	6
12. Hot desert shrub	None	None	None	> 21.0	None	7
13. Cool desert shrub	> 120	None	None	None	None	8
14. Polar desert	None	None	None	None	None	9

available land for agriculture, Cramer and Solomon (1993) overlaid the biome vegetation patterns with a category 'climatologically suitable for agricultural land', which matched fairly well with global maps of non-irrigated crops. This approach is further elaborated upon by linking BIOME with a potential agricultural model (Leemans and Solomon 1993). Both are now an essential part of the terrestrial environment system of the IMAGE 2 model (Alcamo *et al.* 1994; Leemans and van den Born 1994), which determines global vegetation response to changing land use, atmospheric conditions and climate.

16.2.3 Constraints of the different climate–vegetation classifications

The presentation of the different climate classifications and their implementations gives an overview of the progress that has been made during the last decade. The models have achieved significant improvement concerning the mechanisms of the physical relationships between the atmosphere and biosphere. The more recent models (e.g. Box 1981; Woodward 1987; Prentice *et al.* 1992) are all derived from physiological considerations, rather than correlations. This is a very

Table 16.3 Combinations of plant functional types generated by the BIOME model. The area (1000 km^2) is the global extent of each combination

Plant functional types	BIOME name	Area (1000 km^2)
Tropical evergreen trees	Tropical rain forest	7 624
Tropical evergreen trees + tropical raingreen trees	Tropical seasonal forest	7 932
Tropical raingreen trees	Tropical dry forest/savanna	17 179
Warm temperate evergreen trees	Broad-leaved evergreen/warm mixed forest	6 561
Temperate summergreen trees + cool temperate conifers + Boreal summergreen trees	Temperate deciduous forest	5 492
Temperate summergreen trees + cool temperate conifers + boreal evergreen conifers + boreal summergreen trees	Cool mixed forest	4 668
Cool temperate conifers + boreal evergreen conifers + boreal summergreen trees	Cool conifer forest	2 807
Boreal evergreen conifers + boreal summergreen trees	Taiga	11 049
Cool temperate conifers + boreal summergreen trees	Cold mixed forest	759
Boreal summergreen trees	Cold deciduous forest	2 834
Sclerophyll/succulent	Xerophytic woods/scrub	10 636
Warm grasses and shrub	Warm grass/shrub	9 845
Cool grasses and shrub + cold grasses and shrub	Cool grass/shrub	7 117
Cold grasses and shrub	Tundra	11 666
Hot desert shrub	Hot desert	20 699
Cool desert shrub	Semi-desert	5 268
Polar desert	Ice/polar desert	4 024

important development for global change studies, because the more mechanistic models are likely to be more robust under changed climatic conditions.

The major problem for climate change impact assessments based on the first three classifications (life zones, global climate and the biogeographical zones) is their high sensitivity to small changes along the boundaries. These boundaries are very sharp on maps with implementations of these classifications (see e.g. Cramer and Leemans 1993), and small changes could lead to large shifts in the global patterns. In reality, the boundaries are often not so clear, and transient zones or ecotones are abundant across many landscapes. Ecotones are probably also more resilient against climatic change, because they include elements of several zones, which allows for a larger adaptive capability.

The most important improvement of the different approaches is the use of plant functional types. This approach allows for a more realistic response of vegetation to a changing environment. Palaeoecological studies clearly demonstrate that plants react to climate change as individual taxa. Biomes have formed, dissolved and re-formed throughout the Quaternary period (Huntley and Webb 1988). The life zone, global climate and the biogeographical zones classifications are therefore less suitable, because their basic unit is biomes, not a taxon. These models could lead to an inadequate description of vegetation patterns under climatic change, because novel, no-analogue vegetation types per definition cannot occur within these models. The more suitable models, however, are still strongly limited in the dynamic responses of vegetation, taking into account processes like migration, competition and succession.

16.2.4 Comparison of the different data sets and climate classifications

Global land cover data bases and bioclimatic classifications or models have different shortcomings. Nevertheless, a pairwise comparison between data bases from either of these groups may lead to insights into the limitations of either the data base or the model. If so, we can start to derive some general principles which should be included in a model used for global change studies. The comparison presented here is based on the results of all the different data bases and global classifications, implemented into a GIS (Leemans 1992), which is linked to an array of different spatial statistical techniques to compare, overlay and plot different data sets.

The creation of comparable sets from all nine data bases (Table 16.4) for such comparison was not a trivial task. First, we defined a common grid for the comparison of all nine available data sets. We defined this grid as those cells that in all data sets were designated as land cells. We did not consider the large, ice covered land masses, Antarctica and Greenland. The coarser data sets (physiognomic vegetation classification (Küchler, 1949) and world vegetation data base (Matthews 1983)) were overlaid onto the finer common grid of 0.5° longitude and latitude. Incomparable cells, such as those with a specific land use (especially in the ecosystems of the world data base (Olson *et al.* 1985), the category 'cultivation' in the global vegetation data base (Matthews 1983) was empty), had further to be removed from all data sets (cf. Table 16.4). This common grid is used for further analysis and consists of 44 335 cells and includes a surface of 99 239 km^2 (75% of the total terrestrial ecosystems).

Secondly, all data sets had to be aggregated into a comparable and compatible legend by reclassifying and aggregating the original data bases. Legends appear to be compatible using similar labels, but these could mean structurally very

Table 16.4 The aggregation for the different land cover classifications (the numbers listed correspond with the different legend items – see Appendices 16.1 and 16.2)

Cover class	Olson *et al.* (1985)	Matthews (1983)	Küchler (1949)	Melillo *et al.* (1993)	BIOME Prentice *et al.* (1992)	Holdridge (1967)	Köppen (1936)	Budyko (1986)
Agricultural land	12, 13, 16, 17, 18, 19, 34, 35, 36, 37	32						
Ice	1, 44, 45	30[a], 31	32[b]	1	1	1	1	1
Cool desert	31, 33		4, 10, 30	2	2	7, 12	20	
Hot desert	29, 30	30[c]	9, 17, 32[d]	21	13	18, 25, 32, 33	21	16
Tundra	32	20, 22, 29	15, 23, 32[e]	3	3	2, 3, 4, 5, 6	2	2
Cool grass	2, 20	18, 26, 27, 28	16	12[f], 13, 30[f]	6	13, 14	22	5, 7
Warm grass	21	24, 25	11, 12, 20, 21	12[g], 30[g]	12	19, 26	23	10, 15
Xerophytic wood	25, 26, 27, 28	6, 12, 13, 17, 19, 21	2, 3, 6, 8, 18, 29	19, 32	14	20, 21, 27, 28	14, 15, 16	9, 14
Taiga	3, 38, 39, 40	8, 14, 16	14[h], 25, 26	4, 5, 6, 7	9	8, 9, 10, 11	4, 8, 9, 10 11	3, 4
Cool conifer forest	4	10	9	7	15	5		
Cool mixed forest	5	4	24	8	5, 8, 9	16	3, 6, 13	
Temperate deciduous forest	7	11	5, 7	10	10	17	7, 12	6
Warm mixed forest	6, 8, 9, 10	5, 7	13, 14[i], 27, 28, 31	31	11	22, 23, 24	17, 18, 19	8
Tropical dry forest	14, 22	9, 15, 23	19, 22	14	17	34, 35, 36	24	13
Tropical seasonal forest	11	2		18	16	29, 37	25	12
Tropical rain forest	15	1, 3	1	16, 17	15	30, 31, 38, 39	26	11
Wetlands	23, 24, 41, 42, 43, 46			11, 20, 22, 23, 24, 25, 26, 27, 28, 29				
Not used				33, 34				

[a]For grid cells beyond 66° N. [b]For grid cells beyond 60° polewards.
[c]For grid cells below 66° N. [d]For grid cells between −60° and 60° latitude and not between 71° and 92° eastern longitude. [e]For grid cells between −60° and 60° latitude and between 71° and 92° E. [f]For grid cells beyond 42° N. [g]For grid cells below 42° N. [h]For grid cells beyond 37° N and west of 100° W. [i]For grid cells below 37° N and east of 100° W.

different categories. A further difficulty was related to the distinction between actual, potential, or human-induced vegetation in some data bases. This was difficult because the necessary documentation on definitions and sources of the data bases was not always available. We constructed a target classification with only 18 different classes (Table 16.4), which incorporated the major biomes that are needed to obtain an adequate resolution for different C cycle models (e.g. Melillo *et al.* 1993; Leemans and van den Born 1994). The original classes are sometimes split using secondary information on climate, topography or location. If large discrepancies occurred due to peculiar emerging patterns, the whole process was repeated. All rules for the final aggregation are listed in Table 16.4.

The resulting compatible data sets were finally analysed by comparing the extent of different classes separately as well as by analysing the differences in the spatial patterns using the Kappa statistic (Cohen 1960; Monserud and Leemans 1992). This statistic compares cell-to-cell agreement for each category and for the data set as a whole. The Kappa statistic ranges from -1 (total disagreement) to 1 (total agreement) and is very suitable to rank similarities and differences between complex spatial patterns. Monserud and Leemans (1992) suggested that values <0.4 are to be considered poor, 0.4–0.55 fair, 0.55–0.7 good, 0.7–0.85 very good and 0.85–1.0 excellent. A limitation of the original Kappa statistic is that it treats each cell independently and the resolution of the data bases is not suitable to define all scales of interactions. Some patterns emerge on larger scales and there the original Kappa statistic was also applied by Prentice *et al.* (1992) to blocks of aggregated cells. This generalized Kappa index displays an overall fit over larger blocks of cells, taking into account the dependency of neighbouring cells.

The Kappa statistic (Table 16.5) illustrates only a poor to fair agreement between all different maps. This is also indicated by the low correspondence between all extents. The location of the biomes that all data sets agree upon are those with a large extent, such as hot desert, boreal and tropical rain forests. Ecotones and the smaller biomes are characterized by large differences among the different data bases. The biogeographical zones (Budyko 1986) do not translate into the whole range of categories. The mixed categories (Table 16.4) are lacking and their cells are merged with the often broader neighbouring categories. This could lead to an unrealistically high agreement between the major biomes and the overall Kappa statistic and extent figures for the biogeographical zones.

It is clear that the different observational land cover databases are heterogeneous. The Kappa statistic indicated only poor to fair agreement between them (Table 16.5). Up to 60% of all cells may be classified differently. This difference is most visible in the comparison between the global vegetation data base (Matthews 1983) and the major ecosystems of the world data base (Olson *et al.* 1985). The other data bases display a somewhat better agreement among themselves. The discrepancy between the different data sets can perhaps be explained by the large differences in the actual classifications, but we have tried to

Table 16.5 Kappa statistic and extent in common (10^3 km^2; upper half of the matrix) between the different global land cover maps. Wetlands and agricultural lands are excluded from the analysis. The total extent of land cover used for this assessment was 99 239 × 10^3 km^2

Data base	1	2	3	4	5	6	7	8	9
1 Olson *et al.* (1985)		42 069	45 577	44 262	47 713	44 383	41 570	42 221	45 609
2 Matthews (1983)	0.39		47 245	52 688	39 831	36 316	38 801	40 397	38 987
3 Küchler (1949)	0.43	0.43		48 353	45 136	37 518	44 427	46 580	49 496
4 Melillo *et al.* (1993)	0.42	0.48	0.45		41 254	38 982	37 757	38 417	40 421
5 Prentice *et al.* (1992)	0.44	0.36	0.46	0.35		50 650	57 109	58 256	52 514
6 Holdridge (1967)	0.41	0.33	0.35	0.35	0.48		46 751	45 927	46 896
7 Köppen (1936)	0.38	0.35	0.39	0.34	0.54	0.44		69 219	50 592
8 Trewartha (1968)	0.39	0.36	0.42	0.35	0.57	0.45	0.68		52 055
9 Budyko (1986)	0.42	0.35	0.43	0.38	0.47	0.42	0.47	0.48	

be as careful as possible in creating compatible data sets. Our conclusion from this comparison is that we actually have only a very approximate comprehension of actual global land cover. This conclusion is further strengthened by the fact that even in the statistical, tabular data bases based on country censuses, such as FAO agricultural and forestry yearbooks, large differences occur between the data.

There is a clear pattern for the comparison between the different models based on the climate–vegetation classifications and global land cover data bases (Table 16.5). The BIOME model generally provides the more adequate simulation of global vegetation patterns, closely followed by Budyko's biogeographical zones. The global climate classifications display an intermediate fit. The frequently used life zone classification has the poorest fit. Both the Kappa statistic and the extent figures present these trends. The land cover categories that are simulated most accurately by all models are hot desert and tundra, closely followed by tropical rainforest and taiga. All other categories are simulated with a much reduced accuracy.

We have drawn the following conclusions for this emerging pattern of data base comparisons. The life zone classification (Holdridge 1967) is based only on annual mean climatic parameters that do not capture the major driving forces of climate upon vegetation patterns. Major physiognomic patterns are adaptations to specific seasonality in temperature and precipitation regimes. The global climate classifications (Köppen 1936; Trewartha 1968) explicitly include seasonality and therefore give a somewhat better performance. The modified Köppen classification (Trewartha 1968) shows a closer match with most observed distributions of biomes because of the close correlation with the boundaries between arid and moist zones. The BIOME model (Prentice *et al.* 1992) and the biogeographical zones (Budyko 1986) both include a more realistic parameterization for the moisture balance. The biogeographical zone model is only based on potential evapotranspiration, while BIOME incorporates a more elaborate moisture availability scheme, including soil characteristics. However, both models do not strongly consider seasonal aspects of moisture availability and can be improved in this respect. For example, inclusion of seasonality by using the characteristics of a dry or growing period (cf. Leemans and Solomon 1993) could already enhance performance.

16.3 APPLICATIONS OF GLOBAL CLIMATE-VEGETATION MODELS

16.3.1 Predicting future biome redistribution caused by climate change

The earliest approaches that determined the impacts of climate on ecosystems used climate–vegetation classifications. Emanuel *et al.* (1985) used the Holdridge life zone classification to determine the extent of forests and grasslands under different climatic conditions. Their analysis clearly showed that with climatic warming the broad-scale vegetation patterns could shift considerably polewards. Besides these (largely) latitudinal shifts, they predicted that some regions might shift from forested to grassland ecosystems. On a global scale, this scenario seems unreasonable today, because changes in precipitation were assumed to be zero. Consequently, warming could lead to a global decrease in moisture, and the global water balance would not be stable.

Since the time of this far-reaching study, many other impacts studies using climate–vegetation classifications have been conducted. A thorough review is given by Cramer and Leemans (1993), who reimplemented both the life zone classifications and the first computerized plant functional type model (Box 1981). The sensitivities of these models were different under the climate change scenarios using anomalies for both temperature and precipitation. The life zone classification showed large shifts in vegetation zones for most high latitude regions, while in Box's model, tropical forests also showed large decreases in extent. This latter decline may be an artefact of the strong temperature sensitivity in Box's model. In his setting of the climatical limits, tropical trees could not survive at temperatures over 30 °C.

The potential shifts in life zones were used to assess the impact on large nature reserves (Leemans and Halpin 1992). Depending on the actual scenario, between 30 and 60% of all reserves could be severely affected with strong negative consequences for biodiversity.

We have repeated the climate-change impacts analysis for the different climate–vegetation classifications (except for the original global climate classification) with several climate scenarios based on both temperature and precipitation anomalies produced by four general circulation models for the Atmosphere (GCMs): the Oregon State University model (OSU, Schlesinger and Zhao 1989); the Goddard Institute for Space Studies model (GISS, Hansen *et al.* 1988); the Princeton General Fluid Dynamics Laboratory model (GFDL, Manabe and Wetherald 1987) and the United Kingdom Meteorological Office model (UKMO) of Mitchell (1983). The scenarios are listed in order of their global annual mean temperature increase under doubled CO_2 atmospheric conditions. All four scenarios differ in the details of the geographic distribution of climate changes they predict. In general all models predict a temperature increase, which is larger in the winter season and high latitude regions. The simulated precipitation change is geographically more complex, but all models, except OSU, show a significant increase of precipitation (e.g. Mitchell *et al.* 1990). The methodology for defining a future climate scenario is described by Leemans (1992).

Figure 16.2 illustrates the potential shifts in vegetation zones using the life zone classification under a changed climate. Even for the most modest scenario (OSU), few areas remain unaffected. Similar changes can be observed for the other climate–vegetation models (Figures 16.3, 16.5 and 16.7). From these figures it becomes clear that all climate–vegetation classifications show similar responses.

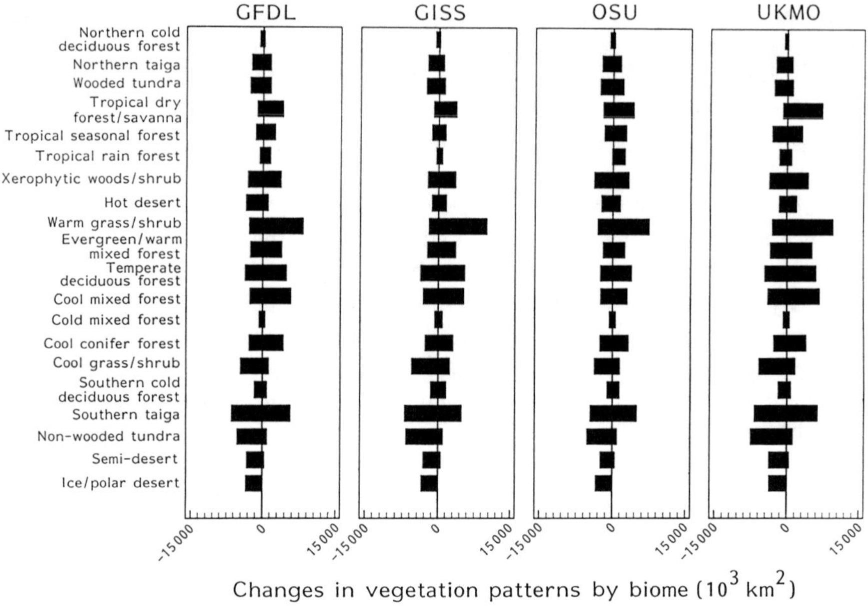

Figure 16.7 Shifts in biomes (Prentice *et al.* 1992) under several doubled CO_2-derived climates. The left part of the histogram presents a decrease in extent, while the right part presents an increase, both in respect to the current extent

Most shifts occur in high latitude regions. The extent of tundra is largely reduced, while the southern part of the tundra is replaced by boreal forests, the southern part of the boreal forest by temperate forests in more maritime climates and by steppe in the more continental climate. The changes in the tropics are less easy to generalize. Deserts and tropical rain forests remain relatively constant, while the vegetation types in more seasonal climates change considerably. Here the largest differences occur between different GCM scenarios.

Looking at these different impact assessments, it is striking how similar the changes unfold for the different GCM scenarios and climate–vegetation models, especially if one accounts for the uncertainties and differences in all models and their simulations. The four GCM scenarios considered here actually span the whole range of 1.5–5.0 °C temperature increase given by IPCC (Houghton *et al.* 1990). The reason for the convergence in these impact assessments can partly be explained by the methodology of creating a scenario. We have used a baseline climatology with monthly values (Leemans and Cramer 1991) and overlaid them with the appropriate climatic anomalies. The underlying baseline pattern is always part of the scenario, and pronounced climatic gradients will remain, especially with a coarser GCM resolution (3–7°) than the baseline (here 0.5°) This method could therefore lead to a greater similarity between scenarios than when the doubled CO_2 GCM simulations are compared together. In interpreting the results, one has to be aware of these limitations, but there are currently no better approaches which are widely accepted (Carter *et al.* 1992).

However, we believe that the climatic anomalies of the different GCMs are much more similar, as experienced by vegetation. The temperature signal between the different GCMs varies considerably, but mostly in the temperature ranges that do not affect plants very much. This is the main explanation for the relatively similar patterns of change in bioclimatic patterns of change in bioclimatic boundaries between different GCM scenarios. For example, the largest differences in the four scenarios occur in January temperatures at high latitudes with increases ranging from 5 °C (OSU) to 25 °C (GFDL). The large temperature increase significantly influences the mean annual temperature increase. Plants are not strongly affected when temperatures remain well below freezing, as is the case for all scenarios.

There are still considerable limitations to such a vegetation–climate model approach. First, it is implicitly assumed that vegetation is in equilibrium with climate. However, the response, adjustment and redistribution of plants require time. Simulations with vegetation–climate models are probably valid for longer time spans, such as millennia (Prentice *et al.* 1991), but they are surely not valid in the decade to century time span for which the climatic change due to the enhanced greenhouse effect is foreseen. Several dynamic models (e.g. Prentice *et al.* 1993; Smith *et al.* 1992b) have been used to determine the time-dependent vegetation response simulating species-specific life-histories and succession processes. From these analyses, it becomes obvious that time lags up to several centuries could easily occur.

Such impact studies further assume that species can reach suitable localities instantaneously. Using palaeoecological data, Davis (1981) has clearly demonstrated that this is not true and that maximum historical migration speeds for trees range from 10 to 100 km per century. This slow rate could be further reduced in the current fragmented landscape, but others have argued that humans also could facilitate higher migration rates. However, with a relatively fast rate of climatic change, conditions could be suitable for any species during the seedling or sapling stage, but not any more for more mature stages. This is especially true for long-lived species, such as trees, and would make an adequate adaptation in sectors such as forestry, extremely difficult.

Secondly, the parameters used in most climate–vegetation models are not the ones that define the actual distribution of vegetation. (We have already discussed that species and not biomes are the actual unit of change.) Together with climate and soil, vegetation history strongly determines the actually occurring vegetation in a given locality and these vegetation patterns should be used for defining a more realistic transient response, not the potential patterns. Determining the response of actual vegetation could be achieved by linking 'plant functional type'-schemes with more dynamic models. The principles of such an approach were recently presented by Smith and Shugart (1993), but the results of their approach were less reliable because they still used the life zone classification.

16.3.2 Determining the global C budget

One of the most important applications of global vegetation or land cover models and data bases is the determination of the role of the terrestrial biosphere in the global C cycle. A straightforward method is to use climate data to derive the C contents of each vegetation type. The first model that used this approach was the Miami model (Lieth 1975). This model is based on a regression between climate indices and net primary productivity (NPP) and is still used in some of the current C cycle models (Esser, 1991). This regression method was further improved by Seino and Uchhijima (1992) who used the indices developed for the biogeographical zones (Budyko 1986) to define NPP and produce global productivity maps. This model, however, has not been used in global change studies.

Prentice and Fung (1990) used the life zone classification (Holdridge 1967) to define the potential C content of the terrestrial biosphere. This is done by: (1) assigning a C density to each life zone; (2) computing the extents of each life zone and (3) multiplying (1) and (2). They did this analysis for the last glacial maximum (LGM: 18 000 BP), current climate and doubled CO_2 climate conditions. The distribution of life zones for the different climate was determined (cf. Figure 16.2) and the specific C content computed. Their conclusion was that the terrestrial biosphere had been acting as a C sink since the LGM and would continue to do so under climatic change. Smith *et al.* (1992a) have redone this analysis using the life zone classification at its full resolution and added soil C densities using the values given by Post *et al.* (1982). Aboveground C densities were assigned using the values given by Olson *et al.* (1985). They did the analyses for several GCM scenarios and concluded that the sink strength of the terrestrial biosphere was not as large as initially predicted by Prentice and Fung (1990). Similar approaches using different models were taken by Adams *et al.* (1990) and Prentice *et al.* (1994; Figure 16.7).

These approaches are all based on the equilibrium between the terrestrial biosphere, climate and the other components of the C cycle. Such an equilibrium may exist for longer time scales, but is not valid for scales ranging from decades to centuries (Prentice *et al.* 1991). The short-term transient dynamics could give a completely different picture of C storage in the terrestrial biosphere. Several analyses (Neilson 1993; Smith and Shugart 1993) have shown that through an increased forest dieback there could be a C pulse to the atmosphere from the biosphere under climatic warming, before the distribution of vegetation belts are readjusted and biospheric C uptake increases again. Such transient dynamics might have large implications for mitigation strategies that aim to reduce CO_2 in the atmosphere. The response of the biosphere would not likely be linear and smooth. Many surprises could occur and our understanding of most processes and interactions is rather incomplete. This is clearly illustrated by the recent sudden decline in atmospheric CO_2 build-up (Sarmiento 1993). No comprehensive explanations have been proposed yet.

Another problem with the global C budget is the inability to balance the global C cycle and most analyses cannot account for a 'missing sink' of about 2 Pg C per year. Siegentaler and Sarmiento (1993) concluded in a recent review that the uptake of CO_2 by the oceans is relatively well understood and that the excess C should be taken up by the terrestrial biosphere. Kauppi *et al.* (1992) reports that boreal and temperate forests are currently sinks of C. This sink function is further elaborated on by Dai and Fung (1993) who assume a large sink in these regions through slightly increasing temperatures. Houghton (1993) challenges this conclusion, stating that these large increases must have been observed in annual growth increments.

Part of this important issue of balancing the global C budget and reducing the missing sink, can well be explained by the inadequacies of the global land cover data sets and models. The current state-of-the-art global C budget models (e.g. Emanuel *et al.* 1984; Goudriaan and Ketner 1984; Esser 1991; Melillo *et al.* 1993) all use different land cover data bases to initialize and parameterize the different ecosystems' productivities and environmental feedbacks. The differences between these data sets are too large to permit a good assessment of the global C cycle. The models therefore can give insights into the processes and their interactions and be very effective in defining trends in the C dynamics under a changing climate, increasing atmospheric CO_2 concentrations, and SOM. But these models and especially their underlying data bases do not allow us to determine the actual size of sources and sinks in the terrestrial biosphere. Discussions on the actual location and characteristic of the missing sink will continue for a while.

16.4 CONCLUSIONS AND SUMMARY

Recently, several global models that simulate one or more aspects of the terrestrial biosphere have been developed (Melillo *et al.* 1993; Prentice *et al.* 1992; Alcamo *et al.* 1994). The analysis presented above stresses the importance of the development of a high-quality global land cover data base, which can be used to initialize, calibrate and/or validate these global models. No reliable and generally accepted data base exists today. This has led to a multitude of approaches, often leading to largely conflicting opinions on the importance of ecological processes.

The current global carbon cycle models (e.g. Melillo *et al.* 1993; Kindermann *et al.* 1993) are not yet able to simulate shifts in vegetation zones caused by climatic change and many other vegetation related feedbacks. To include such important vegetation responses, global climate–vegetation classifications should be applied. Of the currently available models, BIOME is the most appropriate candidate, because it includes the most important aspects of seasonality in temperature patterns, a quasi-realistic moisture balance and its vegetation responses are based on a series of independent plant functional types. However, BIOME still lacks an appropriate seasonality in its definition of the moisture regime. Advancements in modelling global vegetation patterns should start with improving the simple moisture balance model and include a better array of soil characteristics, especially now that an improved version of the FAO world soil resources has been released (Anonymous 1993).

Several global assessments on the impacts of climatic change on terrestrial ecosystems have been accomplished (e.g. Houghton *et al.* 1990, 1992; Izrael *et al.* 1990, 1992). These studies have been limited by the sparsely available data with a comprehensive global cover. Most analyses have therefore been limited to regional and/or sectorial studies and little integration has been achieved. Many global change studies have therefore been very anecdotal. Only recently has a more global ecological theory begun to emerge (e.g. Solomon and Shugart 1993; Kareiva *et al.* 1993; Ehleringer and Field 1993). Currently, a series of integrated models are being developed that aim to assess the most important aspects of global change and will be used to address important policy issues (e.g. Alcamo *et al.* 1994). These models can now be developed because of the improved understanding of global biogeochemical cycles, ecological processes and interactions between climate, soil and land cover. It is probably this kind of model that will most clearly indicate the weaknesses of current data bases and will benefit most from future improvements.

16.5 ACKNOWLEDGEMENTS

We would like to thank I. Colin Prentice and Martin T. Sykes for providing the data of Figure 16.7. Critical readings of earlier versions of this draft by Joe Alcamo and Kees Klein-Goldewijk are appreciated. The research was funded by the Dutch National Programme on 'Global Air Pollution and Climate Change' under contracts NOP 852067 (MAP 481510 and 482507) and the Dutch Ministry of Housing, Physical Planning and the Environment under contract MAP 481507. The study further contributes to core research of the Global Change and Terrestrial Ecosystems project of IGBP.

16.6 REFERENCES

Adams, J. M., Faure, H., Faure-Denard, L., McClade, J. M. and Woodward, F. I. (1990) Increases in terrestrial carbon storage from the Last Glacial Maximum to the present. *Nature* **348**, 711–714.

Alcamo, J., Kreileman, G. J. J., Krol, M. and Zuidema, G. (1994) Modeling the global society–biosphere–climate system, Part 1: Model description and testing. *Water Air Soil Pollut.* **76**, 1–35.

Anderson, J. R., Hardy, E. E., Roach, J. T. and Witmer, R. E. (1976) *A Land Use and Land Cover Classification System for Use with Remote Sensor Data.* Professional Paper 964. US Geological Survey, Washington DC.

Anonymous (1993) *World Soil Resources. An Explanatory Note on the FAO World Soil Resources Map at 1:25 000 000 scale.* World Soil Resources Report 66 rev. 1. Food and Agriculture Organization of the United Nations, Rome.

Bailey, R. G. (1980) *Description of the Ecoregions of the United States.* Miscellaneous Publication No. 1391. US Department of Agriculture, Ogden, Utah.

Bailey, R. G. (1983) Delineation of ecosystem regions. *Environ. Manage.* **7**, 365–373.

Bailey, R. G. (1989) Explanatory supplement to ecoregions maps of the continents. *Environ. Conserv.* **16**, 307–309.

Bartholomew, J. C., Christie, J. H., Ewington, A., Geelan, P. J. M., Lewisobe, H. A. G., Middleton, P. and Winkleman, H. (Eds) (1988) *The Times' Atlas of the World.* Times Books Limited, London.

Bazzaz, F. A. (1990) The response of natural ecosystems to the rising global CO_2 levels. *Ann. Rev. Ecol. Syst.* **21**, 167–196.

Bouwman, A. F., Fung, I., Matthews, E. and John, J. (1993) Global analysis of the potential for N_2O production in natural soils. *Glob. Biogeochem. Cyc.* **7**, 557–597.

Box, E. O. (1981) *Macroclimate and Plant Forms: An Introduction to Predictive Modeling in Phytogeography.* Dr W. Junk Publishers, The Hague.

Braun-Blanquet, J. (1964) *Pflanzensoziologie, Grundzüge der Vegetationskunde.* Springer-Verlag, Vienna.

Budyko, M. I. (1986) *The Evolution of the Biosphere.* D. Reidel Publishing Company, Dordrecht.

Carter, T. R., Parry, M. L., Nishioka, S. and Harasawa, H. (1992) *Preliminary Guidelines for Assessing Impacts of Climate Change.* IPCC Working group 2 report. Environmental Change Unit, Oxford and Center for Global Environmental Research, Tsukuba.

Cohen, J. (1960) A coefficient of agreement for nominal scales. *Educat. Psychol. Measurements* **20**, 37–46.

CORINE (1989) *CORINE database manual, version 2.1.* Commission of the European Communities, Brussels.

Cramer, W. and Leemans, R. (1993) Assessing impacts of climate change on vegetation using climate classification systems. In: Solomon, A. M. and Shugart, H. H. (Eds) *Vegetation Dynamics Modelling and Global Change*, pp. 190–217. Chapman & Hall, New York.

Cramer, W. and Solomon, A. M. (1993) Climatic classification and future global redistribution of agricultural land. *Clim. Res.* **3**, 97–110.

Dai, A. G. and Fung, I. Y. (1993) Can climate variability contribute to the 'Missing' CO_2 sink? *Glob. Biogeochem. Cyc.* **7**, 599–609.

Davis, M. B. (1981) Quaternary history and the stability of forest communities. In: West, D. C., Shugart, H. H. and Botkin, D. B. (Eds) *Forest Succession: Concepts and Application*, pp 132–154. Springer-Verlag, New York.

Ehleringer, J. R. and Field, C. B. (Eds) (1993) *Scaling Physiological Processes: Leaf to Globe.* Academic Press, San Diego.

Emanuel, W. R., Killough, G. E. G. and Olson, J. S. (1981) Modelling the circulation of carbon in the world's terrestrial ecosystems. In: Bolin, B. (Ed.) *Carbon Cycle Modelling*, pp. 335–353. Wiley & Sons, New York.

Emanuel, W. R., Killough, G. E. G., Post, W. M. and Shugart, H. H. (1984) Modeling terrestrial ecosystems in the global carbon cycle with shifts in carbon storage capacity by land-use change. *Ecology* **65**, 970–983.

Emanuel, W. R., Shugart, H. H. and Stevenson, M. P. (1985) Climatic change and the broad-scale distribution of terrestrial ecosystems complexes. *Clim. Change* **7**, 29–43.

Espenshade, E. B. Jr. and Morrison, J. L. (Eds) (1991) *Goode's World Atlas*. Rand McNally & Company, Chicago.

Esser, G. (1991) Osnabrück biosphere model: structure, construction, results. In: Esser, G. and Overdieck, D. (Eds) *Modern Ecology, Basic and Applied Aspects*, pp. 679–709. Elsevier, Amsterdam.

Fitter, A. H. and Hay, R. K. M. (1981) *Environmental Physiology of Plants*. Academic Press, London.

Goudriaan, J. and Ketner, P. (1984) A simulation study for the global carbon cycle, including man's impact on the biosphere. *Clim. Change* **6**, 167–192.

Grabherr, G. and Kojima, S. (1993) Vegetation diversity and classification systems. In: Solomon, A. M. and Shugart, H. H. (Eds) *Vegetation Dynamics and Global Change*, pp. 218–232. Chapman & Hall, New York.

Guetter, P. J. and Kutzbach, J. E. (1990) A modified Köppen classification applied to model simulations of glacial and interglacial climates. *Clim. Change* **16**, 193–215.

Hansen, J., Fung, I., Lacis, A., Rind, D., Lebedeff, S., Ruedy, R., Russell, G. and Stone, P. (1988) Global climate changes as forecast by Goddard Institute for Space Studies three-dimensional model. *J. Geophys. Res.* **93**, 9341–9364.

Holdridge, L. R. (1959) Simple method for determining potential evapotranspiration from temperature data. *Science* **130**, 572.

Holdridge, L. R. (1967) *Life Zone Ecology*. Tropical Science Center, San José, Costa Rica.

Holten, J. I. (1990) *Biologiske og økologiske konsekvenser av klimaforandringer i Norge*. NINA Utreding 11. Norsk Institutt for Naturforskning, Trondheim.

Houghton, J.T., Callander, B. A. and Varney, S. K. (Eds) (1992) *Climate Change 1992. The Supplementary Report to the IPCC Scientific Assessment*. Cambridge University Press, Cambridge.

Houghton, J.T., Jenkins, G. J. and Ephraums, J. J. (Eds) (1990) *Climate Change: The IPCC Scientific Assessment*. Cambridge University Press, Cambridge.

Houghton, R. A. (1993) Is carbon accumulating in the northern temperate zone? *Glob. Biogeochem. Cyc.* **7**, 611–617.

Houghton, R. A., Hobbie, J. E., Melillo, J. M., Moore, B. III, Peterson, B. J., Shaver, G. R. and Woodwell, G. M. (1983) Changes in the carbon content of terrestrial biota and soils between 1860 and 1980: a net release of CO_2 to the atmosphere. *Ecol. Monogr.* **53**, 235–262.

Hulme, M., Wigley, T., Jiang, T., Zhao, Z.-C., Wang, F., Ding, Y., Leemans, R. and Markham, A. (1992) *Climate Change due to the Greenhouse Effect and its Implications for China*. WWF, Gland, Switzerland.

Huntley, B. and Webb, T. III (Eds) (1988) *Vegetation History*. Kluwer Academic Publishers, Dordrecht.

Izrael, Y. A., Canziani, O., Hashimoto, M., Odingo, O. S. and Tegart, W. J. M. (Eds) (1992) *Climate Change 1992: The Supplementary Report to the IPCC Impact Assessment*. WMO and UNEP, Geneva.

Izrael, Y. A., Hashimoto, M. and Tegart, W. J. M. (Eds) (1990) *Climate Change: The IPCC Impact Assessment*. Australian Government Publishing Service, Canberra.

Kareiva, P. M., Kingsolver, J. G. and Huey, R. B. (Eds) (1993) *Biotic Interactions and Global Change*. Sinauer Associates Inc., Sunderland, Mass.

Kaufman, Y. J., Tucker, C. J. and Fung, I. (1990) Remote sensing of biomass burning in the tropics. *J. Geophys. Res.* **95**, 9927–9939.

Kauppi, P., Mielikäinen, K. and Kuusela, K. (1992) Biomass and carbon budget of European forests, 1971 to 1990. *Science* **256**, 70–74.

Kindermann, J., Lüdeke, M. K. B., Badeck, F. W., Otto, R. D., Klaudius, A., Hager, C., Wurth, G., Lang, T., Donges, S., Habermehl, S. and Kohlmaier, G. H. (1993) Structure of a global and seasonal carbon exchange model for the terrestrial biosphere – The Frankfurt Biosphere Model (FBM). *Water Air Soil Pollut.* **70**, 675–684.

Kineman, J. J. (1992) *Global Ecosystems database Version 1.0* (on CDROM) *User's guide.* Key to Geophysical Records Documentation No. 26. USDOC/NOAA National Oceanic and Atmospheric Administration, Boulder, Color.

Klein-Goldewijk, K., van Minnen, J. G., Kreileman, G. J. J., Vloedbeld, M. and Leemans, R. (1994) Simulating the carbon flux between the terrestrial environment and the atmosphere. *Water Air Soil Pollut.* **76**, 199–230.

Köppen, W. (1936) Das geographische System der Klimate. In: Köppen, W. and Geiger, R. (Eds) *Handbuch der Klimatologie*, pp. 1–46. Gebrüder Borntraeger, Berlin.

Küchler, A. W. (1949) A physiognomic classification of vegetation. *Ann. Ass. Amer. Geog.* **39**, 201–210.

Küchler, A. W. (1967) *Vegetation Mapping.* The Ronald Press Company, New York.

Küchler, A. W. and Zonneveld, I. S. (Eds) (1988) *Vegetation Mapping.* Kluwer Academic Publishers, Dordrecht.

Larcher, W. (1980) *Physiological Plant Ecology.* Springer-Verlag, Berlin.

Leemans, R. (1992) Modelling ecological and agricultural impacts of global change on a global scale. *J. Sci. Ind. Res.* **51**, 709–724.

Leemans, R. and Cramer, W. (1991) *The IIASA Database for Mean Monthly Values of Temperature, Precipitation and Cloudiness on a Global Terrestrial Grid.* Research Report RR-91-18. International Institute of Applied Systems Analyses, Laxenburg, Austria.

Leemans, R. and Halpin, P. (1992) Global change and biodiversity. In: Groombridge, B. (Ed.) *Biodiversity 1992: Status of the Earth's Living Resources*, pp. 254–255. Chapman & Hall, London.

Leemans, R. and Solomon, A. M. (1993) The potential response and redistribution of crops under a doubled CO_2 climate. *Clim. Res.* **3**, 79–96.

Leemans, R. and van den Born, G. J. (1994) Determining the potential global distribution of natural vegetation, crops and agricultural productivity. *Water Air Soil Pollut.* **76**, 133–161.

Lieth, H. (1975) Modelling the primary production of the world. In: Lieth, H. and Whittaker, R. H. (Eds) *Primary Productivity of the Biosphere*, pp. 203–215. Springer-Verlag, Berlin.

McGuire, A. D., Melillo, J. M., Joyce, L. A., Kicklighter, D. W., Grace, A.L., Moore, B. III and Vrsmarty, C. J. (1992) Interactions between carbon and nitrogen dynamics in estimating net primary productivity in North America. *Glob. Biogeochem. Cyc.* **6**, 101–124.

Manabe, S. and Wetherald, R. T. (1987) Large-scale changes in soil wetness induced by an increase in carbon dioxide. *J. Atmos. Sci.* **44**, 1211–1235.

Matthews, E. (1983) Global vegetation and land use: new high-resolution data bases for climate studies. *J. Clim. Appl. Meteorol.* **22**, 474–487.

Matthews, E. and Fung, I. (1989) Methane emission from natural wetlands: global distribution, area, and environmental characteristics of sources. *Glob. Biogeochem. Cyc.* **1**, 61–68.

Melillo, J. M., McGuire, A. D., Kicklighter, D. W., Moore, B. III, Vörösmarty, C. J. and Schloss, A. L. (1993) Global climate change and terrestrial net primary production. *Nature* **363**, 234–240.

Mitchell, J.F.B. (1983) The seasonal response of a general circulation model to changes in CO_2 and sea temperatures. *Quart. J. R. Meteorol. Soc.* **109**, 113–152.

Mitchell, J.F.B., Manabe, S., Meleshko, V. and Tokioka, T. (1990) Equilibrium climate change–and its implications for the future. In: Houghton, J. T., Jenkins, G. J. and Ephraums, J. J. (Eds) *Climate Change: The IPCC Scientific Assessment*, pp. 131–172. Cambridge University Press, Cambridge.

Monserud, R. A. and Leemans, R. (1992) The comparison of global vegetation maps. *Ecol. Model.* **62**, 275–293.

Mueller-Dombois, D. and Ellenberg, H. (1974) *Aims and Methods of Vegetation Ecology.* John Wiley & Sons, Chichester.

Mller, M. J. (1982) *Selected Climatic Data for a Global Set of Standard Stations for Vegetation Science.* Dr W. Junk Publishers, The Hague.

Neilson, R. P. (1993) Vegetation redistribution–a possible biosphere source of CO_2 during climatic change. *Water Air Soil Pollut.* **70**, 659–673.

Neilson, R. P., King, G. A. and Koerper, G. (1992) Toward a rule-based biome model. *Landscape Ecol.* **7**, 27–43.

Olson, J., Watts, J. A. and Allison, L. J. (1985) *Major World Ecosystem Complexes Ranked by Carbon in Live Vegetation: A Database.* Report NDP-017. Oak Ridge National Laboratory, Oak Ridge, Tenn.

Parry, M. (1992) The potential effect of climate changes on agriculture and land use. *Adv. Ecol. Res.* **22**, 63–92.

Parton, W. J., Schimel, D. S., Cole, C. V. and Ojima, D. S. (1987) Analysis of factors controlling soil organic matter levels in Great Plains grasslands. *Soil Sci. Soc. Am. J.* **51**, 1173–1179.

Post, W. M., Emanuel, W.R., Zinke, P. J. and Stangenberger, A. G. (1982) Soil carbon pools and world life zones. *Nature* **298**, 156–159.

Prentice, I. C., Bartlein, P. J. and Webb, T. III (1991) Vegetation and climate change in eastern North America since the last glacial maximum. *Ecology* **72**, 2038–2056.

Prentice, I. C., Cramer, W., Harrison, S. P., Leemans, R., Monserud, R. A. and Solomon, A. M. (1992) A global biome model based on plant physiology and dominance, soil properties and climate. *J. Biogeogr.* **19**, 117–134.

Prentice, I. C. and Fung, I. Y. (1990) The sensitivity of terrestrial carbon storage to climate change. *Nature* **346**, 48–51.

Prentice, I. C., Sykes, M. T. and Cramer, W. (1993) A simulation model for the transient effects of climate change on forest landscapes. *Ecol. Model.* **65**, 51–70.

Prentice, I. C., Sykes, M. T., Lautenschlager, M., Harrison, S. P., Denissenko, O. and Bartlein, P. J. (1994) Modelling global vegetation patterns and terrestrial carbon storage at the last glacial maximum. *Biogeogr. Let.* **3**, 67–76.

Raunkiær, C. (1934) *Life Forms of Plants and Terrestrial Plant Geography.* Clarendon Press, Oxford.

Sarmiento, J. L. (1993) Carbon cycle–atmospheric CO_2 stalled. *Nature* **365**, 697–698.

Schlesinger, M. E. and Zhao, Z.-C. (1989) Seasonal climatic changes induced by doubled CO_2 as simulated by the OSU atmospheric GCM/mixed-layer ocean model. *J. Clim.* **2**, 459–495.

Seino, H. and Uchhijima, Z. (1992) Global distribution of net primary productivity of terrestrial vegetation. *J. Agr. Met.* **48**, 39–48.

Siegentaler, U. and Sarmiento, J. L. (1993) Atmospheric carbon dioxide and the ocean. *Nature* **365**, 119–125.

Skole, D. and Tucker, C. (1993) Tropical deforestation and habitat fragmentation in the Amazon: Satellite data from 1978 to 1988. *Science* **260**, 1905–1910.

Smith, T. M., Leemans, R. and Shugart, H. H. (1992a) Sensitivity of terrestrial carbon storage to CO_2-induced climate change: comparison of four scenarios based on general circulation models. *Clim. Change* **21**, 367–384.

Smith, T. M. and Shugart, H. H. (1993) The transient response of terrestrial carbon storage to a perturbed climate. *Nature* **361**, 523–526.

Smith, T. M., Shugart, H. H., Bonan, G. B. and Smith, J. B. (1992b) Modeling the potential response of vegetation to global change. *Adv. Ecol. Res.* **22**, 93–116.

Solomon, A. M. and Shugart, H. H. (Eds) (1993) *Vegetation Dynamics and Global Change.* Chapman & Hall, New York.

Strahler, A. N. and Strahler, A. H. (1987) *Modern Physical Geography.* J. Wiley & Sons, New York.

Tahktajan, A. A. (1973) *Origin and Distribution of the Angiosperms.* Academic Press, New York.

Tchebakova, N. M., Monserud, R. A. and Leemans, R. (1993a) Global vegetation change predicted by the modified Budyko model. *Clim. Change* **25**, 59–83.

Tchebakova, N. M., Monserud, R. A., Leemans, R. and Golovanov, S. (1993b) A global vegetation model based on the climatological approach of Budyko. *J. Biogeogr.* **20**, 219–244.

Thornthwaite, C. W. (1943) Problems in the classification of climate. *Geogr. Rev.* **33**, 233–255.

Thornthwaite, C. W. (1948) An approach toward a rational classification of climate. *Geogr. Rev.* **38**, 55–94.

Thornthwaite, C. W. and Mather, J. R. (1957) Instructions and tables for computing potential evapotranspiration and the water balance. *Publ. Climatol.* **10**, 185–310.

Townshend, J. R. G. (1992) *Improved Global Data for Land Application: A Proposal for a New High Resolution Data Set.* IGBP-Report No. 20. International Geosphere-Biosphere Programme, Stockholm.

Trewartha, G. T. (1968) Köppen's classification of climates. In: Trewartha, G. T. (Ed.) *An Introduction to Climate*, pp. 395–399. McGraw-Hill, New York.

Tuhkanen, S. (1980) Climatic parameters and indices in plant geography. *Acta Phytogeogr. Suec.* **67**, 1–110.

UN-ECE/FAO (1992) *The Forest Resources of the Temperate Zones: 1990 Forest Resource Assessment.* UN-ECE/FAO, Rome.

UNEP/GEMS (1993) *Vegetation Classification.* Report of the UNEP-HEM/WCMC/GCTE preparatory meeting, Charlottesville, Virginia, USA. GEMS Report Series No. 19. United Nations Environment Programme, Nairobi.

UNESCO (1973) *International Classification and Mapping of Vegetation.* Report United Nations Educational, Scientific and Cultural Organization, Paris.

Von Humboldt, F. H. A. (1807) *Ideen zu einer Geographie der Pflanzen neben einem Naturgemälde der Tropenländer.* Tübingen.

Walter, H. (1970) *Vegetationszonen und Klima.* Eugen Ulmer, Stuttgart.

Walter, H. (1985) *Vegetation of the Earth and Ecological Systems of the Geo-Biosphere.* Springer-Verlag, Berlin.

White, F. (1983) *UNESCO Vegetation Map of Africa.* UNESCO, Paris.

Whitmore, T. C. (1984) A vegetation map of Malaysia at the scale 1:5 million. *J. Biogeogr.* **11**, 461–471.

Whittaker, R. H. (1975) *Communities and Ecosystems.* Macmillan, New York.

Woodward, F. I. (1987) *Climate and Plant Distribution.* Cambridge University Press, Cambridge.

16.7 APPENDICES

Appendix 16.1 The legend of the different land cover data bases. The numbers in the left column correspond with the numbers in Table 16.4

	Ecosystems of the world (Olson *et al.*, 1985)	Matthews' (1983) global vegetation	Küchler's (1949) physiognomic classification	Melillo *et al.*'s (1993) potential vegetation
1	Antarctica	Tropical evergreen rain forests and mangrove forest	Broad-leaved evergreen trees (B)	Ice
2	Siberian parks	Tropical/subtropical evergreen seasonal broad-leaved forest	Broad-leaved evergreen trees and shrubs (Bs)	Polar desert/alpine tundra
3	Main taiga	Subtropical evergreen rain forest	Broad-leaved evergreen trees and patchy shrubs (Bsp)	Wet/moist tundra
4	Cool conifer forest	Temperate/subpolar evergreen rain forest	Broad-leaved evergreen, dwarf shrub (BZI, Bz)	Boreal forest
5	Cool mixed forest	Temperate evergreen seasonal broadleaved forest, summer rain	Broad-leaved deciduous trees (D)	Forested boreal wetland
6	Warm mixed forest	Evergreen broad-leaved sclerophyllous forest, winter rain	Broad-leaved deciduous trees, open stand (Di)	Boreal woodland
7	Warm deciduous forest	Tropical/subtropical evergreen needle-leaved forest	Broad-leaved deciduous trees and shrubs (Ds)	Non-forested boreal wetland
8	Broad-leaved evergreen forest	Temperate/subpolar evergreen needle-leaved forest	Broad-leaved deciduous trees with patchy shrub (Dsi)	Temperate mixed forest
9	Warm conifer forest	Tropical/subtropical drought deciduous forest	Broad-leaved deciduous trees with single shrubs (Dsp)	Temperate coniferous forest
10	Tropical montane forest	Cold-deciduous forest with evergreens	Broad-leaved deciduous trees and dwarf shrubs (Dzp)	Temperate deciduous forest
11	Tropical seasonal forest	Cold-deciduous forest without evergreens	Broad-leaved deciduous, shrubs and grasses (DG)	Temperate forested wetland
12	Cool crops	Xeromorphic forest and woodlands	Broad-leaved deciduous trees and grasses (DG)	Tall/medium grassland

Appendix 16.1 (continued)

13	Warm farms	Evergreen broad-leaved sclerophyllous woodland	Broad-leaved deciduous and evergreen trees and shrubs (DBs)	Short grassland
14	Tropical dry forest	Evergreen needle-leaved woodland	Needle-leaved evergreen trees (E)	Tropical savanna
15	Equatorial evergreen forest	Tropical/subtropical drought-deciduous woodland	Needle-leaved evergreen trees in patches (Ep)	Arid shrubland
16	Paddyland	Cold-deciduous woodland	Grass and herbaceous plants (G)	Tropical evergreen forest
17	Warm irrigated drylands	Evergreen broad-leaved shrubland/thicket	Grass and herbaceous plants in patches (Gp)	Tropical forested wetland
18	Cold irrigated drylands	Evergreen needle-leaved or microphyllous shrubland/thicket	Grass, herbs and broad-leaved evergreen trees (GBp)	Tropical deciduous forest
19	Cool irrigated drylands	Drought-deciduous shrubland /thicket	Grass, herbs and broad-leaved deciduous trees (GD)	Xeromorphic forest/woodland
20	Cool grass/shrub	Cold-deciduous subalpine/subpolar shrubland	Grass, herbs and broad-leaved deciduous trees in patches (GDp)	Tropical forested floodplain
21	Warm grass/shrub	Xeromorphic shrubland, dwarf shrubland	Grass, herbs and broad-leaved deciduous shrubs in patches (GDsp)	Desert
22	Tropical savanna	Arctic/alpine tundra, mossy bog	Grass, herbs and broad-leaved semi-deciduous trees (GSp)	Tropical non-forested wetland
23	Bogs, bog woods	Grassland with 10–40% woody tree cover	Herbaceous plants other than grasses (L)	Tropical non-forested floodplain
24	Mangroves	Grassland with < 10% woody tree cover	Mixed, broad-leaved and needle-leaved evergreen trees (M)	Temperate non-forested wetland
25	Low scrub	Grassland with shrub cover	Needle-leaved deciduous trees (N)	Temperate forested floodplain
26	Mediterranean grazing	Tall grassland, no woody cover	Needle-leaved and broad-leaved deciduous trees (ND)	Temperate non-forested floodplain

Appendix 16.1 (continued)

	Ecosystems of the world (Olson *et al.*, 1985)	Matthews' (1983) global vegetation	Küchler's (1949) physiognomic classification	Melillo *et al.*'s (1993) potential vegetation
27	Semi-arid woods	Medium grassland, no woody cover	Semi-deciduous, broad-leaved evergreen deciduous trees (S)	Wet savanna
28	Succulent thorns	Meadow, short grassland, no woody cover	Semi-deciduous, broad-leaved evergreen deciduous shrub (Ss)	Salt marsh
29	Sand desert	Forb formations	Semi-deciduous, broad-leaved evergreen deciduous shrub (SsG)	Mangrove
30	Hot desert	Desert	Semi-deciduous, broadleaved evergreen deciduous shrub (Szp)	Temperate savanna
31	Cool desert	Ice	Semi-deciduous, broad-leaved and needle-leaved trees (SE)	Temperate broad-leaved evergreen forest
32	Tundra	Cultivation	Absent (b)	Mediterranean shrubland
33	Polar desert			Freshwater lakes
34	Cool fields/woods			Saltwater lakes and inlets
35	Warm woods/fields			
36	Cool woods/fields			
37	Warm fields/woods			
38	Southern taiga			
39	Northern taiga			
40	Wooded tundra			
41	Heaths and moors			
42	Marsh and swamps			
43	Coastal edges			
44	Polar desert			
45	Ice			
46	Water			

Appendix 16.2 The legend of the different bioclimatic classifications. The numbers in the left column correspond with the numbers in Table 16.4

	Holdridge life zone classification (1967)	Köppen (1936) climate classification	Budyko's (1986) biogeographical zones	BIOME (Prentice *et al.* 1992)
1	Ice	EF	Ice/polar desert	Ice/polar desert
2	Polar desert	ET	Tundra	Semi-desert
3	Polar dry tundra	Dfd	Taiga	Tundra
4	Polar moist tundra	Dfc	Continental taiga	Taiga
5	Polar wet tundra	Dfb	Continental steppe	Cold deciduous forest
6	Polar rain tundra	DFa	Temperate mixed and deciduous forest	Cool grass and shrubs
7	Boreal desert	Dwd	Temperate steppe	Cool conifer forest
8	Boreal dry bush	Dwc	Subtropical mixed and deciduous forest	Cold mixed forest
9	Boreal moist forest	Dwb	Subtropical xerophytic wood	Cool mixed forest
10	Boreal wet forest	Dwa	Subtropical steppe	Temperate deciduous forest
11	Boreal rain forest	Cfc	Tropical rain forest	Evergreen/warm mixed forest
12	Cool temperate desert	Cfb	Tropical seasonal forest	Warm grass and shrubs
13	Cool temperate desert bush	Cfa	Tropical savanna	Hot desert
14	Cool temperate steppe	Csc	Tropical thorns	Xerophytic woods/shrubs
15	Cool temperate moist forest	Csb	Tropical steppe	Tropical rain forest
16	Cool temperate wet forest	Csa	Desert	Tropical seasonal forest
17	Cool temperate rain forest	Cwc		Tropical dry forest or savanna
18	Warm temperate desert	Cwb		
19	Warm temperate desert bush	Cwa		
20	Warm temperate thorn steppe	BWk		
21	Warm temperate dry forest	BWh		
22	Warm temperate moist forest	BSk		

Appendix 16.2 (continued)

	Holdridge life zone classification (1967)	Köppen (1936) climate classification	Budyko's (1986) biogeographical zones	BIOME (Prentice *et al.* 1992)
23	Warm temperate wet forest	BSh		
24	Warm temperate rain forest	Aw		
25	Subtropical desert	Am		
26	Subtropical desert bush	Af		
27	Subtropical thorn steppe			
28	Subtropical dry forest			
29	Subtropical moist forest			
30	Subtropical wet forest			
31	Subtropical rain forest			
32	Tropical desert			
33	Tropical desert bush			
34	Tropical thorn steppe			
35	Tropical very dry forest			
36	Tropical dry forest			
37	Tropical moist forest			
38	Tropical wet forest			
39	Tropical rain forest			

Index

Note: Page reference in *italics* refer to Figures; those in **bold** refer to Tables

Index compiled by Annette J. Muster